Strategies in Cold:
Natural Torpidity and Thermogenesis

ACADEMIC PRESS RAPID MANUSCRIPT REPRODUCTION

STRATEGIES IN COLD:
Natural Torpidity and Thermogenesis
Jasper Park Lodge, October 3–8, 1977

Strategies in Cold:
Natural Torpidity and Thermogenesis

Edited by

Lawrence C. H. Wang
Department of Zoology
University of Alberta
Edmonton, Canada

Jack W. Hudson
Section of Ecology and Systematics
Cornell University
Ithaca, New York

ACADEMIC PRESS NEW YORK SAN FRANCISCO LONDON 1978
A Subsidiary of Harcourt Brace Jovanovich, Publishers

ACADEMIC PRESS, INC.
111 Fifth Avenue, New York, New York 10003

United Kingdom Edition published by
ACADEMIC PRESS, INC. (LONDON) LTD.
24/28 Oval Road, London NW1 7DX

Main entry under title:

Strategies in cold.

Papers presented at a symposium held in Alberta, Can., Oct. 3-8, 1977.
1. Body temperature–Regulation–Congresses.
2. Cold–Physiological effect–Congresses.
3. Hibernation–Congresses. 4. Dormancy (Biology) –Congresses. I. Wang, Lawrence C.H.
II. Hudson, Jack W. [DNLM: 1. Hibernation–Congresses. 2. Mammals–Physiology–Congresses.
3. Body temperature regulation–Congresses.
4. Adaptation, Physiological–Congresses. 5. Cold climate–Congresses. W3 IN919GM 5th 1977s / QL755 I61 1977s]
QP135.S77 599.5'43 78-8287
ISBN 0-12-734550-7

PRINTED IN THE UNITED STATES OF AMERICA

Contents

List of Participants

B. Abbotts, Department of Zoology, University of Alberta, Edmonton, Alberta T6G 2E9, Canada

R. Agid, Institut de Physiologie, Universite Paul Sabatier, 31 Toulouse, France

L. Ambid, Institut de Physiologie, Universite Paul Sabatier, 31 Toulouse, France

J. S. Ballantyne, Department of Zoology, University of British Columbia, Vancouver, British Columbia V6T 1W5, Canada

D. Balfour, Department of Zoology, University of Alberta, Edmonton, Alberta T6G 2E9, Canada

A. L. Beckman, Department of Physiology, School of Medicine, University of Pennsylvania, Philadelphia, PA 19104

Hans W. Behrisch, Institute of Arctic Biology, University of Alaska, Fairbanks, AK 99701

L. C. Bergmann, Loyola Medical Center, Maywood, IL 60153

V. Bernson, The Wenner-Gren Institute, Norrtullsgatan 16, S-113 45 Stockholm, Sweden

G. L. Bintz, Department of Biology, Eastern Montana College, Billings, MT 59101

M. A. Bisby, Faculty of Medicine, University of Calgary, Calgary, Alberta T2N 1N4, Canada

M. A. Brock, Gerontology Research Center, National Institute on Aging, Baltimore City Hospitals, Baltimore, MD 21224

D. S. Bruce, 321 Brookside Circle, Wheaton, IL 60187

Kurt Brück, Zentrum für Physiologie, Justus-Liebig-Universität-Giessen, Aulweg 129, D-6300 Lahn-Giessen 1, Federal Republic of Germany

R. F. Burlington, Biology Department, Central Michigan University, Mt. Pleasant, MI 48855

Barbara Cannon, The Wenner-Gren Institute, Norrtullsgatan 16, S-113 45 Stockholm, Sweden

C. Carey, Department of Evolutionary Biology, University of Colorado, Boulder, CO 80309

John S. Charnock, Department of Pharmacology, University of Alberta, Edmonton, Alberta T6G 2H7, Canada

R. Christopherson, Department of Animal Science, University of Alberta, Edmonton, Alberta T6G 2E3, Canada

B. Clark, Department of Biology, University of New Mexico, Albuquerque, NM 87131

K. Cooper, Department of Physiology, University of Calgary, Calgary, Alberta T2N 1N4, Canada

W. Cottle, Department of Physical Education, University of Alberta, Edmonton, Alberta T6G 2H7, Canada

J. A. Cranford, Virginia Polytechnic Institute and State University Biology Department, Blacksburg, VA 24061

S. A. Dalton, Department of Zoology, University of Alberta, Edmonton, Alberta T6G 2E9, Canada

Albert R. Dawe, 1050 Knollwood Road, Deerfield, IL 60015

W. R. Dawson, Division of Biological Sciences, University of Michigan, Ann Arbor, MI 48103

D. R. Deavers, Dalton Research Center, University of Missouri, Columbia, MO 65201

B. J. Donovan, Department of Zoology, University of Alberta, Edmonton, Alberta T6G 2E9, Canada

J. R. Feierman, Corrales Mental Health Clinic, Albuquerque, NM 87114

D. D. Feist, Institute of Arctic Biology, University of Alaska, Fairbanks, AK 99701

G. L. Florant, Department of Biology, Stanford University, Stanford, CA 94305

G. E. Folk, Jr., Department of Physiology, University of Iowa, Iowa City, IA 52240

A. R. French, Department of Biology, University of California, Riverside, CA 92521

W. A. Galster, Flammability Research Center, University of Utah, Salt Lake City, UT 84112

S. Gehnrich, 413½ S. 6th Street, Laramie, WY 82070

J. C. George, Department of Zoology, University of Guelph, Guelph, Ontario N1G 2W1, Canada

J. D. Glass, Department of Zoology, University of Alberta, Edmonton, Alberta T6G 2E9, Canada

A. Graham, Department of Animal Science, University of Alberta, Edmonton, Alberta T6G 2E3, Canada

V. Gregg, Department of Animal Science, University of Alberta, Edmonton, Alberta T6G 2E3, Canada

F. Reed Hainsworth, Department of Biology, Syracuse University, Syracuse, NY 13210

D. Harding, Department of Zoology, University of Manitoba, Winnipeg, Manitoba R3T 2N2, Canada

William C. Hartner, Department of Biology, Northeastern University, Boston, MA 02115

J. S. Hayward, Department of Biology, University of Victoria, Victoria, British Columbia V8W 2Y2, Canada

J. Heath, Department of Physiology and Biophysics, University of Illinois, Urbana, IL 61801

G. Heldmaier, Zoologisches Institut, Universität Frankfurt, 6000 Frankfurt, Federal Republic of Germany

H. C. Heller, Department of Biology, Stanford University, Stanford, CA 94305

N. E. Henderson, Department of Biology, University of Calgary, Calgary, Alberta T2N 1N4, Canada

Jean Himms-Hagen, Department of Biochemistry, University of Ottawa, Ottawa, Ontario K1N 9A9, Canada

R. Hissa, Department of Zoology, University of Oulu, 90100 Oulu, 10, Finland

W. Holmes, University of Washington, Seattle, WA 98105

B. A. Horwitz, Department of Animal Physiology, University of California, Davis, CA 95616

Jack W. Hudson, Section of Ecology and Systematics, Cornell University, Ithaca, NY 14853

S. W. Jack, Department of Veterinarian Science, Animal Science Center, University of Maryland, College Park, MD 20742

L. Janský, Department of Comparative Physiology, Faculty of Sciences, Charles University, Prague, Czechoslovakia

B. W. Johansson, Heart Section, General Hospital, S-214 01 Malmö, Sweden

D. L. Jones, Department of Physiology, Faculty of Medicine, University of Western Ontario, London, N6A 5C1, Canada

S. B. Jones, Department of Physiology, Loyola University Medical Center, Maywood, IL 60153

M. Koban, Institute of Arctic Biology, University of Alaska, Fairbanks, AK 99701

J. Koski, Jr., Department of Zoology, Washington State University, Pullman, WA 99163

B. LaGrange, 14 Patrick Street, S. Burlington, VT 05401

Jacques LeBlanc, Department of Physiology, School of Medicine, Laval University, Quebec City, Quebec G1K 7P4, Canada

O. Lindberg, The Wenner-Gren Institute, Norrtullsgatan 16, S-113 45 Stockholm, Sweden

P. Lundberg, The Wenner-Gren Institute, Norrtullsgatan 16, S-113 45 Stockholm, Sweden

Charles P. Lyman, Department of Anatomy, Harvard Medical School, Boston, MA 02115

C. B. Lynch, Department of Biology, Wesleyan University, Middletown, CT 06457

G. R. Lynch, Department of Biology, Wesleyan University, Middletown, CT 06457

R. MacArthur, Faculty of Environmental Design, University of Calgary, Calgary, Alberta T2N 1N4, Canada

C. Martineau, Department of Zoology, University of Montana, Missoula, MT 59801

G. A. Meininger, Dalton Research Center, University of Missouri-Columbia, Columbia, MO 65201

P. E. Meyers, Department of Anatomy, Chicago Medical School, Chicago, IL 60612

G. Michener, Box 99, Picture Butte, Alberta T0K 1V0, Canada

Virginia M. Miller, Department of Psychology, University of Virginia, Charlottesville, VA 22901

J. G. Minor, Department of Biology, University of Missouri-Kansas City, Kansas City, MO 64110

T. W. Moon, Department of Biology, University of Ottawa, Ottawa, K1N 6N5, Canada

P. Morrison, Institute of Arctic Biology, University of Alaska, Fairbanks, AK 99701

S. R. Morton, Ecology and Evolutionary Biology, University of California, Irvine, CA 92717

N. Mrosovsky, Department of Zoology, University of Toronto, Toronto, Ontario M5S 1A1, Canada

A. Muchlinski, The Museum, Michigan State University, East Lansing, MI 48824

X. J. Musacchia, Dalton Research Center, University of Missouri-Columbia, Columbia, MO 65201

M. Néchad, The Wenner-Gren Institute, Norrtullsgatan 16, S-113 45 Stockholm, Sweden

Jan Nedergaard, The Wenner-Gren Institute, Norrtullsgatan 16, S-113 45 Stockholm, Sweden

A. M. Nicol, Department of Animal Science, University of Alberta, Edmonton, Alberta T6G 2E3, Canada

R. C. O'Brien, Department of Anatomy, Harvard Medical School, Boston, MA 02115

P. R. Oeltgen, Lexington VA Hospital, Pathology Service 113, Lexington, KY 40507

G. Parker, Laurentian University, Biology Department, Sudbury, Ontario P3E 2C6, Canada

J. N. Pasley, Department of Physiology, University of Arkansas Medical Center, Little Rock, AR 92201

D. Pehowich, Department of Zoology, University of Alberta, Edmonton, Alberta T6G 2E9, Canada

E. T. Pengelley, Department of Biology, University of California, Riverside, CA 92521

H. Pohl, Max-Planck-Institut für Verhaltensphysiologie, D-8737 Erling-Andechs, Federal Republic of Germany
J. C. Rauch, Department of Zoology, University of Manitoba, Winnipeg, Manitoba R3T 2N2, Canada
W. J. Ray, Department of Physiology, School of Medicine, University of Missouri-Columbia, Columbia, MO 65201
F. Reinhard, Max-Planck-Institut für Biochemistry, Munich, West Germany
M. L. Riedesel, Department of Biology, University of New Mexico, Albuquerque, NM 87131
J. Roberts, 901 Charles River, Needham, MA 02192
Lennart Romert, The Wenner-Gren Institute, Norrtullsgatan 16, S-113 45 Stockholm, Sweden
R. Sable-Amplis, Institut de Physiologie, Universite Paul Sabatier, 31 Toulouse, France
E. Satinoff, Department of Psychology, University of Illinois at Urbana, Champaign, IL 61820
G. Workman Scott, Department of Zoology, University of Toronto, Toronto, Ontario M5S 1A1, Canada
D. M. Secoy, Biology Department, University of Regina, Regina, Saskatchewan S4S 0A2, Canada
R. C. Simmonds, P.O. Box 34551, West Bethesda, MD 20034
B. D. Snapp, Department of Biology, Stanford University, Stanford, CA 94305
W. A. Spurrier, Department of Physiology, Loyola University Medical Center, Maywood, IL 60153
L. L. Stebbins, Department of Biology, University of Lethbridge, Lethbridge, Alberta T1K 3M4, Canada
J. M. Steffen, Department of Biology, University of New Mexico, Albuquerque, NM 87131
J. K. Sullivan, Biology Department, Wesleyan University, Middletown, CT 06457
Ulf Sundin, The Wenner-Gren Institute, Norrtullsgatan 16, S-113 45 Stockholm, Sweden
Jan Svartengren, The Wenner-Gren Institute, Norrtullsgatan 16, S-113 45 Stockholm, Sweden
H. Swan, 6700 W. Lakeridge Road, Lakewood, CO 80227
D. M. Toth, Department of Anatomy, University of New Mexico, School of Medicine, Albuquerque, NM 87131
Janet Twente, Dalton Research Center, University of Missouri-Columbia, Columbia, MO 65201
John W. Twente, Dalton Research Center, University of Missouri-Columbia, Columbia, MO 65201
R. Van Buskirk, Department of Biology, Stanford University, Stanford, CA 94305

W. L. Veale, Division of Medical Physiology, The University of Calgary, Calgary, Alberta T2N 1N4, Canada

F. D. Vogt, Department of Biology, Wesleyan University, Middletown, CT 06457

Lawrence C. H. Wang, Department of Zoology, University of Alberta, Edmonton, Alberta T6G 2E9, Canada

J. M. Ward, Jr., Department of Physiology and Cell Biology, The University of Kansas, Lawrence, KS 66045

R. G. White, Institute of Arctic Biology, University of Alaska, Fairbanks, AK 99701

S. Wickler, Division of Biological Science, University of Michigan, Ann Arbor, MI 48109

J. S. Willis, Department of Physiology and Biophysics, University of Illinois, Urbana, IL 61801

L. C. Wit, Department of Zoology and Entomology, Auburn University, Auburn, AL 36830

Larry L. Wolf, Department of Biology, Syracuse University, Syracuse, NY 13210

D. Woodward, Physiology Programs, Office of Naval Research, Arlington, VA 22217

W. Wünnenberg, Department of Zoology, The University of Kiel, 23 Kiel, Germany

B. A. Wunder, Department of Zoology and Entomology, Colorado State University, Ft. Collins, CO 80523

B. A. Young, Department of Animal Science, University of Alberta, Edmonton, Alberta T6G 2E3, Canada

R. A. Young, Harriet G. Bird Memorial Laboratory, Stow, MA 01775

M. Zatzman, Department of Physiology, University of Missouri Medical Center, Columbia, MO 65201

Eugen Zeisberger, Zentrum für Physiologie, Justus-Liebig-Universität-Giessen, Aulweg 129, D-6300 Lahn-Giessen 1, Federal Republic of Germany

Preface

The broad topics of natural torpidity and thermogenesis constituted the central theme of the Fifth International Symposium on Mammalian Hibernation. The symposium entitled "Strategies in Cold: Natural Torpidity and Thermogenesis" was held at Jasper Park Lodge, Alberta, Canada, October 3–8, 1977. The present volume is a collection of the 20 invited review papers presented at the symposium, selected to review advances made since the 1971 symposium held at Aspen, Colorado, in the areas of molecular, biochemical, and cellular adaptations of natural torpidity and the role of the central nervous system in regulation of natural torpidity. The comparative aspects of natural torpidity, the mechanisms of thermogenesis, and the adaptation of man in cold were also covered.

The funding for this symposium was received from three sources: the Office of Naval Research, Department of Navy, U.S.A., under contract ONR NR 201-288; the National Research Council of Canada; and the University of Alberta. We are grateful for their generous support. The Department of Zoology at the University of Alberta, which was the hosting institution, provided much logistic support for the symposium, and we are particularly thankful to Dr. J. R. Nursall, Chairman of the Department of Zoology, and the secretarial staff headed by L. Overton. The preparation of the manuscripts demanded expert skills and tireless devotion. We are fortunate to have B. Donovan, who spent numerous hours typing and revising all the manuscripts to camera-ready form. Without her invaluable assistance, early publication of this volume would have been impossible. The staff at Academic Press provided valuable advice and assistance in publication of this book. Last, but not least, we wish to thank all individual authors for working closely and cooperatively with us; without their substantial contributions, this symposium and its textual proceedings would not have been possible.

STRATEGIES IN COLD

INTRODUCTORY REMARKS

Lawrence C. H. Wang
University of Alberta
Department of Zoology
Edmonton, Alberta, Canada T6G 2E9

Jack W. Hudson
Section of Ecology and Systematics
Cornell University
Ithaca, New York 14853
U.S.A.

When survival is challenged by cold, homeotherms react with two options: to counter it or to escape from it. To counter cold, physical (insulation of various forms) and behavioral means are employed to reduce heat loss while physiological means, i.e., the various mechanisms of thermogenesis, are activated to increase heat production so that a constant body temperature can be maintained. To escape from cold, animals either migrate to warmer climates or exhibit torpor while remaining on native grounds.

Dealing with the broad topics of natural torpidity and thermogenesis, the Jasper symposium included four review sessions on the following subjects: 1) Comparative Aspects of Natural Torpidity; 2) Central Nervous System Regulation in Natural Torpidity; 3) Cellular and Biochemical Adaptation in Natural Torpidity; and 4) Thermogenesis and Man in Cold. Highlights of these review sessions are presented below to orient the readers as to the scope of coverage in this text.

From the five papers presented in the session on "Comparative Aspects of Natural Torpidity", several conclusions may be drawn. The regularity with which annual hibernation cycles occur in nature has prompted the search for underlying mechanisms which drive and entrain such circannual rhythms. Three possible models exist: a self-sustaining circannual oscillator; a frequency demultiplication mechanism and a sequence-of-events model. It is apparent that species differences

ISBN 0-12-734550-7

and the lack of systematic data for many hibernating species preclude definite conclusions as to which of these models may be the most wide-spread. Future research on the role of external zeitgebers on circannual rhythm of hibernation could contribute important insight to understanding of the phenomenon. There is an impressive number of small mammals which exhibit daily torpor. Particular attention was paid to the inability of these animals to survive body temperatures much below 10-15°C and the possible underlying limiting mechanisms such as function of the heart at low temperature. The realization that only a fraction of the population exhibit torpor suggests torpor could be an adaptation for the population rather than the individual and further substantiates the need for more field observations. The pattern of hibernation and the energetics of hibernation under natural conditions have always been intriguing. Radio-telemetry has revealed that in a population of ground squirrels, the pattern of hibernation is more predictable and the duration of hibernation bouts longer under natural conditions than they are in the laboratory. The energy saved by utilizing hibernation rather than remaining homeothermic, could amount to 88% for the hibernation season. The use of an indirect calorimetric and time-budget approach in such studies also allows the calculation of energetic aspects associated with individual phases of a hibernation cycle; such an approach should prove to be useful in future research on energetics of torpor in other species. A frequently asked question is "if hibernation is so energetically advantageous why is hibernation not more prevalent among small mammals?" One answer is to be derived by viewing torpor from a cost-benefit perspective. This takes into account influences of prevailing environmental factors such as temperature variations, energy availability, cost of activity, predation constraints, and social interactions, which could act as constraints on evolution and occurrence of torpor in a given species. A review of the progress made in the last thirty years on studies of natural torpidity presented clear evidence that physiological changes observed in different species are not identical and that a broad generalization based on data from a single species may confuse rather than clarify the mechanisms underlying natural torpidity.

Five papers were presented in the session on "Central Nervous System Regulation in Natural Torpidity". The various neuronal models for body temperature regulation in euthermia and in hibernation were brought up to date. A model for hibernation was elaborated in which the temperature-response curve of the pre-optic neurons is exponential and that thermogenesis can be initiated or terminated in response to decreasing or increasing body or ambient temperatures, by disinhibition or inhibition of a thermogenic neuronal pool. Recent studies revealed that pre-optic neurons of hibernators remain responsive to hypothalamic temperature below 10°C while similar neurons in non-hibernators become insensitive below a hypothalamic temperature of 30°C. Although neurons in the medial and dorsal hypothalamus are not thermally sensitive in the hibernators (as are their counter parts in the anterior hypothalamus), they are important in integrating thermal inputs and controlling thermoregulatory outputs to effectors. Undoubtedly, future research on single-unit activities will provide much needed data in formulating mechanisms of thermoregulation in the hibernators. Electrophysiological studies on sleep demonstrated that entrance into hibernation is an extension of the non-REM sleep with a reduction of hypothalamic threshold temperature. The reduction of proportional constant for thermogenesis is also typical for both sleep and hibernation. Thus, neurophysiological control of hibernation could have evolved from non-REM sleep. Obviously more studies are needed from a diversity of species before this interesting concept can be firmly established. One paper dealing with the regulatory role of autonomic nervous system on hibernation proposed that the hibernation state is maintained via strong parasympathetic suppression (autonomic hibernation maintenance response, AHMR) of the sympathetic reflex, which shows progressive irritability as hibernation progresses. The origin of the AHMR is presently conjectural and begs the question of its control. The role of biogenic amines in thermoregulation, in particular, that of serotonin, has generated much interest in recent years. The greatly increased brain concentration and turnover rates of serotonin during preparation for hibernation and during hibernation suggest a functional involvement of the central serotonergic pathways in hibernation. Since the destruction rate of 5-HT was observed to be constant, the increased turnover of 5-HT is largely due to increased substrate for 5-HT synthesis. In this regard, it was particularly interesting that enhanced occurrence of hibernation could be demonstrated when animals were fed with a tryptophan-rich diet, i.e., given ample precursor of 5-HT. However, much remains to be learned as to the mechanisms by which the increased 5-HT turnover facilitates hibernation.

In the review session on "Cellular and Biochemical Adaptation in Natural Torpidity", four papers were presented with the aim of elucidating the adaptive modifications at the membrane, biochemical, and cellular levels as well as the endocrine regulation of metabolism in the hibernators. At the membrane level, it was concluded that no unifying concept of the role of lipid phase-transitions in response to hibernation can be developed at the present time since qualitative differences exist depending on the particular membrane-bound enzyme system as well as the subcellular localization of the enzyme. It was clear, however, that mathematical models more powerful than the conventional Arrhenius formula are needed for analysis of effects of temperature on membrane lipid phase-transitions and of the roles of proteins in hibernation. The discovery that isoenzymes of different kinetic characteristics particularly suitable for hibernation or for euthermia dominate, respectively, the preparation for hibernation and the emergence from hibernation suggest that biochemically, adaptive adjustments of certain key regulatory enzymes for energy metabolism are an integral part of hibernation. It remains to be documented how such changes are initiated and how such regulatory enzymes maintain their functions at low body temperatures during hibernation. It is recognized that tolerance of mammalian cells to cold may be dependent on differences of species or tissue, isolation from organism as well as prior experience to cold. Further investigations into the differential effects of damage by cold to the cells is needed. These studies should include a determination of damages affecting the physiological functions and reproductive capacity of the cell. While very little is known about how cold damages the reproductive ability of the cell in an organsism, the damage by cold to physiological functions of the cell seems to act via its effects on energy metabolism or ion regulation. Our knowledge of endocrine functions in hibernation remains fragmentary. The seasonal cyclic activity of the thyroid gland seems to be correlated with lipid metabolism and protein metabolism rather than with thermogenesis. The typical autumnal involution of the gonads can be prevented by continuous cold exposure from April onward and inactivity of the adrenal medulla often precedes the onset of the hibernation season. Clearly, much is needed to be learned about endocrine influences on energy metabolism during the different phases of the annual hibernation cycle.

A special session on "hibernation trigger" was included in this symposium to update research activities in this exciting area. A historical account of past developments was presented by a film on the "hibernation trigger" and this was followed by a theoretical discussion as to the possible mode of action of the "trigger". The session was concluded by a progress report on the chemical characterizations of the "trigger". Experiences using the "trigger" transfusion technique were exchanged among different laboratories. Although rates of success varied, it was felt that progress in isolation and purification of the "trigger" would undoubtedly help to resolve much of the apparent differences.

There were five papers included in the session on "Thermogenesis and Man in Cold". Aspects of biochemical mechanisms for non-shivering thermogenesis in brown adipose tissue and skeletal muscle, the regulation of non-shivering thermogenesis by neurohumors, the neural mechanisms underlying thermoregulatory threshold deviations in thermal adaptation and the adaptation of man in cold were discussed. At the biochemical level, the recently developed isolated cell preparation of hamster brown adipose tissue which shows characteristics of hormonal sensitivity and duration of thermogenesis typical of *in vivo* responses, provides a suitable system for fruitful studies on mechanisms of non-shivering thermogenesis and mitochondrial biogenesis. The mediatory role of norepinephrine in stimulation of non-shivering thermogenesis possibly occurs via altered mitochondrial coupling, enhanced ion transport across the plasma membrane and ion transport across the mitochondrial membrane. The interaction between catecholamines and other hormones (e.g. thyroid hormone, TSH, growth hormone, ACTH) on thermogenesis however, remains to be explored. In man, intermittent exposure to severe cold is more typical than continuous exposure to moderate cold as is the case in experimental animal models. Although the existence of non-shivering thermogenesis in adult man still remains a possibility, decreased sympathetic response, enhanced parasympathetic activity and a diminution of pain sensation seem to be general adaptations to intermittent exposure to severe cold in man. A downward shift of the central threshold for thermogenesis possibly accounts for the reduction of shivering and reduced body temperature level observed in man after intermittent cold exposure. This shift in central threshold for thermogenesis is likely mediated by norepinephrine. Much information is needed however, on autonomic effects of biogenic amines and the adaptive capability for cross adaptations, i.e., protection against one or more stressors after repeated or prolonged exposure to another stressor in man.

In addition to the 20 review papers, there were 55 voluntary papers also presented at the symposium. Abstracts of these papers are to be published in the April, 1978 issue of Journal of Thermal Biology.

Session I

Comparative Aspects of Natural Torpidity

NATURAL TORPIDITY, PROBLEMS AND PERSPECTIVES

Charles P. Lyman
Department of Anatomy
Harvard Medical School
Boston, Massachusetts 02115

and

Museum of Comparative Zoology
Harvard University
Cambridge, Massachusetts 02138

ABSTRACT

The progress during the last thirty years in the study of natural torpidity and hibernation in mammals and birds is presented briefly. It is emphasized that the abandonment of the warm-blooded state as a strategy for survival during unfavorable periods has been developed by many different groups, and that the present evidence indicates that the physiological changes which occur are not identical in each case. Broad generalizations supported by evidence from a single species are risky at best and may actually confuse rather than clarify.

ISBN 0-12-734550-7

When Dr. Wang asked me to open this symposium with a few introductory remarks, my immediate reaction was "Why me?" Although I am probably the only one in this group who has read Raphäel Dubois' masterful Physiologie Comparée de la Marmotte (6) published in 1896, from cover to cover, still this did not in itself seem to be reason enough to pick me for this difficult task. After considerable thought, I decided that my extreme age was the deciding factor and the intent was to capture my pearls of wisdom before I retired to a life of leisure.

The title of my talk is "Natural Torpidity, Problems and Perspectives." I hasten to explain that it was Dr. Wang who chose the title and that I feel in no way adequate to cover all the ground that this title implies. Instead, I would like to continue on the geriatrics theme and suggest that the subtitle should be "How it feels to be in hibernation for thirty years." Taking my theme from the subtitle, I will try to give you a little bit of historical retrospection concerning the research on cold, torpor, and hibernation in mammals and birds, starting at the end of World War II, and then offer a few ideas about future research.

In doing this, I realize that I may say some things which may appear critical and which in fact may be quite wrong, but the purpose of this symposium is to exchange ideas and come out with better ones, so I hope no feelings will be hurt.

Thirty years ago there was very little research on hibernation and torpor being carried out in the New World and the chief centers of study were Charles Kayser's laboratory in France and Paavo Suomalainen's laboratory in Finland. For many years torpor in mammals had been regarded as a manifestation of inferior temperature control, and animals that hibernated were believed to either lose their ability to maintain their body temperature at certain seasons of the year, or else be innately incapable of maintaining the warm-blooded state.

I have never been quite sure why this concept was accepted, for it was recognized that all hibernators could rewarm from the chilled state without heat from exogenous sources, and this would certainly indicate that the animal had not succumbed to the cold because it was incapable of combating it.

The concept of inferior temperature regulation during hibernation nurtured the idea that the animal had lost control of its internal environment and that it was no longer a homeostatic organism. An example of this concept is found in the book of Benedict and Lee on "Hibernation and Marmot Physiology" published in 1938 by the Carnegie Institution of Washington (2). In these studies the measurements of metabolism of woodchucks in natural hibernation and chilled under nembutal anesthesia are lumped together under the chapter heading "The Physiology of the Marmot in Hibernation", and the authors speak of "The lowest respiration rate in hibernation, under the influence of nembutal, is ....." Of course, the degree of homeostasis under various conditions of torpor is still the object of study in many laboratories, but I think it is generally agreed that the animal in natural torpor has not completely abandoned its avian or mammalian heritage of homeostasis.

Thirty years ago it was recognized that there were degrees of torpor among different species, but the multitudes of variation were not known. Bears were supposed to hibernate, but no one had yet dared to take their rectal temperatures. Now it has been shown that they undergo profound metabolic changes during their period of dormancy and that these changes are very different from the changes which take place during starvation (18, 19). Thirty years ago the reports on daily torpor in small rodents were anecdotal and no precise physiological measurements had yet been made. Now we have excellent data on some of the small rodents of North America, and there appear to be gradations of torpor among the various species, from a daily light torpor to a deeply hibernating state which lasts several days. Thirty years ago it was believed that the temperate zone microchiropteran bats were the only ones to indulge in torpor, but it is now clear that daily torpor occurs, not only in some tropical microchiropterans, but also in a small megachiropteran. Thirty years ago there were stories of torpor in birds, and that old lexicographer Samuel Johnson (3) claimed that swallows slept all winter. According to him, "a number of them conglobulate together, by flying round and round, and then all in a heap throw themselves under water and lie in the bed of a river." Now we know that hummingbirds indulge in daily torpor and that birds of the Caprimulgiformes family enter a state of hibernation which may last several days. Thirty years ago there was a good deal of information concerning the deeply hibernating mammals, particularly ground squirrels, marmots and the European hedgehog, but there was virtually no documentation of variation within the species. Now there is a great deal of data on the various types of torpor and hibernation in different species of mammals, so much so, in fact, that many feel it is time to categorize the various physiological states.

The many states of torpor, aestivation and hibernation have not been catalogued and Russell, in his report on temperature regulation of the cheirogaleid lemurs of Madagascar (21), expresses the frustration felt by many when he heads one section of his paper as "Cheirogaleids and Hibernation, Estivation or Whatever." He goes to say "Clearly, many of the fundamental terms and concepts of mammalian thermoregulatory physiology are in need of review." I agree with him that there is a need for review, but I do not think that this is the time to establish precise definitions for each physiological condition, for when one categorizes, one must set up limitations that admit no exceptions and I do not believe that we know enough to do this.

Although our knowledge of temperature regulation in North American and European rodents has advanced, yet we know virtually nothing about South American hystricomorph rodents. Actually Pearson (20) has suggested that one of the cricetid mice (*Phyllotis sublimis*) of the Altiplano of Peru may hibernate, and some of the hystricomorph rodents live under similar conditions in the same area. There is also very little known concerning the mammals of Asia, at least by those of us in the Western world. If torpor is a strategy to avoid arid conditions, study of the small mammals of the semi-Sahara in Africa should provide interesting comparisons with what we now know about animals living under similar conditions in North America.

The primitive fauna of Madagascar has been studied to some extent and it is known that several lemurs and insectivores enter a daily and seasonal torpor, but not enough is known to make physiological comparisons with torpor in the small mammals of North America. For example, it is tempting to inquire why the small tenrecs of Madagascar apparently enter torpor (7) while the shrews of North America do not. Both belong to the Order Insectivora. The shrews are smaller and must be exposed to at least as unfavourable an environment. Is there a physiological stricture which denies these animals the ability to have recourse to torpor as an escape from this environment? The tenrecs have been isolated from mainland Africa for at least 20,000,000 years and have followed their own evolutionary course. Is their type of torpor and hibernation more primitive physiologically than the more "modern" rodent hibernators? Can they, for example, sense the onset of potentially lethal cold and respond with an increased metabolic rate? There is some evidence that the Madagascan "hedgehog" (*Setifer setosus*) in torpor will increase its metabolic rate if its body temperature starts to drop too low (9). Is this the same reaction that one finds in the sophisticated hibernating rodents?

About thirty years ago, while recognizing that there were various degrees of torpor, I attempted to categorize the deepest type of torpor as it occurred in the golden hamster and coined the phrase "deep hibernation" for this state (11). This definition could be moderately precise, because it related the animal's lowest body temperature to the freezing point of water. However, we know now that there are important differences between species in deeply hibernating mammals.

In deep hibernation, the cerebral cortex of *Citellus beecheyi* and other ground squirrels and marmots is electrically active (10, 14, 23), while in the golden hamster it is electrically silent (4). Golden hamsters and golden mantled and thirteen-lined ground squirrels do not sense potentially lethal environmental cold with peripheral receptors (15), but marmots do (22) and dormice, which hibernate in relatively exposed places, apparently use peripheral receptors as a first line of defense (16). Most animals, in waking from deep torpor, undergo profound circulatory changes when arousing (13), but the aestivating Mohave ground squirrel arousing from torpor usually does not (1). Totally curarized bats can warm as rapidly from the hibernating state as can normal animals, presumably by using their brown fat, but curarization of dormice and golden hamsters greatly delays arousal (8) - and the list could go on.

These differences in species can lead to unexpected difficulties. For example, it is well known that many hibernators, particularly of the tribe *Marmotini*, become very fat in the autumn and do not store food for the winter of hibernation. When exposed to the cold in the autumn, they are apt to enter hibernation almost at once. Golden hamsters, on the other hand, store large amounts of food, lose weight when exposed to cold, and eat the food during periods of activity between bouts of hibernation. There is usually a delay of days or weeks after exposure to cold before these hamsters enter hibernation. It is logical to assume that the *Marmotini* hibernators prepare for hibernation in the autumn prior to cold exposure while the golden hamster undergoes a period of adaptation after exposure to cold before it hibernates. Thus it appears that there are two types of rodent hibernators: the preadapted species that fatten in the summer and starve during the hibernating period, and the non-preadapted species that store food and eat. Differences in intermediary metabolism have been found between these species (5, 24) and it is logical to relate these to the different behavioral patterns. However, we have recently completed a study on the Turkish hamster, *Mesocricetus brandti*, which, in contrast to the golden hamster, has a wide geographic range (17). Its distribution apparently surrounds that of the golden hamster in Syria and it is found across Asiatic Turkey over the Caucasus into Daghestan and also in the northern part of Iraq and northwest Iran. These animals, like the golden hamster, do not fatten in the fall, lose weight when exposed to cold, store food, and eat when they are active between bouts of hibernation, but they enter hibernation as readily as do ground squirrels. In this case, then, there appears to be no relationship between autumn fattening, starvation and readiness to hibernate immediately when exposed to cold.

Yet here again we run into trouble, for very little is known about the behavior in the field of either the golden or the Turkish hamster except that both of them store large quantities of food. It has been shown in the laboratory that denying the ability to store food in the cold will delay the onset of hibernation in the golden hamster (12) and this may also influence hibernation in the Turkish hamster (17). Do our laboratory conditions include some subtle factor, such as the amount of food available for storage prior to cold exposure, which is satisfied in the case of the Turkish hamster, but falls short of complete satisfaction in the golden hamster? Alternatively, is there some basic physiological difference between the two species so that one is preadapted for hibernation and the other is not? Only the realization that these are different wild animals living under different ecological conditions will lead to the correct answer to problems such as these.

Turning back again to yesteryear, one must remember that the complicated and useful apparatus of the modern laboratory simply did not exist and many of the techniques which are now so helpful and revealing had not yet been developed.

Instrumentation was relatively primitive, particularly multichannel temperature recorders which were just coming on the market. No techniques for chronic recording of electroencephalography, electromyography or electrocardiography of hibernating animals had been worked out. There were no techniques for implanting longterm vascular cannulae for monitoring circulation, sampling blood, or infusing drugs without disturbing the animal. The elegant use of indwelling thermodes to alter the temperature of specific loci in the brain was not to be perfected for many years, and infusion of minute amounts of drugs into critical areas of the brain was not regarded, at least by me, as being remotely possible. Analyzers which would measure oxygen consumption or CO_2 output of small mammals over long periods were just becoming available commercially and virtually all total body metabolic measurements were accomplished with home-made apparatus.

The development of new techniques and better apparatus has greatly increased our knowledge, but it carries with it some hidden dangers for those who seek the ultimate truth. The early experiments were relatively simple and easy to repeat, but the newer experiments are much more difficult, for they often require surgical preparation of the animal and very expensive apparatus. Furthermore, many of the experiments are long drawn out affairs, taking several days to carry out. When the investigator has obtained significant results with one species, the last thing he wants to do is to start all over again, using the same old, now boring, technique but with a different species as the experimental animal. Furthermore, no other investigator, unless he has some specific reason for doing so, is going to repeat the experiment with the only difference being that he is using another species.

I have no quarrel with experiments on a single species provided that the investigator limits his conclusions to the species investigated. The evidence for the polyphyletic evolutionary origin of the various types of torpor and hibernation appears to me to be overwhelming. These physiological modifications probably developed among the various groups of mammals and birds as methods of escaping from the costs of a tachymetabolic rate during physiologically taxing periods, whether of cold or of lack of food or of water. If this is the case, there is no reason to assume that each new adventurer into hibernation should or would solve its problems as did its remote relative.

Every investigator wants to hit upon a great general truth which will explain a fascinating and complex phenomenon. But remember, you are up against a group of animals which have been fighting for untold thousands of years against unfavourable environmental conditions. Of course, there are biological and biochemical restraints which limit the animal's maneuverability, and it is very important to identify those constraints. But within these strictures, the various species will have tried every trick in the book, and one should not expect, for example, that a bat and a hamster would have ended up with the same solutions. There are more things in heaven and earth, Horatio, than are dreamt of in your philosophy.

REFERENCES

1. Bartholomew, G. A. and J. W. Hudson. Aestivation in the Mohave ground squirrel. Bull. Mus. Comp. Zool. 124: 193-208, 1960.
2. Benedict, F. G. and R. C. Lee. Hibernation and marmot physiology. Carnegie Inst. Wash. Publ. #497: 1-239, 1938.
3. Boswell, J. Life of Johnson. London: Oxford University Press, 1927, Vol. 1, pp. 1-704.
4. Chatfield, P. O., C. P. Lyman and D. P. Purpura. The effects of temperature on the spontaneous and induced electrical activity in the cerebral cortex of the golden hamster. E. E. G. Clin. Neurophysiol. 3: 225-230, 1951.
5. Denyes, A. and J. D. Carter. Utilization of acetate-1-C^{14} by hepatic tissue from cold exposed and hibernating hamsters. Am. J. Physiol. 200: 1043-1046, 1961.
6. Dubois, R. Physiologie Comparée de la Marmotte. Annales de l'Université de Lyon. Masson et Cie, Ed., Paris, pp. 1-269, 1896.
7. Eisenberg, J. F. and E. Gould. The tenrecs: a study in mammalian behavior and evolution. Smithsonian Contrib. to Zoology 27: 1-134, 1970.
8. Hayward, J. S. and C. P. Lyman. Non-shivering heat production during arousal from hibernation and evidence for the contribution of brown fat. Mammalian Hibernation III. Fisher, K.C., A.R. Dawe, C.P. Lyman, E. Schönbaum, and F.E. South (editors). Edinburgh: Oliver and Boyd, 1967, pp. 346-355.
9. Hildwein, G. Evolution saisonnière de la thermorégulation chez l'Ericulus (*Setifer setosus*) (Mammifère Insectivore). C. R. Soc. Biol. 158: 1580-1583, 1964.
10. Kayser, C., F. Rohmer and G. Hiebel. L'E. E. G. de l'hibernant. Léthargie et réveil spontané du spermophile. Essai de reproduction et l'E. E. G. chez le spermophile reveille et le rat blanc. Rev. Neurol. 84: 570-578, 1951.
11. Lyman, C. P. The oxygen consumption and temperature regulation of hibernating hamsters. J. Exp. Zool. 109: 55-78, 1948.

12. Lyman, C. P. Activity, food consumption and hoarding in hibernators. J. Mamm. 35: 545-552, 1954.

13. Lyman, C. P. Circulation in mammalian hibernation. Handbook of Physiology - Circulation III. Ch. 56, pp. 1967-1989, 1965.

14. Lyman, C. P. and P. O. Chatfield. Hibernation and cortical electrical activity in the woodchuck (*Marmota monax*). Science 117: 533-534, 1953.

15. Lyman, C. P. and R. C. O'Brien. Sensitivity to low temperature in hibernating rodents. Am. J. Physiol. 222: 864-869, 1972.

16. Lyman, C. P. and R. C. O'Brien. A comparison of temperature regulation in hibernating rodents. Am. J. Physiol. 227: 218-223, 1974.

17. Lyman, C. P. and R. C. O'Brien. A laboratory study of the Turkish hamster *Mesocricetus brandti*. Breviora, Mus. Comp. Zool. 442: 1-27, 1977.

18. Nelson, R. A., J. D. Jones, H. W. Wahner, D. B. McGill and C. F. Code. Nitrogen metabolism in bears: urea metabolism in summer starvation and in winter sleep and role of urinary bladder in water and nitrogen conservation. Mayo Clin. Proc. 50: 141-146, 1975.

19. Nelson, R. A., H. W. Wahner, J. D. Jones, R. D. Ellifson and P. E. Zollman. Metabolism of bears before, during, and after winter sleep. Am. J. Physiol. 224: 491-496, 1973.

20. Pearson, O. P. Mammals in the highlands of southern Peru. Bull. Mus. Comp. Zool. 106: 117-174, 1951.

21. Russell, R. J. Body temperature and behavior of captive cheirogaleids. In: Lemur Biology. I. Tattersall and R. W. Sussman (editors). New York and London: Plenum Press, 1975, pp. 1-365.

22. South, F. E., W. C. Hartner and R. H. Luecke. Responses to preoptic temperature manipulation in the awake and hibernating marmot. Am. J. Physiol. 229: 150-160, 1975.

23. Strumwasser, F. Some physiological principles governing hibernation in *Citellus beecheyi*. Bull. Mus. Comp. Zool. 124: 285-320, 1960.

24. Whitten, B. K. and G. J. Klain. NADP specific dehydrogenases and hepatic lipogenesis in the hibernator. Comp. Biochem. Physiol. 29: 1099-1104, 1969.

CIRCANNUAL CYCLES IN HIBERNATORS

N. Mrosovsky
Departments of Zoology and Psychology
University of Toronto
Toronto, Ontario, Canada M5S 1A1

ABSTRACT

This review considers three possible ways of generating circannual cycles in constant conditions: i) a self-sustaining oscillator analogous to a circadian oscillator but with a characteristic frequency close to a year, ii) a frequency demultiplication mechanism with a circannual cycle based on approximately 365 circadian cycles, iii) a sequence of linked stages with each stage normally taking a relatively standard amount of time to complete. Data on circannual cycles in different mammalian hibernators are then outlined in turn but in all cases there is insufficient information to be sure what kind of mechanism operates. Widespread application of the concept of circannual cycles persisting but only expressing themselves under limited environmental conditions is considered unsatisfactory. It is suggested that further research on the importance of external factors in circannual cycles may be especially profitable.

ISBN 0-12-734550-7

Ever since Pengelley and Fisher's (62, 63) classic demonstration of the persistence of cycles of body weight and hibernation in golden-mantled ground squirrels kept in constant conditions, the phenomenon of circannual cycles in hibernators has intrigued and puzzled biologists. The existing data have been thoroughly reviewed in previous hibernation symposia and elsewhere (11, 41, 54, 58, 59). The present review, although it attempts at least to cite more recent findings, concentrates on theoretical distinctions. Its specific aims are twofold. First it considers a number of different possible ways of generating circannual rhythms. To help formulate issues, data from non-hibernating species as well as from hibernators are introduced from time to time. Second, after having discussed possible mechanisms for generating circannual cycles, the paper goes through available information on a number of mammalian hibernators.

I. WAYS OF GENERATING CIRCANNUAL CYCLES

A. Self-sustaining Oscillators with Characteristic Frequencies

It is natural that concepts flourishing in the field of circadian rhythms should soon take root in that of circannual rhythms, especially since, at least superficially, both types of rhythms have several properties in common. Among these are persistence in constant conditions, over several cycles, with periodicities different from but approximating those of the external oscillators, a relative compensation of periodicity for temperature and an ability to show cycles without ever having experienced them in nature. For mammals this last point can be demonstrated even more convincingly in the case of circannual rhythms because with animals born into constant conditions in captivity there is no question of them ever having experienced an annual cycle in nature (25, 57, Boshes, unpublished, with *Citellus lateralis*); in the case of mammals exhibiting circadian cycles after developing in constant conditions (73) there remains the possibility that experience of the mother's rhythms *in utero* or soon after birth was important.

To account for the properties of circadian cycles just mentioned, the existence of self-sustaining oscillators (sometimes called pacemakers or endogenous clocks), has been proposed. The oscillators are assumed to have characteristic frequencies that are relatively inflexible but still susceptible to entrainment by light-dark cycles or other Zeitgebers. By analogy then, circannual rhythms might be controlled by generally similar oscillators but with a natural frequency close to a year rather than a day.

There has been some debate on how close the analogy between circadian and circannual rhythms really is. For instance it has been pointed out that in ground squirrels cycles are sometimes 8-9 months rather than close to a year, and that there are suggestions that cycles might be slowed down by low temperature (40, 41). More recent data, however, suggest that the effect of temperature on cycle lengths, if it exists, is not overwhelming, and certainly nowhere near to producing a Q_{10} of 2 (43, 79). On the matter of the cycle lengths much shorter than a year, Pengelley and Asmundson (58) have countered that, given within species variability, some short periods are to be expected. Data on variability in the period of circadian rhythms makes this point even more plausible. Rodent species with circadian periods close to 24 hrs show less variability than those with periods generally further away from 24 hrs (67). Similarly when one compares the variability of circannual periodicity in ground squirrels (43, 56) and warblers (22) to how closely the average period approximates a year, one finds that species with periods further from a year, the ground squirrels, tend to have greater standard deviations than the species with periods generally closer to a year, the warblers. In this context the existence of some very short free-running periods in ground squirrels could be taken not as spoiling the circadian analogy but rather as reinforcing it.

These various views are given to emphasize that the validity of the analogy between circadian and circannual is not a matter of proof or disproof but of how different people look at the same facts. Is a cycle length of 8 months circannual or not? It is a matter of opinion. In trying to evaluate the circadian-circannual analogy then, perhaps we should first ask what we expect of analogies and what their dangers are. "Although analogies prove nothing, they can have great value in providing generative insights....The danger in an analogy is not that it is wrong, but that it is overdrawn" (41). More specifically, "if it is overdrawn, the analogy between circadian and circannual rhythms unproductively biases the kinds of question which people studying the latter ask" (34). "Premature adoption of models devised for circadian rhythms may be restrictive in terms of the experiments they stimulate" (42).

In this context the possibility that temperature compensation may be somewhat poor in certain circannual cycles, or the fact that periods can differ considerably from a year, cannot disprove the idea that there is a circannual oscillator (they could tell one something about the properties of that oscillator, if it exists). However, acceptance of the idea of circannual oscillators analogous to circadian oscillators does tend to lead to experiments designed to reveal or concentrate on oscillator type properties rather than other aspects of circannual cycles. Whether this is desirable or not is again a matter of opinion, and discussion of this is best deferred until we have considered alternative lines of experimentation that other viewpoints might stimulate (sections 2 and 3 below). But one technical criticism of comparisons of properties of circannual and circadian oscillators can be mentioned right away. Menaker (34) has argued that it may be impossible ever to characterize the properties of circannual oscillators adequately because of the long time spans involved. Circadian rhythms sometimes have to go through several cycles before transients have passed, and after-effects of changing lighting schedules can persist through 100 cycles (67). "Formal model building in the study of circadian rhythms has depended almost exclusively on kinds of data which simply will not be available for circannual rhythms" (34). This is a compelling argument, but it is perhaps overly pessimistic. Obviously one cannot perform experiments in which an individual animal is followed through many cycles before and after a manipulation. But if sufficiently large groups of animals are used to overcome individual variation, then it is possible to obtain information about whether phase shifts in circannual rhythms have occurred in response to a manipulation (43). Obtaining at least a crude phase response curve in a few cycles is not necessarily a logical impossibility.

A recent example of the kind of question that can be asked with the circadian analogy as a stimulant is whether Aschoff's rule applies to circannual rhythms. Pengelley *et al.* (61) have investigated this by keeping groups of animals in constant light of different intensities, rather than following changes within animals as intensity is varied. They detected no differences between the free-running circannual periods of ground squirrels kept in 500 lux and 24 lux, or between those and other groups that were enucleated. Perhaps with their sample size individual variation masked any differences but their data gave no hint of any effects.

The absence of light intensity effects does not disprove the idea that there is a circannual oscillator analogous to a circadian one. There are exceptions to Aschoff's rule as applied to circadian rhythms: for instance man does not consistently obey the rule (86). Moreover, since it has not been demonstrated that light is a Zeitgeber for ground squirrels, there is no particular reason to suppose it would have important effects on periodicity (cf. 67). If, as seems more likely, photoperiod is the Zeitgeber, then a better test of Aschoff's rule would be to compare circannual periods of animals kept on short and long days. If a Zeitgeber can be found, then it may well be that its intensity when continously present would have effects similar to those incorporated in Aschoff's rule.

This brings one to the question of Zeitgebers. It is widely recognized that one of the weakest parts of the circannual-circadian analogy for hibernators is that no Zeitgeber for circannual cycles has been satisfactorily demonstrated (34, 40, 41, 59). However, recently some interesting results have been obtained with European hamsters (6). This species has a circannual cycle in constant conditions but on accelerated photoperiodic schedules the cycle can be as short as 7 months. Similar, more dramatic, results have been obtained on cycles of antler growth in sika deer, non-hibernators (18).

Such demonstrations of circannual cycles in constant conditions but much shorter cycles under accelerated photoperiods suggest that photoperiod, or possibly simply changes in photoperiod irrespective of whether increased or decreased (17), can act as a Zeitgeber for circannual cycles. But they have much wider implications than this. They raise the whole question of the distinction between circannual cycles and photoperiodically driven cycles and how photoperiodism acts. A cycle is described as circannual if it persists in constant conditions. If the cycle does not persist in constant photoperiod but can be produced by changing photoperiods, or if the various parts of the natural cycle can be produced in succession by photoperiodic stimulation, then it can be described as a photoperiodic cycle. These are simply descriptive classifications, but most people assume that in cycles described as circannual there is a large measure of endogenous control, and that in cycles described as photoperiodic external controls are of paramount importance. A problem then arises when both photoperiodic effects and circannual cycles can be demonstrated in the same species (eg. sika deer, European hamsters, starlings). Are endogenous or exogenous controls most important in these species? If, as it seems, both types of controls are operating, then what circumstances determine which type of control predominates? What is the relationship between the two kinds of control?

One way of unifying these complexities is to argue that daylength does not act by directly inducing changes such as gonad growth, antler growth or hibernation but rather that it acts by entraining underlying cycles in these variables. In other words the distinction between a circannual cycle and a potoperiodic cycle, useful as it is to describe phenomena, does not necessarily reflect a basic difference in mechanism. The cyclical mechanism in a photoperiodic species might be based on a circannual cycle with photoperiod having a synchronizing rather than an inducing role.

This elegantly unifying idea creates another problem, but one that circadian-circannual analogy is capable of rising to. If the accelerated cycles seen in deer or European hamsters on accelerated photoperiodic schedules are really a reflection of the entraining effects of daylength on an underlying circannual cycle, then without any changes in photoperiod, in constant conditions, one should see that underlying cycle in action in a freerunning state. But in fact circannual cycles are not always evident. For instance, they are not manifest in deer kept on LD 12:12 but do appear when deer are kept on LD 6:18 or LD 18:6 (16); starlings do not display circannual cycles on LD 11:13 or LD 13:11 but do on LD 12:12 (78). The circadian-circannual analogy provides a ready answer: the cycle is present but does not express itself under all these conditions (cf. 16, 19). This is a reputable concept in the field of circadian rhythms where there is evidence that underlying rhythmic processes can persist in the absence of external measurable rhythms. For instance, following electroshock rats become inactive and circadian wheel running rhythms disappear; after a few days when the animals recover from the electroshock the activity rhythms reappear and are phased in a way that would have been expected had no manipulation occurred (71). This indicates the persistence of a rhythm that did not express itself in the few days immediately after the electroshock. Similar conclusions can be drawn from experiments where tying weights onto leaves has prevented the expression of circadian cycles in the form of leaf movements (4).

In the field of circannual rhythms there are similar examples. For instance, temporary restriction of food does not abolish an underlying cycle in a programme for body weight in hibernators because on refeeding the animals return to weights appropriate for that time of year rather than to their pre-deprivation levels (see 44, for a review). Similarly preventing migratory restlessness by placing warblers in total darkness does not abolish circannual activity rhythms but only stops their expression because when the birds are returned to more normal nighttime conditions, activity returns to a value appropriate for that time of year (21).

There are then plenty of precedents for arguing that underlying rhythmicity, even if it is not reflected in an overt rhythm, may still be present.

Given this possibility, the circadian analogy in its most imperialistic form engulfs not only the circannual rhythms of species with both circannual and photoperiodic cycles in the descriptive sense, but also potentially all photoperiodic cycles, even those of species for which no circannual cycles have ever been found when animals are kept in constant conditions. In such cases, presumably, the right conditions for the expression of circannual cycles, which might be quite limited, have not been found.

Surely then, guided by the circannual-circadian analogy with its unifying power and rich source of ideas, the study of Zeitgebers for circannual rhythms and the relationship of circannual to photoperiodic cycles is an exciting line of research.

B. Circadian Cycles Generating Circannual Cycles

Rather than postulating separate slow frequency oscillators to account for circannual cycles, it would be economical if circadian oscillators could somehow be used to generate circannual rhythms, just as they can be put to use in photoperiodism or in sun-compass orientation. The relationship between circadian and circannual cycles is under active investigation at present (7, 14, 20, 23, 47). It would be satisfying if someone could come up with something as elegant as Bünning's hypothesis for photoperiodism, but at present most of the work is essentially descriptive. Perhaps these efforts will provide the data that lead to a good hypothesis.

There is, however, one hypothesis of how circannual and circadian events might be related that is well formulated enough to test, namely the frequency demultiplication hypothesis. Several groups have been interested in this idea independently. As Gwinner (20) puts it: "There are physical devices which generate low-frequency rhythms by frequency demultiplication of high frequency rhythms. An example is the electric clock which is equipped with a mechanism that transforms the commercial periodicity of 50 cycles/sec to 1 cycle/day so as to provide information about the time of day. In an analogous way, organisms may be capable of transforming circadian rhythms into circannual rhythms to provide information about the time of year."

This idea can be investigated by seeing whether animals taking short subjective days in constant conditions also take short subjective years. This has been done by Gwinner (20) with starlings; he found that the average period of the circadian activity rhythm correlated significantly with the circannual period of moulting. This result is consistent with the frequency demultiplication hypothesis, but for various reasons, some mentioned by Gwinner, it is not strong evidence in its favour. First, circannual moult cycles took rather more than the predicted 365 circadian cycles. Second, data on how many subjective circadian cycles the birds actually went through in a circannual cycle were not available; instead estimates of average circadian periodicity were based on those parts of the year when the starlings showed clear activity rhythms. Third, if there was any internal desynchronization between different rhythmic functions, as has been found with circannual rhythms in other avian species (22), then the moult may not have been the circannual rhythm that best related to the number of circadian activity cycles. More important, even when correlations between circannual and circadian cycle lengths are demonstrated, they do not in themselves prove that the circannual is based on the circadian; it may simply be that both circannual and circadian cycles tend to be long in some individuals while in others they tend to be short. Finally, the results of experiments where starlings have been entrained to short or long days do not obviously conform to predictions from the hypothesis (Gwinner, personal communication). Despite such problems in this particular case, frequency demulitiplication remains one possible mechanism of generating circannual rhythms (24) and the cylces of hibernators can be investigated with this in mind.

In Toronto, we tried to do this by looking at the relationship of circadian to circannual periodicty in golden-mantled ground squirrels kept in constant light but were frustrated by lack of clear circadian cycles during much of the year. This is often a problem with species that become torpid during part of the year. However, one animal out of 13 did not become torpid and gave excellent activity records for more than a year. In this individual there was a circannual cycle in the period of the circadian cycle. This cycle of cycles corresponded with the circannual body weight cycle (47).

Data like these showing relationships between circadian and circannaul cycles do not of course establish causal links. It is quite possible that the changes in circadian period, which were not very large in any case, resulted from the different hormonal states that occur in the course of the circannual cycle of reproductive state. On the other hand, it also remains possible that the changes in the circadian cycles were instrumental in producing the circannual cycle. If there are changes in the period of the circadian rhythm in constant conditions, then when entraining agents are present there might also be changes in the phase angle. Such changes could lead to the presence of external light coinciding with critical parts of circadian cycles of sensitivity to light (external coincidence). However, it is more likely, given that ground squirrels show circannual cycles both when blind and when in LL, that there are changes in the temporal relationships between different endogenous circadian rhythms whose periods change differently over the course of the year (internal coincidence). But it must be stressed that neither type of coincidence explains what might cause circadian patterns to vary over the year in constant conditions. Coincidence systems might well be a way through which circannual cycles create different physiological states at different times of year, but the fundamental problem of what generates the circannual rhythmicity is untouched by these explanations or by similar ones offered by Gwinner (20).

Perhaps at this stage more general tests of the proposition that circannual cycles are based on circadian cycles are needed. In birds and mammals circadian rhythmicity can be virtually abolished, or certainly drastically altered, by pinealectomy or by suprachiasmatic lesions respectively (35, 88). With such interventions it would be possible to test whether circadian cycles are necessary at all for long duration rhythmicity. If they were not, it would dampen the search for instructive relationships between circadian and circannual cycles. In ground squirrels sustaining partial damage to the suprachiasmatic nucleus (and its caudal projections) as a side effect of lesions aimed at other areas, there was no sign of loss of circannual body weight cycles (43). But this is only suggestive; complete lesions of the suprachiasmatic nucleus and concomitant monitoring of circadian activity are needed to investigate this matter properly.

C. Sequence of Stages, Chain of Interval Timers

Alongside the circadian-circannual analogy and the idea that circannual cycles may be based on circadian cycles, there has arisen a third class of explanations of circannual cycles. This views them as composed of a series of stages that in normal circumstances take a relatively fixed amount of time to complete. Several versions of this idea with slightly different emphasis have been advanced (13, 24, 39, 40, 83). The common point in these accounts is that, although some people may still wish to use the term oscillator for describing certain formal properties of the rhythm, there is no need to postulate a special oscillator mechanism controlling other variables or a clock to explain circannual cycles. Instead they could be generated by a sequence of events each triggering, on completion, the next event in the chain. As long as the last event was linked back to the first in the series the cycle could persist in constant conditions. By implication, temprqarily blocking or speeding up parts of the chain should lead to alterations in the period of the cycle. More generally, if circannual cycles do not depend on an oscillator with a characteristic frequency it should be possible to alter their frequency over wide ranges.

In one sense, it is unfortunate that the terms "sequence of events" or "sequence of stages" have been used to label this way of looking at circannual cycles because obviously an oscillator must also comprise, in the final analysis, a sequence of events (34, 41, 42). On the other hand, such terms may be valuable because they steer clear of assuming certain oscillatory type properties in the system, especially the idea that there is a characteristic inherent frequency or that the system is necessarily self-sustaining (42). They place the emphasis on attempts to characterise the stages rather than on the formal aspects of the rhythmic mechanisms (34).

Another way of conceptualizing these distinctions is to think of a continuum. At one end are systems that have characteristic, relatively inflexible, frequencies, are self-sustaining and well protected against environmental perturbations (e.g. many circadian cycles). At the other end are systems with more flexible frequencies, requiring input to keep going and susceptible to environmental changes. If one wants to stress fundamental similarities one can call everything on this continuum by the same name. If one wants to stress differences between ends of the continuum (and even quantitative differences require something different in the underlying mechanism) one can use labels to distinguish the ends of the continuum. The sequence of stages approach considers circannual cycles as being toward the flexible end of such a continuum.

If hibernator cycles are best viewed in this way it should be possible to identify these stages and then to arrest or accelerate them and so alter the period of cycles. Early attempts to do this were based on the guess that the stages might consist of particular metabolic states reflected by body weight (39, 83). However, experiments that forced animals off body weights normally prevailing at a given phase of the cycle, although not continued for long enough with large enough groups of animals (40, 41), do not give much support to this idea (see 44, for a review). As mentioned, when the manipulations interfering with weight end, the animals return to weights appropriate for that time of year and the cycle continues, without a phase shift, as if nothing had happened.

Such demonstrations do not undermine the sequence of stages viewpoint but show rather that the actual states of fat reserves are not critical events in the chain. Regardless, it seems, of the actual amount of fat, there is an underlying programme for fatness that cycles on a circannual basis. At each phase of this cycle there is a set-point for fat that is defended, but the value of this set-point changes over the cycle. Perhaps the critical stages are composed of the underlying programme for how fat the animal should be at a given time of year rather than the actual state of the fat reserves.

There is, of course, a potential sterility about this kind of argument if it is invoked too often. It would always be possible to argue one's way out of failure to identify critical parts of a sequence of stages by proposing some underlying process, untouched by the particular experimental manipulations used, and asserting that this comprised the real sequence of events making up the cycle. However, these are early years in research on circannual cycles. Moreover, the suggestion of an underlying cycle of set-points for fat is not something intangible because much is known about how to alter the levels of fat that are defended in non-hibernating mammals. In particular, lesions of the ventromedial hypothalamus lead to elevated weights and regulation around those elevated weights, while lesions of the lateral hypothalamus lead to weight loss and regulation around lowered weights (26, 68). Making similar lesions in hibernators should alter not only the actual weights but also the programmed weights and might therefore also interefere with the progression through a cycle of programmed weights. This idea has been tested by making weight-elevating hypothalamic lesions in ground squirrels during the phase of the circannual cycle when they are losing weight (43). It was found that lesions were associated with shorter body weight cycles. This might have been due only to some temporary disruption in the link between actual weights and programmed weights; more telling was the finding that the lesioned animals continued to be phase-advanced relative to the controls even on the cycle subsequent to the one when the lesion was made. However, the effect was only marginally significant. The sequence of stages formulation leads to the expectation that with the correct manipulations large phase shifts even within a few cycles could be obtained.

A more promising approach may be trying to manipulate the duration of the reproductive stage of the cycle. Several authors have commented on the fact that circannual cycles in ground squirrels, while occasionally longer than a year (43, 79), are far more often shorter than a year. Pengelley and Asmundson (59) report that out of 61 cycles in golden-mantled ground squirrels kept in constant conditions 48 (79%)

were shorter than a year. One possibility is that reduction of breeding activities in capitivity is often responsible for truncating the cycle (41, 83). This idea has a precedent in the 9.6 month cycles of equatorial sea-birds. If these birds lose their eggs or young they seldom lay again but they do return to breed again sooner than birds that raised chicks successfully (82). Coming back to hibernators, similar observations suggest that their cycles can be abbreviated by lack of full reproductive activities: female ground squirrels giving birth after trapping in the spring go into hibernation later in the year than those producing no litter and this difference is maintained even in the second year in captivity (60). This provides circumstantial evidence for a relationship between short cycles and lack of breeding but does not prove the point because the two groups might have differed even before they were trapped. Moreover, even castrated ground squirrels sometimes have cycles longer than a year (64). It therefore has to be conceded that so far there have been no really convincing demontrations of arresting or accelerating stages of circannual cycles in hibernators.

Apart from lack of direct experimental support, it has also been suggested on other grounds that sequence of stages explanations are likely to be inadequate for circannual cycles, at least in some species. Gwinner (22) has argued that internal desynchronization between various different circannual rhythms in birds runs counter to the idea of one stage ushering in the next. However, even if internal desynchronization were demonstrated in hibernators it would not dismiss the sequence of stages viewpoint. If one can postulate several self-sustaining oscillators with characteristic frequencies to account for internal desynchronization then one can just as reasonably postulate several sequences of stages running together like threads in a skein of multicoloured wool but capable of sliding relative to each other if not anchored by appropriate environmental stimuli. This brings up the matter of how external influences are viewed from the sequences of stages perspective.

In its purest form a sequence of stages could operate without environmental input at all if the last stage were linked back to the first. Perhaps this occurs in the case of the 9.6 month cycle in tropical birds. But in the case of hibernators in temperate zones there must be some mechanisms for insuring that in nature the durations of the various stages add to a year. Environmental stimuli could act by stretching or compressing one or more of the component stages.

Another possibility is that, rather than only stretching or compressing stages of the cycle that would still occur anyway, environmental stimuli are actually essential at certain stages for the cycle to proceed at all (cf. 74). This alternative is worth careful consideration because it relates to the problem of why circannual cycles do not appear under various experimental conditions. This has sometimes been ascribed to a failure of the cycles to express themselves under certain conditions (see above). Saying that cycles do not express themselves is a way of acknowledging the importance of external factors while retaining the notion of a self-sustaining endogenous oscillator. But if a failure of cycles to appear under a variety of conditions--conditions that differ moreover for different species--is a widespread phenomenon, as seems to be the case with circannual rhythms, then surely we need some more specific formulation of the role of environmental stimuli than simply saying the cycle does not express itself. One possibility is that there are hour-glass mechanisms (cf. 14) with constraints on the external stimuli that can tip the hour-glass. Some specific examples may help illustrate how such a mechanism might encompass the lack of expression of cycles.

Many years ago it was reported that thirteen-lined ground squirrels kept in the cold remained in reproductive condition much longer than those kept in the warm (85). More recently it has been found that with the same species kept in constant conditions at different temperatures a fairly high proportion of the animals did not show persistent body weight rhythms, especially in the colder of two rooms (48; Fig. 1). These animals, having gained weight their first fall in captivity and then lost it over the subsequent winter, remained thereafter at low spring weights. However, when temperatures were raised later, weight

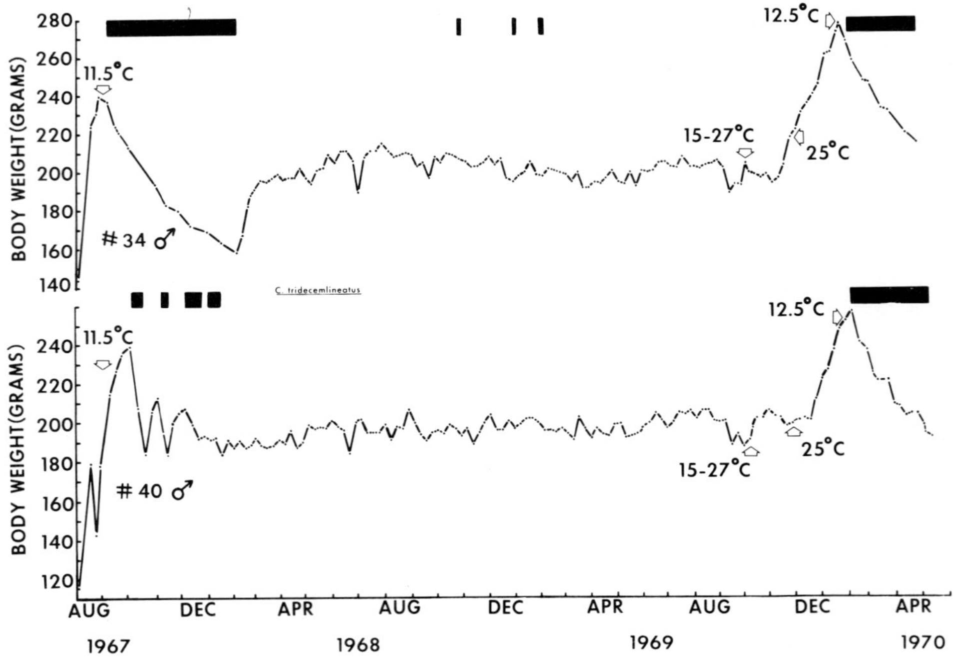

Fig. 1. *Body weights of individual thirteen-lined ground squirrels, illustrating loss of circannual cycles in a cold room. Black bars indicate that the animal was found hibernating on at least 2 days out of a week. Arrows show room temperature (data from 48).*

gain followed within a number of weeks. Two points are emphasized. First, it is not that the animals were incapable of gaining weight or hibernating under the conditions of the experiment because they could do so in the first season in captivity (e.g. No. 40, Fig. 1). Second, the squirrels did not simply become arrhythmic but their rhythms stopped at a specific phase of the annual cycle, the spring phase associated with low body weights and reproductive proclivities. These points suggest that a certain degree of warmth in the spring phase of the cycle, the exact amount varying among individuals, is necessary for regression of reproductive glands and subsequent fattening. The cycle can be thought of as a series of stages that require environmental input at one particular stage. If this input, warmth in the present example, is always present, then the cycle would persist in constant conditions and give the appearance of sufficient endogenous control whereas in fact it might be that external stimuli were necessary. If the animals were in the cold, then the cycle would proceed until the spring phase and then stick there until adequate input of warmth occurred. Consideration of the control of cyclicity in another hibernating species may help elaborate this type of explanation and show it to be plausible.

Golden hamsters are photoperiodic: when kept in short days the testes regress within a few weeks. However, without further changes in photoperiod, there is spontaneous recrudescence about 4-5 months later (69). Unless they now receive long photoperiods for a while they remain in reproductive condition, even though still kept on short days. Long photoperiods are necessary to break a refractory phase to short photoperiods. Golden hamsters therefore need two environmental inputs to produce a full cycle in reproductive state. There is nevertheless a certain amount of endogenous control because, once stimulated with short photoperiods, changes proceed for about 6 months until the hamsters stick in the reproductive phase.

Now one could easily imagine a system with the endogenous control more highly developed so that changes initiated by short photoperiods last close to a year rather than 6 months. If these changes brought the animals back to their original state, then the still present short photoperiods would reinitiate regression of the testes and a circannual rhythm would be seen. But if the animals were kept continously on photoperiods incapable of initiating the whole chain of events, then a circannual cycle would not be seen (or, as some would say, the circannual cycle would not express itself). Thus circannual cycles in some hibernating species may not be radically different from the photoperiodic cycles in hamsters where the role of external stimuli is clearly recognizable. It might be that the essential difference is that only one rather than two environmental inputs are necessary. The consequences of such a modification however could make the control systems look much more different than they really are. If only one input is necessary, the cycle will persist when the necessary environmental stimuli are always present. The range of constant conditions under which the circannual cycle appears will depend on how broad a set of environmental stimuli are capable of initiating the chain of events that lasts about a year and brings the animal back into a state where these events can be reinitiated. If such an hour-glass mechanism is easily tipped over, then circannual cycles will appear in many different conditions; if it is exclusive, then they may appear in only very limited conditions. In summary then, the sequence of stages viewpoint certainly allows for the importance of external stimuli, and it offers a fairly specific means of conceptualizing the puzzling way that circannual cycles occur in some conditions but not others.

II. CIRCANNUAL CYCLES IN HIBERNATORS

There follows a review of circannual cycles in different mammalian hibernators. A cycle is accepted as circannual if in constant conditions at least one full cycle occurs with the animal returning to its original phase. Obviously demonstration of persistence through several cycles would be stronger evidence but this paper will risk errors of commission on this matter. Cycles are not considered as proven to be circannual if only data on animals grouped together are presented (5, 52) even though they may well have been occurring in individuals. One problem in reviewing this subject is that heroism in data collection is not always matched by completeness in data presentation. It is a pity when all the work that goes into studying animals over such long periods does not culminate in the tables or graphs of periodicities and other parameters for individual animals that are so necessary to an understanding of this relatively new phenomenon. Grouping of cycles with different periodicities can be very hard to interpret. Also statements on temperature levels without any indications of variation should be viewed with caution; even small temperature fluctuations might be important for some species.

Circannual cycles in hibernators have been measured in a variety of ways. The only general statement that can be made is that, paradoxically, hibernation itself is one of the less reliable indices. This is because some animals may become torpid one year and not the next (6, 43), or, if they do, then only for a short time. Also the onset and cessation of hibernation are sometimes poorly defined with occasional short-lived torpor occurring an appreciable amount of time before long-lasting torpor develops (79). Body weight, food or water intake, even if not checked as frequently, may be better measures of circannual cycles in hibernators.

Because of lack of information it is impossible to progress with any certainty from description to deducing what kind of mechanisms underly circannual cycles in various species. However, to meet a reviewer's obligation to summarize the state of a field, Table 1 lists the species considered below together with the mechanism that seems most consistent with the available data. Apart from noting the suggestion of considerable diversity in mechanism, the reader should treat this table with caution: it is only an interim assessment, just 20 years after circannual cycles were first adequately demonstrated, of data that have many gaps, contradictions and possibilities for alternative interpretations.

A. *Cricetus cricetus* (European hamster)

An extensive study by Canguilhem *et al.* (6) has demonstrated circannual cycles in European hamsters. They followed body weights of their animals in LL, DD (.2 lux), LD 12:12 and on an accelerated photoperiod

Table 1. Circannual cycles in hibernators and possible mechanisms

Species	Available data most consistent with the following mechanism	Important external factors
Cricetus cricetus	circannual self-sustaining oscillator	? photoperiod as Zeitgeber
Marmota monax	"	"
Citellus tridecemlineatus	sequence of stages	temperature
Citellus lateralis	circannual self-sustaining oscillator	? temperature as Zeitgeber
Citellus richardsonii	?	? temperature
Citellus undulatus	?	? photoperiod
Other *Citellus sp.*	?	?
Eutamias sp and *Tamias striatus*	?	?
Erinaceus europeaus	?	?
Glis glis	sequence of stages	temperature
Eliomys quercinus	?	?
Perognathus logimembris	sequence of stages	food supplies

with two long-day phases occurring each year. On each of these lighting schedules, groups were studied at 7°C and 20°C; LL and DD groups were also studied under naturally varying temperatures. In addition some animals were kept under both natural temperature and lighting conditons.

The main results were that circannual cycles were evident on LD 12:12 at 20°C, and probably also in LD 12:12 at 7°C, though poor survival in the latter group raises the question of whether the second weight peak shown really reflects the maximum of circannual cycle or the sudden onset of sickness followed by steep weight loss and death. At 7°C in LL, 3 out of 8 animals showed clear circannual cycles. In general in LL, weight changes were erratic and cycles were not clearly discernible. In DD, on the other hand, the progression of weight changes was smoother, but on reaching high weights after about a year in capitivity further cyclical changes stopped and the hamsters tended to remain at stable elevated weights. The results from the accelerated photoperiod group have been alluded to above; at all three temperatures there was evidence of weight cycles related to the photoperiodic schedule. In summary, it was established that circannual cycles occur in *C. cricetus*, but their appearance depends greatly on lighting conditions.

There are nevertheless a number of problems in interpretation in addition to those steming from the poor survival rate. For instance, some hamsters on LL had cycle lengths of only 165 days, as short as those of groups on the accelerated photoperiods. Even the hamsters on natural photoperiods had cycles a few weeks shorter than a year. Finally, there was much individual variation and no assurances given that the groups were well matched before the experiment. Follow up work is certainly needed. The finding of circannual cycles on LD 12:12 together with responses to accelerated photoperiods in European hamsters also makes it tempting now to look for circannual cycles in golden hamsters where photperiodic control has been firmly established. Perhaps on photoperiods just between long and short days, such as LD 12:12, they too would show circannual cycles. The behaviour of European hamsters on LD 18:6 and LD 6:18 would also be worth sutdying now.

B. *Marmota monax* (Woodchuck)

Circannual cycles of body weight occur in woodchucks kept in LD 16:18, both at 20°C and 6°C (9). More recently Davis and Finnie (10) have proposed that photoperiod may be the Zeitgeber for these cycles. They sent woodchucks from the United States to Australia and noted phase advances in the time of peak weight in subsequent years. Others remain skeptical about the Zeitgeber interpretation because the animals were only yearlings when flown to Australia (44); even woodchucks staying in the United States can show some phase advances in time of peak weight during their first three years of life (87; see 44, for further discussion). Larger samples and more controls are necessary. Furthermore, if photoperiod is important, then it is not clear why earlier experiments did not implicate it: for instance, there were no consistent differences in food intake between groups of woodchucks kept from June 21 into September on LD 18:6, LD 16:8 and LL; nor were diferences in weight gain detected between groups kept from April (for an unstated length of time) on LD circa 1.5:22.5 (6°C ambient temperature) and those on naturally increasing daylengths (variable temperature, 0-15°C); finally woodchucks kept from June 18-September 28 on naturally decreasing photoperiods were not obviously different in the timing of their main weight gains from those kept in a comparable situation on LD 16:8, though they did seem to start to lose weight earlier (9). Perhaps the different photoperiods were not present for long enough (11) or sample sizes were too small in these experiments to reveal any clear effects. If photoperiod is a Zeitgeber, then differing reactions to different photoperiods should be demonstrable at least at some phases of the circannual cycle.

Despite these hesitations, that daylength is a Zeitgeber remains one reasonable interpretation of the Davis and Finnie (10) experiment. If follow up work is done it might take less long to identify a Zeitgeber by driving circannual rhythms at frequencies different from a year, as did Goss (16) and Canguilhem *et al.* (6), than by phase shifting. At the same time the limits of entrainment could be studied and this would give information on the flexibility of any oscillator.

C. *Citellus tridecemlineatus* (Thirteen-lined ground squirrel)

As mentioned above, a high percentage of individuals of this species do not show persistent cycles when kept in a cold room (11°C and 4°C, LD 12:12) but become stuck in their reproductive phase (48; Fig. 1). In order to evaluate the idea that warm temperatures in the spring are needed to initiate an hour-glass process that runs for about a year, it is necessary to know what happens to animals kept continously warm. According to the proposed explanation they should mostly show persistent cycles. Wheel running and body weight rhythms can occur in blinded thirteen-lined ground squirrels kept 24-28°C (71, 72) but data on the number of animals showing such cycles have not been given. Twenty animals kept in LD 12:12 at 19-22°C were studied more recently by Armitage and Schulenberger (1) and circannual cycles of body weight occurred. They only present group data but presumably most or all individuals showed these cycles since there is no mention to the contrary. It has also been reported that thirteen-lined ground squirrels kept at 25°C display circannual cycles in the amount of torpor they show when transferred for a test period of a few days to a colder environment (36). In these experiments the squirrels were kept on photoperiods changing progressively between LD 9:15 and LD 19:5 and back again once a year. For some of the animals the daylengths were increasing over a given period while for others the lighting schedule was reversed so that the daylengths were decreasing. However, the lighting conditions were without effect; the animals in both groups showed similar persistence of cycles for more than a year. Evidently rhythmic changes in this species are not dependent on having photoperiodic stimulation, a conclusion that fits with earlier work (27). These experiments then again suggest that thirteen-lined ground squirrels show persistent cycles when kept in warm conditions, but unfortunately information on sample size was not given and the graphs apparently showed data for groups of animals, making it impossible to be sure that individuals were displaying circannual rather than annual cycles, and if so what percentage. Therefore, the available data suggest rather than establish that cycles persist more often in warm than cold environments in this species. Moreover, *C. tridecemlineatus* has a wide distribution and different populations in experiments may not be comparable. But enough has already been found out to encourage carefully planned studies of the role of temperature for this species.

D. *Citellus lateralis* (Golden-mantled ground squirrel)

Circannual cycles have been more intensively studied in this animal than any other species of hibernator. There are demonstrations of the cycle persisting in constant conditons for 3 and even 4 years (3, 43, 56). The cycles also have been observed in several different laboratories and in a wide range of environmental conditions. They occur in LD 12:12 (25, 43, 63, 79) in LL (47, 61) and in blinded animals (57). They occur in temperatures ranging from 0-35°C (63, 79) and any effect of temperature on period is relatively minor over most of this range, if it exists at all (see above). Circannual cycles also appear in animals born in captivity (57; Boshes, unpublished). They persist in food deprived animals (25) in castrated animals (64) and in animals with medial or lateral hypothalamic lesions [(43); Mrosovsky and Hallonquist, unpublished].

Such resistance to interference is consistent with the operation of a self-sustaining oscillator compensated against temperature and other disturbances as is the circadian system (66). If only a Zeitgeber could be identified, it would not merely strengthen the analogy to the circadian system but would also open up a way of tackling the physiological basis of the circannual rhythm. Input in the relevant modality could be traced into the central nervous system and lead one to the pacemaker, just as, with circadian rhythms, knowledge that light is the most important Zeitgeber has been followed by tracing visual projections into the suprachiasmatic nucleus and identifying this area as of major importance (88).

In the case of circannual rhythms in golden-mantled ground squirrels temperature is a better candidate for a Zeitgeber than light. Various experiments with altering photoperiodic schedules have failed to affect the time of onset of hibernation (63), although it is possible that these treatments were begun too late, in August, to influence the results that year (54). With temperature on the other hand, there is some evidence consistent with Zeitgiber function from experiments where animals were transferred from a warm to a cold room at different phases of their cycles; the timing of the subsequent cycle depended on the phase at which the transfer took place (59, 63). If temperature is the main Zeitgeber this would be consistent with lack of evidence that circannual rhythms in this species fit with Aschoff's rule (61) and would also be consistent with any minor effects of temperature levels on the period of the cycle (40, 41).

All in all, a strong case can be made that circannual cycles in *C. lateralis* depend on self-sustaining oscillators. Nevertheless, despite the number of studies on this species, other possibilities have hardly been investigated, and one or two observations suggest that acceptance of the oscillator explanation could be premature. In particular, occasionally golden-mantled ground squirrels do not show persistent cycles in constant conditions but become stuck in the phase of low body weight and breeding state (Fig. 2) just as do a much larger proportion of thirteen-lined ground squirrels in similar circumstances (48). It is possible therefore golden-mantled have the same basic mechanism as thirteen-lined ground squirrels, but that most of them are more sensitive to warm temperatures in the spring. Persistence of cycles would then be more frequently seen.

E. *Citellus richardsonii* (Richardson's ground squirrel)

This species shows a circannual rhythm of hibernation in LD 12:12, its period however is considerably shorter than a year, only about 7 1/2 months on average, both at 0°C and 18°C (79). Another possible indication of lability in frequency control is that young males go into hibernation considerably later than the adult males. Differences in time of entry to hibernation among age classes is widespread in ground squirrels but is especially marked in *C. richardsonii*.

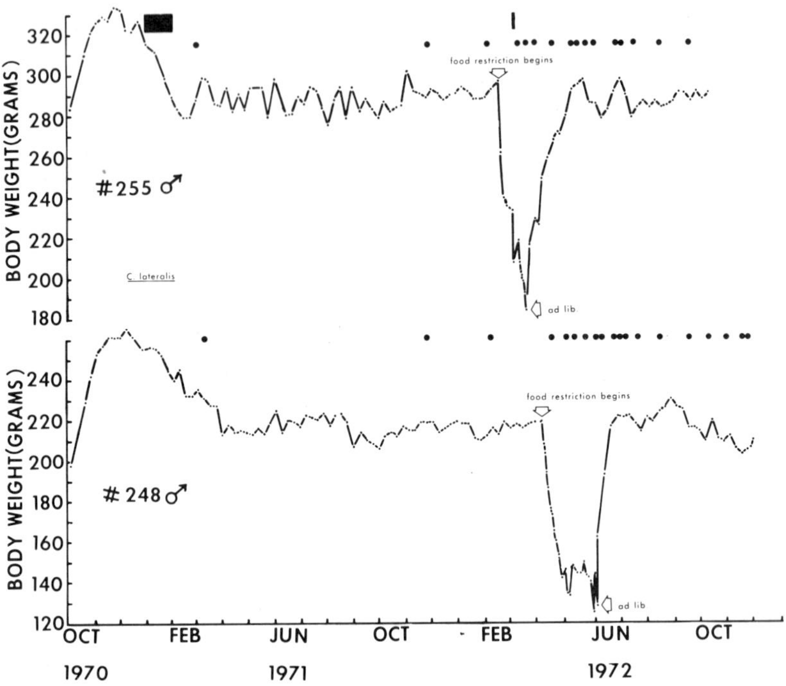

Fig. 2. *Loss of circannual cycles of body weights in two golden-mantled ground squirrels kept in cold room at 12.5°C (for details see 43). At least 22 other golden-mantled ground squirrels kept in the same room showed circannual body weight cycles (43). Black bars indicate the animal was hibernating for at least 2 days out of a week. Solid circles mark times when testes were descended. Because loss of cycles had not been expected, testes condition was not checked frequently at first, but only at the times shown by circles; the animals may have been in reproductive condition at other times. Note that the animals were in reproductive condition in November 1971 and for about a year thereafter. Restriction of food and refeeding was done to check if the animals were regulating weights, as appeared to be the case, rather than being unable to gain weight. Shortly after the period covered by the records shown, both animals became noticeably unhealthy and lost weight.*

For instance, in Alberta the young disappear in mid September while the adult males are already mostly gone by the end of July. This implies that the time between the onset of the first to the second hibernation is about 10 months, while the time between the onset of the second to the third hibernation is about 12 months; a circannual oscillator with a characteristic frequency would not be expected to vary so much from year to year (Wang, personal communication). However, because the time that animals disappear into their burrows does not necessarily reflect the state of an oscillator, further studies are needed; monitoring captive animals given plenty of food over their first year might be instructive.

Although *C. richardsonii* can show circannual cycles of hibernation both at 0 and 18°C (79), it has also been noted that weight cycles are absent in a number of individuals kept in constant conditions (Wang, personal communication). In the latter work 14 out of 44 animals (32%) kept at 12 or 5°C for more than 1 year did not have weight cycles. These animals also tended to hibernate less, and in this respect the findings are similar to those on thirteen-lined ground squirrels (48). However, unlike thirteen-lined ground squirrels, they did not even show changes in weight and hibernation at the start of their first year in captivity.

F. *Citellus undulatus* (Arctic ground squirrel)

Arctic ground squirrels hibernate more and earlier if kept in naturally decreasing photoperiods rather than in constant illumination (12) or when kept in darkness rather than in natural photoperiods (37). These data tell one about factors that predispose the animals to hibernate but they should not be taken as demonstrating a Zeitgeber action of photoperiod. To show that would require studying the animals for an additional year and finding the differences still present then. It is just as likely, however, that photoperiods alter thresholds for becoming torpid without affecting the timing mechanisms that make torpor more likely at one time of year than another: in one of these experiments, although the total amount of hibernation was different between the groups, the hibernation bout lengths peaked at about the same date

(12) and in the other experiment, although again the amount of hibernation was different between the groups, its onset was only about a week apart (37).

G. *Citellus sp.* (Other Ground Squirrels)

Circannual hibernation rhythms occur in several other species of hibernating ground squirrels, in particular in *C. columbianus* in LD 12:12 at 0°C (79; Scott, personal communication), *C. variegatus*, *C. beecheyi*, *C. mohavensis* in LD 12:12 at 12°C and 3°C and in *C. tereticaudus* at 12°C (64). Unfortunately long-term survival was not good at 3°C so that comparisons of the rhythms in the two temperatures cannot be made. *C. mohavensis* also have circannual cycles at 18°C (55). *C. beldingi* in LD 12:12 at 16°C shows a circannual rhythm in reproductive condition but hibernation often does not occur at this temperature (25).

H. *Eutamias sp.* and *Tamias striatus* (Chipmunks)

Chipmunks of the genus *Eutamias* kept at 16°C have circannual cycles of breeding condition and also of phases of low water intake associated with inactivity, though not necessarily with hibernation itself (25). Hibernation in chipmunks is variable and much dependent on the conditions (52). When they are transferred from warm to cold rooms more pronounced torpor occurs. This has been found both for *Eutamias sp.* moved from 16 to 5°C (25) and for *Tamias striatus* moved from 18 to 0°C (80). Although hibernation was more evident initially, in both these studies it decreased the second year after the transfer. A continuous low ambient temperature appears to hinder hibernation (80). This had been attributed to adaptation to the cold (25) but it might be possible that a block of the cycle similar to that shown by thirteen-lined ground squirrels in the cold is involved (Fig. 1). Some of the chipmunks kept at 5°C remained in reproductive condition many months; however it was not stated how often breeding was checked and this appears to have been less frequent before the transfer (25). If there is any tendency to remain in a non-hibernating breeding phase of the cycle, then this does not affect all individuals equally because chipmunks can show circannual cycles at 5 as well at 16°C (25) and at 0 as well as 18°C (80). No differences in period at different temperatures were evident but sample sizes were small. Effects of different photoperiods have not been studied; LD 12:12 prevailed in both the investigations cited.

I. *Erinaceus europaeus* (Hedgehog)

Kristoffersson and Suomalainen's (30) data are generally cited as supporting the existence of circannual cycles in hedgehogs (41, 59). They showed a persistence of weight changes in animals kept at 20-22°C, but lighting conditions were not completely constant as the animal room had a window in it (Kristoffersson, personal cummunication). If there is a circannual cycle in hedgehogs it might be illuminating to make comparisons in constant conditions between New Zealand hedgehogs and the British stock from which they derived, because in the wild the milder New Zealand conditions are associated with a shorter hibernation and longer breeding season (53).

J. *Glis glis* (Edible dormouse)

Cycles in this species are so unusual that they have been classed as a new kind of endogenous rhythm (45). In a warm environment there is a cycle of body weight that lasts very approximately 6 weeks (38, 45, 75, 76, 77). Associated with body weight are changes in food intake, meal patterns, water intake, activity, body temperature and thresholds for hypothalamically elicited feeding (Mrosovsky, Melnyk, Hallonquist, Boshes and Lang, unpublished). But when dormice are hibernating in a cold or cool environment, then the cycle is much longer, over a year sometimes (14, 45, 81), with the period correlated to how much the animals hibernate (Fig. 3). If there is an endogenous oscillator, then clearly it is not well compensated for temperature. In this respect the control system in dormice differs from that in ground squirrels and woodchucks; the latter show circannual weight cycles whether they actually hibernate or not (9, 43).

Temperature compensation is a widespread feature of timing systems with obvious advantages (8, 66); the value of a non-temperature-compensated system in dormice is puzzling. Field studies of this small

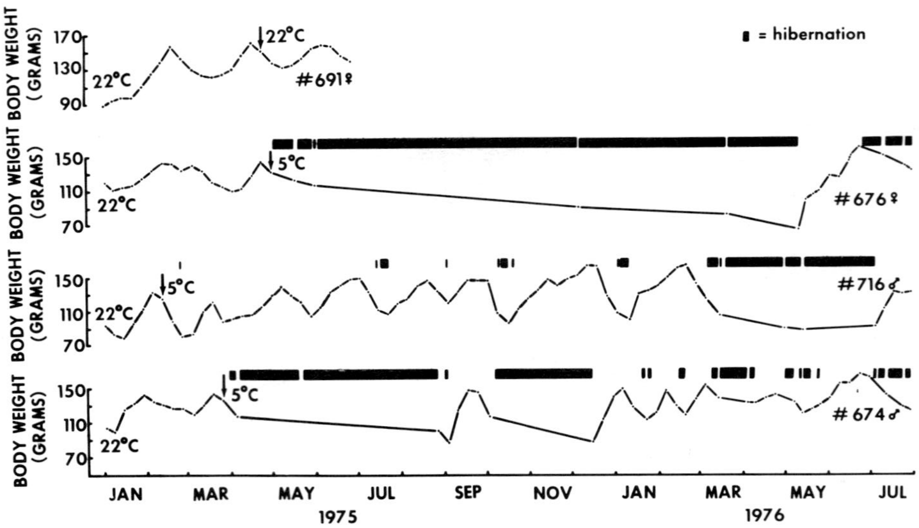

Fig. 3. *Body weight and hibernation of individual dormice, Glis glis, illustrating cycles of only a few weeks duration in animals in a warm room or in non-hibernating animals in a cold room. Arrows show when the animals were transferred to a different room, together with the temperature of that room. Periodic arousals from hibernation and return to hibernation in a day are not shown (adapted from Mrosovsky, 1977).*

arboreal nocturnal rodent have not been extensive enough to illuminate this matter. However, it has been noted that dormice enter hibernation at such different times of year and in such different weather conditions that some measure of endogenous control seems to be involved (84); differing states in different individuals with respect to cycles of a few weeks duration might well be one factor. Possibly temperature dependence of these cycles reflects some form of opportunism that enables dormice to take advantage of mild winters.

As puzzling as the role of temperature is that of light. Cycles of a few weeks duration in dormice have been seen not only in LD 12:12 (45, 76) but also under natural photoperiod, temperature and humidity (75). Yet there are also reports that dormice are photoperiodic, with respect both to weight gain (29) and to ease of inducing hibernation when transferred to a cold room (36). The persistence of cycles of few weeks in naturally changing photoperiods in some individuals (75) seems incompatible with photoperiodic effects. However, there are data suggesting that when daylength is not completely constant, cycles in dormice diminish in amplitude over the summer (38, 45, 75, 76). Since with animals kept under a strict LD 12:12 for more than a year no circannual cycle in the 6 weeks cycles was obvious (Mrosovsky *et al.*, unpublished), it is likely that exogenous rather than endogenous factors are responsible. Perhaps long days attenuate the 6 week cycles, in some individuals at least; these cycles would then show up more strongly over the winter months. Another possibility is that some of the effects attributed to photoperiod might have occurred spontaneously. When dormice are kept in the laboratory their weight patterns vary considerably; some animals hold stable weights for many months and then suddenly begin to fatten and cycle thereafter without any deliberate or noticeable change in the conditions. Such individual variability and spontaneous changes mean that extensive controls are desirable in experiments involving altered daylengths.

In summary then, while more studies on rhythmicity in dormice are needed, especially on the role of photoperiod and the realtionship of the short cycles seen in captivity to the annual cycles in wild animals

(29, 84), there is already enough information to argue against the operation of an oscillator with a characteristic frequency. Nor does frequency demultiplication seem likely because there are no striking relationships in this species between circadian and circannual periods in experiments where both have been monitored (12°C ambient temperature and an environment where the animals could select a dark sleeping place out of a generally illuminated larger area, ref. 14). This leaves the sequence of stages approach as the most plausible one to follow.

K. *Eliomys quercinus* (Garden dormouse)

Given the remarkable rhythms in edible dormice, it is of interest to know whether other members of the family Gliridae display similar rhythms. When garden dormice are kept in a cold environment, 5°C, hibernation can continue for more than 10 months; this was found with animals deprived of food and water (32). More interesting than the prolonged hibernation, which also occurs in food deprived ground squirrels (28, 46, 63), was the finding that the rate of weight loss and interarousal intervals did not show progressive changes as hibernation continued. It must be emphasized that the animals were food deprived, because progressive changes do occur in garden dormice with access to food (see 7, for a review). Such progressive changes in hibernation bout length are characteristic of ground squirrels and many hibernators (see 41, for a review) and they occur even in food deprived ground squirrels (28, 46). The decreasing bout lengths toward the end of the hibernation season presumably reflect the operation of endogenous mechanisms that normally bring hibernation to an end altogether. Their absence in garden dormice kept hibernating at 5°C suggest that temperature is important in determining the duration of hibernation and so of the whole cycle (32). This is consistent with the fact that garden dormice kept in constant conditions that are cool enough, 12°C, to support prolonged torpor, display circannual cycles (measured by the duration of spontaneous torpor, ref. 7). To assess whether the circannual period in the latter case reflects the characteristic frequency of the system or the operation of a temperature dependent flexible frequency system, it is important to know what happens when garden dormice are kept in conditions warm enough to prevent deep or prolonged torpor. Lachiver and Boulouard (31) state that the weights of garden dormice kept in a warm room over the winter rise steadily. However, as weighings were relatively infrequent and group data are graphed, their paper cannot be taken as proof that cycles of a few weeks duration do not occur in warm conditions.

It should be noted that all the data mentioned above refer to populations of garden dormice from France. In contrast, garden dormice from Finland have more highly developed endogenous controls; the Finnish animals do show progressive declines in hibernation bout length toward the end of the winter season, and also changes in oxygen consumption and susceptibility to disturbance when kept without food and water at 4°C (49, 50, 51). The only indication of endogenous controls in French dormice that has been detected under such conditions is a decrease in hibernation bout length at the end of the season, but only at certain temperatures, not at 0°C, and then only in male animals. Further comparisons of closely related populations at the interface of endogenous and exogenous controls are likely to be instructive.

L. *Perognathus longimembris* (Pocket mouse)

Studies on pocket mice further illustrate the limited range of conditions under which circannual cycles occur (15). At 16°C in LD 12:12 there were cycles of above-ground activity of about 10 months period but at 31°C these were absent or unobvious; no circannual cycles were detected in the duration of torpor at 18°C (LD 9:15 and LD 15:9) or at 8°C (LD 9:15). Heating the soil with a lamp towards the end of the phase of underground existence brought the mice above ground but the animals were not studied long enough thereafter to determine if a phase shift had occurred. There is no strong evidence therefore that altering temperature gradients in the soil acts as a Zeitgeber.

More recent work (French, personal communication) suggests that there may be no need to invoke an endogenous circannual oscillator for these pocket mice because the time they stay underground is much influenced by the amount of food they have stored away. With no changes in soil temperature, it seems they stay underground until their supplies are almost exhausted. If provided with large amounts of food when in their active aboveground phase, they may later stay underground for more than a year. The rate their hoards are used up depends greatly on how often they are torpid. The cycle in pocket mice therefore has similarities to the cycle in dormice described above; in both cases the duration of one of the phases is much influenced by the amount of hibernation. It would be interesting to learn now how much altering the duration of the underground phase in pocket mice through food supplies affects the date when they next go underground.

III. CONCLUSIONS

It is obvious that there are species differences in circannual cycles in hibernators with external control being clearest in *Glis glis* and least implicated in *Citellus lateralis*. It is also obvious that data on any single species are insufficient to tell us even what kind of mechanism generates its circannual cycles. The safest course would be to refrain from making further conclusions. As Renan (70) said "Qui sait si la finesse d'esprit ne consiste pas à s'abstenir de conclure". But the experimental scientist cannot afford the wisdom of such abstinence. At least at the level of choosing experiments that are likely to be instructive, an especially important matter with those that take so long, some guiding viewpoints and intuitions must play a role. Lest the reader has detected the dull note of impartiality in what has been written above, some personal viewpoints, or prejudices, will now be made explicit.

One overworked concept in chronobiology today is that of underlying rhythms persisting but not expressing themselves. Certainly there are many demonstrable cases where this occurs, but there are other cases where this explanation hovers between the unsatisfactory and the unlikely. With circannual rhythms that do not appear in some environmental conditions, there is nothing particular to suggest the persistence of unexpressed rhythms. It is just as plausible to assume that certain environmental conditions are important in the generation of the rhythmicity itself. Processes that last several months can be initiated by environmental events in golden hamsters. Such processes might last close to a year in other species with re-initiation depending on the presence of appropriate external stimuli. To the extent that this is true, more attention should be given to understanding why circannual rhythms do not appear in some conditions and how they can be restarted, rather than studying their properties in conditions where they do occur. Species such as *C. tridecemlineatus* or *E. quercinus*, in which both exogenous and endogenous controls seem readily detectable, might be subjects of choice initially, though to discover if this viewpoint has wide applicability, it would be challenging to search for conditions under which cycles to not occur in species whose rhythms appear to be most firmly entrenched in endogenous mechanisms, *C. lateralis* for example.

It is pertinent that in the field of circadian rhythms interest in the conditions under which cycles do not occur, such as bright LL, is reviving (2,33). It has been found that when house sparrows that are arrhythmic in LL are transferred to DD they begin to become active about 12 hours later, no matter what time the transfer took place, and their subsequent circadian rhythms in DD follow on from this point (2). This suggests that LL stops oscillations and darkness sets them going again. However, an alternative explanation of such results is that the oscillations were not ever blocked but were not expressing themselves (2, 65). This interpretation requires the additional assumptions that transitions from LL to DD not only phase shift the unexpressed oscillation but also that such LL-DD transitions are especially efficient in phase shifting. Given these elaborations it seems more productive that the onus of proof should fall on the side of demonstrating the existence of unexpressed hypothetical invisible rhythms than on demonstrating that they are not there. As a working hypothesis it is more reasonable to accept that oscillations have stopped in LL and that darkness is necessary to re-initiate them.

This type of explanation is analogous to that discussed above for the role of warm springtime temperatures in the weight cycles of thirteen-lined ground squirrels. In the latter case it is much easier to see that the rhythm has stopped at a particular phase of the cycle, whereas with circadian activity rhythms of sparrows this is not readily apparent. Moreover, with circannual time spans, slow acting manipulations that interfere with cycles can still be applied at a relatively precise phase, but in the experiments mentioned on sparrows in LL, for instance, it takes several or more days for the animals to become arrhythmic and by the time this happens the phase of any underlying cycle can no longer be specified at a given time. Circannual cycles may therefore offer some advantages in working with this kind of phenomenon.

It would certainly be surprising if research on circannual cycles, with its difficulties of time span, could change or even assist in changing ideas about circadian cycles. And yet it remains conceivable that the concept of self-sustaining oscillators that do not express themselves in certain conditions will give way in both fields (at least in some cases in some species .....there is no reason to expect uniformity) to explanations in terms of hour-glasses that are reset by a limited range of environmental inputs, and the circannual-circadian analogy be reborn transformed.

ACKNOWLEDGEMENTS

The author thanks B. Rusak, J.T. Enright, K. Hoffmann, R. Melnyk and L. Wang for comments. The experimental work was supported by the National Research Council of Canada.

REFERENCES

1. Armitage, K.B., and E. Schulenberger. Evidence for a circannual metabolic cycle in *Citellus tridecemlineatus*, a hibernator. Comp. Biochem. Physiol. 42A: 667-688, 1972.
2. Binkley, S. Constant light; effects on the circadian locomotor rhythm in the house sparrow. Physiol. Zool. 50: 170-181, 1977.
3. Blake, B. H. The annual cycle of fat storage in two populations of golden-mantled ground squirrels. J. Mammal. 53: 157-167, 1972.
4. Bünning, E. The Physiological Clock. (3rd ed.) New York: Springer-Verlag, 1973.
5. Canguilhem, B., and R. Bloch. Évolution saisonnière de l'élimination des hormones surrénaliennes chez un hibernant, *Cricetus cricetus*. Arch. Sci. Physiol. 21: 27-44, 1967.
6. Canguilhem, B., J.-P. Schieber, and A. Koch. Rhythme circannuel pondéral du hamster d'europe (*Cricetus cricetus*). Influences respectives de la photopériode et de la température externe sur son déroulement. Arch. Sci. Physiol. 27: 67-90, 1973.
7. Daan, S. Periodicity of heterothermy in the garden dormouse, *Eliomys quercinus* (L.). Neth. J. Zool. 23: 237-265, 1973.
8. Daan, S., and C. S. Pittendrigh. A functional analysis of circadian pacemakers in nocturnal rodents. III. Heavy water and constant light: homeostasis of frequency? J. Comp. Physiol. 106: 267-290, 1976.
9. Davis, D.E. The annual rhythm of fat deposition in woodchucks (*Marmota monax*). Physiol. Zool. 40: 391-402, 1967.
10. Davis, D. E. Hibernation and circannual rhythms of food consumption in marmots and ground squirrels. Quart. Rev. Biol. 51: 477-514, 1976.
11. Davis, D. E., and E. P. Finnie. Entrainment of circannual rhythm in weight of woodchucks. J. Mammal. 56: 199-203, 1975.
12. Drescher, J. W. Environmental influences on initiation and maintenance of hibernation in the arctic ground squirrel, *Citellus undulatus*. Ecology 48: 962-966, 1967.

13. Enright, J. T. Ecological aspects of endogenous rhythmicity. Ann. Rev. Ecol. Syst. 1: 221-238, 1970.
14. Fischer von K., H.-W. Butschke, and D. Mahlert. Untersuchungen zur circadianen und circannualen Rhythmik bei Siebenschläfern *(Glis glis* L.) im Selbstwählversuch. Z. Säugetierkunde 40: 65-74, 1975.
15. French, A.R. Circannual rhythmicity and entrainment of surface activity in the hibernator, *Perognathus longimembris*. J. Mammal. 58: 37-43, 1977.
16. Goss, R.J. Photoperiodic control of antler cycles in deer II. Alterations in amplitude. J. Exp. Zool. 171: 223-234, 1969.
17. Goss, R.J. Photoperiodic control of antler cycles in deer III. Decreasing versus increasing day lengths. J. Exp. Zool. 197: 307-312, 1976.
18. Goss, R.J., C.E. Dinsmore, L.N. Grimes, and J.K. Rosen. Expression and suppression of the circannual antler growth in deer. In: Circannual Clocks: Annual Biological Rhythms. Pengelley, E.T. (editor). New York: Academic Press, 1974, p. 393-422.
19. Gwinner, E. A comparative study of circannual rhythms in warblers. In: Chronobiology. Menaker, M. (editor). Washington, D.C.: National Academy of Sciences, 1971, p. 405-427.
20. Gwinner, E. Circannual rhythms in birds: their interaction with circadian rhythms and environmental photoperiod. J. Reprod. Fert., Suppl. 19: 51-65, 1973.
21. Gwinner, E. Endogenous temporal control of migratory restlessness in warblers. Naturwissenschaften 61: 405, 1974.
22. Gwinner, E. Adaptive significance of circannual rhythms in birds. In: Physiological Adaptation to the Environment. Vernberg, F.J. (editor). New York: Intext Educational Publishers, 1975, p. 417-433.

23. Gwinner, E. Effects of season and external testosterone on the free running circadian activity rhythm of euorpean starlings (*Sturnus vulgaris*) J. Comp. Physiol. 103: 315-328, 1975.

24. Hamner, W.M. On seeking an alternative to the endogenous reproductive rhythm hypothesis in birds. In: Chronobiology. Menaker, M. (editor). Washington, D.C.: National Academy of Sciences, 1971, p. 448-462.

25. Heller, H.C., and T.L. Poulson. Circannian rhythms - II. Endogenous and exogenous factors controlling reproduction and hibernation in chipmunks *(Eutamias)* and ground squirrels (*Spermophilus)*. Comp. Biochem. Physiol. 33: 357-383, 1970.

26. Hoebel, B.G., and P. Teitelbaum. Weight regulation in normal and hypothalamic hyperphagic rats. J. Comp. Physiol. Psychol. 61: 189-193, 1966.

27. Johnson, G.E., and E.L. Gann. Light in relation to the sexual cycle and to hibernation in the thirteen-lined ground squirrel. Anat. Rec. Suppl. 57: 28, 1933.

28. Kayser, C., and C. Aron. Rôle du jeûne et de la thyroide dans l'hibernation. Étude fait sur le spermophile *(Citellus citellus)* C. R. Séance Soc. Biol. 146: 1376-1379, 1952.

29. König, C. Einflüsse von Licht und Temperatur auf den Winterschlaf des Siebenschläfers *Glis g. glis* (Linnaeus 1766). Z. Morph. Ökol Tiere 48: 545-575, 1960.

30. Kristoffersson, R., and P. Suomalainen. Studies on the physiology of the hibernating hedgehog 2. Changes of body weight of hibernating and non-hibernating animals. Ann. Acad. Sci. Fenn. Ser. A. IV. 76: 1-11, 1964.

31. Lachiver, F., and R. Boulouard. Évolution du poids corporel chez un hibernant le Lérot (*Eliomys quercinus* L.). durante la période hivernale et au cours de l'exposition au froid en été. Mammalia 29: 343-366, 1965.

32. Lachiver, F., and R. Boulouard. Évolution de la périodicité des phases de sommeil et de reveil chez le Lérot *(Eliomys quercinus)* au cours du sommeil hivernal et de la léthargie induite en été. J. Physiol. Paris 59: 250-251, 1967.
33. McMillan, J.P., J.A. Elliott, and M. Menaker. On the role of eyes and brain photoreceptors in the sparrow: arrhythmicity in constant light. J. Comp. Physiol. 102: 263-268, 1976.
34. Menaker, M. Circannual rhythms in circadian perspective. In: Circannual Clocks: Annual Biolgical Rhythms. Pengelley, E.T. (editor). New York: Academic Press, 1974, p. 507-520.
35. Menaker, M. and N. Zimmerman. Role of the pineal in the circadian system of birds. Amer. Zool. 16: 45-55, 1976.
36. Morrison, P.R. Adaptation of small mammals to the arctic. Fedn. Proc. 23: 1202-1206, 1964.
37. Morrison, P., and W. Galster. Patterns of hibernation in the arctic ground squirrel. Can. J. Zool. 53: 1345-1355, 1975.
38. Mrosovsky, N. Acceleration of annual hibernating cycle to 6 weeks in captive dormice. Can. J. Zool. 44: 903-910, 1966.
39. Mrosovsky, N. Comment. In: Mammalian Hibernation III. Fisher, K.C., A.R, Dawe, C.P. Lyman, E. Schönbaum, and F.E. South (editors). Edinburgh and London: Oliver and Boyd, 1967, p. 27-28.
40. Mrosovsky, N. Mechanism of hibernation cycles in ground squirrels: circannian rhythm or sequence of stages. Penn. Acad. Sci. 44: 172-175, 1970.
41. Mrosovsky, N. Hibernation and the Hypothalamus. New York: Appleton-Century-Crofts, 1971.
42. Mrosovsky, N. Comment. In: Circannual Clocks: Annual Biological Rhythms. Pengelley, E.T. (editor). New York: Academic Press, 1974, p. 161-163.
43. Mrosovsky, N. The amplitude and period of circannual cycles of body weight in golden-mantled ground squirrels with medial hypothalamic lesions. Brain Res. 99: 97-116, 1975.

44. Mrosovsky, N. Lipid programmes and life strategies in hibernators. Amer. Zool. 16: 685-697, 1976.
45. Mrosovsky, N. Hibernation and body weight in dormice: a new type of endogenous cycle. Science, 196: 902-903, 1977.
46. Mrosovsky, N. and D. S. Barnes. Anorexia, food deprivation and hibernation. Physiol. Behav. 12: 265-270, 1974.
47. Mrosovsky, N., M. Boshes, J. D. Hallonquist, and K. Lang. Circannual cycle of circadian cycles in a golden-mantled ground squirrel. Naturwissenschaften 63: 298-299, 1976.
48. Mrosovsky, N., and K. Lang, Disturbances in the annual weight and hibernation cycles of thirteen-lined ground squirrels kept in constant conditions and the effects of temperature changes. J. Interdiscipl. Cycle Res. 2: 79-90, 1971.
49. Pajunen, I. Body temperature, heart rate, breathing pattern, weight loss and periodicity of hibernation in the Finnish garden dormouse, *Eliomys quercinus* L. Ann. Zool. Fenn. 7: 251-266, 1970.
50. Pajunen, I. A comparison of oxygen consumption and respiratory quotients in Finnish and French garden dormice *Eliomys quercinus* L., hibernating at 4.2 ± 0.5°C. Ann. Zool. Fenn. 13: 161-173, 1976.
51. Pajunen, I. Hibernation in Finnish and French garden dormice, *Eliomys quercinus* L. at 4.2 ± 0.5°C. Ph.D. thesis, University of Helsinki. 1976.
52. Panuska, J.A. Weight patterns and hibernation in *Tamias striatus*. J. Mammal. 40: 554-566, 1959.
53. Parkes, J. Some aspects of the biology of the hedgehog (*Erinaceus europaeus* L.) in the Manawatu, New Zealand. New Zealand J. Zool. 2: 463-472, 1975.
54. Pengelley, E.T. The relation of external conditions to the onset and termination of hibernation and estivation. In: Mammalian Hibernation III. Fisher, K.C., A.R. Dawe, C.P.

Lyman, E. Schönbaum, and F.E. South (editors). Edinburgh and London: Oliver and Boyd, 1967, p. 1-29.

55. Pengelley, E.T. Influence of light on hibernation in the Mojave ground squirrel *(Citellus mohavensis)*. In: Physiological Systems in Semiarid Environments. Hoff, C.C., and M.L. Riedesel (editors). Albuquerque, New Mexico: University of New Mexico Press, 1969, p. 11-16.

56. Pengelley, E.T. and S.J. Asmundson. Free-running of endogenous circannian rhythms in the golden-mantled ground squirrel, *Citellus lateralis*. Comp. Biochem. Physiol. 30: 177-183, 1969.

57. Pengelley, E.T., and S. J. Asmundson. The effect of light on the free running circannual rhythm of the golden-mantled ground squirrel, *Citellus lateralis*. Comp. Biochem. Physiol. 32: 155-160, 1970.

58. Pengelley, E.T. and S.J. Asmundson. An analysis of the mechanisms by which mammalaian hibernators synchronize their behavioral physiology with the environment. In: Hibernation and Hypothermia, Perspectives and Challenges. South, F.E., J.P. Hannon, J.R. Willis, E.T. Pengelley, and N.R. Alpert (editors). Amsterdam: Elsevier, 1972, p. 637-661.

59. Pengelley, E.T., and S.J. Asmundson. Circannual rhythmicity in hibernating mammals. In: Circannual Clocks: Annual Biological Rhythms. Pengelley, E.T. (editor). New York: Academic Press, 1974, p. 95-160.

60. Pengelley, E.T. and S.J. Asmundson. Female gestation and lactation as zeitbegers for circannual rhythmicity in the hibernating ground squirrel, *Citellus lateralis*. Comp. Biochem. Physiol. 50A: 621-625, 1975.

61. Pengelley, E.T., S.J. Asmundson, B. Barnes, and R.C. Aloia. Relationship of light intensity and photoperiod to circannual rhythmicity in the hibernating ground squirrel, *Citellus lateralis*. Comp. Biochem. Physiol. 53A: 273-277, 1976.

62. Pengelley, E.T., and K.C. Fisher. Onset and cessation of hibernation under constant temperature and light in the golden-mantled ground squirrel *(Citellus lateralis)*. Nature 180: 1371-1372, 1957.
63. Pengelley, E.T., and K.C. Fisher. The effect of temperature and photoperiod on the yearly hibernating behaviour of captive golden-mantled ground squirrels *(Citellus lateralis tescorum)*. Can. J. Zool. 41: 1103-1120, 1963.
64. Pengelley, E.T., and K.H. Kelly. A "circannian" rhythm in hibernating species of the genus *Citellus* with observations on their physiological evolution. Comp. Biochem. Physiol. 19: 603-697, 1966.
65. Pittendrigh, C.S. Circadian clocks: what are they? In: The Molecular Basis of Circadian Rhythms. Hastings, J.W., and H.-G. Schweiger (editors). Berlin: Dahlem Konferenzen, 1976, p. 11-48.
66. Pittendrigh, C.S., and P.C. Caldarola. General homeostasis of the frequency of circadian oscillations. Proc. Nat. Acad. Sci. U.S.A. 70: 2697-2701, 1973.
67. Pittendrigh, C.S., and S. Daan. A functional analysis of circadian pacemakers in nocturnal rodents. I. The stability and lability of spontaneous frequency. J. Comp. Physiol. 106: 223-252, 1976.
68. Powley, T.L. and R.E. Keesey. Relationship of body weight to the lateral hypothalamic feeding syndrome. J. Comp. Physiol. Psychol. 70: 25-36, 1970.
69. Reiter, R.J. Comparative physiology: pineal gland. Ann. Rev. Physiol. 35: 305-328, 1973.
70. Renan, E. (1852) quoted in Arber, A. The Mind and the Eye. Cambridge University Press, 1954, p. 113.
71. Richter, C.P. Biological Clocks in Medicine and Psychiatry. Springfield, Illinois: Charles C. Thomas, 1965.

72. Richter, C.P. Comment. In: Mammalian Hibernation III. Fisher, K.C., A.R. Dawe, C.P. Lyman, E. Schönbaum, and F.E. South (editors). Edinburgh and London: Oliver and Boyd, 1967, p. 27-28.

73. Richter, C.P. Inborn nature of the rat's 24-hour clock. J. Comp. Physiol. Psychol. 75: 1-4, 1971.

74. Rusak, B., and I. Zucker. Biological rhythms and animal behavior. Ann. Rev. Physiol. 26: 137-171, 1975.

75. Schaefer, A., F. Piquard, D. Kalinkov, and P. Haberey. Variations circannuelles de la régulation pondérale chez le Loir *Glis glis*. C. R. Séance Soc. Biol. 168: 1428-1431, 1974.

76. Schaefer, A., F. Piquard, and P. Haberey. Évolution saisonière de la régulation alimentaire du poids chez le loir, *Glis glis*. J. Physiol., Paris, 70: 649-658, 1975.

77. Schaefer, A., F. Piquard, and P. Haberey. Food self-selection during spontaenous body weight variations in the dormouse (*Glis glis* L.) Comp. Biochem. Physiol. 55A: 115-118, 1976.

78. Schwab, R.G. Circannian testicular periodicity in the european starling in the absence of photoperiodic change. In: Biochronometry. Menaker, M. (editor). Washington, D.C.: National Academy of Sciences, 1971, p. 428-447.

79. Scott, G.W., and K.C. Fisher. The lengths of hibernation cycles in mammalian hibernators living under controlled conditions. Penn. Acad. Sci. 44: 180-183, 1970.

80. Scott, G.W., and K.C. Fisher. Hibernation of eastern chipmunks (*Tamias striatus*) maintained under controlled conditions. Can. J. Zool. 50: 95-105, 1972.

81. Scott, G.W., and K.C. Fisher. Periodicity of hibernation of dormice (*Glis glis*) maintained under controlled conditions. Can. J. Zool. 54: 437-441, 1976.

82. Snow, D.W., and B.K. Snow. The breeding cycle of the swallow-tailed gull, *Creagrus furcatus*. Ibis 109: 14-24. 1967.

83. Strumwasser, F., F.R. Schlechte, and J. Streeter. The internal rhythms of hibernators. In: Mammalian Hibernation III. Fisher, K.C., A.R. Dawe, C.P. Lyman, E. Schönbaum, and F.E. South (editors). Edinburgh and London: Oliver and Boyd, 1967, p. 110-139.
84. Vietinghoff-Riesch, A. von. Der Siebenschläfer (*Glis glis* L.) Monogr. Wildsäugetiere 14: 1-196, 1960.
85. Wells, L.J., and M. Zalesky. Effects of low environmental temperature on the reproductive organs of male mammals with annual aspermia. Am. J. Anat. 60: 429-447, 1940.
86. Wever, R. Autonome circadiane Periodik des Menschen unter dem Einfluss verschiedener Beleuchtungs - Bedingungen. Pflügers Arch. 306: 71-91, 1969.
87. Young, R.A. The woodchuck, *Marmota monax*, as a biomedical model for the study of obesity. Ph.D. Thesis, University of Vermont, 1975.
88. Zucker, I., B. Rusak, and R. G. King. Neural bases for circadian rhythms in rodent behavior. In: Advances in Psychobiology, volume 3 Riesen, A.H., and R.F. Thompson (editors). New York: Wiley, 1976, p. 35-74.

SHALLOW, DAILY TORPOR: A THERMOREGULATORY ADAPTATION

Jack W. Hudson
Section of Ecology and Systematics
Cornell University
Ithaca, New York 14853

ABSTRACT

Shallow daily torpor is now known to occur naturally in a large number of small marsupial and placental mammals, including shrews. It is a means of reducing energy requirements at a time food is less available. Torpor is only a few hours in length and its duration is proportionate to the available food. It is not a loss of thermoregulatory control. The cardiovascular system remains under autonomic control during torpor. Each species seems to have a "critical body temperature", below which it can not spontaneously arouse. The inability of these small mammals to tolerate low body temperatures may be due to a loss of adequate membrane function, particularly in the heart, as demonstrated by K^+ loss. Typically only a fraction of a population seems to exhibit shallow torpor. Assuming that hibernation (or shallow torpor) is genetically determined, natural selection functions so that a small fraction of the population possesses the gene for torpor. Thus, only a small fraction of the population survives the winter when food is severely limited. There only need be a sufficient number of individuals who are in adequate condition for rebuilding the population prior to the spring production of renewed food resources. By coupling high fecundity with differential survival, the population *per se* survives. These small mammals do not assume the horizontal, curled posture typical of hibernators, but remain upright, so that without careful observation of breathing rates or measurement of body temperature, torpor can easily be missed, particularly when relatively high body temperatures permit typical locomotory behavior. There is evidence that shallow, daily torpor is seasonal and may require acclimatization.

ISBN 0-12-734550-7

I. INTRODUCTION

The difficulties of computer searches for titles dealing with the comparative aspects of hibernation is convincing evidence that simple words such as hypothermia, heterothermia, dormancy, torpor, estivation and hibernation have been used in so many ways that these terms have come to mean different phenomena to different biologists. As a result, the complete retrival of data dealing with the basic physiology of hibernation is an impossible task. One would not dare use as a key word "hypothermia," for it has become too extensively involved in the biomedical literature. Part of the difficulty in reaching agreement on the use of key words is the unexpressed and ill-defined concept of what constitutes a hibernator or an estivator or hypothermia. Are these terms which have only physiological meaning, or do we use them only in an ecological context, or have we tended to use them both ways?

Davis (9) restricts the use of hibernation and estivation to its ecological context, viz., if torpor occurs during the summer, the mammal is an estivator, but if it occurs during the winter, the mammal is a hibernator. Does this mean that any mammal which becomes torpid before September 21st, but remains intermittently torpid throughout the winter, is both an estivator and a hibernator, but one which does not become torpid until September 22nd is a hibernator? Most of us studying the physiology of torpor wonder if these terms have a specific physiological meaning. Is estivation shallow torpor and hibernation deep torpor? There seems to be implicit in many statements about estivation the assumption that estivation is a physiological response to heat and aridity, yet to my knowledge there is no unequivocal evidence that dormancy can be induced by heat or water deprivation (although water deprivation will lead to self-imposed starvation, so that if dormancy occurs, it is probably the starvation rather than the dehydration which is the causative factor). If estivation and hibernation are to be differentiated on the basis of depth of body temperature, where is the dividing line? Is it at 15°C? Regardless of the term to be applied to the form of dormancy or torpidity, it is in the final analysis a response to a limitation of energy, whether the response is anticipatory of future deprivation or a consequence of an immediate withdrawal of food signalling ultimate deprivation. And because mammals live in different habitats with differences in availability of food, both with respect to timing and abundance, it strikes me as incredible that we have spent so much effort forcing a wide variety of mammals into our concept of hibernation as deep and prolonged torpor. There are over 4000 different species of mammals living in many different environments. Since summer in the northern hemisphere is a time of warmth, it follows that any mammal torpid at that time (i.e., one in which the body temperature has fallen below some level such as 31°C, admittedly an arbitrary delineation but one which is generally acceptable), tend to have relatively high body temperatures. Thus, estivation has probably come to mean dormancy or torpidity at a high body temperature. The difficulty of attaching much significance to body temperature alone is the paradoxical circumstance in which a hibernator, i.e., a mammal which can cope with low body temperatures, may also be dormant at high body temperatures. But there are many examples of species which may be dormant at high body temperatures but cannot tolerate low body temperatures (exclusive of the biomedical application of hypothermia). Thus in a sense, a hibernator can be an estivator, but the converse is not necessarily true.

Obviously words such as "high" or "low" are imprecise and not particularly useful for the delineation of categories, though probably most of us tend to think of low as "5°C" and high as "15°C." And to many of us the marmot, the hedgehog and the bat, which become torpid for extended periods of time during the winter often letting their body temperature approach air temperature which may be near 0°C are "classical hibernators." Such species arouse infrequently throughout the winter and as spring approaches, they come out of hibernation and resume a normothermic body temperature in conjunction with their above-ground activities. (Although not all of these hibernators have their hibernacula in the ground; some may construct their nests on the forest floor, in crevices of walls, hollow logs and other localities, so that the extent to which the temperature and light of their hibernacula remain constant varies tremendously.). Why should we assume that all hibernators have endogenous rhythms, when in fact many are exposed to marked changes in both temperature and photoperiod throughout the year including their period of hibernation? Compare the common dormouse, *Muscardinus arvenallaris* with the marmot, *Marmota monax*. Surely these two species must be subjected to vastly different ambient conditions of temperature and light fluctuation.

Only a few families of six of the nineteen orders of mammals include hibernators (*Monotremes, Marsupials, Insectivores, Chiropterans, Primates and Rodents*) (25). But within the past twenty-five years we have discovered a large number of small mammals which exhibit daily, shallow torpor. At one time this torpor was thought to be a state of morbidity induced by the stress of confinement in traps on cold nights or due to a severe reduction in food. We now know that daily torpor is a physiological and ecological adaptation for coping with limited energy resources and it is a historical review of these insights on which I would like to focus my review.

In addition to reviewing the historical progress of our understanding of shallow, daily torpor and its distribution in various families predominantly from the marsupial, insectivore and rodent orders, some attention will be given to the chipmunks (*Tamias striatus* and *Eutamias spp.*), a group which seems to be intermediate between the "classical hibernator" and the "shallow hibernator" with respect to their depth and length of torpor.

II. SHALLOW DAILY TORPOR

Although there are multitudinous reports of hypothermia in various kinds of mammals, unless a species spontaneously, or at least during a restriction of food and/or a lowering of the ambient temperature becomes dormant, i.e., lets its body temperature drop markedly below 31°C, and can spontaneously rewarm to normothermia at the ambient temperature at which it was dormant, true dormancy is not considered to have occurred. This differentiates dormancy or torpidity from artificial hypothermia and keeps the thermoregulatory performance of the species in an ecological context. While it is possible there may be a progressive warming of the ambient temperatures to serve as a source of exogenous heat for arousal, this can not under natural conditions be very great, and certainly is not nearly the same magnitude as occurs when animals are taken from refrigerators at an ambient temperature of 5°C and then aroused at room temperature (i.e., 22°C).

In attempting to present a historical review to show how our thinking has shifted from viewing starvation-induced torpor as thermoregulatory inadequacy to interpreting it to be a natural response to proximate or ultimate limitations of food, inevitably I shall fail to give proper credit in some instances and improperly give credit in others. This cannot be helped.

With the advantage of hindsight, it now seems perfectly obvious that there should be a wide variety of hibernation patterns, since species occupy different niches where their physical and biotic problems vary. Why should a small desert rodent, able to survive indefinitely on a dry diet, living in subterranean burrow permitting escape from, or avoidance of, prolonged exposure to extremely low temperatures, exhibit the same thermoregulatory pattern as a temperate or tropical bat, which must contend with a seasonal food supply while living in a relatively constant physical environment? While there may be valid taxonomic and evolutionary reasons for expecting the physiological features of hibernators from a single family to be approximately the same, there is little reason to expect all families of hibernators to be similar. Certainly the golden-mantled ground squirrel, *Citellus (Spermophilus* or *Callospermophilus) lateralis* which lives at high elevations where much of the time the ground is usually covered with a heavy snow pack, but an environment where food is probably routinely abundant but often inaccessible, would be subjected to intense selection pressure to remain dormant for extended periods of time. The pocket mice, *Perognathus spp.*, on the other hand, which live in deserts or semi-arid habitats may, for the most part, be faced with the consequence of extended periods of drought and forced to cope with an ever dwindling supply of seeds. In fact some pocket mice do not hibernate during times of abundant food (27).

My basic assumption in the evaluation of hibernation as an adaptive response to limits of food is that it is probably advantageous for an animal to be torpid as little as possible while balancing its energy requirements with the energy available. Thus, for some species, it may suffice to have a reduction in body temperature for only a few hours each day, because ambient conditions are sufficiently mild so that daily foraging at a time food resources are dwindling yields sufficient energy to meet metabolic demands. For other species, daily foraging is out of the question and it becomes necessary to lower the body temperature for extended periods of time.

The identification of shallow torpor due to a lack of food and exposure to low temperatures probably had its genesis in the 1951 paper by Howard in which he reported the effects of low temperature, abundance of nesting material and food on the survival of various species of *Peromyscus* (principally *P. leucopus noveboracensis* and *P. maniculatus bairdi*). At various places throughout his paper Howard uses the term torpor, though he never exactly specifies how he differentiated torpid animals from those simply dying of starvation. He concluded his paper with a statement which remains essentially true after some 26 years of studying torpor in a large number of small species of mammals: "Apparently less food is required to maintain *Peromyscus* at freezing temperatures, when they can become torpid during the daytime, than is needed at warmer temperatures when the animals cannot become torpid and can only huddle and sleep by day." With the advantages of hindsight, it is clear that Howard implicitly recognizes spontaneous torpor as something different from torpor induced by starvation. And while he did not measure rectal body temperature, his measurements of surface body temperatures from the middle of a huddled group of *Peromyscus* suggests, along with some other observations on core to surface gradients of normothermic animals, that his torpid mice had body temperatures of 15-16°C.

Dawson (11) observed body temperatures as low as 22.0 and 17.8°C, respectively in *Dipodomys merriami* and *Dipodomys panamintinus*, in a general thermoregulatory study of a variety of desert rodents. He interpreted this as hypothermia and a result of their inability in a post-absorptive state to mobilize fat reserves rapidly enough to permit adequate heat production. It is interesting that this viewpoint tended to prevail for a considerable period of time, i.e., the assumption that starvation-induced torpor represents an inadequacy of thermoregulation rather than an adaptation of the thermoregulatory process.

Perhaps the work which essentially initiated the series of studies which ultimately led to the conclusion that starvation-induced torpor was a controlled physiological response rather than a breakdown in thermoregulation, was the work of Tucker (48) in which he was able to regulate the incidence and length of spontaneous torpor in *Perognathus californicus* by rationing their food. He also noted that animals could not arouse from torpor if their body temperature fell below 15°C. Correlated with this was his observation that the torpor cycle was irregular at ambient temperatures between 10 and 12°C, evidence that there was some optimum temperature at which regular daily torpor occurred.

III. TORPIDITY IN CRICETIDS

MacMillen (31) systematically studied temperature regulation and torpor in the cactus mouse, *Peromyscus eremicus*, a study engenderd by his observation that *P. eremicus* disappeared from his study plot during a 6-8 week period of early summer when water was least abundant. In only one instance did he observe spontaneous torpor in the laboratory in a group of eight animals, all of which could be starved into torpor. However, MacMillen believed that spontaneous torpor occurred in nature, since the same individuals reappeared in the trapping quadrant after their prolonged absence, though he could not obtain a significant degree of spontaneous torpor in his laboratory. MacMillen's study demonstrates one of the problems typical of hibernation studies, *viz.*, the extent to which the conditions of captivity interfere with the natural responses of the animals? Since moisture was conspicuously absent in the habitat of *P. eremicus* during the period the animals disappeared, MacMillen (31) sought to relate torpor in this species to a lack of water.

However, even though *P. eremicus* could maintain itself indefinitely on a dry diet at 20°C and a relative humidity of 60-80%, animals were not in positive water balance at other temperatures without free water to drink. Thus, water withdrawal ultimately led to self-imposed starvation, and leaves in doubt the basic stimulus causing torpor. Hudson (22) observed daily torpor in *Baiomys taylori*, a 7 gram cricetid, by limiting the food. Morhardt and Hudson (34) reported starvation (i.e., food rationed) induced torpor in several species of *Peromyscus (P. crinitus, P. boylei, P. eremicus, P. maniculatus* and *P. leucopus). P. truei* and *P. californicus* could also be induced to exhibit daily torpor by rationing their food (23). In all of these species the progressive reduction of food led to torpor not only at room temperature, but also at temperatures below 15°C. Under these conditions the body temperature was anywhere between approximately 15 and 25°C and the episodes of torpor lasted only a few hours each day. Arousal was spontaneous (i.e., did not require exogenous heat) and the length of their torpor periods (Fig. 1) were related to the amount of daily food rations (22). Whether these animals were starved is a moot point. If starvation means the absence of food at the time animals enter torpor, then all of these various species of small cricetid rodents were starved. But this is also true of many "classical hibernators", e.g. bats. It is likely, as Davis (9) points out, that even if food is present, starvation may be self-imposed, as the absence of eating always precedes the induction of torpor. That small rodents such as various species of *Peromyscus* become torpid with a body weight reduction of 9 to 27% is suggestive of a depletion of body fat which has preceded torpor, although this figure may be spuriously high since animals tend to fatten in the laboratory and it is quite possible they were only reaching their field weights at the time they became torpid (32).

If daily torpor in *Peromyscus* is not simply a breakdown in thermoregulatory control due to starvation, there should be evidence that these species maintain control over their cardiovascular, respiratory and nervous systems during episodes of torpor. Morhardt (33) examined the performance of the heart of *Peromyscus*

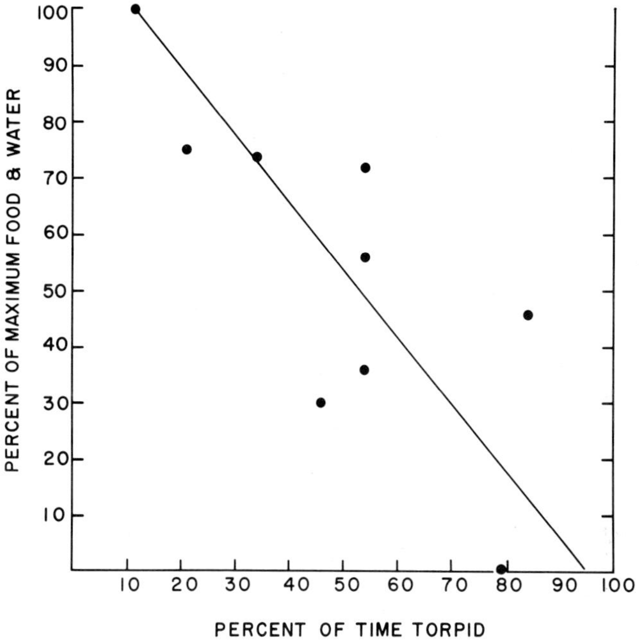

Fig. 1. *Relation between percentage of time during 24 hr animal was torpid and ration of food it was given. Ration of food is expressed as a percentage of 3.7710 g of seed and potato, an amount inadequate for normothermia throughout a 24-hr period yet eliciting only a minimum period of daily torpor. From (22), by permission of University of Chicago Press.*

spp. during torpor and by using both abdominal and jugular cannulae to inject atropine into torpid deer mice. The ensuing cardio-acceleration (Figs. 2 and 3) was evidence that the heart was under vagal inhibition. He also observed that for a particular body temperature, the heart rate was greater during arousal from torpor than it was during entry into torpor (Fig. 4), further evidence of control over heart rate. Also, the oxygen consumption at a particular body temperature was much greater during arousal than it was during entry into torpor, evidence that at least one of two physiological effects occur: i) there is suppression of metabolism during entry into torpor, or ii) there is an

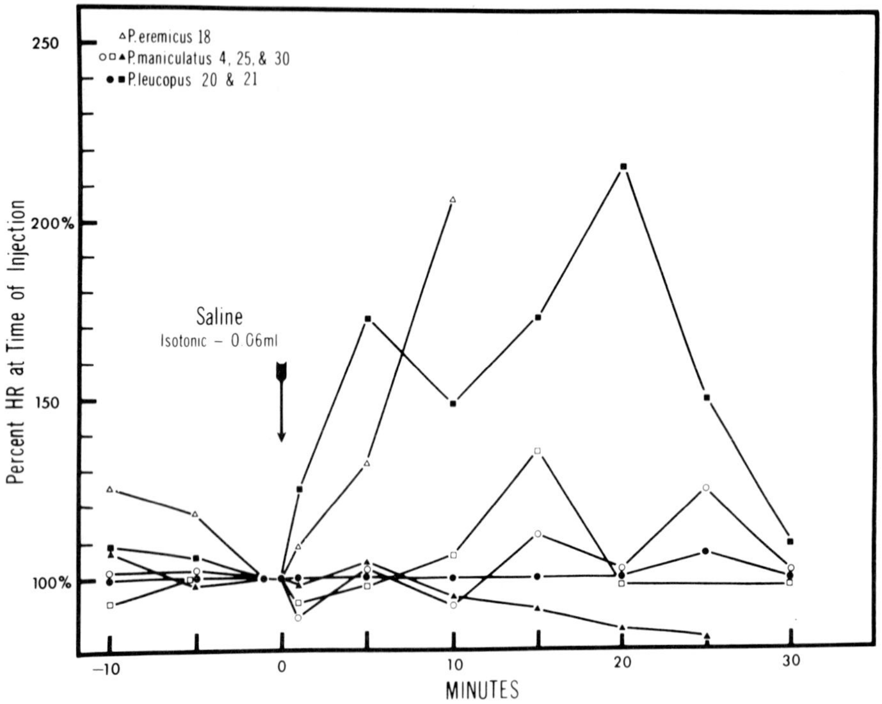

Fig. 2. *The heart rates (plotted as percentage of the heart rate at the time of infusion) before and after infusion of isotonic saline solution into the peritoneal cavities of torpid Peromyscus through indwelling catheters. All animals were fully torpid except P. eremicus No. 18 which was completing entrance. From (33).*

intensification of metabolism greater than a simple Q_{10} effect during arousal from torpor. These metabolic effects are accompanied by corresponding changes in respiration. All of the observed cardiovascular, respiratory, and metabolic effects Morhardt observed in *Peromyscus* are similar to those Strumwasser (46) reported for *Citellus (Spermophilus) beecheyi*, with one major exception, the lowest body temperature from which *Peromyscus* could successfully arouse spontaneously was much higher than in *S. beecheyi*. And in fact, it appears as if most of these small cricetids capable of

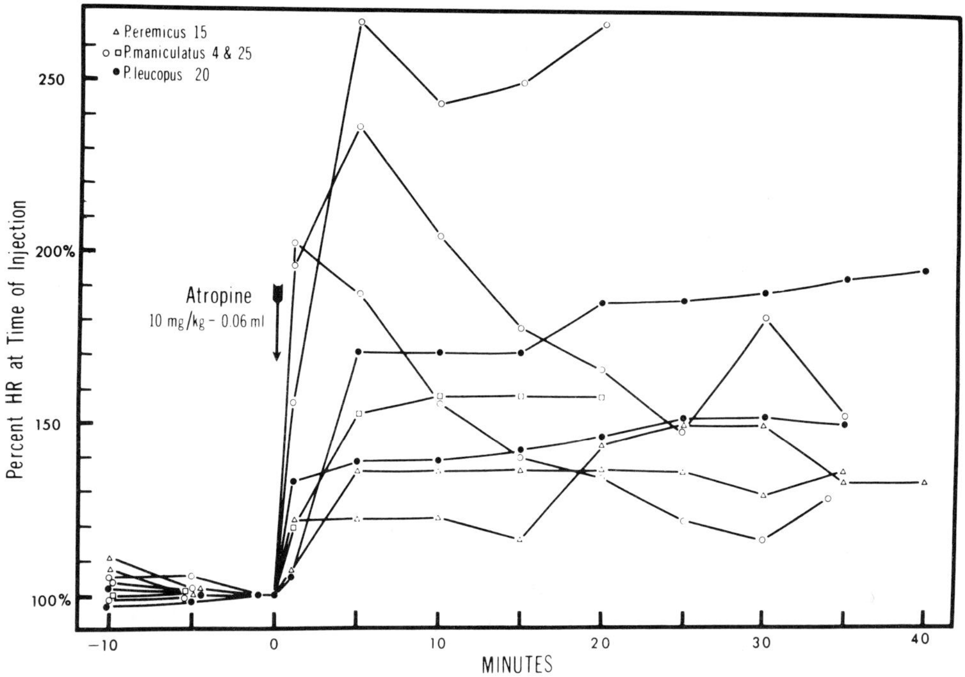

Fig. 3. *The heart rates (plotted as percentage of the heart rate at the time of infusion) before and after the infusion of atropine sulfate solution into the peritoneal cavities of fully torpid* Peromyscus *through indwelling catheters. From (33).*

daily torpor have relatively high "critical body temperatures", i.e., body temperatures below which they are incapable of spontaneously arousing (though it may be possible to provide sufficient exogenous heat to restore normothermia, but this is highly artificial).

In an effort to determine what limited the depth (lowest body temperature) of torpor, Hudson and Eller (26) examined the ability of isolated perfused hearts of *Peromyscus leucopus* to maintain tension and electrolyte balance as a function of perfusion temperatures (Figs. 5, 6, and 7). The maximum tension and the temperature at which the isolated hearts began to lose ^{42}K at an increased rate corresponded well with the lowest body temperature from which daily spontaneous arousals occurred (Tables 1 and 2).

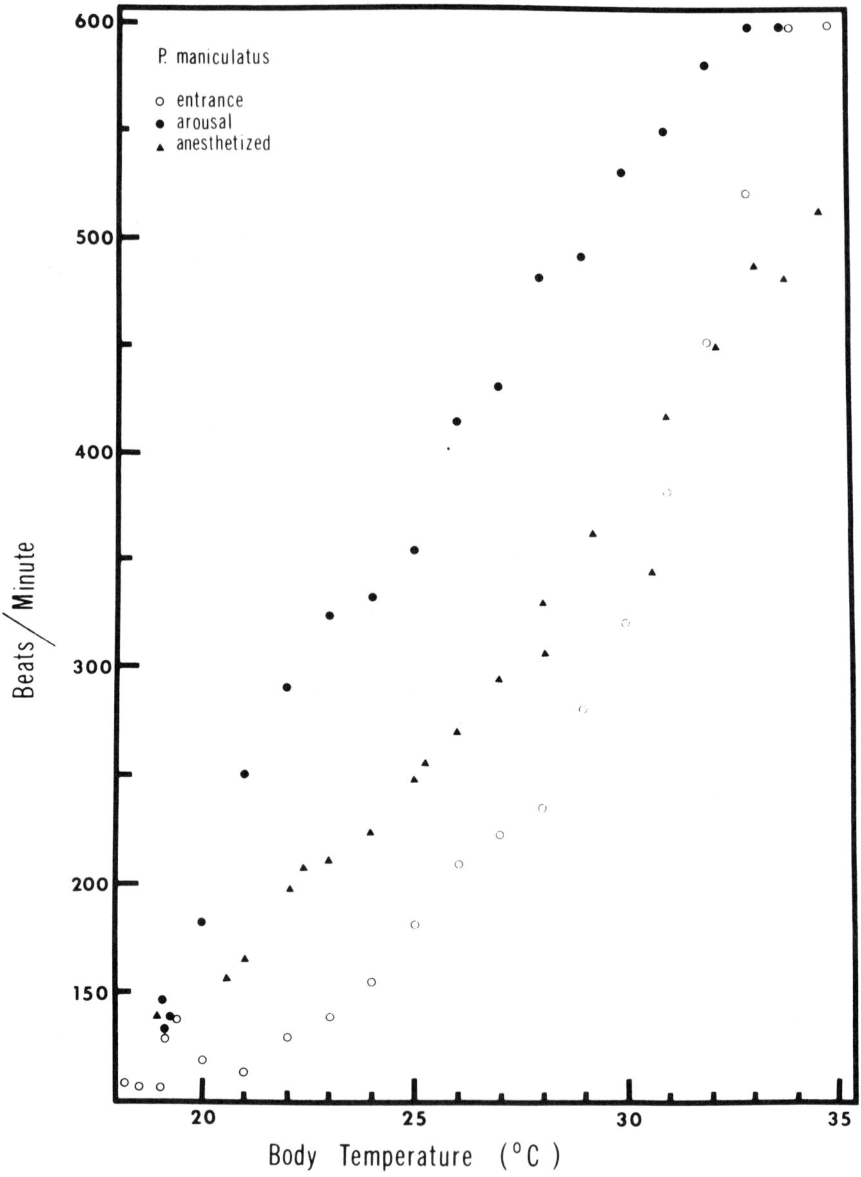

Fig. 4. *The heart rates of P. maiculatus No. 6 during torpor plotted as a function of the body temperature during entrance (open circles) and arousal (closed circles). The closed triangles represent the pooled heart rates of two P. maniculatus anesthetized with sodium pentobarbital and cooled. From (33).*

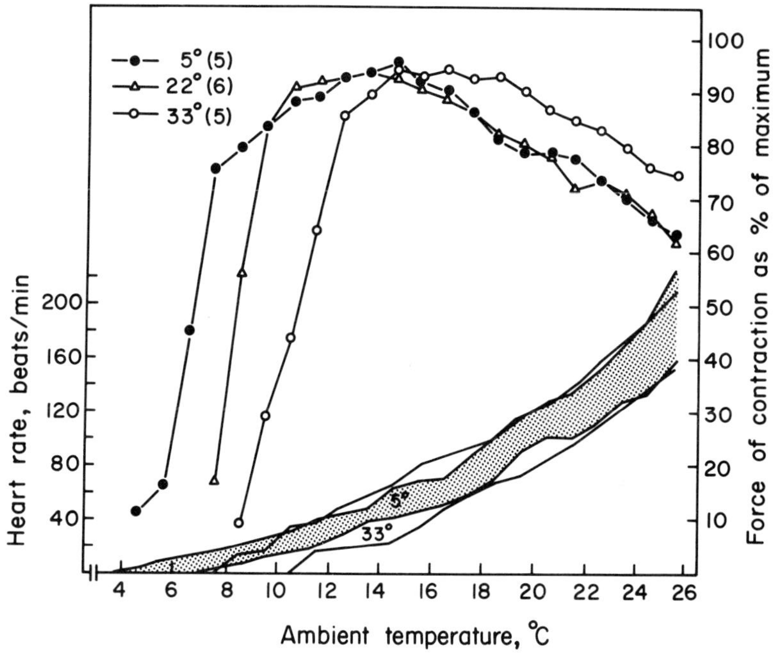

Fig. 5. *The relationship between ambient temperature during perfusion and the average heart rate or average force of contraction of isolated hearts from five animals acclimated to 5°C (stippled area in lower part of graph or ●); 22°C (△) or 33°C (open area demarcated by upper and lower lines in lower area of graph or ○). A resting tension of 200 mg was applied to the heart during cooling. From (26).*

Up to this point we have been considering a phenomenon which is primarily laboratory, with evidence that it is a natural thermoregulatory response to limited availability of food, there was little evidence that it occurred in the field. However, Stebbins (45) says: "*P. maniculatus* underwent periods of daily torpor in winter at Heart Lake. They had the lowest level of winter activity of the three species tested. During early January and early March there was almost no activity during the day and little at night. The deer mice were removed from their nest box while torpid and found to fit the description of torpid *Peromyscus*

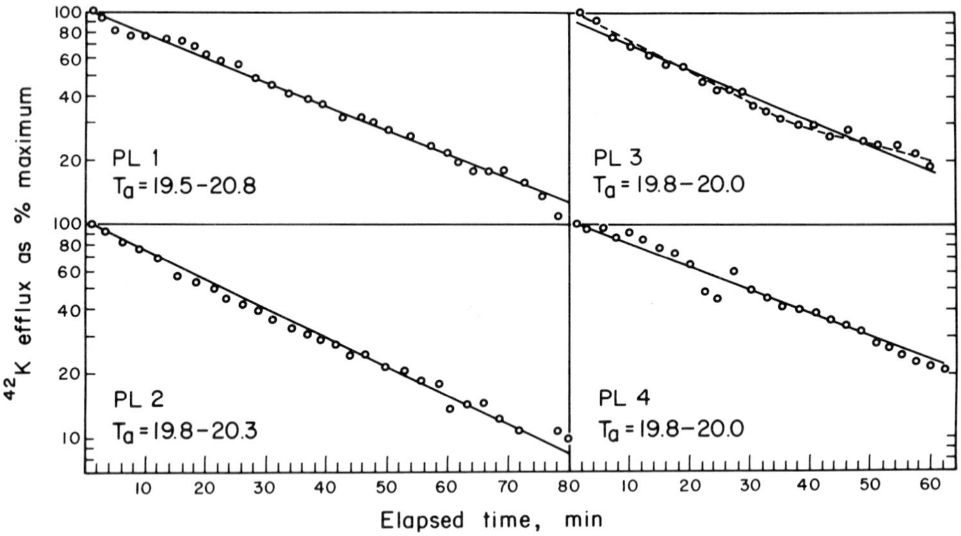

Fig. 6. *^{42}K in the perfusate from isolated hearts of four animals perfused at the constant temperature indicated below each line. A straight line has been fitted by eye to the points. The dashed line in the upper right corner represents an alternative fit of lines to these points. From (26).*

spp. as described in nature by Howard (21). They had a slow breathing rate, moved their limbs slowly when walking, let their tails drag and made no attempt to run or escape. When the temperature moderated somewhat in late March and early April, *P. maniculatus* were no longer observed in torpor, and their general level of activity increased considerably." In three separate recent studies: Gaertner, *et al.*, (16); Hill (19) and Lynch *et al.*, (30) have supplied additional evidence that daily torpor in *Peromyscus* might be a naturally occurring phenomenon. In all three studies, *P. leucopus* became torpid in the pesence of food, i.e., torpor occurred without the necessity of restricting food. Gaertner, *et al.*, (16) compared the physiological responses of *P. leucopus* forced into torpor by food deprivation with *P. leucopus* in which torpor occurred spontaneously in the presence of food. The starved animals exhibited a slower spontaneous arousal than

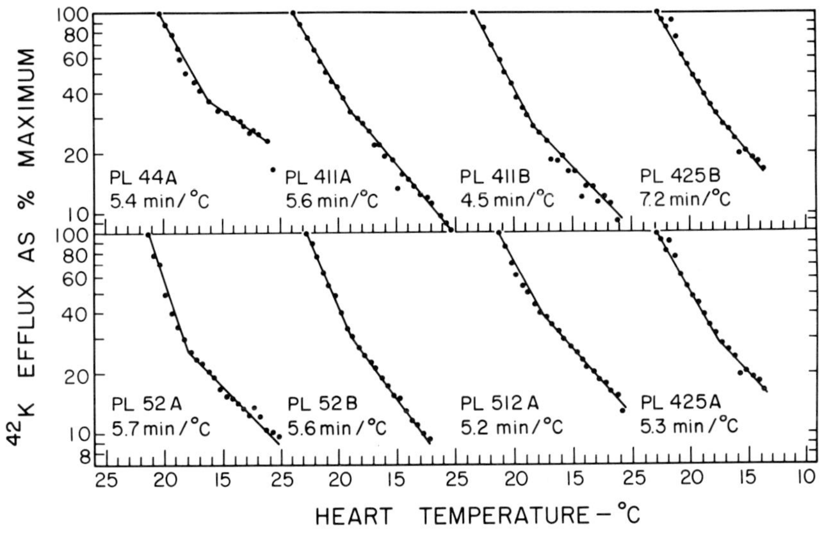

Fig. 7. *^{42}K in the perfusate from isolated hearts of animals acclimated to 5°C perfused at decreasing ambient temperatures. The two straight lines have been fitted by eye to all of the points for each individual animal. Cooling rate for each experiment given below the animal number. From (26).*

those entering torpor in the presence of food, a response analogous to that observed by Bartholomew and Cade (1) in their study of *Perognathus longimembris*, one of the smallest of the pocket mice (Heteromyidae). Also, there was a striking difference in the performance of the heart as recorded by telemetry. Unlike the heart rate of *P. leucopus* studied by Morhardt (33), Gaertner, *et al.*, (16) did not observe skipped beating during torpor in their animals. It is, of course, possible that this difference merely reflects methodology: telemetry versus the restraining effects of EKG leads. However, when this observation is coupled with the differences in arousal rates of starved and non-starved *Peromyscus*, it appears as though there must be some subtle differences in the manifestations of control mechanisms utilized by *Peromyscus* (and *Perognathus*) depending upon the availability of food. Gaertner *et al.*, (16) also noted that acclimatization achieved by

Table 1. TEMPERATURE AND TIME CHARACTERISTICS OF PERIODS OF TORPOR OF WHITE-FOOTED MICE

Species	No. of periods of torpor	Lowest body temperature (°C)	Range of ambient temperatures (°C)	Maximum arousal rate (°C/min)	Maximum duration of torpor (hr)
P. eremicus	17	19.0	9.5-22.1	0.43	11.2
P. boylei	23	18.0	11.0-21.0	0.54	8.8
P. leucopus	19	17.0	11.0-21.5	0.42	11.3
P. crinitus	28	17.0	12.2-26.5	0.51	9.5
P. maniculatus sonoriensis	25	13.4	-0.5-27.5	0.64	10.8

TABLE 2. AMBIENT TEMPERATURE OF CHANGE IN ACTION POTENTIAL AND ^{42}K EFFLUX OF PERFUSED, ISOLATED HEARTS

	Animals acclimated to 30°C			Animals acclimated to 5°C	
Animal No.	T_a of maximum action potential (°C)	T_a of ^{42}K efflux change (°C)	Animal No.	T_a of maximum action potential (°C)	T_a of ^{42}K efflux change (°C)
418 A	16.0	20.8	44 A	-----	16.0
418 B	21.0	20.0	411 A	17.5	19.0
59 A	18.0	18.0	411 B	12.0	18.5
59 A	21.0	19.5	425 A	20.0	17.5
523 A	19.0	19.5	425 B	15.5	18.0
523 B	22.5	19.5	52 A	11.0	18.0
66 A	17.5	22.5	52 B	18.0	19.0
66 A	16.0	20.0	516 A	17.0	17.5
			516 B	14.0	16.0
$\overline{X} \pm 2$ S.E.	18.9 ± 1.7	20.0 ± 0.9		15.6 ± 2.2	17.7 ± 0.7

exposure of *P. leucopus* to the natural conditions of a reducing photoperiod and ambient temperature in the fall appeared to be essential for spontaneous torpor to occur. Whatever role this acclimatization played was retained even after animals were subjected to laboratory conditions; mice brought into the laboratory after mid-December exhibited a rate of oxygen consumption indicative of spontaneous torpor after they were returned to outdoor nest boxes in late Febraury. This spontaneous torpor in *Peromyscus* was seasonal since no torpor occurred durring the months of May and June, and in late April and May the ability to enter and recover from torpor was reduced. Hill (19) studied torpor in a laboratory colony of *P. leucopus* and observed regular daily torpor in the presence of food *ad libitum* in seven of seventeen animals, while four others displayed sporadic episodes of torpor. (Unfortunately, Gaertner *et al.*, (16) did not specify what fraction of their population exhibited spontaneous torpor). Hill attributed these individual differences in proclivity for spontaneous torpor to genetic differences and suggested that his results might portray "an example of mammalian torpor in a potential state of contemporary evolutionary flux." It is not unusual for some individuals of species considered to be "classical hibernators" to fail to undergo torpor in the laboratory. No doubt many of us have assumed that this is a consequence of laboratory conditions. But there is evidence that the capacity for hibernation in hibernators such as the hamster *Mesocricetus auratus*, is absent in some individuals and that this may be genetically determined (8).

That daily torpor is widely prevalent in nature has been convincingly demonstrated by the recent study of Vogt and Lynch (51) who placed 36 nest boxes in each of two 1.7 hectare study plots and provided approximately 25 grams of cotton in each box. There were eight food hoppers in one of the study plots while the other plot was without a food supplement. All nest boxes were examined between 8 and 11 a.m. at 7-10 day intervals between November and March. Evidence of torpor was based predominantly on behavior, though in every case in which an animal judged to be torpid and had its body temperature measured, the body temperature was 25°C or lower. Of 143 observations, there were 34 instances of torpor in Site I (where food was provided), while only 17 of 205 observations in Site II (no food provided) were mice found to be torpid. These observations, which were corroborated by telemetered body temperatures from animals confined to outdoor cages, also demonstrated a higher incidence of torpor during the winter (December, January and February). Thus, the capstone of an observation which was initiated in 1951 by Howard (21) has culminated in the conclusive proof that torpor is a natural event in the life of *P. leucopus* and that it is not a breakdown of the thermoregulatory machinery.

Further evidence that torpor can be daily in its frequency and shallow in its physiology in many small cricetids is the observations by Fisler (14). In his general study of the adaptations of harvest mice to salt marshes he observed spontaneous daily torpor in *Reithrodontomys raviventris raviventris* and chill induced torpor in *R. megalotis*, although he was unable to induce torpor in *R. r. halicoetes* even by exposure to 4.8°C for 14.5 hrs. Perhaps the temperature was too low and this subspecies resisted torpor because of the potentially lethal consequences should it fail to keep its body temperature from falling below some critical level. The lowest body temperature Fisler (14) observed for *R. r. raviventris* was 21.5°C and at no time did the body temperature fall to the ambient temperature which varied between about 17 and 20°C. When handled, these torpid mice aroused within a half-hour or so. Since Fisler considered *R. r. raviventris and R. r. halicoetes* to have been derived from *R. megalotis*, he interprets his results to mean that selection intensified the trait in *raviventris* and eliminated it entirely from *halicoetes*. This interpretation implies considerable liability to the evolution of torpor.

IV. TORPOR IN HETEROMYIDAE

A similar history of our understanding of torpor in *Peromyscus* which has been repeated for the kangaroo rat, *Dipodomys merriami*, began with Dawson's observation of hypothermia after 12 hours of exposure of fasting, post-absorptive, *D. merriami* to an ambient temperature of approximately 15°C. Subsequently, Carpenter (7) elicited torpor in *D. Merriami* by progressively restricting their food. However, mortality was high in these experiments, though the animals survived for over a week and had body temperatures of 16-17°C at an ambient temperature of 15°C.

In a study of torpor in *D. merriami*, Yousef and Dill (53) observed daily torpor in kangaroo rats maintained at 13-15°C with a ration of 2.0 grams of rolled oats, which was 45% of their *ad libitum* food consumption at that temperature. Of the eight of eighteen animals in which the T_{re} dropped to 14°C or below only one survived the 7 day experimental period. No measurements were made of spontaneous rewarming at 13-15°C, but these authors note that spontaneous arousals must have occurred since "...they enter hibernation periodically every day." Among animals rewarming at an ambient temperature of 27-30°C, nine whose body temperatures were above 15°C at the beginning of arousal survived, while six which died during arousal had body temperatures slightly below 15°C. Of the thirteen animals which were on a restricted food regime, the fat content of those surviving was 1.7% while the fat content of those which died was only 1.0%. These levels of fat are markedly below those of animals fed *ad libitum* - 3.7% for animals kept at 25 ± 2°C and 7.8% for animals kept at 13-15°C. As yet there is no evidence of spontaneous torpor in any *D. merriami* either in the laboratory or more importantly, in the field. But based on the history of our studies of *Peromyscus spp.*, it may be only a matter of time until someone makes the critical observations. It may well be that a less severe reduction in food, or a more gradual reduction, coupled with sufficient acclimatization could show that torpor in this heteromyid is not qualitatively different than that of another heteromyid, *Microdipodops pallidus*. Brown and Bartholomew (5) report that *M. pallidus* does not become torpid at either 17 or 6°C when given seeds in excess each day, but when the daily ration is reduced to 1.5 g, all individuals become torpid, an observation at odds with the report of Bartholomew and MacMillen (3) that "Under laboratory conditions we found that kangaroo mice at any time of year will spontaneously become dormant at ambient temperatures ranging at least from 5 to 26."

With the advantage of hindsight, one wonders to what extent an inadvertent starvation imposed torpor was induced. The role of food restriction in this family appears to play a significant role in the induction of torpor. Kenagy (27) was able to trap *Perognathus longimembris* (the little pocket mouse) throughout the winter of 1969-70, a time when the seed crop was abundant, whereas in 1970-71 and 1971-72, when the seed was less abundant, no animals appeared in his traps. Furthermore, during the year of abundance, there was the greatest survival rate of his population over the winter -- 82%. French (15) noted that captive *P. longimembris* maximize their time of euthermia by selecting high environmental temperatures at all times of the year.

V. TORPOR IN DIPODIDAE

The ecological equivalent of the New World heteromyids in the Old World Deserts are the dipodids, both families including hibernators as well as non-hibernators. Most of the information about the hibernation of dipodids up to 1961 is in the Russian literature and has been summarized by Cade (6).

Cade (6) describes *Jaculus jaculus* as exhibiting lethargy and hypothermia in captivity with irregular periods of inactivity in nature. Hooper and El-Hilali (20) observed torpor in captive jerboas, *Jaculus orientalis* (Body Weight = 132-142 g) and *Jaculus deserti*, (Body Weight = 63-78 g), but only when food was withheld. These species do not store food. "Deep hypothermic lethergy" (their term) was observed in 15 animals (7 *J. orientalis* and 8 *J. deserti*), and in all of the 10 known instances of rewarming (arousal), the process took place without the application of external heat at ambient temperatures of 17-21°C. At least three animals aroused spontaneously, i.e., without disturbance, at the ambient temperature at which they had been torpid. El-Hilali and Veillat (13) observed torpor in two *Jaculus orientalis* which had been maintained at 5°C for more than two months. The rectal temperatures of these animals was 10.8 and 9.5°C, and while both animals aroused, they did so at an ambient temperature of 24°C, so that they had been subjected to considerable exogenous heat. There was no indication as to the availability of food during dormancy. These same authors (*op. cit.*) excavated forty-one plugged burrows in November at which time the soil temperature at a depth of 30-45 cm ranged from 9.5 to 10°C. They found 38 animals (23 females and 15 males), which had an average body weight of 145.9 ± 9.9 g ($\bar{x}$ ± s.e.) torpid with body temperatures between 10 and 11°C. One animal was found to be active and two were dead. Interestingly, the torpid animals were all in an upright but crouched position, with the nose on the forepaws, the hind legs hidden beneath the body, and the tail running in a sagittal plane between the fore and hind paws. The head was bent downwards with the eyes closed and the upper half of the pinna folded back. This posture is similar to that observed in torpid *Baiomys taylori* (22), but it is not typical of hibernators. This contrasts with the "classical hibernator" which assumes a horizontal, curved posture when torpid. If *Jaculus* does not store seeds, as has been noted (13), either it has deposited considerable fat, or its torpor must be daily, in which case the animals would periodically forage for food. Their arousal rate is incredibly slow, being approximately 0.10 to 0.15 degrees/minute (my calculations); a response reminiscent of animals forced into hypothermia. While they were initially apneic, breathing soon became regular during arousal and they began to shiver when their body temperature reached 17°C. Although these authors describe *Jaculus orientalis* as a "true hibernator," much remains to be learned about the physiology of torpor in this species and to what extent it is similar to the "classical hibernator." The data at present argues for daily torpor.

VI. TORPOR IN MURIDAE

Many of the small rodents which engage in daily, shallow torpor are cricetids, a group considered to be ancestral to the murids (44). With the exception of Petter's unverified account of summer lethargy in *Mus musculus spretus* as described by Cade (6) and Robert's (40) description of estivation in *Steatomys*, murids are not considered to be hibernators, although Hudson (23) observed a thermal sensitivity in the perfused isolated hearts from *Mus musculus* var albino slightly above that of other small rodents which exhibited shallow daily torpor. Recently, Morton (36) observed torpor in *Mus musculus* located in nests underneath large rocks in Victoria, Australia. They had body temperatures of 18.2 to 21.5 at air temperatures of 11 to 12°C. These observations represent a third example of a phenomenon occurring under natural conditions though there were hints from their isolated hearts (23) that shallow, daily torpor could have been considered to be a laboratory phenomenon. Morton's observations strengthens the case for presuming that shallow, daily torpor probably occurs widely in nature, even though its naturalness remains to be documented in most cases.

VII. TORPOR IN INSECTIVORES

Although there are many species in the families Tenrecidae and Erinaceidae known to hibernate or estivate, the special thermoregulatory feature shallow, daily torpor only recently been reported for the family Soricidae, a family which includes many species which weigh less than ten grams. Vogel (50) describes cold resistance and reversible hypothermia in *Suncus etruscus*, the smallest species of mammal. The females weigh 1.953 g and the males weigh 1.761 g. Torpor occurred intermittently at a T_a of 16 - 18°C for 1.5 to 7.5 hrs with a body temperature 2°C above the ambient temperature when these animals were subjected to a restricted food supply. Well fed animals resisted temperatures below 0°C. Three specimens were able to survive without food for a 25 hr. trial. These observations contrast markedly with the results from the systematic study of thermoregulation in three species of shrews, *Sorex minutus*, *S. araneus* and *Neomys fodiens*, averaging 3.7 g; 7.0 and 17.1 g respectively (17). Gebczynski (17) looked unsuccessfully for torpor in freshly captured animals during all seasons of the year at ambient temperatures ranging from 3 to 25°C and was unable to observe a significant reduction in metabolism which did not result in death when food was removed. But, of course, he did not do studies in which food was progressively restricted, and since his animals were freshly captured, surely their behavior must have been disturbed under these conditions. In view of what we now know about torpor in a variety of small mammals as it relates to food supply and ambient temperatures, it would be worth repeating these experiments using animals which had become accustomed to conditions of captivity; the metabolic measuring devices; and a daily progressive restriction of food.

VIII. TORPOR IN PRIMATES

Russell* (see page 193-206, reference in footnote) studied nine *Microcebus murinus* and three *Cheirogaleus medius* which he kept under relatively constant temperature and light conditions for a year. These species, which ranged in size from about 80 to 300 g exhibited prehibernal fattening and spontaneous daily torpor in the austral winter-spring (March - April), but did not show daily torpor at other times of the year.

IX. TORPOR IN MARSUPIALS

Not only do small placental mammals exhibit the special pattern of daily, shallow torpidity, but it has been observed in several species of marsupials. We have, of course, known for some time about torpor in the phalangerids (2).

However, it now appears that shallow, daily torpor is probably widespread amongst the small dasyurids; carnivorous marsupials. Wallis (52) studied daily torpor in *Antechinus stuartii* by means of a thermistor probe implanted in the interscapular region and connected with a transmitter. He also made periodic measurements of rectal temperature and found these to be within 1°C of the subcutaneous temperature. Eight animals captured in December and January (the austral summer) did not become torpid, even at a T_a of 15°C, yet 7 out of 9 animals captured between April and September (austral fall and winter) and given only 2 grams of food (Harpers Puppy Chow) had body temperatures which briefly fell below 30°C, with the animals entering torpor between 1 and 12 times per experiment. In one experiment, the total oxygen consumption over a 25 h.period at a T_b of 23.8°C was 41% of the resting, normothermic level of *A. stuartii*. As with other species of small mammals, these torpid marsupials assumed a crouched posture during torpor with the head tucked under the abdomen. Respiration was reduced from 60-180/minute to 40-80/minute, and there were periods of apnea. The T_b never fell below 18-21°C in any of the animals which survived torpor. Dawson (10) reports torpor in three species of planigales, *Planigale ingrami*, *P. tenuirostris* and *P. gilesi*, which weigh 5.8-9.3 g; 6.6-7.3 g; and 9.4 to 10.0 respectively. These marsupials (closely related to *Antechinus*) are considered ecological equivalents of the placental shrews (39). The body temperatures of the planigales were as low as 14 °C, the lowest temperature which occurred in the unheated animal house. They entered torpor as soon as the lights were turned on and the torpid period persisted for 2-4 hours under experimental conditions.

The body temperatures during torpor in two other dasyurids, *Sminthopsis crassicaudata* (N = 5) and *S. larapinta* (N = 5) were measured by Godfrey (18) who studied laboratory bred animals which nested in a Perspex tube fitted with a thermocouple soldered to a thin copper plate cemented to the floor of the tube so as to record skin temperature of resting animals. The ambient temperatures in the laboratory were 21°C., while a few animals were kept outdoors where the ambient temperature ranged from 9 to 16°C. Starvation in these species is not likely to be reflected in a marked depletion of body fat because of the heavy deposits of fat in their tails. Fully fed *S. carassicaudata* did not become torpid, but when food was limited to half the daily *ad libitum* consumption, torpor occurred more readily, particularly at low temperatures. The lowest skin temperature varied from 17-23°C, but since the skin temperature is probably at least 1°C less than the rectal temperature, this species probably has a "critical T_b" of 18°C, which is similar to *A. stuartii*. Animals completely deprived of food aroused normally from torpor, but did not feed when food was offered and subsequently died. The skin temperatures of *S. larapinta* at T_a ranging between 15.0 and 23.9°C were between 18 and 27°C. Based on Morton's observation (36) it is now known that *S. crassicaudata* can be hand captured in the field during episodes of torpor in which the body temperature ranges between 17 and 23°C at a T_a of 9 to 11°C, and that it arouses within 30 minutes after capture at the prevailing air temperature, indicative of the spontaneous nature of their torpidity. This is the fourth case in which long standing laboratory observations have only recently been corroborated by field studies.

In conjunction with a general study of water turnover, Kennedy and McFarlane (28) observed the body temperature of *Dasycercus crassicaudata* to be 5-10°C below its normal value of 35-37°C at ambient temperatures between 20 and 30°C, though at a T_a of 20°C the body temperature remained 35-37°C. [Morrison and McNab, (35) report this too but as an unpublished observation using the old generic name *Chaetocercus*.] *D. crassicaudata* deprived of food for 7 days exhibited a 50-60% decrease in oxygen consumption with a body temperature which steadily dropped from 35-36 to 22-24°C, i.e., to a level within 2°C of ambient temperature. It is unknown what these authors meant by "steadily." On refeeding, the body temperature rose to normal by the following morning. The body temperatures of *Dasycercus* maintained on minimal food rations were 10°C below normal, or within 3°C of ambient temperature. Significantly, *Dasycercus* is primarily a desert dasyurid, whereas *Sminthopsis* is most common in the more temperate regions, and in conformity, *Dasycercus* has a "critical body temperature" which is probably about 23°C, whereas the more temperate *Antechinus* and *Sminthopsis* have critical body temperatures of 18°C.

Shallow, daily torpor is not exclusive to the marsupials of Australia. Morrison and McNab (35) report torpor, which is daily and shallow in the murine opossum, *Marmosa spp.*, a 13 gram didelphid marsupial from South America. This species (thought to be *M. microtarsus*) exhibited diurnal cycles of torpor lasting 5-6 hours with a slow (3 hour) spontaneous arousal at a T_a of 16°C (T_b not measured during this particular experiment, but based on other measurements it was probably somewhere between 16 and 18°C). These authors also report arousal in 30 minutes by an animal which had a T_b of 18°C (as estimated from their figure 7). The depressed metabolism that Morrison and McNab (35) observed at 11°C in *Marmosa elegans*, another small oppossum which weighs about 30-48 g suggests torpor, but its exact nature remains unknown.

X. DISCUSSION - DAILY TORPOR

The most consistent feature to the torpor exhibited by all of these small marsupials, insectivores and rodents is that there appears to be a "critical body temperature," below which spontaneous arousals are lost. The daily episodes of torpor last for only a few hours, usually beginning about the time that their foraging activity ceases. Presumably the length of torpor is adjusted to the food supply, though this remains to be critically studied in most of the species known to engage in daily torpor. To what extent the "critical body temperature" is a reflection of natural selection in response to ambient temperature encountered during torpor in their respective habitats is unknown, though the evidence is suggestive that there is a match between ambient temperature and "critical body temperature." It is important to carefully consider the temperature of their microhabitats, for some species are fossorial while others are not. For example, even though *Baiomys taylori* lives at the same latitudes as the *P. leucopus* we studied (26) it has a "critical body temperature" at least 5 to 7°C above that exhibited by *Peromyscus*, probably because it is not fossorial and is exposed to higher ambient temperatures at certain times of the year. It is not known why these species cannot or will not let their body temperatures drop below the "critical level." However, there is evidence which suggests that they lack the capability for the maintenance of normal cardiac function below the critical threshold, perhaps because of an increased membrane permeability to potassium; a failure to maintain adequate electrolyte balance; and a decreased ability to produce adequate ventricular force of contraction. Also unknown is the extent to which acclimatization may facilitate not only the onset and utilization of daily torpor, but the level of the critical body temperature.

It is now clear that at least a fraction of the population of small mammals which utilizes this daily torpor does so under natural conditions. If one ignores the survival of the individual and focuses on the survival of the population, daily torpor may be an adaptive response assuring that a sufficient fraction of the population survives a period of dwindling food resources until more favourable conditions for reproduction and a rebuilding of the population density occur. This fraction is a genetically selected group in which the gene frequency is the result of opposing selective forces. If too many animals are torpid, the disadvantages of torpor, such as predation act to reduce gene frequency. If an insufficient number of individuals become torpid, the selection pressure will function to increase gene frequency for torpor by favoring those individuals with the least energy requirement.

XI. CHIPMUNKS

A group of hibernators which seem to be somewhat intermediate between the "classical hibernators" such as hedgehogs, marmots, bats, and ground squirrels and the small insectivores, rodents and marsupials utilizing shallow, daily torpor, are the chipmunks, *Tamias striatus* and *Eutamias spp*. While the "critical body temperature" of the classical hibernator may be near 0°C and 15-20°C for the daily hibernator, it seems to be about 5-7°C in the chipmunks. In none of the studies of *T. striatus* has the body temperature of torpid animals been observed to approach 0°C (37, 38, 41, 43, and 54).

In none of these studies has anyone reported torpor to occur in all of the animals studied in the laboratory. Wang and Hudson (54) deafened animals which did not exhibit torpor, and 4 of the 6 animals which had not previously become torpid, did so spontaneously within one month after deafening. These results suggest that reluctant hibernators may in some cases be individuals easily disturbed by the ambient noise in their laboratory environment. Most of the studies in which there has been a systematic measurement of body or skin temperature of torpid chipmunks at ambient temperatures below 10°C report considerable variation in the lowest body temperature and the length of the torpid period of individual chipmunks. Wang and Hudson (54) found body temperatures between 5.0 and 28.5°C in 330 measurements of 52 torpid chipmunks, but in only 23 of the 163 measurements at a T_a below 9.5°C was the T_b less than 10°C. The lowest T_b from which an animal spontaneously aroused was 5.0°C and this was at a T_a of 4.0°C. Scott and Fisher (43) found skin temperatures of torpid chipmunks to range between 5.5 and

28.7 at ambient temperatures of either 0 ± 2°C or 18 ± 3°C. Pivorun (38) reports minimum body temperatures approximately 1°C above ambient temperatures of 3-16°C, but body temperatures of 6 - 14.5°C above an ambient temperature of 0°C. In general most investigators report a duration of from 1 to 6 days for the individual bouts of torpor. Based on information about body temperature transmitted by telemetry, it appears that the length of the individual bouts of torpor is directly related to the ambient temperature at T_a's between 3 and 16°C (38), but this relationship breaks down at an ambient temperature of 0°C. Furthermore, entry into hibernation at the termination of the non-hibernating period involved test drops in which the depth and length of torpor sequentially increased, but after the so-called "plateau" stage had been reached, animals would return to their lowest body temperatures and longest duration of torpor following subsequent arousals (38).

Pivorun (38) does not specify the ambient temperatures at which he measured his test drops, though it can be assumed to be the 8°C he reports for one of his experiments in which the "plateau" body temperature was 10°C. Pivorun reports the sympatric *Eutamias minimus* to be a deeper hibernator than *T. striatus*, but he defines deeper as "less prone to arousal" and not exhibiting "test drops." However, he did not expose *E. minimus* to ambient temperatures below 8°C.

The distribution of the capacity for hibernation in the family Sciuridae is fascinating, since *Tamias* [Black (4) treats *Tamias* and *Eutamias* as a single group] is thought to represent the stem of the family (4) with the flying tree squirrel branch (subfamily Petauristinae and subfamily Sciurinae, tribe Sciurini) not possessing any ability whatsoever for torpor, while the other branch, which includes the ground squirrels (subfamily Sciurinae, tribe Marmotini) representing the classical hibernation pattern. Black (4) has undertaken a thorough and systematic analysis of fossil representatives of this family and derived a phylogenetic tree (Fig. 8) in which he considers Sciurids to have originated from Paramys, an extinct group he describes as "..presumably ground living forms with a rather generalized, scampering type skeleton." Black speculates that the paramyine line reached the sciurid level of organization in the late Eocene or Oligocene, and that the tribal groups Tamini, Sciurini and Marmotini differentiated in the late Oligocene.

If hibernation is a conservative feature coupled with primitive morphological patterns (6), then it must have been present in *Tamias spp.* as far back as the Oligocene and was lost in the Sciurini tribe, but

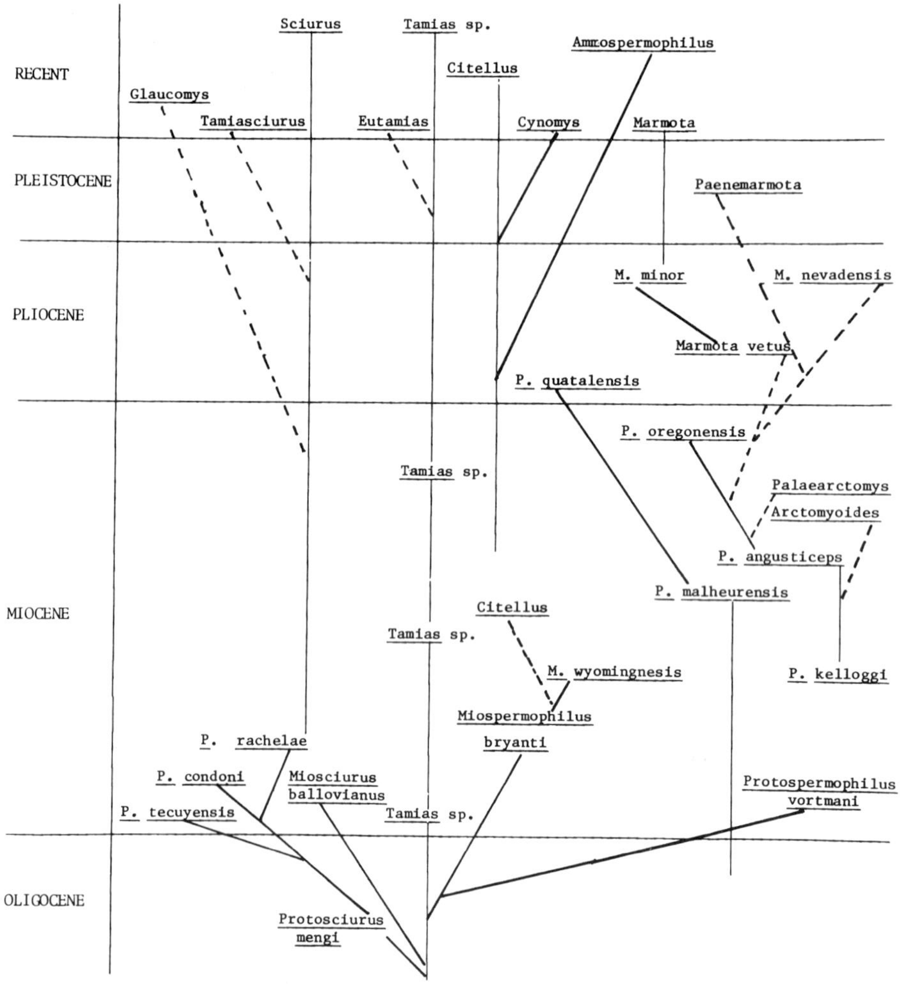

Fig. 8. *Phylogenetic tree of North American Sciuridae. Solid lines indicate probable relationships, broken lines possible relationships. From (4).*

further developed in the Marmotini tribe. The alternative interpretation would be to assume that hibernation is so polyphyletic and specifically adaptive for each different ecological niche, that it arose independently within the various sciurid lineages, exhibiting different patterns in the Tamini and Marmotini tribes. One of the most puzzling contradictions is the observation that the antelope ground squirrel, *Ammospermophilus leucurus*, which is a non-hibernator, possess a heart which has the ability to beat at nearly 0°C just as the heart from the hibernator, *Citellus mohavensis* can beat at 0°C or less (29), while the isolated, perfused hearts of the desert ground squirrel, *Citellus (Spermophilus) tereticaudus* stops beating at perfusion temperatures below 5°C and the thirteen-lined ground squirrel, *Spermophilus tridecemlineatus* can beat at perfusion temperatures below freezing (but not below the freezing point of the perfusion medium), [Table 3, (24)]. Lyman (29) assumes that *Ammospermophilus leucurus* belongs to a lineage of hibernators and has retained some of the features which characterize hibernators, while Hudson (25) assumes that hibernation represents an adaptive response to a particular set of ecological circumstances and that it can be repeatedly evolved depending upon the selection pressure. This does not, of course, explain why a non-hibernator possesses the heart of a hibernator. There is, of course, no resolution of these contrasting views, though a systematic study of the characteristics of perfused isolated hearts from the ground squirrels (including marmots), chipmunks and tree squirrels from North America should provide us with some substantial information as to what extent this thermal sensitivity is a taxonomic feature, i.e. to what extent it reflects the lowest body temperature to which the hibernator is exposed. It seems strange that the groups which have been around the longest (*Tamias* and *Eutamias*) are intermediate between the "classical hibernator" and the daily shallow hibernator and that the fraction of the population which utilizes torpor in the laboratory seems also to be less than typically observed with the "classical hibernator."

One group of sciurids which has been little studied with respect to their capacity for hibernation is the

Table 3. Body weight, heart weight, and heart rate from intact animals and isolated hearts of *C. tereticaudus* and *C. tridecemlineatus* (± 2 S.E.)

Species	N	Weight		Heart rate (beats/minute) at different ambient temperatures (Ta)			
		body (G)	heart (MG)	intact at 23-25°C	isolated		
					at 20°C	Ta of slowest rate	slowest rate
C. tereticaudus	19	217.2 ± 5.6	819.8 ±75.5	255 ±47[a]	77.6 ±5.5	5.7 ±0.7	2.1 ±0.6
C. tridecemlineatus	15	224.3 ±19.9	951.1 ±108.4	406 ±41[b]	94.6 ±14.7	-0.7 ±0.9	2.3 ±0.8

[a] N = 6

[b] N = 10

prairie dogs, and consequently their status as hibernators remains obscure and contradictory. Egoscue (12) studied the Utah prairie dog, *Cynomys parvidens*, which is closely allied to *C. leucurus* in captivity at the U.S. National Zoo. He reports that all activity above ground ceased by mid-July, the burrow entrances deteriorated, and the food remained untouched. The animals began to reappear in early January and all were active by 10 March (two males and four females), a schedule which was repeated in 1972, 1973 and 1974. This contrasts with the zoos' previous colony of black-tailed prairie dogs, *C. ludovicianus* which were active year-long in the same enclosure. On the other hand, Thomas and Riedesel (47), acknowledging reports of naturalists that the black-tailed prairie dog, *C. ludovicianus*, disappears during the winter months, which suggests hibernation, undertook a study of the body temperatures of five animals implanted with telemeters and kept in outdoor cages. The body temperatures of four animals ranged from 25.5 to 40.6°C, while the fifth animal had a body temperature which was between 13 and 40.2°C. When these animals were placed in a 12°C hibernaculum, one animal had a body temperature near ambient, while the other four had body temperatures above 27.7°C. While the data available is too skimpy to warrant conclusions, it may be that *Cynomys spp.*, is more similar to chipmunks than to ground squirrels and marmots in their hibernation patterns.

XII. EVOLUTION OF TORPIDITY

Although most authors of recent papers dealing with hibernation tend to view it as polyphyletic (25), perhaps precluding the value of even reviewing its evolution at this time, there are some interesting aspects of daily, shallow torpor which can be usefully considered.

Our preoccupation with the "classical hibernator" and questions of maintenance and arousal from hibernation with respect to the cardiovascular, nervous and endocrine systems as well as those special preparative cellular and biochemical adaptations permitting survival at body temperatures near freezing for extended periods of time, has led us to ignore the multitudinous variations on this central theme.

Hibernation has been viewed as a state of prolonged depression of body temperature to levels near 0°C. At the second hibernation symposium, Cade (6) suggested that hibernation was best developed in those rodent families which were at the base or their particular phyletic lines, describing hibernation as associated with mammals which were primitive. Hibernation arose in mammals which were not precise homeotherms (maintaining a constitutional eurythermism), but evolved a specialized thermoregulation in which hibernation was a precisely controlled thermoregulatory phenomenon. In that same symposium, Twente and Twente (49) proposed that bats which hibernate were derived from homeothermic ancestors, (implying that precise homeothermy developed first), which was followed by heterothermia and then developed hibernation. Cade (6) suggested that mammals can be forced into hypothermia if they were originally hibernators, by virtue of their constitutional eurythermism but have reached the evolutionary state where they have lost part of the ability typical of the "classical hibernator and that this advancement is commensurate with their morphological evolution.

I have proposed (25) that there is yet another way of viewing the evolution of hibernation, though I have no data to support my contention, making speculation rather easy. Is it not possible that most small, altricial neonates are not already in a sense hibernators and from their physiology could have repeatedly (for different species) evolved hibernation or daily, shallow torpor as we know it? It may not really be much of an evolutionary feat to become a hibernator if you are small enough and born prematurely (altricial). Certainly the heart should be able to function at low body temperature, which frequently occurs within the nest due to parental absence. Brown fat is widely distributed among neonates. There must be some correlation between the biochemical adaptations for anoxia of neonates and that of hibernators (though this latter point has been contested). Scholander (42) has suggested that selective ischemia is an important asphyxial defense in newborn; a regional distribution of blood similar to that of hibernators. Thus we have a number of features common to both hibernators and neonates, so all we need is the central nervous system control coupled to these in the right way at the right time and hibernation can evolve. Any small generalized mammal which is born in the altricial state and subsequently acquires homeothermy with the development of its thermoregulatory system and other supporting physiological systems, represents a physiological substrate from which hibernation and/or daily, shallow torpor could easily have been selected.

XIII. SUMMARY

Over the last 25 years it has been discovered that the lethargy or hypothermia frequently observed in small mammals exposed to combinations of low temperature and an inadequate supply of food is a natural thermoregulatory response which occurs in the field, probably as a means of reducing energy requirements at a time food is less available. The torpor episodes are usually only a few hours in length, though it is likely their duration is related to the quantity of food obtained during the previous evening of foraging. Torpor is not very deep, possibly because most of the small animals in which this has been observed to date, live in environments where they are not exposed to particularly low temperatures (if one allows for the more equable conditions prevailing in an underground nest or nesting logs or other structures providing considerable insulation). The inability of these small mammals to tolerate low body temperatures (i.e., temperatures much below 10 - 15°C) may be due to the absence of adequate membrane function of their hearts (and other organs) so as to prevent the loss of potassium at low temperatures, resulting in an inadequate electrolyte balance. At present it appears that only a small fraction of a population can be induced into torpor by forced starvation (i.e. food restriction). Similarly only a small fraction of a population in nature seems to exhibit shallow torpor spontaneously at any one time. Whether all members of the population exhibit this in nature at one time or another, is not known. The adaptive significance of this response is apparently for the population rather than the individual. So long as an adequate number of individuals survive periods of unfavorable conditions to serve as a breeding nucleus for re-establishing the population, the welfare of the population is secure. The ultimate objective of survival is to maintain the gene pool. As we have so frequently observed in the animal world, there are multitudinous strategies for achieving that goal. In this example perhaps it can be reasoned that were all the individuals to survive unfavourable winters, breeding could not occur since, i) the animals might be greatly weakened by their winter ordeal, and ii) until plants had initiated growth, bloomed and seeded, there still might not be much food around. Thus, at any time the fraction of the population surviving winter in a condition suitable for breeding exceeds the accessible food the entire population dies off. By coupling high fecundity with differential survival, determined by the ability to utilize torpor, the population as a whole survives and is ready to rebuild following one or more seasons of unfavourable conditions. What selection mechanism operates to keep the entire population from becoming hibernators? Presumably disadvantages such as predation balance advantages such as a reduced energy requirement to keep a genetic balance at a low gene frequency for torpor.

The number of documented examples of daily, shallow torpor among marsupials, insectivores and rodents has increased substantially within the last 25 years. The documentation of its occurrence in nature has lagged significantly behind our laboratory demonstrations of its existence. Unlike the "classical hibernators" these "shallow, daily hibernators" do not assume the horizontal, curled posture, but remain upright, so that without careful observation of breathing rates or measurement of body temperature, it could easily be missed, expecially if the body temperature were relatively high, permitting typical locomotory behavior.

It must be emphasized that true "shallow, daily torpor" is spontaneous (i.e., can occur in the presence of food, albeit restricted) and that animals can restore their body temperature to normal levels without the application of external heat. Moving animals from a refrigerator to room temperature to arouse is not satisfactory evidence. If they are exposed to temperatures which are too low, they may resist entry into torpor, so that it is important in testing suspected cases that the animals not be subjected to ambient temperatures much below 15 or much above 20°C. Because not all individuals tend to exhibit such torpor, it is important to examine sufficient numbers. And since it may be seasonal at least in some species, one must consider the possibility of circannial rhythms or exogenous cues such as photoperiod and temperature. Also, it may well be that it occurs in different seasons (e.g. summer dormancy) in different species, for reasons not well understood, but hidden under the rubric of estivation.

ACKNOWLEDGEMENT

The preparation of this manuscript was supported by a grant from the National Science Foundation (U.S.A.) BM 74-007789.

FOOTNOTE

*Subsequent to the completion of this review my attention was called to the chapter in Lemur Biology on "Body temperatures and behavior of captive cheirogaleids by Robert Jay Russell, I. Tattersall and R.W. Sussman (editors). Plenum Press, New York and London, 1975.

REFERENCES

1. Bartholomew, G. A., and T. J. Cade. Temperature regulation, hibernation, and aestivation in the little pocket mouse. *Perognathus longimembris*. J. Mammal. 38 (1):60-72, 1957.
2. Bartholomew, G. A. and J. W. Hudson. Hibernation, estivation, temperature regulation, evaporative water loss, and heart rate of the pigmy possum, *Cercaertus nanus*. Physiol. Zool. 35(1): 94-107, 1962.
3. Batholomew, G. A., and R. E. Macmillen. Oxygen consumption, estivation, and hibernation in the kanagaroo mouse, *Microdipodops pallidus*. Physiol. Zool. 34(3):177-183, 1961.
4. Black, C. C. A review of the North American tertiary sciuridae. Bull. Mus. Comp. Zool., Harvard University 130(3): 109-248, 1963.
5. Brown, J. H., and G. A. Bartholomew. Periodicity and energetics of torpor in the kangaroo mouse, *Microdipodops pallidus*. Ecology 50(4): 705-709, 1969.
6. Cade, T. J. The evolution of torpidity in rodents. Annales academiae scientiarum fennicae. Series A. IV. Biologica 71/6: 78-112, 1964.
7. Carpenter, R. E. A comparison of thermoregulation and water metabolism in the kangaroo rats *Dipodomys agilis* and *Dipodomys merriami*. Univ. Calif. Publ. Zool. 78: 1-36, 1966.
8. Chaffee, R. R. J. On experimental selection for super-hibernating and non-hibernating lines of Syrian hamsters. J. Theoret. Biol. 12: 151-154, 1966.
9. Davis, D. E. Hibernation and circannual rhythms of food consumption in marmots and ground squirrels. The Quarterly Review of Biology (54)(4):477-514, 1976.
10. Dawson, T. J. Metabolism, thermoregulation and torpor in shrew sized marsupials of genus *Planigale*. (In Manuscript).
11. Dawson, W. R. The relation of oxygen consumption to temperature in desert rodents. J. Mammal. 36(4):543-553.

12. Egoscue, H. J. Abnormal juvenile pelages and estivation in the Utah prairie dog, *Cynomys parvidens*. Southwest Nat. 20(1): 139-142, 1975.
13. El-Hilali, M. and J. P. Veillat. *Jaculus orientalis:* A true hibernator. Mammalia 39(3): 401-404, 1975.
14. Fisler, G. F. Adaptations and speciation in harvest mice of the marshes of San Francisco bay. Univ. Calif. Publ. Zool. 77:1-108.
15. French, A. R. Selection of high temperatures for hibernation by the pocket mouse, *Perognathus longimembris*: ecological advantages and energetic consequences. Ecology 57: 185-191, 1976.
16. Gaertner, R., J. S. Hart, and O. Z. Roy. Seasonal spontaneous torpor in the white-footed mouse, *Peromyscus leucopus*. Comp. Biochem. Physiol. 45A: 169-181, 1973.
17. Gebczynski, M. Oxygen consumption in starving shrews. Acta Theriologica 16, 18: 288-292, 1971.
18. Godfrey, G. K. Body-temperatures and torpor in *Sminthopsis crassicaudata* and *S. larapinta* (Marsupialia-Dasyuridae). J. Zool., Lond. 156: 499-511, 1968.
19. Hill, R. W. Daily torpor in *Peromyscus leucopus* on an adequate diet. Comp. Biochem. Physiol. 51A: 413-423, 1975.
20. Hooper, E. T., and M. El-Hilali. Effects of starvation on body weight and temperature in captive jerboas, Genus *Jaculus*. Mammalia 37: 105-121, 1973.
21. Howard, W. E. Relation between low temperature and avilable food to survival of small rodents. J. of Mammal. 32(3): 300-312, 1951.
22. Hudson, J. W. Temperature regulation and torpidity in the pygmy mouse, *Baiomys taylori*. Physiol. Zool. 38(3):243-254, 1965.
23. Hudson, J. W. Variations in the patterns of torpidity of small homeotherms. In: Mammalian Hibernation III. Fisher, K. C., A. R. Dawe, C. P. Lyman, E. Schönbaum, and F. E. South, Jr. (editors). London: Oliver & Boyd Ltd., 1967, p. 30-46.

24. Hudson, J. W. Thermal sensitivity of isolated-perfused hearts from the ground squirrels, *Citellus tereticaudus* and *Citellus tridecemlineatus*. Z. vergl. Physiologie 71: 342-349, 1971.

25. Hudson, J. W. Torpidity in mammals. In: Comparative Physiology of Thermoregulation, Vol. III. Whittow, G. C. (editor). New York and London: Academic Press, 1973, p. 97-165.

26. Hudson, J. W. and R. R. Eller. ^{42}K efflux, EKG and tension in isolated perfused hearts of white-footed mice. Comp. Biochem. Physiol. 49A: 743-755, 1974.

27. Kenagy, G. J. Daily and seasonal patterns of activity and energetics in a heteromyid rodent community. Ecology 54(6): 1201-1219, 1973.

28. Kennedy, P. M. and W. V. McFarlane. Oxygen consumption and water turnover of the fat-tailed marsupials *Dasycercus cristicauda* and *Sminthopsis crassicaudata*. Comp. Biochem. Physiol. 40A:723-732, 1971.

29. Lyman, C. P. The effect of low temperature on the isolated hearts of *Citellus leucurus* and *C. mohavensis*. J. Mammal. 45(1):122-126, 1964.

30. Lynch, G. R., S. E. White, R. Grundel, and M. S. Berger. Effects of photoperiod, melatonin administration and thyroid block on spontaneous daily torpor and temperature regulation in the white-footed mouse *Peromyscus leucopus*. J. Comp. Physiol. (In press)

31. MacMillen, R. E. Aestivation in the cactus mouse, *Peromyscus eremicus*. Comp. Biochem. Physiol. 16: 227-248, 1965.

32. Morhardt, J. E. Body temperatures of white-footed mice *(Peromyscus spp.)* during daily torpor. Comp. Biochem. Physiol. 33: 423-439, 1970.

33. Morhardt, J. E. Heart rates, breathing rates and the effects of atropine and acetylcholine on white-footed mice *(Peromyscus spp.)* during daily torpor. Comp. Biochem. Physiol. 33: 441-457, 1970.

34. Morhardt, J. E., and J. W. Hudson. Daily torpor induced in white-footed mice (*Peromyscus spp.*) by starvation. Nature 212 (5066): 1046-1047, 1966.
35. Morrison, P., and B. K. McNab. Daily torpor in a brazilian murine opossum (*Marmosa*). Comp. Biochem. Physiol. 6:57-68, 1962.
36. Morton, S. R. Torpor and nest-sharing: Their ecological significance in an insectivorous marsupial, *Sminthopsis crassicaudata*, and the house mouse, *Mus musculus*. J. Mammal. (In press).
37. Panuska, J. A. Weight patterns and hibernation in *Tasmias striatus*. J. of Mammal. 40(4):554-565, 1959.
38. Pivorun, E. B. A biotelemetry study of the thermoregulatory patterns of *Tamias striatus* and *Eutamias minimus* during hibernation. Comp. Biochem. Physiol. 53A: 265-271, 1976.
39. Ride, W. D. L. A guide to the Native Mammals of Australia. Melbourne, Oxford University Press, xiv + 249, 1970.
40. Roberts, A. The mammals of South Africa. South Africa: Central News Agency, 1951, 700 pp.
41. Schaffer, R. E., and R. W. Bullard. Comparisons of rewarming from natural torpidity and induced hypothermia in chipmunks (*Tamias striatus*) with reference to heart rate and temperature relationships. Proc. Indiana Acad. Sci. 79: 476-483, 1970.
42. Scholander, P. F. The master switch of life. Scientific American, 209: 2-10, 1963.
43. Scott, G. W. and K. C. Fisher. Hibernation of eastern chipmunks (*Tamias striatus*) maintained under controlled conditions. Canad. J. Zool. 50(1): 95-105, 1972.
44. Simpson, G. G. The principles of classification and a classification of mammals. Bull. Amer. Mus. Nat. Hist. 85: 1-350, 1945.
45. Stebbins, L. L. Seasonal and latitudinal variations in the circadian rythms of three species of small rodents in Northern Canada. Ph. D. dissertation, University of Alberta, 1969.

46. Strumwasser, F. Thermoregulatory, brain and behavioral mechanisms during entrance into hibernation in the squirrel, *Citellus beecheyi*. Am. J. Physiol. 196(1): 15-22, 1959.
47. Thomas, T. H., and M. L. Riedesel. Evidence of hibernation in the blacktailed prairie dog, *Cynomys ludovicianus*. Cryobiology 12(6): 559, 1975.
48. Tucker, V. A. Oxygen consumption, thermal conductance, and torpor in the California pocket mouse, *Perognathus californicus*. J. Cell. and Comp. Physiol. 65(3):393-404, 1965.
49. Twente, J. W., and J. A. Twente. An hypothesis concerning the evolution of heterothermy in bats. Annales academiae scientiarum fennicae. Series A. IV, Biologica 434-442, 1964.
50. Vogel, P. Kalteresistenz und reversible Hypothermie der Etruskerspitzmaus (*Suncus etruscus, Soricidae, Insectivora*). Z. Saugetierkunde 39: 78-88, 1974.
51. Vogt, F. D., and G. R. Lynch. Seasonal study of spontaneous daily torpor in free-ranging white-footed mice (*Peromyscus leucopus*). Manuscript.
52. Wallis, R. L. Torpor in the dasyurid marsupial, *Antechinus stuartii*. Comp. Biochem. Physiol. 53A:319-332, 1976.
53. Yousef, M. K., and D. B. Dill. Daily cycles of hibernation in the kangaroo rat, *Dipodomys merriami*. Cryobiology 8:441-446, 1971.
54. Wang, L. C-H., and J. W. Hudson. Temperature regulation in normothermic and hibernating eastern chipmunk, *Tamias striatus*. Comp. Biochem. Physiol. 38(1A): 59-90, 1971.

ENERGETIC AND FIELD ASPECTS OF MAMMALIAN TORPOR: THE RICHARDSON'S GROUND SQUIRREL

Lawrence C. H. Wang
University of Alberta
Department of Zoology
Edmonton, Alberta, Canada T6G 2E9

ABSTRACT

Previous studies on field aspects of mammalian torpidity concern mainly with the timing of "immergence" and "emergence" in seasonal torpor. Little is known during the torpor season itself. The present report summarizes results from a 3-year telemetric study on pattern of torpidity in the Richardson's ground squirrel (*Spermophilus richardsonii*) under natural conditions. The evaluation of energetics for seasonal torpor adopts an "indirect calorimetry and time-budget" approach; the energetics associated with different phases of a torpor cycle are compartmentalized in the laboratory and the time-budget derived from patterns of torpidity observed in the field. The results indicate that the torpor season commences in mid-July for adult *S. richardsonii* and mid-September for juveniles, but terminates synchronously for both groups in mid-March. Duration of torpor averages 3 and 4.3 days in July and August and gradually lengthens to 19.1 days in January; after which it shortens to 14.2 and 6.0 days, respectively, in February and March. The duration of inter-torpor homeothermy ranges from 300 to 1500 min. The amount of time spent in torpor ranges from 43.1% in July and 60.4% in March to 90.1-98.4% between August and February. The energetic cost averages 12.8% for entry into torpor, 16.6% for deep torpor, 19.0% for arousal from torpor and 51.6% for the period of inter-torpor homeothermy, throughout a torpor season. The monthly energy savings by utilization of torpor range from 38.0% in July to 51.2% in March but are 81.6% to 96.0% between August and February. For the torpor season as a whole the energy saving comes to 87.8%.

ISBN 0-12-734550-7

I. INTRODUCTION

The study of mammalian torpor (estivation, hibernation and daily torpor) has made impressive progress in the last two decades as evidenced by the previous four international symposia on the subject and the many publications in major scientific journals. The general concensus is that physiologically, estivation and hibernation are similar phenomena, differing only in the timing of seasonal occurrence and the depth of body temperature (Tb) depression which is dictated by the ambient temperature (Ta) at which torpor occurs (16). Physiological studies on daily torpor in small cricetids and heteromyids (15, 26, 28, 40, 44) have also indicated that qualitatively it is similar to hibernation but the depression of Tb is generally shallow and the duration of torpor is typically less than 24 hr. The relatively shallow seasonal torpor in the Chipmunks (*Tamias* and *Eutamias*) also showed similarities physiologically to those observed in hibernation (37, 45). Ecologically, the different patterns of torpidity are considered adaptive in coping with seasonal environmental stresses for energy and/or water conservation as well as avoidance of daily or seasonal extremes in temperature (15).

The majority of field studies on mammalian torpidity have been concerned primarily with the timing of seasonal torpor and the duration of the torpor season (e.g. 1, 27, 31, 35). This information is usually obtained through regular census and/or systematic trapping. The disappearance of animals from above the ground is generally referred to as the time of "immergence" for torpidity while the re-appearance of the animal as "emergence". The period between "immergence" and "emergence" is the torpor season. A typical study illustrating such techniques is found in Michener (27), in which she reported a mean hibernating season of 222 and 249 days respectively, for adult *Spermophilus richardsonii* and *S. columbianus*, and 157-190 and 250-255 days, respectively, for the juvenile males and females of the two species. A recent review by Davis (6) summarized to date the timing of estivation and hibernation in ground squirrels and marmots under field conditions with the following generalizations: (i) the timing for immergence and emergence in a given species is rather predictable but varies between species; (ii) sex and age differences exist on timing of immergence and emergence with males immerging and emerging before the females and adults immerging and emerging before the young (6). For obvious reasons of inaccessibility to the animal, these studies can not provide any information on the underground activity of the animals during the period between immergence and emergence. It seems paradoxical that although the torpor season proper is the focus of attention, it is the least understood aspect amongst a plethora of information concerning natural torpidity. With the advancement of biotelemetry it has become possible to measure the Tb of an animal in it's burrow under natural conditions (2, 46). Using this method, we have been conducting field studies recording Tb in animals during the torpor season, aiming to fill this gap in our knowledge on the subject (47). In this chapter, more recent information will be presented to characterize the pattern of torpidity under natural conditions in a population of Richardson's ground squirrels.

Equally perplexing is the fact that although hibernation is viewed as an effective means for energy conservation in overwintering, there has been no study which documents the energy expenditure of a hibernator in its hibernating season under natural conditions. It is perhaps not difficult to understand why this should be the case. The lack of such a study is a direct consequence to the lack of specific information on patterns of hibernation under natural conditions. To estimate energy consumption during the hibernating season, one must know (i) the timing of onset and termination of the hibernating season; (ii) the duration of each hibernating bout and its seasonal variation; (iii) the energy consumption for entry into and arousal from hibernation; (iv) the energy consumption during hibernation at different Tb's; and (v) the period of homeothermy between successive hibernation bouts and the energy consumption for each such period and at different Ta's. Since none of this information was available for hibernation under natural conditions, it has not been possible to construct an energy budget for the hibernation season. Our recent field studies however, have provided pertinent information on these aspects to allow for an estimation of such a budget.

Methods currently available for studies of energetics in free-living small mammals have been reviewed by Gessaman (9), and the ecological aspects of energetics by Grodzinski and Wunder (11). Only a brief summary will be presented here with reference to relevance of the method in application to study of natural torpidity: (i) the use of double-labeled water ($D_2{}^{18}O$) for estimation of rate of CO_2 production (21) has been successfully applied in birds (20, 42, 43) and small rodents (22, 32, 33, 34). However, since this method requires at least two samples of body water from the same animal, one at the time of release and one at recapturing, its usefulness is limited to estimation of energetics of the whole torpor season but not for compartmentalization of seasonal differences in energy expenditures within the torpor season due to the absence of surface activity after immergence; (ii) the use of telemetered heart rate as a predictor for oxygen consumption ($\dot{V}O_2$) based on empirical correlation between heart rate and $\dot{V}O_2$. Although significant correlations have been observed in many species of rodents (23, 29), such relationships appear to be stable only for one to a few days in the same individual and may vary between individuals. Since the change of $\dot{V}O_2$ is reflected not only in heart rate but also in stroke volume and especially the arterial to venous difference in O_2 content, the usefulness of the method is probably limited to steady-state conditions at relatively low metabolic rates where heart rate increases linearly with the increase of $\dot{V}O_2$. Since both the entry and arousal phases of a torpor episode are not steady states in so far as changes of cardiovascular parameters are concerned (25, 45), the use of heart rate for $\dot{V}O_2$ prediction in natural hibernation is probably not suitable; (iii) the estimation of energetics with the incorporation of the "climatic space" concept (14) also does not seem to be applicable in the study of natural torpidity although such approaches have been employed in evaluation of energetics in the ground squirrel in its non-torpor season (30); (iv) the classical method of food consumption and assimilation analysis (36) is probably not suitable in the study of torpor energetics as most animals use fat reserves acquired in the fall as the energy source during the torpor season. However, it has been demonstrated that by manipulation of food supply in the laboratory an inverse relationship was observed between the amount of food available and the amount of time spent in torpor in some heteromyids (3, 7, 41). Apparently, the animal could adjust its energy budget for torpor or homeothermy, according to the amount of food present (7). Thus by knowing the assimilation coeffient (e.g. 7, 11 Table 8.2) and the amount of food hoard consumed and in nature, it is possible that the energy consumption in a torpor season can be quantified; (v) the combination of indirect calorimetry ($\dot{V}O_2$ or CO_2 production, $\dot{V}CO_2$) for estimation of resting and active metabolism and a time-budget for daily activities of different types (feeding, flying, resting, grooming, aggresion etc.) under natural conditions has been utilized for the construction of an energy budget in many birds and mammals (see 9 for review). This method is suitable for the study of energetics in torpidity. Although tedious, one can measure the $\dot{V}O_2$ and Tb simultaneously in the laboratory throughout different phases of a torpor cycle. Having established the energy expenditure during entry into torpor, deep torpor, arousal from torpor, and the period of homeothermy between torpor bouts at Tb's and Ta's similar to those observed in nature, it becomes possible to compartmentalize energy costs associated with each of these phases as well as the sum of them. For construction of an energy budget for the torpor season, it remains necessary however, that field data on duration of torpor and that of homeothermy between torpor bouts be made available. The energetic study to be presented here adopted this approach in methodology.

Before one can justify the extrapolation of lab energetic data to field conditions, the question "how accurate is it quantitatively to extend metabolic measurements made in the lab to field conditions?" must be answered. Unfortunately, there is no direct answer to this. Judging from changes of telemetered Tb under both field and lab conditions, the rates of cooling and rewarming in a torpor bout are comparable between the two conditions (47). Since the measurement of Tb in the lab was made with the animal in a typical metabolic chamber for simultaneous $\dot{V}O_2$ measurement (47), it seems reasonable to assume that the $\dot{V}O_2$ changes under field conditions would also be similar. With present technology, direct measurement of $\dot{V}O_2$ in a natural burrow is not possible, thus the verification of quantitative equivalency between field and lab metabolic measurements must await future evaluations.

To reiterate, the objectives of this presentation are to provide information on: (i) patterns of torpidity under natural conditions in the Richardson's ground squirrel *S. richardsonii* as studied by radiotelemetry; (ii) for comparative purpose, the pattern of torpidity under constant conditions (total darkness and 5°C) in the laboratory; (iii) energetic costs of different phases of a torpor cycle and their seasonal variations; and (iv) energy savings derived from exhibition of torpidity in *S. richardsonii* in overwintering. Although somewhat specific in presentation, it is hoped that the present study illustrates an approach that is suitable for the study of energetics of mammalian torpor in general.

II. FIELD ASPECTS OF NATURAL TORPIDITY

Field aspects of the present study were conducted on a private farm located 22 km east of the city of Edmonton, Canada. The study area is an open, short grass field 600 m long and 50 m wide (7.4 acres). It is surrounded by other grass and wheat fields approximately 1.2 km^2 (300 acre) in total area. Richardson's ground squirrels are abundant in the study area and surrounding fields. The characteristics of the Tb transmitter used in the study have been described in detail elsewhere (46). Using an automatic recording system (47), signals from the transmitters were received once every 10 min. In addition, daily recordings of Tb were also made by walking close to the site of the burrow with a receiver/tape-recorder (46). The transmitter was surgically implanted in the peritoneal cavity of the ground squirrel and implanted animals were allowed 1-3 days to recover in the laboratory before being released to their original sites of capture. A continuous surveillance of each released animal was made for at least 2 days with the aid of color tags sutured on the animal and binoculars. This procedure was necessary because animals could roam freely in and out of the study area yet it was essential to ascertain the particular burrow that the animal occupied for torpidity. Unfortunately, the ground squirrels rarely stayed in any given burrow for more than a few days prior to exhibiting torpor. This, coupled with the fact that the Tb transmitter had only short ranges of transmission (3-10 m), made the tracking of individuals extremely difficult. Of the 64 animals which had been released over a three year span, only 12 animals were recovered. These provided data on torpor. The relatively low recovery rate could be attributed to: (i) death of animals in their burrows (for unknown reasons) during their first epidsode of hypothermia (10 animals); (ii) malfunctions of transmitters; (iii) implanted animals wandering off the study area; and (iv) natural predation by badgers, weasels and hawks in the study area.

The burrow temperature was estimated by Tb of transmitter-implanted animals which were found dead after release (failure to rewarm spontaneously from a depressed Tb in the first episode of decreased Tb), and by a Telethermometer probe (No. 402, Yellow Spring Instruments) buried approximately 1 m below the ground surface (similar depth of natural burrow). Readings were made daily during the summer months (June-August), and bi-weekly during the fall and winter. However, since no animal was actually present in the burrow, the measured temperature was probably a low estimation of the burrow with the animal present.

The earliest episode of torpor observed in *S. richardsonii* was July 17, in an adult female (No. L-1) which was released on June 20. The first episode of torpor in this animal lasted two days with a minimum Tb during torpor at 16.4°C. Regular torpor episodes were observed subsequently throughout fall and winter until its emergence from torpor in the following March (Table 1). Another female adult (No. 101) which was released on July 17 was observed in torpor on July 25; the first episode lasted also two days at a minimum Tb of 15.8°C (Table 1). This animal was subsequently predated by a badger (judging from the size of excavation leading to the squirrel's burrow) sometime on August 14 while in torpor. Six other adult females were also observed to exhibit frequent torpor during August (Table 1) while the juvenile males usually did not commence torpor until mid-September (Table 1). The juvenile male (No. 15)

Table 1. Seasonal variations of duration of torpor in *S. richardsonii* under natural conditions*

Animal No.	JUL	AUG	SEP	OCT	NOV	DEC	JAN	FEB	MAR	Date of 1st Torpor
				Duration of individual torpor episodes (days).						
L-1 Adult ♀	2,4,6	5,7,7, 6,7	8,7,8, 9	9,11, 10	18,19	21	19,17	16,12	8,6	July 17
101 Adult ♀	2,2,2	3,4,4, 5								July 25
L-2 Adult ♀		2,2,3, 5,6	6,7,8, 9	8,11 15						Aug 7
L-14 Adult ♀		3,4,5 6	5,7,7, 6,6	7,9 10						Aug 17
L-5 Adult ♀		4,6	7							Aug 19
2 Adult ♀		3,4	4,4,5, 7,7,5							Aug 21
95 Adult ♀		3,3,3	5,5,6, 8,8	9,11, 11	9,12 13	16,14	12,19	19,9	7,6,6	Aug 21
78 Adult ♀		3	4,5,5, 6,5,4	5,8, 10,8	8,13					Aug 29
33 Juvenile ♂			4,4,5	6,6,7, 7,8	14,8 8	20	24,24	15	3	Sept 12
L-17 Juvenile ♂			2,5							Sept 21
27 Juvenile ♂			4,6	4,5,5	12,10 10	13,11	22,17			Sept 22
15 Juvenile ♂					6,7,8	11,15	17,20			Nov 5
$\bar{x}$ =	3	4.3	6.0	8.3	11.0	15.1	19.1	14.2	6.0	

which was released on August 7 did not become torpid until November 5, probably due to a delayed fattening prior to onset of torpor (47). The timing of onset of the torpor season observed in the present study generally coincided closely with the dates of immergence reported for this species according to trapping records (27). Since trapping records also indicated that adult males are the first to immerge in this species (27), it is possible that adult males would exhibit torpor even before mid-July. However, the difficulty encountered in the present study in trapping and tracking adult males precludes a confirmation on that aspect.

The duration of torpor in July and August was typically short, lasting from 2 to 7 days and averaging 3 and 4.3 days respectively (Table 1). The burrow temperature in these months ranged between 13-17°C in different localities and different years. The duration of torpor gradually lengthened as the season progressed and the burrow temperature gradually decreased from 13 to near 0°C in January when peak duration of 19.1 days was reached (Table 1; Fig. 1). Subsequently, the duration of torpor shortened to 14.2 days in February and 6 days in March (Table 1; Fig. 1) even though burrow temperature for these months remained at the lowest values for the season at 0 to -2°C. The torpor season terminates in mid to late March in this species, with the earliest observation of emergence on March 18. A complete torpor season in a juvenile male is presented in figure 2 to illustrate the seasonal variations in duration of torpor as well as the minimum Tb during torpor, which was usually higher but within 1-4°C of the burrow temperature.

The dynamics of Tb changes in a torpor cycle observed in mid-winter is illustrated in figure 3. The entry into torpor typically took approximately 24-36 hr before a stabilized minimum Tb was reached. This was observed regardless of the level of burrow temperature. The rate of cooling ranged between 0.012-0.024°C/min, and averaged 0.017°C/min (N=5, 20 measurements). The difference in rates of cooling was mainly due to the greater temperature differential between the body and the burrow during the winter while the time to reach the minimum Tb remained fairly constant. Arousal from torpor, on the other hand, was rapid, usually completed

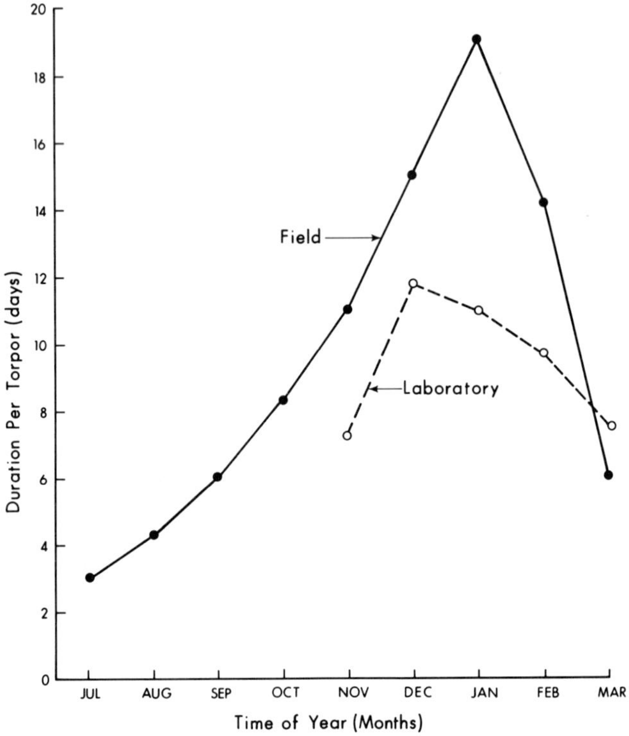

Fig. 1. *Seasonal variations of duration of torpor in* S. *richardsonii under field and laboratory conditions (5 ± 1°C, total darkness).*

in 2-6 hr with the difference largely determined by the initial Tb from which arousal began. The rate of re-warming ranged between 0.093-0.260°C/min, and averaged 0.174°C/min (N=5, 24 measurements).

The period of homeothermy between successive torpor bouts ranged between 300 to 1500 min (Table 2). There seems to be a trend towards shortening of this period from 680 min in September to 450 min in December, and after which, a lengthening to 840 min in February and 1500 min in March (Table 2). However, the scarcity of data precludes a definite statement on this pattern of change at the present time. The significance of this period of homeothermy is unknown; it has been suggested

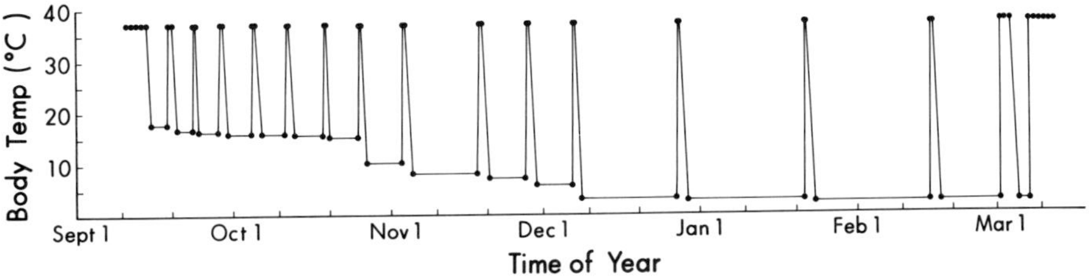

Fig. 2. *A complete torpor season in a juvenile male S. richardsonii under field conditions. Note the seasonal variations in duration of torpor and the decrease of minimum body temperature during torpor.*

that it may be associated with certain metabolic activities such as gluconeogenesis (8) or the redistribution of ions (49). We have recently demonstrated that central injection of 5-HT (serotonin) in *S. richardsonii* suppresses heat production in cold (5°C) if the animal is in its torpor season but not if it is in its non-torpor season (10). We suspect that 5-HT may act to maintain the depressed metabolic state during torpor (Wang and Glass, unpublished). Kudryavtseva and Popova (19), Popova (38) and Janský and Novotná (17) have reported cyclic changes in concentration and turnover rates of 5-HT in the brain throughout a hibernation cycle. The highest turnover rate occurred during hibernation (17) and an increase in concentration occurred shortly after arousal (19). In view of our study and the studies by these authors (17, 19, 38), we suspect that this period of homeothermy between torpor could be significant in that it allows the resynthesis of 5-HT in the CNS prior to onset of the next bout of torpor. Regardless of its functional significance, it is worth noting that this period of homeothermy represents the greatest drain on energy consumption in the torpor season (see section on energetics). Since urination and defecation, feats typically observed after arousal in the laboratory, can be completed in relatively short time, and while excavation of natural burrows in the present study found no stored food present, which eliminates the possibility of feeding, one wonders why the

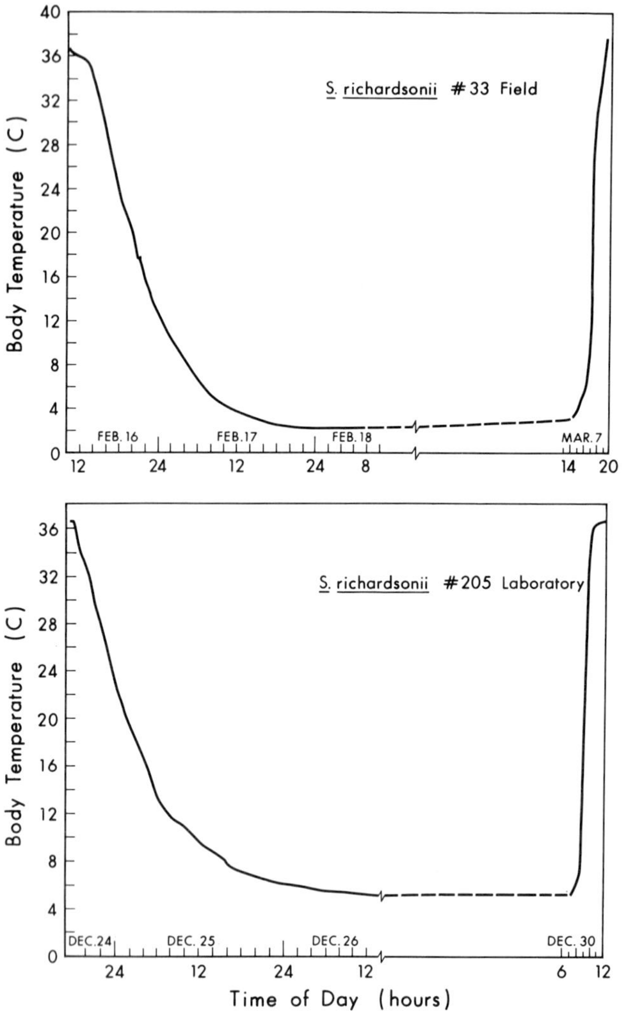

Fig. 3. *Typical dynamic changes of body temperature (telemetered) during a torpor episode in S. richardsonii under field (upper graph) and laboratory (lower graph) conditions. Note the relatively prolonged period during entry into torpor (24 hr or longer) and the relatively rapid rise of body temperature during arousal from torpor. Note also the similarities between the two recordings both qualitatively and quantitatively.*

Table 2. Period of homeothermy (min) between completion of arousal from torpor (Tb > 36°C) and re-entry into torpor (onset of Tb decline) in *S. richardsonii* under natural conditions.*

Animal No.	SEP	OCT	NOV	DEC	JAN	FEB	MAR
L-1	750	700 450	560				
L-2	720 340	510 660	630				
15			570 660	420			
27	840	300 600 450	360	540			
33	750	540 1200 660		390		840	1500
95			330				
$\bar{x}$ =	680	607	518	450		840	1500

*Based on periodic telemetric body temperature recording at 10 minute intervals. Lack of data in any month due to malfunctioning of transmitter or other technical difficulties.

animal would spend a minimum of 5 hrs (shortest time observed, Table 2) burning up precious energy reserves in the midst of an energy conservation campaign!

Based on the average duration of torpor (Table 1) and the period of homeothermy between torpors (Table 2), it is possible to construct a time-budget for torpidity at different months throughout the seasons. The result of such calculations is presented in Table 3, and in figure 4. The amount of time spent in torpor was highest in December and January, being 98 and 98.4% respectively, and lowest in July (43.1%) and March (66.4%), with the rest of the months (i.e. August to November, and February) ranging from 90.6 to 96.8% (Table 3).

III. LABORATORY ASPECTS OF NATURAL TORPIDITY

For observation of over-winter torpidity in the laboratory, 10 ground squirrels, of both sexes, were housed individually in plastic cages measuring 43 x 20 x

Table 3. Seasonal variations in estimated time budget for torpor in *S. richardsonii* under natural conditions.*

Month	$\bar{x}$ day per torpor	$\bar{x}$ hr per torpor	$\bar{x}$ hr per inter-torpor homeothermy	Total hr per torpor cycle	Total hr available for torpor	No. of possible torpor cycles	Total time in torpor (hr)	% time of month spent in torpor
JUL	3	72.0	11.3[a]	83.3	372[b]	4.46	321.1	43.1
AUG	4.3	103.2	11.3[a]	114.5	744	6.49	670.5	90.1
SEP	6.0	144.0	11.3	155.3	720	4.64	667.6	92.7
OCT	8.3	199.2	10.1	209.3	744	3.55	708.0	95.2
NOV	11.0	264.0	8.6	272.6	720	2.64	697.3	96.8
DEC	15.0	360.0	7.5	367.5	744	2.02	728.8	98.0
JAN	19.1	458.4	7.5[c]	465.9	744	1.60	732.0	98.4
FEB	14.2	340.8	14.0	354.8	672	1.90	645.5	96.1
MAR	6.0	144.0	25.0	169.0	528[d]	3.12	449.9	60.4

*Calculated from data in Tables 1 and 2.

a Assumed to be same as in September.
b Since torpor was not observed until July 15, only half of July was counted as available for torpor.
c Assumed to be same as in December.
d Allow 22 days in March before termination of torpor season in this species.

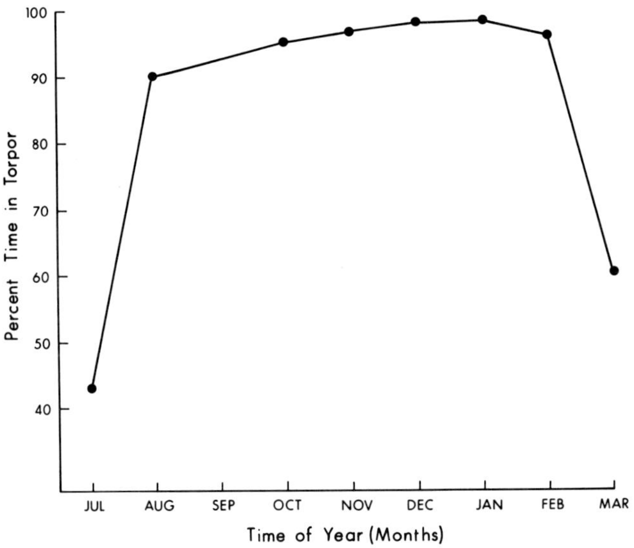

Fig. 4. *Seasonal variations of time-budget for torpor in S. richardsonii under field conditions. Data from Table 3.*

23 cm, with cotton and pine shavings and food and water *ad libitum* in a walk-in environmental chamber (2 x 2.7 x 2.3 m) at Ta = 5 ± 1°C and total darkness. These animals were transferred from warmth (22°C) and 12L:12D photoperiod on October 17 and were allowed 4 weeks to adjust to their new conditions without disturbance except for weekly cleaning (only those which were not torpid at the time). The animals were inspected daily for episodes of torpor using a red-light and the "saw-dust" method. The experiment was terminated on March 31, the following year.

The duration of torpor in these animals ranged between 2 to 20 days and averaged 7.4 to 11.8 days through November to March (Table 4). These values are generally lower than those observed under field conditions (Table 1). With respect to occurrence of torpor, it was not as regular and predictable as that observed in field animals. The torpor season also seemed to be shorter

Table 4. Seasonal variations of duration of torpor in *S. richardsonii* under laboratory conditions (Ta=5 ± 1°C, total darkness).*

Animal No.	NOV	DEC	JAN	FEB	MAR
201	5,6	9,8,10 11	13,12	13,11,4	
202	8,6	11,15,9	8,6,4, 2	3	
205	2,12	11,11,6	9,17	11	
208		20,16	11,12, 5		
301		19,16	15,13	14,13	9
303	7,13	9,13	10,13, 8	3	
305		11,8	8,10, 8		
306		12,12	13,16	15,17	11
308	6,8	6,17	15,18	14,15	8,2
309	6,8	9,13	16,13	7,3,3	
$\bar{x}$	7.3	11.8	11.0	9.7	7.5

*When dates of torpor extended over two calendar months, the torpor was awarded to the month where majority of duration occurred. Absence of data indicates absence of torpor.

than that observed in the field. Furthermore, the seasonal variation in duration of torpor under laboratory conditions was not as pronounced as that observed in nature (Fig. 1). Taken together, it appears that our laboratory conditions had a disturbing effect on the pattern of torpidity in this species; whether this is a consequence of the somewhat noisy physical environment in the hibernaculum (due to fan and selenoid switching) or the disturbing effect of close proximity between individuals (an aroused animal could be quite vocal upon inspection) or a combination of these factors is not known.

For a comparison of changes in Tb during a torpor cycle between lab and field conditions, animals were implanted with transmitters identical to those used for field studies. The animals were placed either in the walk-in environmental chamber at 5°C and total darkness for recording of Tb only, or in an open-circuit system for simultaneous measurement of $\dot{V}O_2$ and $\dot{V}CO_2$ in addition to Tb. The open-circuit system consisted of the following components: (i) a 4.5 l (1 gallon) metabolic chamber (made out of a paint can or acrylics) with ports for air inlet and outlet and housing a thermocouple; (ii) calibrated flow meter (Matheson, 603) for controlled flow-rate of 600 - 1500 cc dry air/min (STP); (iii) constant temperature cabinet with an accuracy of ± 0.5°C; (iv) O_2 analyzer (Beckman G-2) and CO_2 analyzer (Beckman 864); (v) potentiometer for thermocouple sensing (Leeds and Northrup, Speedomax W) and (vi) strip chart recorders for O_2 and CO_2 analyzers. Analog outputs from Tb receiver, O_2 and CO_2 analyzers and thermocouples were fed into respective amplifiers and converted to digital forms (A-D converter, Xincom 3150) for on-line data acquisition by a minicomputer (Texas Instrument, 980A) and print-out by an electronic terminal (Texas Instrument, 733). The rate of data analysis was once per second, and the instantaneous rates of Tb changes, $\dot{V}O_2$ and $\dot{V}CO_2$ as well as the integrated total for these latter parameters were displayed usually once every 5 minutes. The open-circuit measuring system is diagrammatically illustrated in figure 5.

To evaluate the finer details of thermal and energetic dynamics during spontaneous entry and arousal from torpor, a single ground squirrel was chronically implanted with a thermocouple re-entrant tube in the anterior hypothalamus (Glass and Wang, unpublished). The animal was placed in a gradient-layer calorimeter (Thermonectics, SECA-1201) for simultaneous measurments of brain temperature, total dry-heat loss and $\dot{V}O_2$ in spontaneous torpor. Data acquisition was similar to that described above. These limited experiments were done at Ta's of 5 and 15°C.

A typical change of Tb in a torpor cycle is illustrated in figure 3. It is apparent that both qualitatively and quantitatively, the changes of Tb are very similar under lab and field conditions at a comparable Ta. The cooling of Tb during entry in all lab animals ranged between 0.010 - 0.018°C and averaged 0.017°C/min (N=4, 10 measurments). This was almost identical to that observed in the field. The rate of rewarming ranged between 0.111 - 0.222°C and averaged 0.151°C/min (N=4, 10

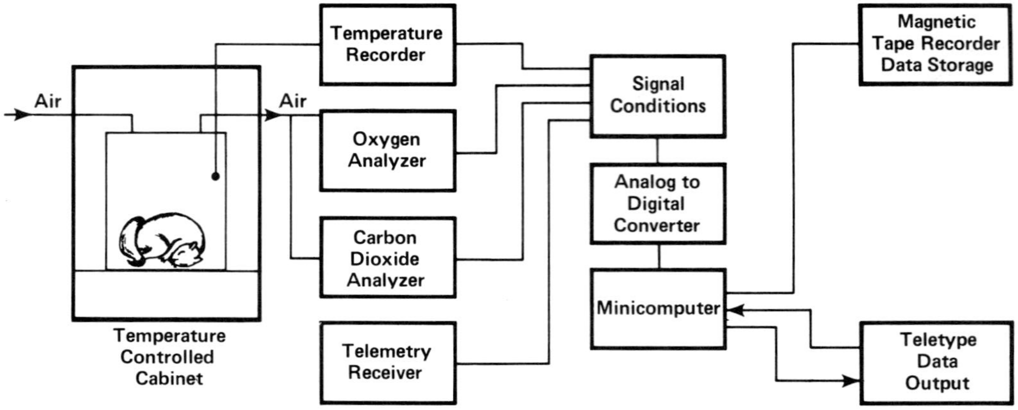

Fig. 5. *Diagrammatic representation of the open-circuit system used for continuous measurement of body temperature (telemetered), oxygen consumption and carbon dioxide production in S. richardsonii throughout a torpor cycle. Components for on-line data acquisition are also shown.*

measurements). This was slightly lower than the 0.174°C/min observed in the field probably due to greater convective heat loss in the metabolic chamber under constant air flow.

A typical change of $\dot{V}O_2$, $\dot{V}CO_2$ and Tb during entry into torpor is shown in figure 6. The decreases of $\dot{V}O_2$ and $\dot{V}CO_2$ preceeded the decrease of Tb such that both metabolic parameters were near their minima while the Tb was still above 30°C. Simultaneous recordings of $\dot{V}O_2$, heat loss and brain temperature yield similar conclusions (Fig. 7). In this case, the $\dot{V}O_2$ was near minimum while the brain temperature remained high at 33°C. The subsequent decrease of brain temperature seemed to be regulated by the rate of heat loss and intermittent pulsatile increases of $\dot{V}O_2$ (Fig. 7). These minute fluctuations between heat loss and heat production and the resultant fluctuations of brain temperature were evident as illustrated on the minute-to-minute changes of these three parameters between 1415 - 1615 hr during the entry period (Fig. 7). It seems probable that these changes also occur in the field animals as their Tb typically fluctuated in a similar fashion but with lesser magnitude as it declined during the entry process (47).

The changes of $\dot{V}O_2$, $\dot{V}CO_2$, R.Q. and Tb during spontaneous arousal from torpor are presented in figure 8. The increases of both metabolic parameters were rapid and continuous, resulting in a smooth rise in both curves. The R.Q. increased from approximately 0.74 during torpor to approximately 0.80 at the early stages of arousal as well as after completion of arousal. After peaking, both metabolic parameters showed constant fluctuations as their rates declined gradually towards the pre-torpor levels. The rise of abdominal Tb lagged behind the increases of both metabolic parameters such that the Tb was only 15°C (a rise of 9.2°C from torpid level) when the metabolic rates peaked. As the metabolic rates were declining, Tb continued to rise and reached its normothermic level of 36°C approximately 4.5 hr after the initiation of arousal (Fig. 8). When $\dot{V}O_2$, heat loss and brain temperature were measured simultaneously (Fig. 9) the peaking of $\dot{V}O_2$ coincided fairly closely with that of the brain temperature. The total heat loss, on the other hand, continued to rise

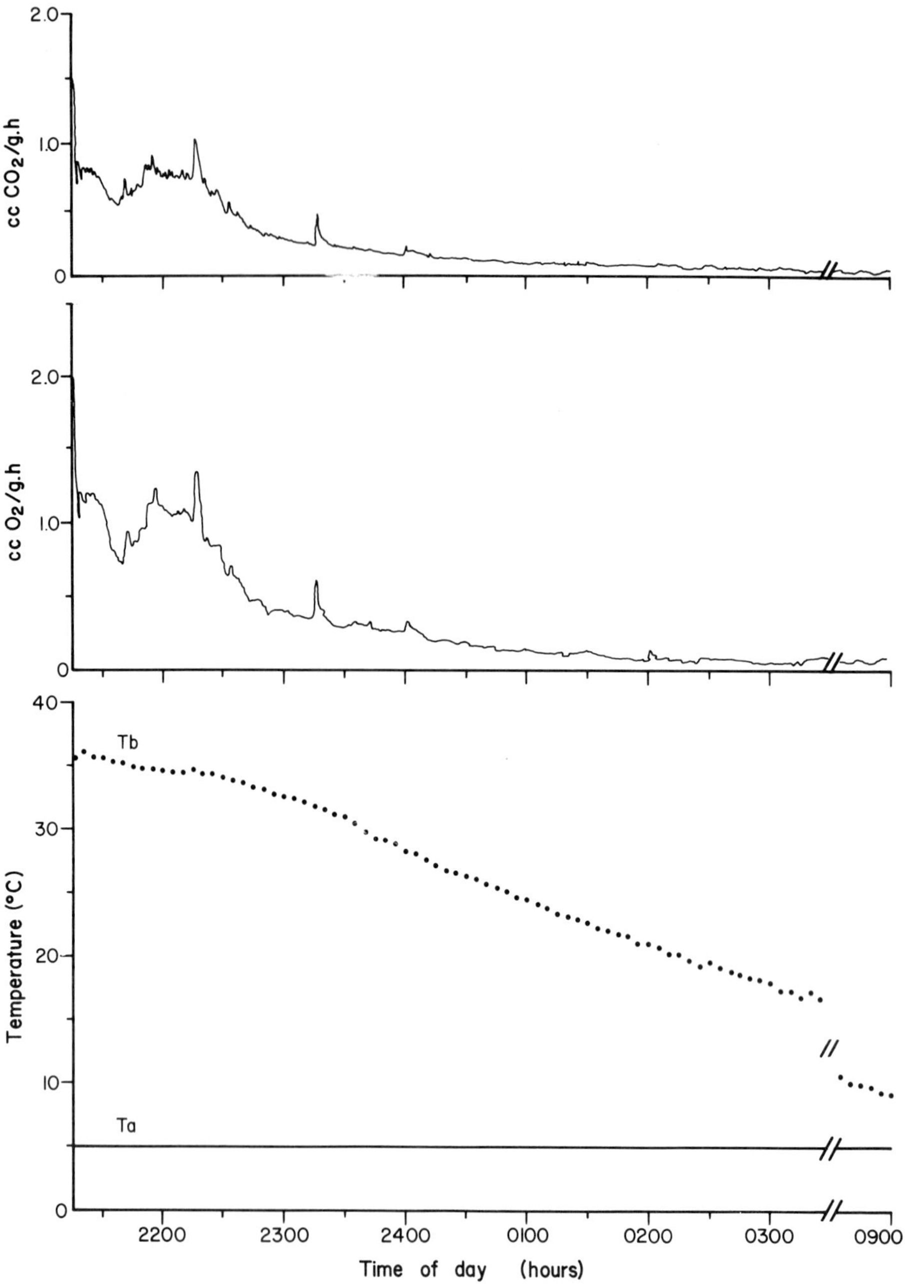

Fig. 6. *Simultaneous recordings of oxygen consumption (cc O_2/g·h), carbon dioxide production (cc CO_2/g·h), and body temperature (Tb, °C; telemetered) during entry into torpor in* S. richardsonii *at an ambient temperature (Ta) of 5°C. Note the rapid declines of both metabolic parameters and the delayed decrease of Tb (abdominal).*

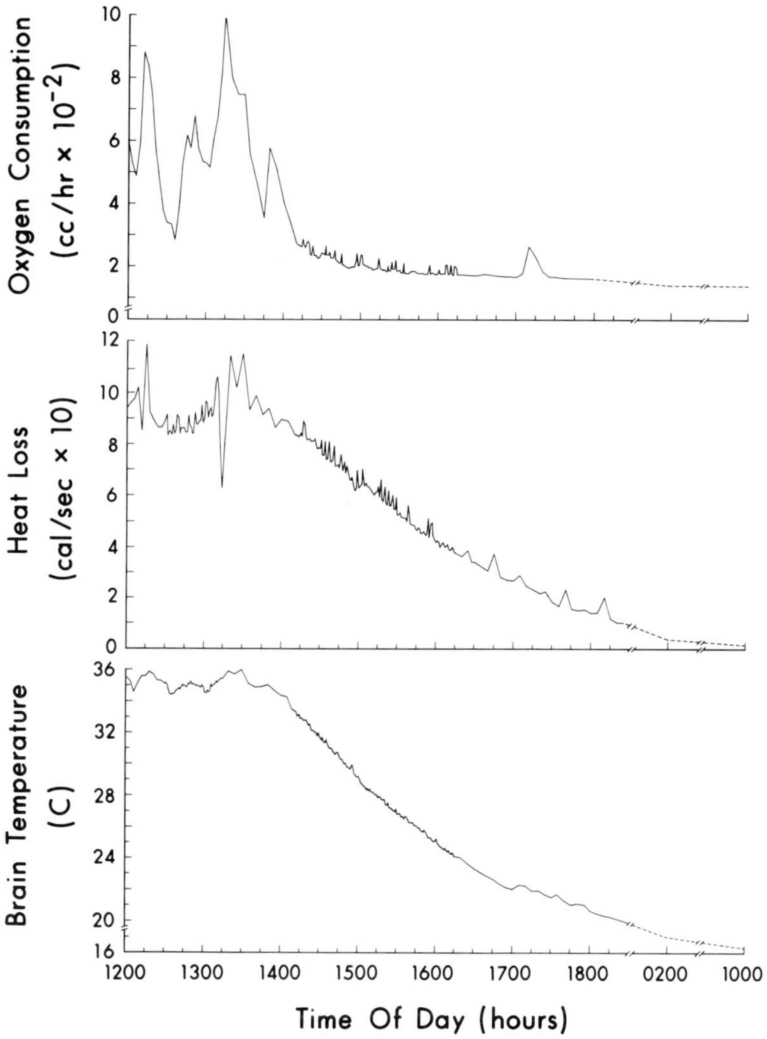

Fig. 7. *Simultaneous recordings of oxygen consumption (cc O_2/hr), dry heat loss (cal/sec), and brain temperature (°C; by thermocouple) during entry into torpor in S. richardsonii at an ambient temperature of 15°C. Recordings between 1415 to 1615 hr are minute to minute changes in all three parameters to illustrate the constant fluctuations of heat production, heat loss and brain temperature during the entry process. Note also the rapid decline of oxygen consumption and slower decreases of heat loss and brain temperature.*

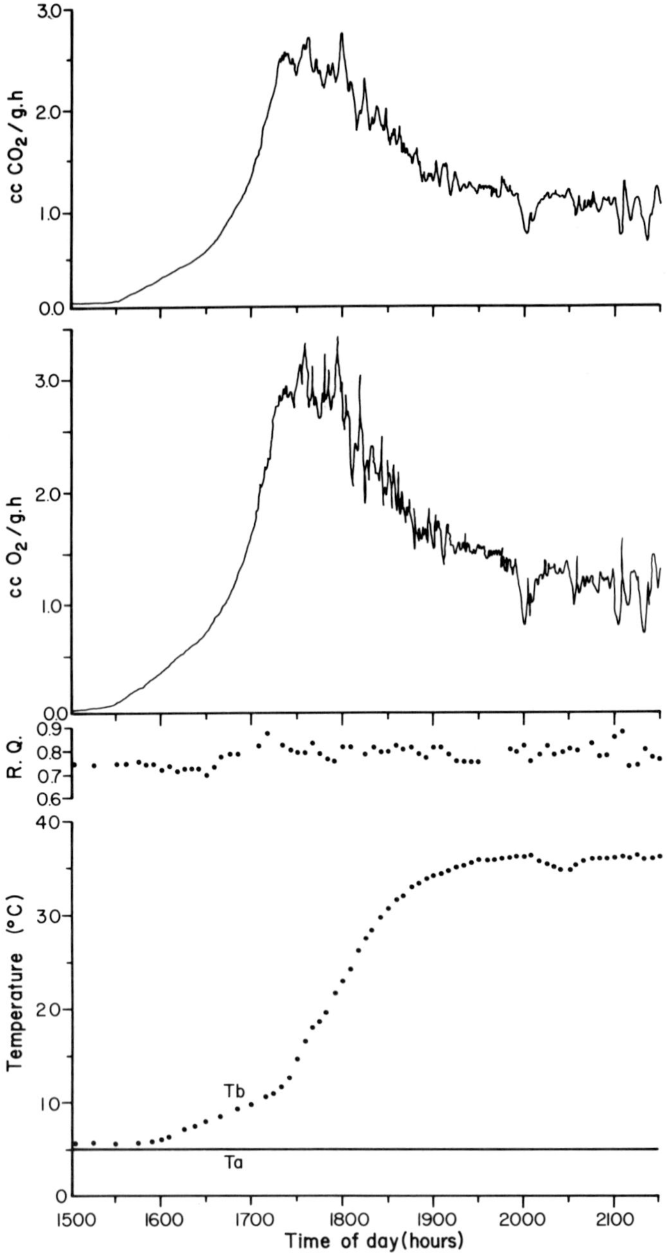

Fig. 8. *Simultaneous recordings of oxygen consumption (cc O_2/g·h), carbon dioxide production (cc CO_2/g·h), respiratory quotient (R.Q.) and body temperature (Tb, °C; telemetered) during arousal from torpor in S. richardsonii at an ambient temperature (Ta) of 5°C. Note the relatively rapid and smooth increases in both metabolic parameters and the delayed increase of Tb (abdominal).*

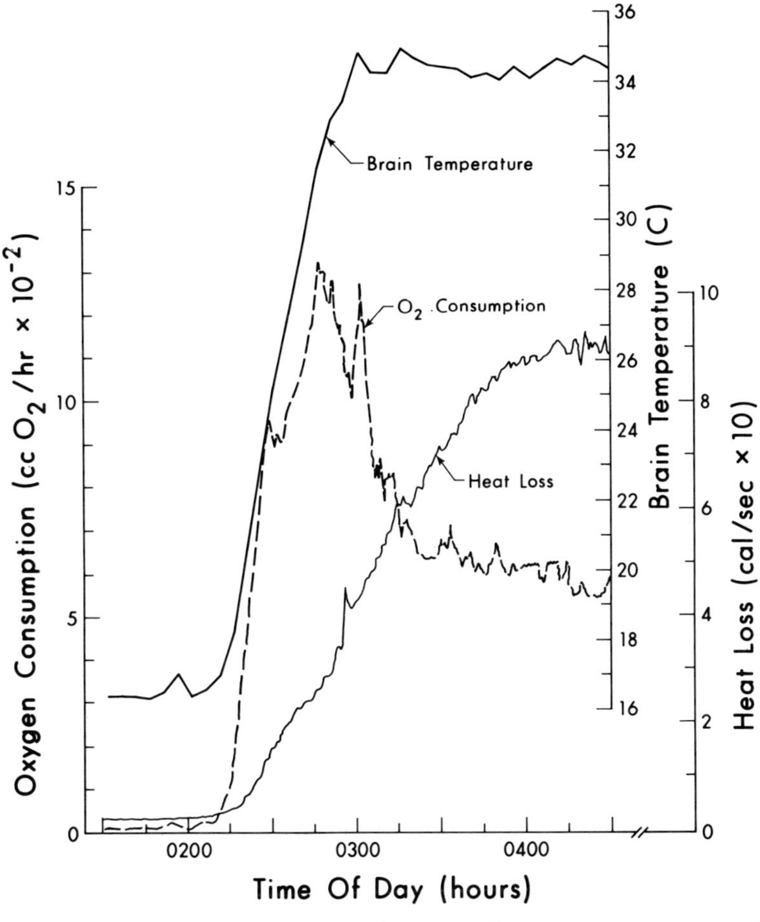

Fig. 9. *Simultaneous recordings of oxygen consumption (cc O_2/hr), dry heat loss (cal/sec), and brain temperature (°C; by thermocouple) during arousal from torpor in S. richardsonii at an ambient temperature of 15°C. Note the coincidence of peaking between oxygen consumption and brain temperature while heat loss was only 1/3 of maximum. Note also the continuous increase of heat loss with concurrent decrease of heat production while the brain temperature remained relatively constant.*

long after the stabilization of brain temperature at its normothermic level; it did not reach its plateau until $\dot{V}O_2$ was close to its minimum (Fig. 9). These changes indicated that the magnitude of heat production during arousal from torpor is closely related to the temperature of the brain which is being rewarmed rapidly after the initiation of arousal. Although this seems to suggest a "set-point" or "threshold" temperature for regulation of heat production by brain temperature during arousal, there remains the question as to why heat production decreased after the stabilization of brain temperature? It is speculated that perhaps the gradually increasing temperature from the rest of the body exerts inhibition on the total heat production when signals from internal thermoreceptors are incorporated at the temperature regulating center in preoptic-anterior hypothalamus. The lower value of heat loss (1/3 of plateaued) observed at the peakings of $\dot{V}O_2$ and brain temperature (Fig. 9) and the subsequent increase of heat loss as the rest of the body was rewarmed suggests that heat loss was confined to only certain portions of the body when peak metabolism occurred. The abrupt increase of heat loss at approximately 0300 hr (Fig. 9), seems to signify the opening of added avenues for heat loss and a steady increase was thereafter observed until the completion of arousal. These observations are similar to those reported by Daudova in the suslik (5), by Hammel *et al.* in *S. lateralis* (12) and are consistent with the now classic observation of differential rewarming of Tb by Lyman (24). It is generally agreed that the selective distribution of blood during the arousal process (4, 18, 39, 48) is mainly responsible for the observed temperature gradient between the anterior and posterior portions of the body. The energetic savings resulting from this differential rewarming will be discussed in a later paragraph.

The $\dot{V}O_2$ at different phases of a torpor cycle at different Ta's is summarized in Table 5. Although there were individual variations, the energy cost for entry into torpor was relatively constant at different Ta's. The 24 hr (a typical time period per entry process) accumulated total in $\dot{V}O_2$ ranged between 1332 - 1845 cc O_2. For this reason, the values were averaged

Table 5. Oxygen consumption at different phases of a spontaneous torpor cycle in *S. richardsonii* at different ambient temperatures (Ta).

Ta(°C)	Entry into torpor (cc O_2/entry)*	During torpor (cc O_2/hr)	Arousal from torpor (cc O_2/arousal)	Inter-torpor homeothermy (cc O_2/hr)
13-15°	1332	16	1607	365
	1845	14.8	1387	480
	1470	17.3		460
				470
				400
		$\bar{x}$=16.0	$\bar{x}$=1497	$\bar{x}$=435
8-10°	1403	12.1	2280	435
		12.1		572
		10.6		460
		13.6		703
				421
				420
		$\bar{x}$=12.1	$\bar{x}$=2280	$\bar{x}$=501
2-6°	1413	9.7	3605	542
	1837	9.1	2909	757
		6.1		802
		9.1		683
		6.1		859
		7.1		
		8.2		
		$\bar{x}$=8.0	$\bar{x}$=3257	$\bar{x}$=728.5
-1-0°				740
				750
				750
	$\bar{x}$=1550			$\bar{x}$=747
	(N=4)	(N=8)	(N=3)	(N=7)

*Based on 24 hour integrated oxygen consumption values.

(N is number of individuals used.)

to provide a single value of 1550 cc O_2 as the cost for entry into torpor when calculating field energetics. The $\dot{V}O_2$ during torpor varied according to Ta and Tb, and ranged between 6.1 - 16 cc O_2/hr between Ta of 2 - 15°C (Table 5). A Q_{10} of approximately 2 was observed for $\dot{V}O_2$ during torpor. The cost of arousal was inversely proportional to the Ta at which torpor occurred. The cost for arousal was more than twice at 2 - 6°C (3252 cc O_2/arousal) than it was at 13 - 15°C (1497 cc O_2/arousal). By comparison, the energetic costs for homeothermy between torpor episodes was high, ranging from 435 cc O_2/hr at 13 - 15°C to 728.5 cc O_2/hr at 2 - 6°C (Table 5); in fact a 4.5 hr homeothermic period would have an energy expenditure exceeding that of an arousal event at the same temperature. It will become apparent that the period of homeothermy is indeed the most energetically costly phase in the final analysis of energetics in natural torpidity.

The energetics of a typical spontaneous arousal as illustrated in figure 9 demonstrates the efficiency of differential rewarming. The total heat loss during arousal between 0202 - 0412 hr, after $\dot{V}O_2$ and heat loss had stabilized (Fig. 9), was 4383.3 cal. The total $\dot{V}O_2$ during the same period was 1607 cc O_2; based on an assumed R.Q. of 0.80 and 4.85 cal/cc O_2, the total heat production was 7793.9 cal. The ground squirrel weighed 406 g, and with a specific heat of 0.83 cal/g·°C for animal tissue (13) the excess heat production should increase the Tb by 10.1°C. But the actual increase was at least 18.6°C (from 16.4 to 35.0°C). Although the difference can be explained by anaerobic metabolism, it is most unlikely as no sustained increase of $\dot{V}O_2$ was evident following arousal which would indicate that an O_2 debt had incurred during arousal. Therefore, one must consider whether the use of 0.83 cal/g·°C as specific heat of the animal was accurate. Based on the excess heat production, the actual rise of Tb and the weight of the animal, the specific heat can be calculated as being 0.45 cal/g·°C. We measured the cooling of a Richardson's ground squirrel carcass, of similar weight, in the calorimeter from 23 to 5°C and obtained a specific heat figure of exactly 0.83 cal/g·°C (Glass and Wang, unpublished). The logical conclusion from the arousal experiment then is that somehow the "functional specific heat" of an arousing ground squirrel is reduced to approximately half of its normal value. Since physically only drastic changes in tissue composition can vary the relationships governing mass, specific heat, temperature and heat capacity, it is highly unlikely that the observed reduction of "functional specific heat" during arousal was due to such physical modifications. A likely possibility could be that the rapidly produced metabolic heat during the early stages of arousal is somewhat trapped inside the animal through vascular adjustments rather than carried to the periphery and lost to the surroundings. An analogous model would be to consider a series of concentric cylinders as representing the demarcation of the animal from core to periphery. The inner-most cylinder is warmed first by metabolic heat but strong vasoconstriction from core to periphery reduces the temperature gradient between periphery and the ambient thus minimizing heat loss to the surroundings. As the next layer is to be warmed up, circulation is selectively open to it so that only a proportional increase in metabolic heat production is necessary to overcome the mass and specific heat of the second layer. At this time heat loss to the surrounding is increased by a proportional amount due to the rise in temperature of the second layer but still would be far less than if the entire periphery were opened to circulation. This process can continue until the outer-most layer is warmed up and the total heat loss by that time would be the plateaued level as observed in figure 9 at the completion of arousal. It is a highly speculative scheme and lacks direct experimental evidence to substantiate it. But the circumstantial evidence on differential rewarming of different parts of the body suggests strongly that this is a definite possibility. The point of emphasis, however, from the energetic point of view is that through the conservation of heat loss the total heat production necessary to raise the Tb from the torpid to the normothermic level can be correspondingly reduced during arousal from torpor. Based on the experiment described here, the energy saving is close to 46% due to the reduction of "functional specific heat". We have conducted additional experiments using 7 more individuals by similar methods but the animals were disturbed to arouse by insertion of brain (through brain re-entrant tube) and rectal thermocouples (Glass and Wang, unpublished). In these latter cases, the "functional specific heat" ranged between 0.48 - 0.69 cal/g•°C and averaged 0.59 cal/g•°C, which represents a reduction of 29% from the normal 0.83 cal/g•°C, and a similar amount in energy saving for arousal.

IV. ENERGETICS OF NATURAL TORPIDITY

By interpolation of laboratory measurements on $\dot{V}O_2$ during episodes of torpor at Ta's comparable to field conditions, the energetics of natural torpidity can be estimated. Based on the energetic data for different phases in a torpor cycle (Table 5), and the time budget for torpor under natural conditions (Table 3), the energy expenditure in seasonal torpor is derived (Table 6). The lowest energy consumption occurs in January (22222.7 cc O_2) when the duration of torpor is the longest and the duration of inter-torpor homeothermy is the shortest. The highest energy consumption occurs in March (76611.6 cc O_2 when the inter-torpor period is the longest. Excepting July when only 15 days are in the torpor season, monthly energy cost gradually declines as the torpor season progressed from August to January; from thereon, the cost begins to rise until the termination of the torpor season in March (Table 6). This trend is mostly due to the gradual lengthening and thereafter, gradual shortening of the duration of torpor (Fig. 1) which, with its low energy requirement has apparently more than sufficiently compensated for the increased energy expenditure for arousal from a low Tb during the winter months (Fig. 3).

With respect to the compartmentalization of energy expenditure in torpor, i.e., the energy cost associated with entry into torpor, during torpor, arousal from torpor, and during inter-torpor homeothermy, the seasonal variations are illustrated in Table 7. These data are calculated from Table 6. Without including the inter-torpor period, the cost of arousal is usually the highest (40.5% of total cost); the cost of entry is also substantial (26.8%) because it usually lasts 24 hr or longer. The cost of deep torpor (32.8%) is directly

Table 6. Estimation of monthly energy expenditure in seasonal torpor in *S. richardsonii* under natural conditions.*

Month	Ta(°C)	$cc0_2$ per entry	$cc0_2$ during torpor	$cc0_2$ per arousal	Total $cc0_2$ per torpor	$cc0_2$ per homeothermy	Monthly total[a] $cc0_2$
JUL	15	1550	736.0	1497	3783.0	4915.0	38793.1[b]
AUG	15	1550	1235.0	1497	4282.2	4915.0	59689.8
SEP	13	1550	1888.0	1497	4935.0	4915.0	45704.0
OCT	10	1550	2238.6	1889	5677.6	4939.9	37692.2
NOV	5	1550	2483.3	2768	6801.3	5160.0	31577.8
DEC	2	1550	2988.0	3257	7795.0	5463.8	26782.7
JAN	0	1550	3439.2	3350	8339.2	5550.0	22222.7
FEB	-1	1550	2182.6	3450	7182.6	10500.0	33596.9
MAR	-1	1550	805.0	3450	5805.0	18750.0	76611.6[c]
						Total for torpor season (July 16 - March 22) =	372670.8

*Calculated from data presented in Tables 3 and 5.

a. By adding total $cc0_2$ per torpor and per homeothermy and multiplying number of torpor episodes in month.
b. Only half of July is considered as torpor season.
c. Only up to March 22 for torpor season.

Table 7. Compartmentalization of monthly energy expenditure for different phases of a torpor cycle in *S. richardsonii* under natural conditions.*

	Within each torpor cycle			Including inter-torpor homeothermy			
Month	Entry, % total	Torpor, % total	Arousal, % total	Entry, % total	Torpor, % total	Arousal % total	Inter-torpor homeothermy, % total
JUL	40.97	19.45	39.57	17.82	8.46	17.21	56.50
AUG	36.19	28.84	34.95	16.85	13.42	16.27	53.44
SEP	31.40	38.25	30.33	15.73	19.16	15.19	49.89
OCT	27.30	39.42	33.27	14.59	21.07	17.79	46.52
NOV	22.78	36.51	40.69	12.95	20.76	23.14	43.14
DEC	19.88	38.33	41.78	11.69	22.53	24.56	41.21
JAN	18.59	41.24	40.17	11.15	24.76	24.11	39.95
FEB	21.57	30.38	48.03	8.76	12.34	19.51	59.38
MAR	26.70	13.86	59.43	6.31	3.27	14.05	76.35
$\bar{x}$=	26.84	32.63	40.50	12.78	16.62	19.01	51.55

*Calculated from data presented in Table 6.

related to the duration of such a state, as well as the Tb during torpor (Table 7). When including the inter-torpor homeothermic period, the entry into torpor accounts for 12.8%, deep torpor 16.6%, arousal from torpor 19.0% and the inter-torpor period 51.6% of the total energy expenditure (Table 7). It is apparent that the period of inter-torpor homeothermy represents the highest energy drain (40-76% of total) regardless of the season (Table 7). As discussed under field studies, it seems appropriate to ask the question once again why this period of homeothermy is necessary, particularly in view of its high energy cost relative to the other phases of a torpor cycle.

To demonstrate the ultimate aspect of energy conservation by the use of natural torpidity, a seasonal budget is calculated (Table 8) based on data from Table 5 and Table 6. The use of torpor saves approximately 38% of an estimated maintenance energy reuqirement in July and 51% in March, the two months with only fractions of time included in the torpor season. In the remaining months, the savings amount to 81 - 96% (Table 8). When the overall benefit is considered, the torpor season conserved an energy expenditure of 87.8% in comparison to being homeothermic for the same duration. This is an impressive accomplishment, especially when one considers that the torpor season in adult *S. richardsonii* covers approximately 8.5 to 9 months. Thus only in the remaining 3 - 3.5 months do the animals need to exert full energy expenditure while completing other important life functions such as reproduction, lactation, growth and the deposition of fat for the next season. The combined effect is that only 34% of energy is spent (12% for nine months, 100% for 3 months) as compared to being homeothermic constantly. In prairies of the north where winters are long and cold, it is not surprising that the Richardson's ground squirrels have evolved physiological adaptations for the exhibition of torpor. Their somewhat unique pattern of torpidity, however, further extends the energetic advantages that natural torpidity exemplifies as a strategy in cold.

Table 8. Energy conservation by natural torpidity in *S. richardsonii*.

Month	Total energy expenditure with torpor[a] (cc O_2)	Total energy expenditure if remain homeothermic[b] (cc O_2)	% energy used by exhibiting torpor[c]	% energy saved by exhibiting torpor
JUL	38793.1	161820	24.0	38.0[d]
AUG	59689.8	323640	18.4	81.6
SEP	45704.0	313200	14.6	85.4
OCT	37692.2	372744	10.1	89.9
NOV	31577.8	432000	7.3	92.7
DEC	26782.7	542004	4.9	95.1
JAN	22222.7	555768	4.0	96.0
FEB	33596.9	501984	6.7	93.3
MAR	76611.6	394416	19.4	51.2[e]
			$\bar{x}$=12.2	87.8

a. From Table 6.
b. Number of hours x cc O_2/h for homeothermy as listed in Table 5.
c. Ratio of torpor/homeothermy.
d. Since torpor commenced after July 15, half of July is homeothermic.
e. Torpor season ends on March 22, therefore 9 days homeothermic.

ACKNOWLEDGEMENTS

The study was supported by a National Research Council of Canada grant No. A6455. I thank M. Kowalchuk for permission to use his farm for the field study, W. Staines, C. Leedham, G. Toews and S. Ruben for field and lab assistance.

REFERENCES

1. Armitage, K.B., J.F. Downhower and G.E. Svendsen. Seasonal chages in weights of marmots. Am. Midland Nat. 96(1):36-51, 1976.
2. Augee, M.L., E.H.N. Ealey and H. Spencer. Biotelemetric studies of torpor in the Echidna. J. Mammal. 51:561-570, 1970.
3. Brown, J.H. and G.A. Bartholomew. Periodicity and energetics of torpor in the Kangaroo mouse, *Microdipodops pallidus*. Ecology 50(4):705-709, 1969.
4. Bullard, R.W. and G.E. Funkhouser. Estimated regional blood flow by Rubidium 86 distribution during arousal from hibernation. Am. J. Physiol. 203(2): 266-270, 1962.
5. Daudova, G.M. Heat production in the suslik *Citellus pygmaeus* during arousal from hibernation. Zh. Evol. Biokhim. Fiziol. 11(2):147-152, 1975.
6. Davis, D.E. Hibernation and circannual rhythms of food consumption in marmots and ground squirrels. Quart. Rev. Biol. 51(4):477-514, 1976.
7. French, A.R. Selection of high temperatures for hibernation by the pocket mouse, *Perognathus longimembris*: ecological advantages and energetic consequences. Ecology 57(1):185-191, 1976.
8. Galster, W. and P. Morrison. Carbohydrate reserves of wild rodents from different latitudes. Comp. Biochem. Physiol. 50A:153-157, 1975.
9. Gessaman, J.A. Methods of estimating the energy cost of free existence. In: Ecological Energetics of Homeotherms. J.A. Gessaman (editor). Logan: Utah State University Press, 1973, Chapter 2, pp. 3-31.
10. Glass, D. and L.C.H. Wang. Thermoregulatory effects of intracerebroventricular injection of serotonin and a monoamine oxidase inhibitor in a hibernator, *Spermophilus richardsonii*. (Abstract). J. Therm. Biol., 1978 (In press).
11. Grodzinski, W. and B.A. Wunder. Ecological energetics of small mammals. In: Small Mammals: Their Productivity and Population Dynamics.

International Biological Programme, Cambridge: Cambridge University Press, 1975, vol. 5, pp. 173-204.

12. Hammel, H.T., T.J. Dawson, R.M. Abrams and H.T. Andersen. Total calorimetric measurements on *Citellus lateralis* in hibernation. Physiol. Zool. 41(3): 341-357, 1968.

13. Hart, J.S. Calorimetric determination of average body temperature of small mammals and its variation with environmental conditions. Can. J. Zool. 29: 224-233, 1951.

14. Heller, H.C. and D.M. Gates. Altitudinal zonation of chipmunks (*Eutamias*): Energy budgets. Ecology 52(3): 424-433, 1971.

15. Hudson, J.W. Variations in the patterns of torpidity of small homeotherms. In: Mammalian Hibernation. Fisher, K.C., A.R. Dawe, C.P. Lyman, E. Schönbaum, and F.E. South, Jr. (editors). Toronto: Oliver and Boyd, vol. III, 1967.

16. Hudson, J.W. and G.A. Bartholomew. Terrestrial animals in dry heat: estivators. In: Handbook of Physiology. Adaptation to the Environment. Washington D.C.: Am. Physiol. Sox., 1964, Sect. 4, Chapt. 34, p. 541-550.

17. Janský, L. and R. Novotná. The role of central aminergic transmission in thermoregulation and hibernation. In: Regulation of Depressed Metabolism and Thermogenesis. Janský, L. and X.J. Musacchia (editors). Springfield: Charles C. Thomas, 1976, Chapter 4.

18. Johansen, K. Distribution of blood in the arousing hibernator. Acta. Physiol. Scand. 52: 379-386, 1961.

19. Kudryavtseva, N.N. and N.K. Popova. Serotonin concentration in various parts of the brain during hibernation and waking. Translated from Byulleten' Eksperimental' noi Biologii i Meditsiny, 75(4): 44-47, 1973.

20. LeFebvre, R.A. The use of D_2O^{18} for measuring energy metabolism in *Columba livia* at rest and in flight. Auk, 81: 403-416, 1964.

21. Lifson, N., G.B. Gordon and R.M. McClintock. Measurement of total carbon dioxide production

by means of D_2O^{18}. J. Appl. Physiol. 7: 704-710, 1955.

22. Little, W.S. and N. Lifson. Validation study of D_2O^{18} method for determination of CO_2 output of the eastern chipmunk (*Tamias striatus*). Comp. Biochem. Physiol. 50A: 55-56, 1975.
23. Lund, G.F. and G.E. Folk, Jr. Simultaneous measurements of heart rate and oxygen consumption in black-tailed prairie dogs (*Cynomys ludovicianus*). Comp. Biochem. Physiol. 55A: 201-206, 1976.
24. Lyman, C.P. The oxygen consumption and temperature regulation of hibernating hamsters. J. Exp. Zool. 109: 55-78, 1948.
25. Lyman, C.P. Circulation in mammalian hibernation. In: Handbook of Physiology. Circulation. Washington, D.C.: Am. Physiol. Soc., 1965, Sect. 2, vol. III, chapt. 56, pp. 1967-1989.
26. MacMillen, R.E. Aestivation in the cactus mouse, *Peromyscus eremicus*. Comp. Biochem. Physiol 16: 227-248, 1965.
27. Michener, G.R. Effect of climatic conditions on the annual activity and hibernation cycle of Richardson's ground squirrels and Columbian ground squirrels. Can. J. Zool. 55: 693-703, 1977.
28. Morhardt, J.E. and J.W. Hudson. Daily torpor induced in white-footed mice (*Peromyscus spp.)* by starvation. Nature 212: 1046-1047, 1966.
29. Morhardt, J.E. and S.S. Morhardt. Correlations between heart rate and oxygen consumption in rodents. Am. J. Physiol. 221(6): 1580-1586, 1971.
30. Morhardt, S.S. and D.M. Gates. Energy-exchange analysis of the Belding ground squirrel and its habitat. Ecological Monographs 44(1): 17-44, 1974.
31. Morton, M.L. Seasonal cycles of body weights and lipids in Belding ground squirrels. Bull. South Calif. Acad. Sci. 74(3): 128-143, 1975.
32. Mullen, R.K. Respiratory metabolism and body water turnover rates of *Perognathus formosus* in its natural environment. Comp. Biochem. Physiol. 32: 259-265, 1970.

33. Mullen, R.K. Energy metabolism and body water turnover rates of two species of free living kangaroo rats, *Dipodomys merriami* and *Dipodomys microps*. Comp. Biochem. Physiol. 39A: 379-380, 1971.
34. Mullen, R.K. Note on the energy metabolism of *Peromyscus crinitus* in its natural environment. J. Mammal. 52: 633-635, 1971.
35. Murie, J.O. Population characteristics and phenology of a Franklin ground squirrel (*Spermophilus franklinii*) colony in Alberta, Canada. Am. Midland. Nat. 90(2): 334-340, 1973.
36. Petrusewicz, K. and A. Macfadyen. Productivity of Terrestrial Animals. Oxford: Blackwell Scientific Publications, 1970, p. 97-111.
37. Pivorun, E.B. A biotelemetry study of the thermoregulatory patterns of *Tamias striatus* and *Eutamias minimus* during hibernation. Comp. Biochem. Physiol. 53A: 265-271, 1976.
38. Popova, N.K. and N.P. Voitenko. Serotonin metabolism during hibernation. Dokl. Akad. Nauk. SSSR Ser. Biol. 218(6): 1488-1490, 1974.
39. Rauch, J.C. Sequential changes in regional distribution of blood in *Eptesicus fuscus* (big brown bat) during arousal from hibernation. Can. J. Zool. 51(9): 973-981, 1973.
40. Tucker, V.A. Oxygen consumption, thermal conductance and torpor in the california pocket mouse, *Perognathus californicus*. J. Cell. Comp. Physiol. 65(3): 393-404, 1965.
41. Tucker, V.A. Diurnal torpor and its relation to food consumption and weight changes in the California pocket mouse *Perognathus californius*. Ecology 7(2): 245-252, 1966.
42. Utter, J.M. Daily energy expenditures of mockingbird Bull. Ecol. Soc. Am. 52: 39, 1971.
43. Utter, J.M. and E.A. LeFebvre. Energy expenditure fo free flight by the purple martin (*Progne subis*). Comp. Biochem. Physiol. 35: 713-719, 1970.
44. Wang, L.C.H. and J.W. Hudson. Some physiological aspects of temperature regulation in the normothermic and torpid hispid pocket mouse, *Perognathus hispidus*. Comp. Biochem. Physiol. 32: 275-293, 1970.

45. Wang, L.C.H. and J.W. Hudson. Temperature regulation in normothermic and hibernating eastern chipmunk, *Tamias striatus*. Comp. Biochem. Physiol. 38(1A): 59-90, 1971.
46. Wang, L.C.H. Circadian body temperature of Richardson's ground squirrel under field and laboratory conditions: A comparative radio-telemetric study. Comp. Biochem. Physiol. 43A:503-510, 1972.
47. Wang, L.C.H. Radiotelemetric study of hibernation under natural and laboratory conditions. Am. J. Physiol. 224(3):673-677, 1973.
48. Wells, L.A. Circulatory patterns of hibernators. Am. J. Physiol. 221(5):1517-1520, 1971.
49. Willis, J.S., S.S. Goldman and R.F. Foster. Tissue K concentration in relation to the role of the kidney in hibernation and the cause of periodic arousal. Comp. Biochem. Physiol. 39A:437-445, 1971.

THE ECONOMICS OF TEMPERATURE REGULATION AND TORPOR IN NONMAMMALIAN ORGANISMS

F. Reed Hainsworth and Larry L. Wolf
Department of Biology
Syracuse University
Syracuse, New York 13210

ABSTRACT

We view temperature regulation and torpor in terms of simple benefit/cost models applied to biological systems. Organisms should maximize their fitness by optimizing relationships between both benefits and costs. Torpor has traditionally been viewed as one mechanism to reduce energetic costs, but its use by an organism also depends importantly on concurrent benefits. An economic interpretation considers homeothermy as a contributor to maximizing benefits relative to costs, so torpor would be predicted only when environmental constraints prevent homeothermy. Several environmental factors that are considered to contribute to a benefit: cost rationale for torpor include: energy availability relative to expenditures (influenced by food availability, costs for activity, constraints from prey selection and digestive processing), predation, social interactions (aggressive and associative), and the nature of environmental temperature variation for organisms with little option for behavioral or physiological regulation. The pertinent literature on temperature regulation and torpor is consistent with a benefit:cost interpretation although current information may not discriminate alternative hypotheses for major constraints generating torpor.

ISBN 0-12-734550-7

I. INTRODUCTION

We wish to identify principles to explain torpor in non-mammalian organisms and temperature regulation in general. We will use simple economic models applied to biological systems. This approach has received attention from ecologists to explain several phenomena (see 82 and references therein), and it is equally applicable to the specific case of body temperature regulation.

When economic models are applied to biological systems, we seek to identify important parameters influencing relative genetic contributions by individuals to future generations, or "fitness", which in the short-term involves survival and reproduction. The basic premise is that organisms have evolved characteristics that approach a maximum fitness within environmental constraints.

Body temperature (Tb) is potentially an important parameter influencing fitness. This is illustrated generally in figure 1. The specific shape of this curve is arbitrary, but its general form is typical (see below). It suggests that there is a unique Tb for maximizing fitness (a, Fig. 1). However, this temperature may not always be realized, and our thesis with

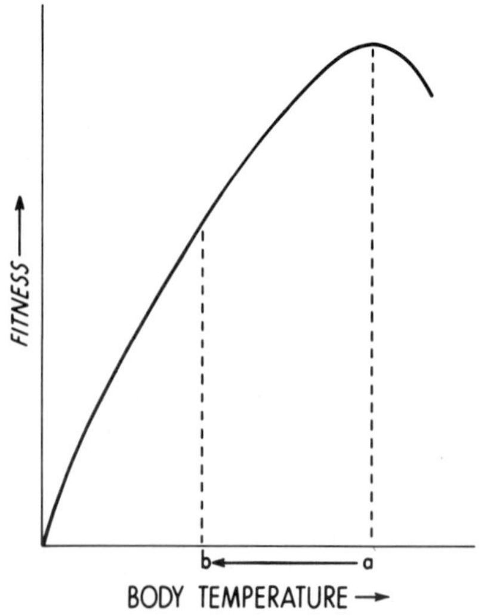

Fig. 1. *A general relationship for "fitness", or biological success from survival and reproduction, as a function of body temperature.*

respect to torpor will indicate environmental constraints where a decrease in Tb (b, Fig. 1) could be adaptive.

Economic models of biological systems attempt to identify variables that influence both benefits and costs associated with an organismal trait. This is illustrated generally in figure 2. Maximizing fitness, then, generally is equated with maximizing benefits above costs (a, Fig. 2), both of which are variable, and this defines the optimum Tb. For a given set of benefit and cost curves, decreasing or increasing Tb relative to the optimum would decrease net benefit (b, Fig. 2) which would be equivalent to shifting from *a* to *b* in figure 1. In economic terms the "currencies" of costs and benefits are not always measurable or interchangeable (e.g., a benefit from predator avoidance may be difficult to quantify in energetic terms, see below), and this presents a major problem. However, with sufficient information about costs and benefits under natural circumstances it is possible to formulate testable hypotheses. We will examine some hypothesized influences on temperature regulation after we provide a cost:benefit definition of homeothermy and torpor.

II. AN ECONOMIC DEFINITION OF HOMEOTHERMY AND TORPOR

The classical definition of homeothermy is maintenance of a constant Tb independent of environmental fluctuations (see 81). This can be accomplished

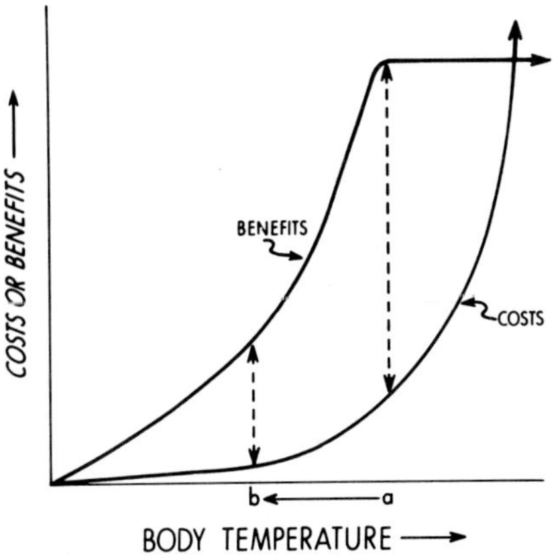

Fig. 2. *A general relationship for benefits and costs as a function of body temperature. The shapes of the functions assume a Q_{10} effect. The benefit curve further assumes some upper limit such as for rate of consumption.*

physiologically (endothermy) and/or behaviorally (ectothermy) (6, 10, 11, 12, 68, 70, 81, 87, 104). Torpor generally is defined as a reduction in Tb from the homeothermic level. Rather than use these operational definitions, we will define a homeothermic temperature as one that results in a maximization of benefits relative to costs when environmental limitations are minimal. This temperature should be homeothermic since regulation by physiology and/or behavior would maintain maximum fitness. The homeothermic temperature may be maintained physiologically, it may be selected from available microenvironments, or it may be available only from temporal variations in temperature (see below). The important feature is that it is that temperature which provides a maximum net benefit for an organism.

The evidence we consider below suggests that there will be a unique temperature for producing a maximum net benefit for any given species (but not necessarily the same temperature among species). Once the homeothermic temperature is identified, torpor would represent a change in Tb as the result of added environmental limitations on costs and benefits. The change in Tb would decrease the absolute benefit to an organism (Fig. 2), but the temperature of torpid organisms should still contribute to maximizing net benefits within the context of the added environmental limitations.

For some organisms (but not all), Tb is maintained in a characteristic range near 35-40°C by physiological and/or behavioral means. These include some plants (76, 81), some insects (5, 49), some fish (22), a variety of reptiles (26, 95, 98), and most birds (27) and mammals. Calloway (20) suggested that the characteristic 35-40°C range of Tb is governed by thermodynamic characteristics of water. Since cells are composed of 80-92% water, the thermodynamic characteristics of the medium (which are invariant) should dictate the efficiency of cellular processes. Calloway argues that within the range of the freezing and boiling points of water, the point of least "stress" would be the logarithmic mean (100°C/e = 36.79°C). Since thermodynamic functions of water change exponentially with temperature, this is the point where water would have characteristics least like either ice or vapor. For example, thermodynamic characteristics of water such as minimum compressibility and specific heat have minimum values at temperatures in the characteristic 35-40°C range (20).

Calloway's hypothesis for an absolute basis for maximizing net benefits through homeothermy has more heuristic value than other hypotheses. The hypothesis that a 35-40°C temperature represents an evolutionarily "advanced" condition (most recently reiterated in 105) offers no explanation for why 35-40°C may be "advanced" and generally neglects the possibility that other temperatures may be adaptive (see below; 54, 73). Other hypotheses rely to some degree on an inherent upper limit to Tb close to 35-40°C. These include the "maxi-therm" hypothesis of Hamilton (45) and hypotheses based on temperatures for protein denaturation (e.g., 19). Calder and King (19), for example, argue that since proteins of birds and mammals denature at high rates at temperatures above 40°C this sets the upper temperature limit, and they set the characteristic range for regulation near the upper limit on the basis of water economy (related to Tb - Ta). However, numerous organisms have cellular temperatures far in excess of 40°C without comparably increased rates of protein denaturation (76°C or higher for some species, 81). This raises the question of why other organisms have not evolved to have a higher Tb than 35-40°C which would provide even greater water economy (88) or heat loss during activity (50).

All organisms do not have a Tb between 35-40°C (for example, microorganisms have environmentally determined temperatures that may be outside this range, see 81). Also, an interesting specialized aspect of homeothermy in organisms that may have a Tb of 35-40°C is the maintenance of a high temperature at low Ta only in critical portions of the body. For example, the few homeothermic plants regulate only the temperature of reproductive organs. For both philodendron (*Philodendron selloum*, 76) and skunk cabbage (*Symplocarpus foetidus*, 62) the maintenance of high temperature of reproductive tissue probably is associated with pollination. Bluefin tuna (*Thunnus thynnus*, 22) maintain high temperatures in swimming muscles and the eye, and bumblebees show elevated thoracic temperatures during feeding, permitting flower exploitation at low ambient temperatures (49).

That all organisms do not have a maximally efficient temperature of 35-40°C or that many only maintain this temperature in certain regions of the body can be explained in terms of environmental limitations on costs and benefits. For some species the costs associated with a Tb of 35-40°C may outweigh the benefits of absolute cellular efficiency. Nevertheless, within limits set by environment, behavior, and physiology, there should be an optimum temperature which maximizes benefits relative to costs. Torpor would represent a reduction in Tb from the optimum as the result of short-term additional environmental constraints. The identification of the environmental conditions that can lead to a departure from homeothermy should help us understand both temperature regulation and torpor.

In the following sections we consider some limitations that influence temperature regulation and torpor for non-mammalian organisms. We have not attempted a comprehensive review (for an extensive survey see 81), but we have selected studies that we feel are most pertinent to a cost:benefit interpretation. Section III deals with the prevalent hypothesis of an energetic basis for temperature regulation; section IV deals with the less prevalent hypothesized role of predation; section V deals with social effects; and section VI discusses the importance of environmental variability. These do not exhaust all possibilities; hopefully, generalities can be extended to a variety of other ecological situations.

III. THE ENERGETIC HYPOTHESIS

This hypothesis assumes that fitness can be equated with rate of net energy gain, or that the costs and benefits in figure 2 are both in energetic currency. For many organisms, maximizing rate of new energy gain results in maximizing growth rate, and this could have a high selective advantage if it increased fecundity because of larger size or decreased age to first reproduction. Maximizing rate of net energy gain could also have selective advantage independent of growth by providing a maximum energy input to cover energy expenditures for a variety of important activities (e.g., courtship, predator avoidance, etc.).

The energetic cost:benefit hypothesis states that energetic expenditures should be minimized relative to energetic intakes per unit time. It may appear that the advantage of torpor in this context is obvious since it is always accompanied by reduced energy expenditures. However, a reduction in cost is only part of a cost:benefit function (Fig. 2). The influence of reduced costs on net benefits per unit time will depend on concurrent benefits. Since a homeothermic temperature would always provide a greater net benefit (see Introduction), torpor should only occur when environmental constraints prevent homeothermy.

To identify environmental limitations that make torpor advantageous we need to appreciate the homeothermic condition in energetic terms as a point of reference. Physiologists have identified specific tissue temperatures associated with maximizing energetic rate processes (81). Dawson (26) argued that the physiological significance of preferred Tb in reptiles is that they result in maximum rate processes. Similarly, Brett (13) noted that rates of oxygen consumption and swimming for sockeye salmon (*Oncorhynchus nerka*) are maximum at 15°C. The temperatures at which maximal rate processes occur vary among species as a result of evolution in environments with different thermal characteristics (38, 81, see Introduction). However, for any species there generally will be a specific temperature in their environment that results in a unique maximum in physiological rate processes.

The temperatures selected or maintained were believed to be determined primarily by these maximum physiological rate responses (26, 38, 81). However, this hypothesis is an incomplete formulation with respect to a cost:benefit function (Fig. 2) since it assumes no limitations or costs in maximizing benefits under natural circumstances. Maintaining a maximum physiological rate depends on the availability of sufficient food; costs associated with obtaining a food benefit (from activity); differences between foods in benefits and costs for acquisition (generating a constraint from prey selection); and costs associated with processing ingested food.

A. Torpor in Ectotherms

1. The Constraint of Food Availability

The most detailed evidence for a potential effect of food availability on temperature regulation in ectotherms is that of Brett (13) for sockeye salmon (*Oncorhynchus nerka*) and Elliot (31) for brown trout (*Salmo trutta*). The effect of food availability on temperatures for maximum rate of growth is illustrated in figure 3 from the data of Brett (13). As food ration is reduced below that resulting in maximum rate of growth at the optimum acclimation temperature of 15°C, the temperature for maximum rate of growth also decreases. Brett (13) discusses the implications of this for energetic costs and benefits. If fish obtained less than the maximum possible food, they should select a temperature which would reduce their metabolic costs (functional torpor) *relative* to the energy from food to generate maximum rate of growth.

The processes of feeding and temperature selection could yield an effect on metabolic rate as illustrated in figure 4 (13). The activity of feeding

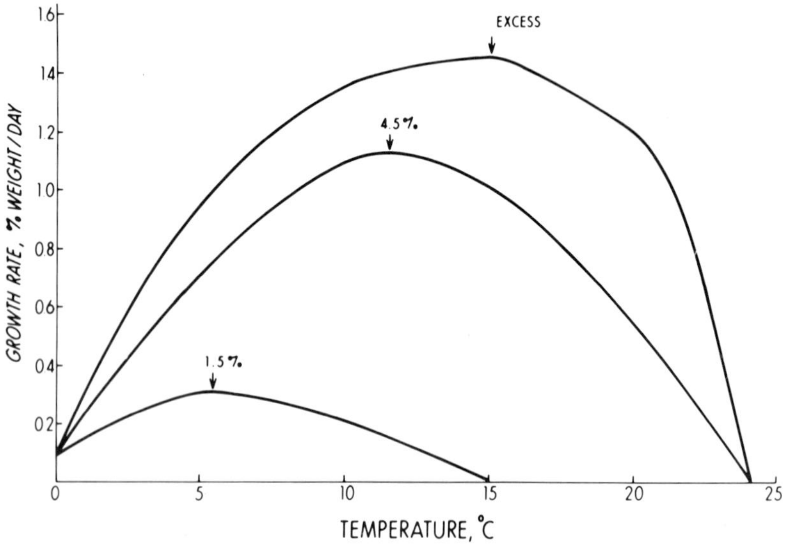

Fig. 3. *Growth rate for young salmon as a function of temperature and food ration. Food rations represent percent of body mass. Modified from Brett* et al. *(14).*

at dawn and dusk in relatively warm surface temperatures could result in relatively high metabolic rates. Depending on food intake, selection of lower temperatures during the day (when feeding does not occur) could reduce metabolic rate and contribute to maximum rate of growth (13). Such an integration of behavior and physiology would employ torpor through temperature selection flexibly with respect to costs and benefits. Of course, if a fish had a choice with respect to food consumption it should select an excess and not enter torpor (remain at the optimum temperature of 15°C for the salmon) if no other added environmental constraints

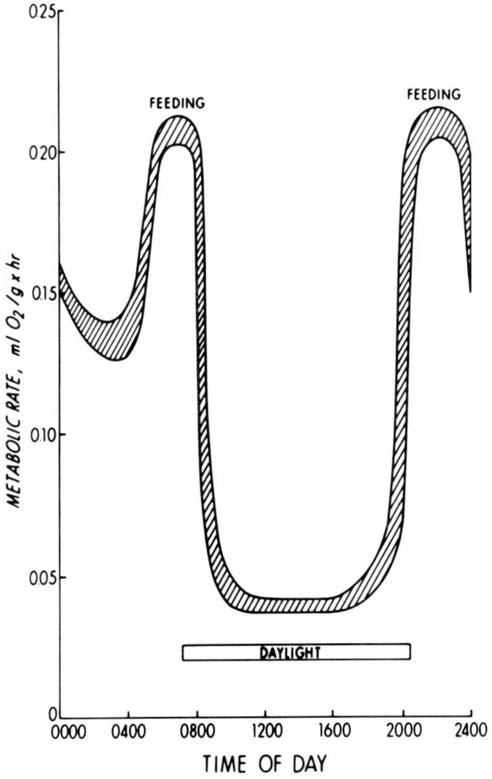

Fig. 4. *Metabolic rate as a function of time of day as a result of observed temperature selection by salmon in a lake environment. Modified from Brett (13).*

influenced temperature selection (see below).

Do fish do this? A direct test would be to examine temperature preferences in the laboratory as a function of growth rate by manipulating food availability (Fig. 3). This has not been done over a range of food availabilities, but Javaid and Anderson (59) noted that starvation influenced temperature selection in salmonids. Groups of 5-6 brook trout (*Salvelinus fontinalis*) and rainbow trout (*Salmo gairdneri*) selected lower temperatures during starvation than when food was available, but groups of atlantic salmon (*Salmo salar*) selected higher temperatures. The atlantic salmon is extremely interesting; for this species the energetic hypothesis predicts that costs would be *less* at a higher than a lower temperature since there are no benefits during starvation. Thus, the atlantic salmon may have a basic difference in underlying physiological processes since most organisms reduce costs by lowering Tb.

A more direct experimental test of the energetic hypothesis is given by Lillywhite *et al.* (69) for a toad (*Bufo boreas*). Figure 5 shows gross conversion efficiencies for growth of toads maintained with excess food at different temperatures; there is a maximum at 27°C. This corresponds to values selected by toads in a thermal gradient with excess food. Starved toads selected more variable temperatures with means of 20°C or lower. This suggests torpor occurred only when energy availability was limited. A selective advantage occurred for toads with limited energy availability since they survived longer at lower environmental temperatures (69).

It is not clear if the temperature for maximum rate of growth shifts with reduced food ration for toads. However, Lillywhite *et al.* (69) suggest toads could benefit from selecting lower temperatures with a reduced level of food intake. Toads maintained at a low temperature (14°C) with unlimited food but given access to a heat source for 5 hours a day achieved significant increases in growth, while 1 hour access to the heat source was insufficient. Swift (93) noted a similar effect for the larvae of an aquatic invertebrate (*Chaoborus trivitatus*), and other studies indicate an effect of temperature fluctuations on rate of growth

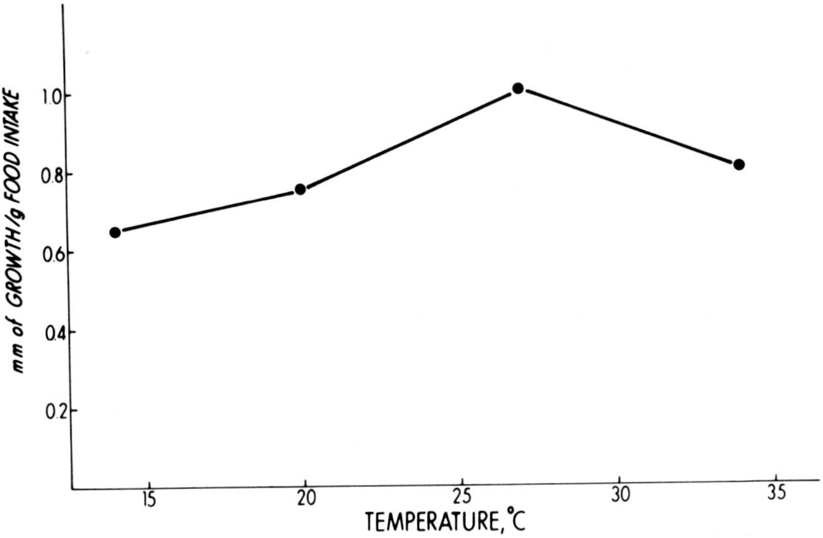

Fig. 5. *Gross conversion efficiencies for linear growth as a function of temperature for toads (Bufo boreas) maintained at the indicated temperatures for four weeks with excess food. From Lillywhite et al. (69), by permission of Ecological Society of America.*

(see 81). Thus, rate of growth may be maximized by the length of time (above some minimum) that an optimum temperature is selected depending on food intake. Such a phenomenon could also occur in fish and may explain selection of an optimum temperature by salmon during the period between dusk and dawn when feeding does not occur (Fig. 4; 13). It does not explain the lower temperatures selected by these salmon during the day which may be related to predation relative to time required for food processing (see below).

Experiments by Regal (84) suggest that temperature selection by reptiles could also have an energetic cost:benefit basis. He found the Tb pattern shown in figure 6 for two *Klauberina riversiana* that had been food deprived for three days. There was a nocturnal selection of lower temperatures which Regal (84) termed "voluntary hypothermia". The pattern suggests minimization of energy expenditures overnight with torpor and selection of higher temperatures during the day when feeding usually occurs.

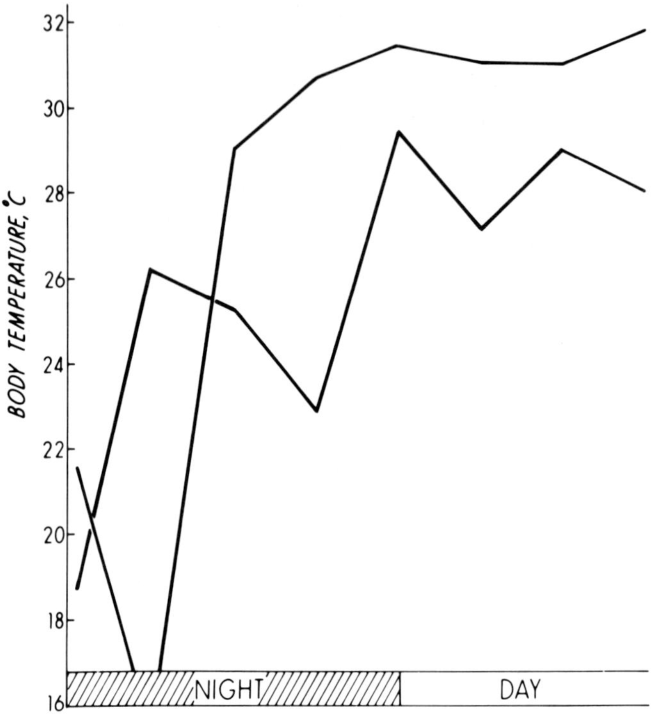

Fig. 6. *Body temperatures selected by two Klauberina riversiana in a thermal gradient after food deprivation for three days. From Regal (84), by permission of American Association for the Advancement of Science.*

It is difficult to place this behavioral torpor for reptiles in a complete cost:benefit context since data apparently are not available that permit relating it directly to rate of growth, survival, or reproduction, and it would be interesting to define optimal situations relative to food availability. For example, selection of high temperatures during the day may be modified under extreme energy restriction, or nocturnal torpor may not occur if the organism has sufficient energy reserves. Cogger (25) recently interpreted variations in observed Tb of Mallee Dragons (*Amphibolurus fordi*) relative to variations in food availability under natural conditions and emphasized the importance of assessing energetic state since a lowering of Tb "can be induced at any time by even mild starvation." Many of the data for reptiles concerning preferred Tb (e.g., 26) could be reassessed profitably if the impact of energy availability were known. For example, perhaps the "surprising" result (95) described by Wilhoft and Anderson (109) of a shift to a lower preferred Tb by *Sceloporus occidentalis* after high temperature acclimation would not be so surprising if interpreted relative to energetic costs and benefits.

2. The Constraint of Activity

Net energy benefits are influenced by energy expenditures for activity as well as food availability since many animals move to feed. Ware (100) analyzed the impact of swimming activity in fish as a cost:benefit problem from data of Ivlev (58) for the bleak (*Alburnus alburnus*). The cost of swimming to feed could influence growth rates (rates of net energy gain), but food intake would increase as a function of swimming speed and food availability. Thus, there could be an optimum expenditure associated with activity to maximize the rate of net energy gain. Ware (100) examined relationships between costs and benefits taking into account information for various metabolic rates. For the data of Ivlev (58) he calculated a maximum net energetic benefit at a swimming speed of 107 m/hr. Observed swimming speed was 111 m/hr. Also, there are theories suggesting that schooling behavior could reduce costs for locomotion by reducing drag on individual fish (8, 101, 102). To the extent that schools are involved in feeding, this could influence rates of net benefit from food.

3. The Constraint of Prey Selection

If different prey are located in different thermal environments, food selection by ectotherms could also influence Tb, energy expenditures, and rates of net energy gain. The costs of prey selection in different environments must be weighed relative to the different benefits of prey. Most economic studies of prey selection by ectotherms deal with prey quality independent of costs determined by environmental temperature. Werner and Hall (103) demonstrated that bluegill sunfish (*Lepomis macrochirus*) selected sizes of prey that maximized benefits relative to foraging time. Eggers (30) found similar relationships from re-analysis of data on alewives (*Alosa pseudoharengus*). It would be interesting to examine this behavior in situations where environmental temperatures influenced rate of net benefit from food selection.

4. The Constraint of Food Processing

Physiological processes involved in food assimilation also could influence rates of net energy gain and could form a basis for temperature selection. Food remains external to an organism until it is assimilated, and the assimilation rate should be temperature dependent. For example, Harlow *et al.* (47) indicated that assimilation is strongly influenced by temperature for the lizard *Dipsosaurus dorsalis* with zero assimilation at 28°C but 70% assimilation at 41°C (the preferred Tb for this species). Regal (83) described temperature dependent behavior of reptiles that also suggests an importance for food processing. For example, a constrictor moved with respect to a heat lamp so the bolus of food remained below a heat lamp during its passage through the digestive tract. In general, processing food at an optimum temperature will contribute to maximizing rate of net energy gain. The quantity of food to be processed may have an impact on duration of homeothermy and torpor.

B. Dinosaur Homeothermy and Extinction

Interpretations of a functional significance for torpor imposed by environmental limitations in ectotherms that regulate temperatures behaviorally have interesting implications relative to recent re-interpretations of thermal relations for dinosaurs. Bakker (2, 3) considered evidence that suggests some dinosaurs may have been endothermic. He divided this into cellular evidence based on bone histology suggesting high, endothermic-like vascularization; gross anatomical evidence from reconstructions of locomotor apparatus suggesting potentially high levels of activity; respiratory and possible heart anatomy suggesting a potential for high aerobic metabolism, and reconstructions of predator/prey relationships in large fossil deposits suggesting a potential similarity of dinosaur communities with modern endotherm communities.

Bakker's arguments led to a flurry of comments. Feduccia (35) provided criticism on structural grounds, but Dodson (28) and Ostrum (78) defended Bakker's hypothesis. Bennett and Dalzell (9) criticized the argument of endothermy related to posture. Spotilla *et al.* (92) developed a model which suggested that large dinosaurs could have been homeothermic without endothermy due to thermal inertia. Farlow (33) examined predator/prey relationships for the Late Cretaceous Oldman Formation community and discussed limitations of this analysis but reached no definitive conclusions concerning endothermy or ectothermy. Recent examinations of *Stegosaurus* suggest a potentially important role for their large, dorsal plates in forced convective heat loss (34), implying a necessity for increased heat loss. Also, Seymour (90) suggested that the large distance between the heart and head of some dinosaurs would have necessitated large hearts to produce pressures sufficient to pump blood to the brain, and this may have required four chambered hearts with a potential for an endothermic-like circulation.

Axelrod and Bailey (1) suggested that environmental temperature changes contributed to dinosaur extinction. This implies an excessive cost associated with endothermy relative to a possibly diminished food availability. The thermal inertia argument of Spotilla *et al.* (92) suggests that large dinosaurs could not have entered torpor behaviorally. Thus, absence of ability to enter torpor either physiologically or behaviorally may have contributed to extinction through excessive energetic costs relative to benefits.

C. Torpor in Endotherms

1. The Constraint of Food Availability

Students of torpor in endotherms have assumed that reduced expenditures could be generally beneficial if they reduced maintanence costs and contributed to a greater net benefit from ingested energy (e.g., 54, 73, 110). Yet the preceding suggests that torpor is to be expected only when energy availability is reduced relative to costs. Evidence on the functional significance of torpor in birds is consistent with this although an energetic interpretation for both endoderms and ectotherms is only one possibility (see below).

Most information comes from studies of hummingbirds, so we will review briefly the energetic characteristics of torpor for these organisms (see 27 for a recent general review). Typical metabolic responses of homeothermic and torpid hummingbirds to Ta are illustrated in figure 7. The "thermostatic" heat production that occurs during torpor results in maintenance of a controlled Tb around a reduced "set point". This pattern has been observed in five species from a variety of habitats (*Eulampis jugularis*, 41; *Panterpe insignis* and *Eugenes fulgens*, 110; *Oreotrochilus estella*, 23; *Archilochus alexandri*, 42). The lack of observation of this phenomenon in earlier studies (64, 65) can be attributed to methodological problems (42), so regulation in torpor with its indication of precise, physiological control probably is a general characteristic of torpor in hummingbirds.

The temperature at which regulation occurs varies between species (from as high as 20°C for *Eulampis* to as low as 5°C for *Oreotrochilus*). Wolf and Hainsworth (110) suggested that this was related to different minimum environmental temperatures, and subsequent information supports this (23, 42). This suggests environmentally determined optimum temperatures for torpor similar to optimum temperatures for other species where environmental constraints do not allow a Tb of 35-40°C. Also, regulation in torpor is not restricted to hummingbirds. For example, brook trout and rainbow trout (59) select lower temperatures during starvation that are greater than the minimum available, constant from day to day, and different between species. Differences between species in temperature for regulation in torpor perhaps could be related to distinct temperature optima for enzyme function set by environmental temperatures (see 91).

Various observations indicate that hummingbirds do not necessarily enter torpor (18, 53, 64). Lasiewski (64) noted that food deprivation was necessary to induce torpor in the laboratory, and Calder and Booser (18) related torpor for an incubating broadtailed hummingbird (*Selasphorus platycercus*) to inclement weather that reduced foraging time. These observations suggest that torpor occurs only after depletion of energy reserves, but, as Dawson and Hudson (27) note, it is difficult to verify this hypothesis without an independent measure of the energetic state of the birds.

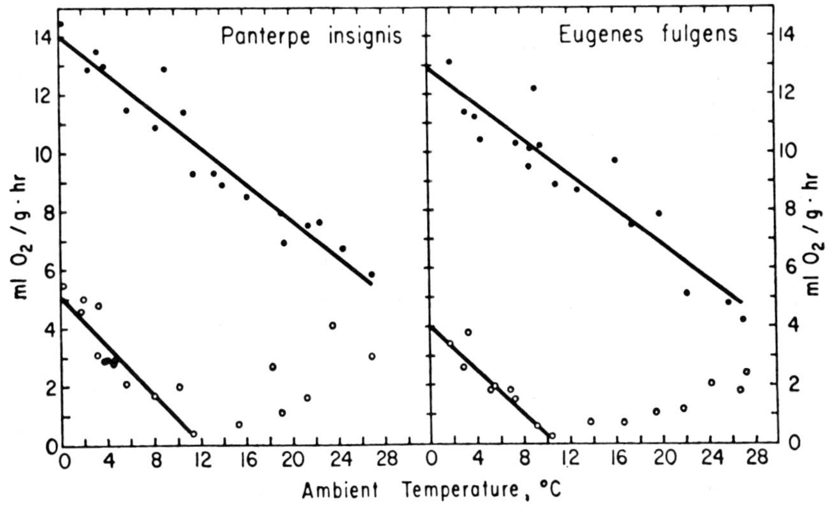

Fig. 7. *Metabolic rates for two hummingbird species from a highland area of Costa Rica as a function of Ta. Closed symbols denote homeothermic condition and open symbols denote torpid condition. From Wolf and Hainsworth (110).*

We have attempted this by using body mass an an index of energy storage (43). We monitored oxygen consumption overnight and varied energy expenditures during the day and overnight using various photoperiods and temperatures. Measurements of body mass just prior to light onset indicated that torpor occurred only in association with low mass. There appeared to be a characteristic lower "threshold" of energy storage, and the extent of energy reserves predicted if and for how long torpor occurred overnight for one species (43). These results support the hypothesis that torpor occurs only when energy availability (in terms of stored reserves) is restricted relative to expenditures. A similar interpretation applies to daily torpor for small mammals (15, 35, 97) and, with suitable timing factors, to hibernation (36).

2. Other Constraints

Constraints from activity, prey selection, and food processing also can apply with some modifications to endotherms. To the extent that energy storage is influenced by expenditures to feed, activity will influence costs, rate of net benefit (111) and ultimately temperature regulation. Differences in prey quality will also influence rates of net energy gain primarily by affecting benefits (111). Food processing would least influence rate of net energy gain for a given prey type and availability, although differences in prey quality can influence rate of net benefit from digestive processing (111).

D. The General Constraint of Body Mass

To the extent that mass determines energy expenditure, mass could also influence temperature regulation and torpor by affecting rate of net energy gain through use of available energy. Due to mass related effects on heat loss, rate of expenditure per gram at a constant Ta increases with decreasing body mass (e.g., 71, 88). With a mass related effect on energy storage for non-feeding periods this could result in a temperature during these periods lower than 35-40°C. But, some organisms of small mass (such as hummingbirds) may be able to store sufficient energy under some conditions to meet energetic costs during non-feeding periods (see above; 112).

At the other extreme, large body mass may limit ability to influence rates of net energy gain from torpor since the cost in time to heat a large mass could be prohibitive and could restrict torpor at least to seasonal time (79). However, due to differences in mass-related metabolic rate [cal/time, scaled to $(\text{mass})^{0.75}$] and mass related digestive capacity [scaled to $(\text{mass})^{1.0}$] (17), larger animals perhaps could achieve energy storage for non-feeding periods more easily than small species. Fisher and Mannery (36) presented a similar argument to explain the frequency of periodic arousals for some relatively large species of hibernators. When applied to non-hibernators this suggests that large size could permit a greater rate of net benefit such that energy processing could take place at an efficient endothermic temperature of 35-40°C. Of course, these processes ultimately have to be viewed within limitations set by environmental variables such as food availability, predation, and interspecific interactions and the requirements relative to these that are dictated by mass (89, 111).

IV. THE PREDATION HYPOTHESIS

The assumption for this hypothesis is that physiology and behavior should be sensitive to predation since it has a dramatic effect on fitness. In general, it is difficult to directly assess the impact of predation on costs and benefits. The best comparison is between individuals that do not minimize predation and those that do to determine the impact of the difference on rates of net energy gain. However, predation can be such a potent selective force that the behaviors or functions involved may have a strong genetic basis, and this makes observation of the null hypothetical condition very difficult.

In an economic discussion of lizard thermoregulation, Huey and Slatkin (56) recognized the possible general importance of predation for behavioral temperature regulation. They related the potential impact of predation to the costs for thermoregulation (generally measured as distance to suitable basking sites 55, 57), but they did not consider any evidence. Most evidence comes from application of economic models to natural conditions where observed rates of net energy gain are less than predicted; the difference is usually attributed to predation (30, 93).

Swift (93) examined the hypothesis that vertical migration patterns of aquatic invertebrates (associated with temperature stratifications) contribute to maximizing rates of net energy gain due to different effects of temperature on energetic costs and benefits (60, 72). He constructed an energy budget for *Chaoborus trivitatus* larvae based on temperature effects on respiration and rate of growth, energetic costs for migration, and benefits from feeding which were estimated for a variety of prey densities. From a computer simulation Swift compared rate of growth using parameters associated with natural migrations and with values associated with staying near the surface where temperatures were higher. Both were examined as a function of prey density near the surface. Swift discounted the energetic hypothesis since larvae would always have had greater rates of net energy gain if they stayed in warmer temperatures near the surface. Their selection of lower temperatures (torpor) through migration was attributed to predation.

We do not believe Swift is justified in accounting for this torpor only on the basis of a predator effect. For example, the extent of the energetic benefit depended importantly on prey density. If larvae obtained a low food ration and remained near the surface until it was processed, they could optimize rates of net energy gain within a predation constraint by migrating to lower temperatures. The data of Swift (93) suggest this since a fluctuation between 5°C (18 hours) and 20°C (6 hours) resulted in increased rate of growth compared with a constant 5°C condition at high and low food availability. Thus, temperature selection behavior of these larvae might best be interpreted with respect to both the energetic and the predation hypotheses.

A similar integrative interpretation was drawn by Eggers (30) from a detailed examination of the energetic hypothesis of Brett (13) for salmon. From a computer simulation of natural parameter effects on costs and benefits, Eggers found that salmon selected temperatures and environmental conditions that did not necessarily maximize rates of net energy gain. He also noted evidence indicating that vertical migration can occur in the absence of temperature stratification. Eggers considered predation as a possible explanation, but he interpreted the effect within the constraint of energy requirements primarily on the basis of selection of optimum temperatures by salmon for part of a day (13).

Eggers also considered schooling behavior in salmon which could influence both predation and rates of net energy gain from feeding. A school could provide a benefit in reducing the probability of predation on an individual (30, 80). To the extent that schools occupy specific temperature environments, this behavior could also influence temperature regulation and torpor. Eggers (30) analyzed the effect of schooling on rates of net energy gain for individuals in a feeding assemblage at uniform prey density. Although he did not consider potential energy savings from schooling, he persuasively argued that individual rates of net energy gain in schools would decrease with decreased distance to nearest neighbor (from overlapping fields of response to prey) and increased distance from the front of the school (from reduced prey availability). Predictions indicated that natural school sizes and densities for salmon would drastically lower individual rates of net energy gain if prey were randomly distributed. This strongly suggests a predation effect on schooling, although schooling still could provide net energetic benefit for individuals if prey were non-randomly distributed (21, 24, 30, 96).

The predation hypothesis applies to torpor in a variety of organisms although available evidence may not permit distinction from the energetic hypothesis. For example, a threshold phenomenon for torpor relative to energy reserves (such as in hummingbirds), can be interpreted according to a cellular energetic efficiency hypothesis (20) or relative to a predation risk associated with torpor (43). Due to a possible genetic effect for the phenomenon it may be difficult to distinguish the hypotheses.

V. SOCIAL INTERACTIONS

Much of the preceding is cast as isolated individuals making "decisions" about appropriate temperatures from an available range. However, no organism operates in a biological vacuum but continually is influenced positively and negatively by others, both conspecifics and different species. It is important to the understanding of temperature regulation in natural situations to determine how these interactions may change individual response. Predators already were considered, primarily in how they might drive an individual away from optimal conditions. Other interactions among individuals are equally possible and all can be considered relative to the question of optimization with respect to Tb. We will discuss two principal types of intraspecific interactions: aggression and group formation. Each also could occur interspecifically and individual identity makes little difference to our discussion.

Aggression often takes the form of displacement of an individual by another. It potentially could drive an individual away from a location where it could maintain an optimum temperature. We do not address why one individual displaces another; we only point out that the displaced individual may be forced to drop off its optimality peak as a result.

Beitinger and Magnuson (7) showed that an adult bluegill sunfish (*Lepomis macrochirus*) was able to change the position of juveniles relative to temperature selected by each juvenile when alone. The juveniles individually selected 31°C, but when an adult was added (and restricted) to the 31°C portion, the young only made short movements into the 31°C area to feed, spending nearly 100% of their time at 27°C or 34°C, whichever was the available alternative. Other workers (e.g., 4) have also discussed the effect of aggressive interactions among fish on realized temperature preferences. In general, the outcome is for the dominant individual to drive subordinates away from their optimum temperature.

Another type of social interaction is for individuals to associate for benefits accruing to each as a result. This could have a variety of influences on Ta selection. For example, group formation could enhance rate of food acquisition in a patchy environment (24,

46, 99). Since temperature choice should be related to levels of food availability and rates of energy intake, group foraging may be an important component in predicting optimal temperature selection. Given variation among environmental sites in the quality and/or quantity of food, it is possible that an organism may have a choice to join a group at other than a homeothermic temperature to increase its rate of net energy gain. Ultimately this temperature environment should be one at which the increased rate of food intake compensates for the costs due to selecting a non-optimal temperature relative to that for solitary feeding. As far as we know, there is no experimental information on this possibility.

Alternatively, a group could result in a decreased rate of individual energy expenditure and a potentially increased rate of net energy gain (assuming rate of intake remains constant or decreases less than expenditures). For example, huddling has been reported to produce energy savings (e.g., 40, 81), primarily for organisms maintaining a high Tb. This behavior also could reduce losses from individuals with variable Tb at ambient temperatures below an optimum (6). There is evidence that a similar phenomenon occurs in young birds and mammals that have not developed a regulated Tb. Rate of energy loss per young is reduced at a low Ta as number of young per nest increases (29, 40, 51, 74, 77), at least to a potential limit set by nest structure. (The converse is the restricted ability to facilitate heat loss as Ta increases; 77). Reduced heat loss by young means that the internal temperature of each may approximate more closely to an optimum with less energy input from the adult from brooding and, probably more importantly, a greater proportion of food can be channelled into growth by young (see 85). This external temperature regulation may be important in determining the optimum clutch size of altricial young as the rate of food delivery by an adult does not increase equally for each additional young (assuming growth rates remain approximately constant) (77, 85, 86).

Finally, an individual might be able to use others as cues for locations of preferred temperatures. This has been examined in a number of contexts associated with "cuing" for feeding, such as areas in which to feed (61, 63), parasitizing individuals as they feed [either stealing food (e.g., 39) or displacing individuals from good feeding locations (21, also, pers. obs.)] and which food items to take from an available array (75). It may be less likely that an individual would be unable to ascertain independently locations of preferred temperature sites, although movements through a gradient may be speeded by the presence of individuals at preferred temperature locations (6). In contrast to the situation described earlier in which aggressive interactions limited available choices, cuing behavior assumes few restrictions on possible locations used by an individual.

VI. TEMPORAL ENVIRONMENTAL VARIATION

We have been discussing primarily organisms that attempt to maximize some benefit:cost function by selecting positions in the environment where Tb is optimum. This selection depends on the organism's mobility and the availability of alternative spatial temperature conditions. We assume that the choice of Ta influences Tb and generally reduces the cost term in the benefit to cost function. We implicitly assumed that an organism, by making a selection, reduced its rate processes associated with energetic expenditures, with relatively less impact on energetic uptake.

For many organisms the opportunity to modify Tb by actively selecting among alternative environments is not available, either because the organism is sedentary or because the degree of mobility does not permit exploitation of locations with different available temperatures. For mobile organisms we have already indicated that there are constraints on selecting among alternatives either because of the costs for movement or because of differential benefits in alternative locations. Many organisms are faced with temporal changes in temperature where they have no choice, either on a long term (e.g., seasonal) or a short term (e.g., daily) basis.

The ecological concept of environmental "grain" influences our view of the role of temporal variation in environmental temperature. Grain relates to the probability of use of environmental patches (in this case, temperature conditions through time) in relation to their frequency of occurrence (66, 106). A fine-grained environment is one that is utilized by an individual essentially randomly so the frequency of patch use is similar to the frequency of occurrence. A coarse-grained environment is one that has a marked differential between the frequency of occurrence and use of patches. An organism that theoretically has access to several temperatures but remains in one would view the environment as coarse-grained. Grain becomes an important aspect of physiological responses for organisms that have no short-term choice in terms of habitat selection. We can imagine long-lived temperate zone organisms that encounter regularly, but gradually, the full range of variations. These would view the environment as fine-grained. On the other hand short-lived organisms could occur in one temperature condition for a generation, but subsequent generations may occur in different conditions, a more coarse-grained view of the environment. The critical question then is: how does a stationary organism cope with the environmental fluctuations?

In fine-grained environments organisms are likely to respond by allowing Tb to fluctuate with Ta if that regularly maximizes the net benefit associated with determining Tb. However, this does not mean that all organisms faced with short-term fluctuations will show a variable Tb. As noted earlier, the probability of short-term shifts in Tb is related importantly to body size and the potential benefits to be derived when costs are reduced.

Temporal fluctuations in temperature allow organisms a choice of temporal pattern of the types and magnitudes of their functional activity. This is conceptually very similar to choosing among available spatial locations of differing temperatures and should be governed by the same principles as outlined earlier. In the short-term, we certainly would expect organisms to have higher levels of functional activity at times when they can maximize their rate of net energy gain (or some other index of fitness) and that may not always occur when Ta would appear to be at a level that maximizes rate processes that influence both benefits and costs.

Seasonal fluctuations in Ta also can provide an opportunity to choose what levels or magnitudes of functional activity to maintain. For organisms with very short generation times the environment becomes more and more coarse-grained and requires less seasonal choice. However, this does not eliminate shorter term temperature cycles or the imposition of a short-term fine-grained environment on a longer term coarse-grained environment. Organisms with longer generation times presumably can selectively utilize portions of the seasonal temperature cycle in an attempt to maximize net benefits. Which portion of the temperature cycle maximizes benefits relative to costs will depend on the relative ability of the organisms to function at various temperatures. Torpor, hibernation, and diapause, among others, are mechanisms for reducing costs when benefits are likely to be low while other organisms function at reasonably high levels of activity even in the face of fluctuating temperatures. Which technique is used will depend on the evolutionary assessment of the benefits and costs of the alternatives.

Organisms that have evolved to utilize the temporally fluctuating temperature environment while maintaining high levels of function have approached the problem in two rather distinct ways: 1) maintenance of a constant, generally high body temperature, and 2) adjustments of the controls of rate processes so that high levels of function are maintained even in the face of varying Tb. We have discussed the first alternative, and we will now consider the second. We shall discuss non-behavioral adaptations; also, for brevity we will not consider in detail such important components of environmental variation as the influence of the variation pattern, both temporally and in relation to the mean value. It must be remembered that these will play an important role in determining which adaptations maximize fitness for these organisms.

While a variety of levels of physiological organization are amenable to interpretation of this general question, we will consider what may be the crux of the energetic rate processes consideration: the ability of enzymes to catalyze reactions associated with energy processing and acquisition. Rather than attempt to deal with many substrate-enzyme systems in detail, we will focus on general aspects of the interaction as they relate to the ability of an organism to optimize temperature regulation in a temporally fluctuating environment.

In general, this view of temperature can be associated with a mechanistic explanation if we assume that there is a single temperature at which each enzyme has a maximal activity. In general, peak enzyme activity seems to be at or near preferred Tb (e.g., for lizards, see 26, 67); the enzyme activity usually falls off rapidly at higher temperatures and more gradually at suboptimal temperatures. A lower or a higher temperature than the optimum slows rates of enzymatic reactions (52) or perhaps other rate limiting mechanisms (107, 108). Adopting a suboptimal rather than a supraoptimal temperature generally puts the organism on the less steep portion of the temperature-activity curve and presumably makes it less susceptible to death from further variations in environmental temperature and enzyme degradation. Selecting supraoptimal temperatures could also increase costs by speeding degradation of enzymes which would have to be replaced.

Four general enzyme-substrate responses are possible to maintain rate processes in the face of varying environmental temperature: 1) changes in enzyme concentrations, perhaps mediated through induction or arrays of modifier genes; 2) production of different general configurations of a single enzyme as a function of temperature, called modulation by Hochachka and Somero (52); 3) changes in substrate availability; and 4) production of different enzymes from different allelic forms at one or more loci. Clearly, these are not mutually exclusive, but are general classes of possible responses.

The fourth mechanism is perhaps the most important, especially if conditions vary predictably with sufficient time to turn structural genes on and off. The organism has adapted to different temperature conditions by having several forms of a gene which may differ in the temperature of peak activity. We can speculate that genetic information for two forms of an enzyme may be a more efficient way of handling predictable temporal changes in temperature of large magnitude than mechanisms 1-3. Small magnitude or unpredictable changes of short duration probably are more efficiently dealt with via changes within a structural gene rather than by multiple structural genes. The primary consideration is that the organism maintains approximately equivalent rates of enzyme activity even though it is forced into different temperature environments. If the difference among environmental conditions is large, reduced fitness associated with allozyme production and maintenance of the genetic heterozygosity may be sufficiently low energetically and genetically that allozymes can be adaptive in many situations. However, as differences among environmental conditions decrease, the efficiency of mixed forms of an enzyme presumably also decreases. This will be strongly dependent on the activity versus temperature curves for each enzyme, a characteristic that varies, presumably adaptively, among enzymes.

Heterozygosity within individuals and polymorphism within the population can be maintained by balancing selection. In addition, a variety of models attempt to predict levels of genetic polymorphism at a single locus as a function of temporal environmental variation (e.g., 32, 44, 48). In these models the maintenance of polymorphism depends strongly on the pattern and predictability of variation (Fig. 8). Numerous models have pointed out that environmental variation can maintain population polymorphisms by selection that favors one or the other homozygote alternatively through time. However, in general, genetic variation within a population is tied more importantly to spatial variation than to temporal variation (48).

In fine-grained environments, heterozygotes often will be able to respond better to the fluctuations, while in coarse-grained environments either homozygosity or heterozygosity might be advantageous (66, 94). Homozygosity could yield high short-term fitness, but decrease long-term fitness if the environment changes. Heterozygosity is a compromise. Which is better in the long-term will depend on rate and variance of environmental change and the autocorrelation (Fig. 8). However, selection presumably acts only on the short-term fitness; foresight in genetic constitution would be impossible to produce with individual selection.

A major remaining problem of enzyme adaptation to varying environments is the issue of mechanisms of adaptation and any apparent correlation of mechanism

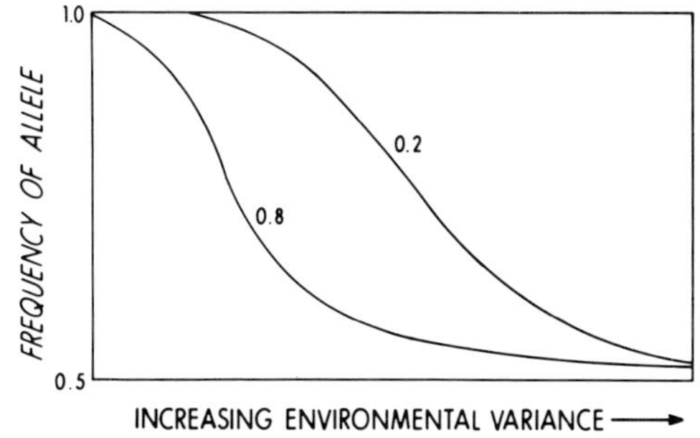

Fig. 8. *A hypothetical relationship between the frequency of a particular gene in a two allele system and the amount of environmental variation. The two curves represent different degrees of environmental predictability estimated by the level of autocorrelation of the environment. A very predictable environment would have a value of 1.0 while a completely unpredictable environment would have a value of 0.0. Modified from Wiens (106) and Bryant (16).*

with patterns of environmental variability, both in terms of predictability and the level of variation about the mean environmental value. For example, while activity levels may differ with temperature for different enzyme forms, it is not clear how much is due to modulation or to simply structural differences that are coded by two alleles. To resolve these questions will require further efforts to refine the actual functional aspects of the enzyme-substrate interaction. K_m or V_{max} values do not provide sufficient information to discriminate among alternative adaptive responses to shifting environmental conditions. Until these details are available it will be difficult to test hypotheses about the adaptive roles of the four possible shifts in enzyme activity in relation to environmental variation. Unfortunately, to understand the mechanistic underpinnings of the benefit:cost model, these details will be required ultimately.

ACKNOWLEDGEMENTS

The preparation of this paper was supported by grants from the National Science Foundation (U.S.A.). We thank Georgia Ventura, Stephanie Petrarca, and Terre Mercier for their assistance, and Drs. P. DeBenedictis, R. Huey, P. Licht, L. Wang and J. Hudson for comments on an earlier draft of the manuscript.

REFERENCES

1. Axelrod, D. I., and H. P. Bailey. Cretaceous dinosaur extinction. Evolution 22: 595-611, 1968.
2. Bakker, R. T. Dinosaur physiology and the origin of mammals. Evolution 25: 636-658, 1971.
3. Bakker, R. T. Anatomical and ecological evidence of endothermy in dinosaurs. Nature 238: 81-85, 1972.
4. Barans, C. A., and R. A. Tubb. Temperatures selected seasonally by four fishes from western Lake Erie. J. Fish. Res. Bd. Can. 30: 1697-1703, 1973.
5. Bartholomew, G.A. Aspects of timing and periodicity of heterothermy. In: Hibernation and Hypothermia, Perspectives and Challenges. South, F. E., J. P. Hannon, J. R. Willis, E. T. Pengelley, and N. R. Alpert (editors). New York: Elsevier, 1972, p. 663-680.
6. Beiswenger, R. E. Diel patterns of aggregation behavior in tadpoles of *Bufo americanus*, in relation to light and temperature. Ecology 58: 98-108, 1977.
7. Beitinger, T. L., and J. J. Magnuson. Influence of social rank and size on thermoselection behavior of bluegill (*Lepomis macrochirus*). J. Fish. Res. Bd. Can. 32: 2133-2136, 1975.
8. Belyayev, V. V., and G. V. Zuyev. Hydrodynamic hypothesis of school formation in fishes. J. Ichthyol. 9: 578-584, 1972.
9. Bennett, A. F., and B. Dalzell. Dinosaur physiology: A critique. Evolution 27: 170-174, 1973.
10. Bogert, C. M. Thermoregulation in reptiles, a factor in evolution. Evolution 3: 195-211, 1949.
11. Brattstrom, B. H. Body temperature of reptiles. Am. Midl. Nat. 73: 376-422, 1965.
12. Brattstrom, B. H. Amphibia. In: Comparative Physiology of Thermoregulation. Whittow, G. C. (editor). New York: Academic Press, Vol. I, 1970, p. 135-166.

13. Brett, J. R. Energetic responses of salmon to temperature. A study of some thermal relations in the physiology and freshwater ecology of sockeye salmon (*Oncorhynchus nerka*). Amer. Zool. 11: 99-114, 1971.
14. Brett, J. R., J. E. Shelbourn, and C. T. Shoop. Growth rate and body composition of fingerling sockeye salmon, *Oncorhynchus nerka*, in relation to temperature and ration size. J. Fish. Res. Bd. Can. 26: 2363-2394, 1969.
15. Brown, J. H., and G. A. Bartholomew. Periodicity and energetics of torpor in the kangaroo mouse, *Microdipodops pallidus*. Ecology 50: 705-709, 1969.
16. Bryant, E. H. Habitat selection in a variable environment. J. theor. Biol. 41: 421-429, 1973.
17. Calder, W. A. Consequences of body size for avian energetics. In: Avian Energetics. Paynter, R. A., Jr. (editor). Cambridge: Publ. Nuttall Ornithol. Club No. 15, 1974, p. 86-144.
18. Calder, W. A., and J. Booser. Hypothermia of Broad-tailed hummingbirds during incubation in nature with ecological correlations. Science 180: 751-753, 1973.
19. Calder, W. A., and J. R. King. Thermal and caloric relations of birds. In: Avian Biology. Farner, D. S. and J. R. King (editors). New York: Academic Press, Vol. IV, 1974, p. 260-413.
20. Calloway, N. O. Body temperature: Thermodynamics of homeothermism. J. theor. Biol. 57: 331-344, 1976.
21. Caraco, T. Ecological regulation of social system stochastic dynamics. Ph.D. Thesis, Syracuse Univ., 1977.
22. Carey, F. G., and J. M. Teal. Regulation of body temperature by the bluefin tuna. Comp. Biochem. Physiol. 28: 205-213, 1969.
23. Carpenter, F. L. Torpor in an Andean hummingbird: Its ecological significance. Science 183: 545-547, 1974.
24. Cody, M. L. Finch flocks in the Mojave desert. Theor. Pop. Biol. 2: 142-158, 1971.

25. Cogger, H. G. Thermal relations of the Mallee Dragon *Amphibolurus fordi* (Lacertilla: Agamidae). Aust. J. Zool. 22: 319-339, 1974.

26. Dawson, W. R. On the physiological significance of the preferred body temperatures of reptiles. In: Perspectives of Biophysical Ecology, Ecological Studies. Gates, D. M. and R. B. Schmerl (editors). New York: Springer-Verlag, Vol. 12, 1975, p. 443-473.

27. Dawson, W. R., and J. W. Hudson. Birds. In: Comparative Physiology of Temperature Regulation. Whittow, G. C. (editor). New York: Academic Press, Vol. 1, 1970, p. 223-310.

28. Dodson, P. Dinosaurs as dinosaurs. Evolution 28: 494-497, 1974.

29. Dunn, E. H. The relationship between brood size and age of effective homeothermy in nestling House Wrens. Wilson Bull. 88: 478-482, 1976.

30. Eggers, D. A synthesis of the feeding behavior and growth of juvenile sockeye salmon. Ph.D. Thesis, Univ. Wash., 1975.

31. Elliot, J. M. The growth rate of brown trout (*Salmo trutta* L.) fed on reduced rations. J. Anim. Ecol. 44: 823-842, 1975.

32. Ewing, E. P. Genetic variation in a heterogeneous environment. V. Temporal and spatial heterogeneity in infinite populations. Amer. Nat. (In press).

33. Farlow, J. O. A consideration of the trophic dynamics of a Late Cretaceous large dinosaur community (Oldman Formation). Ecology 57: 841-857, 1976.

34. Farlow, J. O., C. V. Thompson, and D.E. Rosner. Plates of the dinosaur *Stegosaurus*: Forced convection heat loss fins? Science 192: 1123-1125, 1976.

35. Feduccia, A. Dinosaurs as reptiles. Evolution 27: 166-169, 1973.

36. Fisher, K. C., and J. F. Mannery. Water and electrolyte metabolism in heterotherms. In: Mammalian Hibernation III. Fisher, K. C., A. R. Dawe, C. P. Lyman, E. Schönbaum, and F. E. South, Jr. (editors). New York: American Elsevier, 1967, p. 235-279.

37. French, A. R. Selection of high temperatures for hibernation by the pocket mouse, *Perognathus longimembris*: Ecological advantages and energetic consequences. Ecology 57: 185-191, 1976.
38. Fry, F. E. J., and P. W. Hochachka. Fish. In: Comparative Physiology of Thermoregulation. Whittow, G. C. (editor). New York: Academic Press, Vol. 1, 1970, p. 79-134.
39. Fuchs, E. Kleptoparasitism of Sandwich Terns *Sterna sandvicensis* by Black-headed gulls *Larus ridibundus*. Ibis 119: 183-190, 1977.
40. Glaser, H., and S. Lustick. Energetics and nesting behavior of the northern white-footed mouse, *Peromyscus leucopus noveboracensis*. Physiol. Zool. 48: 105-113, 1975.
41. Hainsworth, F. R., and L. L. Wolf. Regulation of oxygen consumption and body temperature during torpor in a hummingbird, *Eulampis jugularis*. Science 168: 368-369, 1970.
42. Hainsworth, F. R., and L. L. Wolf. Regulation of metabolism during torpor in "temperate" zone hummingbirds. Auk 95: 197-199, 1978.
43. Hainsworth, F. R., B. G. Collins, and L. L. Wolf. The function of torpor in hummingbirds. Physiol. Zool. 50: 215-222, 1977.
44. Haldane, J. B. S., and S. D. Jayakar. Polymorphism due to selection of varying direction. J. Genet. 58: 237-242, 1963.
45. Hamilton, W. J. III. Life's Color Code. New York: McGraw-Hill, 1973, 238 pp.
46. Hamilton, W. J. III, and K. E. F. Watt. Refuging. Ann. Rev. Ecol. Syst. 1: 263-297, 1970.
47. Harlow, H. J., S. S. Hillman, and M. Hoffman. The effect of temperature on digestive efficiency in the herbivorous lizard, *Dipsosaurus dorsalis*. J. comp. Physiol. 111: 1-6, 1976.
48. Hedrick, P. W., M. E. Ginevan, and E. P. Ewing. Genetic polymorphism in heterogeneous environments. Ann. Rev. Ecol. Syst. 7: 1-32, 1976.
49. Heinrich, B. Temperature regulation in the bumblebee, *Bombus vagans*. Science 175: 185-187, 1972.

50. Heinrich, B. Why have some animals evolved to regulate a high body temperature? Amer. Nat. 111: 623-640, 1977.
51. Hill, R. Ontogeny of homeothermy in neonatal *Peromyscus leucopus*. Physiol. Zool. 49: 292-306, 1976.
52. Hochachka, P. W., and G. Somero. Strategies of Biochemical Adaptations. Philadelphia: W. B. Saunders, 1973, 358 pp.
53. Howell, T. R. and W. R. Dawson. Nest temperatures and attentiveness in the Anna hummingbird. Condor 56: 93-97, 1954.
54. Hudson, J. W. Torpidity in mammals. In: Comparative Physiology of Thermoregulation. Whittow, G. C. (editor). New York: Academic Press, Vol. III, 1973, p. 97-165.
55. Huey, R. B. Behavioral thermoregulation in lizards: Importance of associated costs. Science 184: 1001-1003, 1974.
56. Huey, R. B., and M. Slatkin. Cost and benefits of lizard thermoregulation. Quart. Rev. Biol. 51: 363-384, 1976.
57. Huey, R. B., and T. P. Webster. Thermal biology of *Anolis* lizards in a complex fauna: The *cristatellus* group of Puerto Rico. Ecology 57: 985-994, 1976.
58. Ivlev, V. S. On the utilization of food by planktophage fishes. Bull. Math. Biophys. 22: 371-389, 1960.
59. Javaid, M. Y., and J. M. Anderson. Influence of starvation on selected temperature of some salmonids. J. Fish. Res. Bd. Can. 24: 1515-1519, 1967.
60. Kerfoot, W. B. Bioenergetics of vertical migration. Amer. Nat. 104: 529-546, 1970.
61. Kiester, A. R., and M. Slatkin. A strategy of movement and resource utilization. Theor. Pop. Biol. 6: 1-20, 1974.
62. Knutson, R. M. Heat production and temperature regulation in eastern skunk cabbage. Science 186: 746-747, 1974.
63. Krebs, J. R., M. H. MacRoberts, and J. M. Cullen. Flocking and feeding in the Great Tit *Parus major* - an experimental study. Ibis 114: 507-530, 1972.

64. Lasiewski, R. C. Oxygen consumption of torpid, resting, active, and flying hummingbirds. Physiol. Zool. 36: 122-140, 1963.
65. Lasiewski, R. C., and R. J. Lasiewski. Physiological responses of the Blue-throated and Rivoli's hummingbirds. Auk 84: 34-48, 1967.
66. Levins, R. Evolution in changing environments. Monogr. Popul. Biol. Princeton: Princeton Univ. Press, No. 2, 1968, 120 pp.
67. Licht, P., W. R. Dawson, and V. H. Schoemaker. Heat resistance of some Australian lizards. Copeia 1966: 162-169, 1966.
68. Licht, P., and A. G. Brown. Behavioral thermoregulation and its role in the ecology of the red-bellied newt, *Taricha rivularis*. Ecology 48: 598-611, 1967.
69. Lillywhite, H. B., P. Licht, and P. Chelgren. The role of behavioral thermoregulation in the growth energetics of the toad, *Bufo boreas*. Ecology 54: 375-383, 1973.
70. Lucas, E. A., and W. A. Reynolds. Temperature selection by amphibian larvae. Physiol. Zool. 40: 159-171, 1967.
71. May, M. L. Warming rates as a function of body size in periodic endotherms. J. comp. Physiol. 111: 55-70, 1976.
72. McLaren, I. A. Effects of temperature on growth of zooplankton and the adaptive value of vertical migration. J. Fish. Res. Bd. Can. 20: 685-727, 1963.
73. McNab, B. K. The economics of temperature regulation in neotropical bats. Comp. Biochem. Physiol. 31: 227-268, 1969.
74. Mertens, J. A. L. The influence of brood size on the energy metabolism and water loss of nestling Great Tits, *Parus major major*. Ibis 111: 11-16, 1969.
75. Murton, R. K. Why do some bird species feed in flocks? Ibis 113: 534-536, 1971.
76. Nagy, K. A., D. K. Odell, and R. S. Seymour. Temperature regulation by the inflorescence of *Philodendron*. Science 178: 1195-1197, 1972.

77. O'Conner, R. J. The influence of brood size upon metabolic rate and body temperature in nestling Blue tits *Parus caeruleus* and House sparrows *Passer domesticus*. J. Zool. Lond. 175: 391-403, 1975.
78. Ostrom, J. H. Reply to "Dinosaurs as reptiles." Evolution 28: 491-493, 1974.
79. Pearson, O. P. Torpidity in birds. Bull. Mus. Comp. Zool. Harvard Univ. 124: 93-103, 1960.
80. Peterson, C. H. Cruising speed during migration of the striped mullet (*Mugil cephalus* L.): An evolutionary response to predation? Evolution 30: 393-396, 1976.
81. Precht, H., J. Christophersen, H. Hensel, and W. Larcher. Temperature and Life. New York: Springer-Verlag, 1973, 779 pp.
82. Rapport, D. J., and J. I. Turner. Economic models in ecology. Science 195: 367-373, 1977.
83. Regal, P. J. Thermophilic response following feeding in certain reptiles. Copeia 1966: 588-590, 1966.
84. Regal, P. J. Voluntary hypothermia in reptiles. Science 155: 1551-1553, 1967.
85. Ricklefs, R. E. Energetics of reproduction in birds. In: Avian Energetics. Paynter, R. A. (editor). Cambridge: Publ. Nuttall Orn. Club No. 15, 1974, p. 152-297.
86. Royama, T. Factors governing feeding rate, food requirement and brood size of nesting Great Tits, *Parus major*. Ibis 108: 313-347, 1966.
87. Rozin, P. N., and J. Mayer. Thermal reinforcement and thermoregulatory behavior in the goldfish, *Carassius auratus*. Science 134: 942-943, 1961.
88. Schmidt-Nielsen, K. Animal Physiology. London and New York: Cambridge Univ. Press, 1975.
89. Schoener, T. W. Size patterns in West Indian *Anolis* lizards: 1. Size and species diversity. Syst. Zool. 18: 386-401, 1969.
90. Seymour, R. S. Dinosaurs, endothermy and blood pressure. Nature 262: 207-208, 1976.
91. Somero, G. N., and P. W. Hochachka. Biochemical adaptation to the environment. Amer. Zool. 11: 157-165, 1971.

92. Spotila, J. R., P. W. Lommen, G. S. Bakken, and D. M. Gates. A mathematical model for body temperatures of large reptiles: Implications for dinosaur ecology. Amer. Nat. 107: 391-404, 1973.
93. Swift, M. C. Energetics of vertical migration in *Chaoborus trivitatus* larvae. Ecology 57: 900-914, 1976.
94. Templeton, A. R., and Rothman, E. D. Evolution in heterogeneous environments. Amer. Nat. 108: 409-428, 1974.
95. Templeton, J. R. Reptiles. In: Comparative Physiology of Thermoregulation. Whittow, G. C. (editor). New York: Academic Press, Vol. 1, 1970, p. 167-223.
96. Thompson, W. A., I. Vertinsky, and J. R. Krebs. The survival value of flocking in birds: A simulation model. J. Anim. Ecol. 43: 785-820, 1974.
97. Tucker, V. A. Diurnal torpor and its relation to food consumption and weight changes in the California pocket mouse *Perognathus californicus*. Ecology 47: 245-252, 1966.
98. Vinegar, A., V. H. Hutchison, and H. G. Dowling. Metabolism, and thermoregulation during brooding of snakes of the genus *Python* (Reptilia, Boidae). Zoologica 55: 19-48, 1970.
99. Ward, P., and A. Zahavi. The importance of certain assemblages of birds as information centres for food finding. Ibis 115: 517-534, 1973.
100. Ware, D. M. Growth, metabolism, and optimal swimming speed of a pelagic fish. J. Fish. Res. Bd. Can. 32: 33-41, 1975.
101. Weihs, D. Hydromechanics of fish schooling. Nature 241: 290-291, 1973.
102. Weihs, D. Optimal fish cruising speed. Nature 245: 48-50, 1973.
103. Werner, E. E., and D. J. Hall. Optimal foraging and the size selection of prey by the bluegill sunfish (*Lepomis macrochirus*). Ecology 55: 1042-1052, 1974.
104. Whittow, G. C. (editor). Comparative Physiology of Thermoregulation. New York: Academic Press, Vol. I, 1970, 333 pp.

105. Whittow, G. C. Evolution of thermoregulation. In: Comparative Physiology of Temperature Regulation. Whittow, G. C. (editor). New York: Academic Press, Vol. III, 1973, p. 201-258.
106. Wiens, J. A. Population responses to patchy environments. Ann. Rev. Ecol. Syst. 7: 81-120, 1976.
107. Wieser, W. Temperature relations of ectotherms: a speculative review. In: Effects of Temperature on Ectothermic Organisms. Wieser, W. (editor). New York: Springer-Verlag, 1973, p. 1-23.
108. Wieser, W. (editor). Effects of Temperature on Ectothermic Organisms. New York: Springer-Verlag, 1973, 298 pp.
109. Wilhoft, D. C., and J. D. Anderson. Effect of acclimation on the preferred body temperature of the lizard, *Sceloporus occidentalis*. Science 131: 610-611, 1960.
110. Wolf, L. L., and F. R. Hainsworth. Environmental influence on regulated body temperature in torpid hummingbirds. Comp. Biochem. Physiol. 41: 167-173, 1972.
111. Wolf, L. L., and F. R. Hainsworth. Energy: Expenditures and intakes. In: Chemical Zoology. Brush, A. (editor). New York: Academic Press, Vol. 10, 1978, p. 307-358.
112. Wolf, L. L., and F. R. Hainsworth. Temporal patterning of feeding in hummingbirds. Anim. Behav. 25: 976-989, 1977.

Session II

Central Nervous System Regulation in Natural Torpidity

NEURONAL MODELS OF TEMPERATURE REGULATION IN EUTHERMIC AND HIBERNATING MAMMALS: AN ALTERNATE MODEL FOR HIBERNATION

Frank E. South[1]
Virginia M. Miller[2]
William C. Hartner[3]

ABSTRACT

Several neuronal models have been proposed that describe the thermoregulatory phenomena observed in euthermic mammals. Opinion and experimental evidence, as to whether the mammalian regulatory systems operate as constant gain, adjustable threshold proportional controllers, or as multiplicative controllers with both adjustable gain and threshold is sharply divided. Studies on single units of the pre-optic/anterior hypothalamus suggest that neurons exist whose thermal sensitivities would be consistent with either system. Available evidence indicates that during euthermia hibernators do not differ significantly from other mammals in thermoregulatory mechanisms. During hibernation, however, models for euthermic regulation are inadequate to account for, or to predict, responses to thermal stimuli whether they be presented centrally or peripherally. A neuronal model, whose temperature-response curve may be fitted by an exponential equation is proposed. The response to drop in body temperature occurs by disinhibition of a thermogenic neuronal pool. Increase in body or ambient temperature facilitates the inhibitory neurons and heat production is minimized. The model requires the existence of at least two interactive neuronal pools which are functional only during the hibernating state and which supplement those concerned with euthermic regulation. It is proposed that such a system would account for most, or all, of the thermoregulatory responses characteristic of the hibernating state.

ISBN 0-12-734550-7

I. INTRODUCTION

Models of physiological processes represent attempts to place experimental observations and concepts into an ordered context that will make their relationships more understandable to the modeler and his colleagues. The form may be verbal-pictorial, analog, mathematical, or neuronal. If the modeler has been successful he will have provided a structure which is an improved mirror to reality. But just as, or more, important in judging success, to paraphrase Hardy (28), is the degree to which it stimulates experimental and theoretical challenge--and the birthing of improvements or alternates.

Thermoregulatory models of all varieties have been reviewed extensively in recent years. These include those of Hardy (26, 27, 28), Hammel (20, 21, 22) and several recent collections of writings on this and related subjects (4, 5, 29, 53, 63, 64). We will attempt to place the ideas presented in some of these writings into contexts of emergent information and, especially, the thermoregulatory puzzles which confront those of us studying animals capable of undergoing hibernation. We have not included a specific discussion of the possible roles played by neurotransmitters since their functions require, and deserve, separate treatment.

II. BASIC CONCEPTS - EUTHERMIC MAMMALS

Neuronal models which have been proposed over the last decade or so have attempted to define the neuronal network which initiates, maintains, and controls responses resulting in the maintenance of stable body temperature. The basic features of these models (3) are: i) two main pathways sensing temperature changes and affecting temperature responses to warm and cold stimuli, ii) crossed inhibition between these two main pathways, and iii) afferent convergence of receptor pathways onto the effector pathways via central control. The differences among the models are found in the anatomic site and nature of the converging signals, in the mechanism of crossed inhibition and in their applicability to a wide variety of physiological conditions.

Of all the models perhaps none has had the impact and generated as many experiments as has the "hypothalamic proportional control with an adjustable set-point" as first proposed by Hammel and his colleagues (20, 21, 22, 25).

Based on results from hypothalamic temperature manipulation in resting and exercising dogs at various ambient temperatures, Hammel (20, 21, 24, 25) proposed a neuronal model which consists of 5 classes of neurons (Fig. 5): two primary sensors responding differentially to relative warmth or cold and three primary effector neurons (vasomotion, shivering, panting). Considering the single cell recordings from the hypothalamus of Nakayama *et al.* (52), Hammel defined the central high-Q_{10} neurons as those neurons in which spontaneous firing rate increased markedly with increasing hypothalamic temperature and the low-Q_{10} neurons as those in which the spontaneous firing rate increased relatively little with hypothalamic temperature alterations.

Of importance in Hammel's model is the assumption that high-Q_{10} sensory neurons facilitate heat loss mechanisms, and conversely that the low-Q_{10} neurons facilitate heat conserving and generating mechanisms while inhibiting the thermolytic. Thermal input from the periphery would facilitate or inhibit directly the high- or low-Q_{10} neurons appropriately. Another assumption associated with this model is that the primary effector neurons possess no inherent thermosensitivity; their output depending upon the relative degrees of facilitation or inhibition received.

The relationship between a thermoregulatory response and a change in the central temperature from a set level is described by the linear equation (20):

$$R-R_0 = \alpha\ (T_H-T_{set}) \quad \text{or} \quad R = \alpha\ (T_H-T_{set}) + R_0 \qquad \text{(eqn. 1)}$$

where,

$R-R_0$ = The thermoregulatory response in appropriate units or change in effector output.

R_0 = The basal level of the response, usually prior to stimulation.

T_H-T_{set} = The difference between the actual hypothalamic temperature and the set-point for the response, R. This is directly analogous to the load error of a proportional controller (26, 27). It also would be the difference between the hypothalamic temperature and the threshold for a particular response.

α = A fixed value proportionality constant or slope of the line for the response, R. It is equivalent to the sensitivity of the system. Whether it has a positive or negative value depends upon the physiological response (output) being considered, e.g., panting in response to heat or shivering in response to cold.

The T_{set}, or threshold for a response, is represented by the intersection of the firing rates of the high- and low-Q_{10} neurons (Fig. 1A). Alterations in T_{set} or threshold points for initiation of shivering or panting result when the firing rates of either the high-Q_{10} or low-Q_{10} neurons are changed. For example during cold exposure, peripheral cold receptors would facilitate the low-Q_{10} neurons, inhibit the high-Q_{10} neurons (Fig. 1B), thus increasing T_{set}, facilitating shivering and vasoconstriction, and inhibiting panting. It should be noted that, according to this model, facilitation (or inhibition) of the low-Q_{10} (or high-Q_{10}) neurons results in a displacement of the firing rate regression rather than a change in the slope of that regression. Therefore, for a given change in temperature the magnitude of the initiated response is always the same; only the temperature at which the response is initiated varies. The metabolic responses to the interaction of T_H and ambient (or skin) temperature are seen in figure 3 (lower right). As the T_{air} increases the T_H at which metabolism would be increased is shifted to lower temperatures. The lines defining the responses are parallel; that is, the α values (slopes) at each T_{air} are identical. Hammel's model has also been characterized as being additive (13). Theoretically, in an additive model, a metabolic response could be elicited at any T_{air}

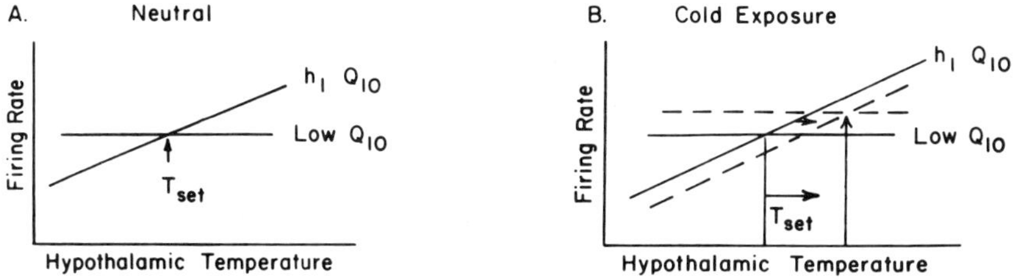

Fig. 1A. *Generation of T_{set} and change of T_{set} according to model of Hammel (19). The intersection of the firing rates of the low- and high-Q_{10} neurons determines the set point or reference signal around which body temperature is regulated. B. Afferent signals from peripheral cold receptors facilitate the low-Q_{10} neurons while inhibiting the high-Q_{10} units. T_{set} is shifted to the right; shivering initiated. Adapted from (20).*

if the T_H were sufficiently altered. It also was proposed (20, 21) that T_{set} would be altered by sleep and exercise. Connections from other brain structures (e.g. the reticular activating system) would synapse directly with the high- and low-Q_{10} neurons and, through facilitatory or inhibitory mechanisms, change the T_{set}.

The aesthetic simplicity of the scheme is such that one is tempted to accept it and to leave it unchallenged. However, that there may be exceptions to it or that other models may be tenable is quite apparent. Integral to the Hammel model is an internal, physiologically generated reference signal--the set-point. Deviation of the controlled variable from the set-point constitutes the load error--which activates the controller, which initiates the controlled response until the load error is corrected. Mitchell *et al.* (50, 51) investigating febrile responses proposed that an inherent reference signal, or fixed set-point, does not exist as such but that the error load, which activates the controller, is brought about by the relative balance between the inputs from warm and cold sensors, whether central or peripheral. Mean body temperature would be the controlled variable, whereas in the Hammel model T_H is the controlled variable--as a representative of mean body temperature. An important characteristic of Mitchell's model is that the relation between input and output for a given response need not be constant.

Evidence from studies involving pyrogen effects on thermosensitive neurons indicates that the discharge frequency of certain neuronal types to a constant input can

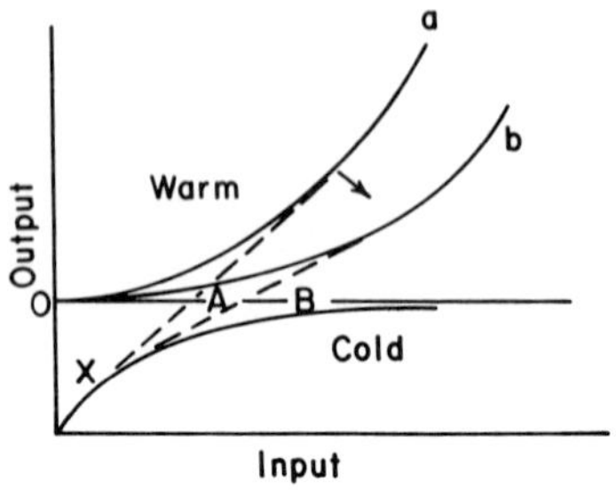

Fig. 2. *Temperature and change of T_{set}. A change in thermosensitivity of warm sensitive neurons (curve a) due to the action of pyrogens (curve b) results in a change in apparent set point (A → B) while actual reference temperature would remain unaltered (X). From Mitchell, Snellin and Atkins (50).*

be altered while the thermoinsensitive neurons remain unaffected (15, 16, 61). Contrasted to the Hammel model, wherein firing rates of neurons at various hypothalamic temperatures are linear functions of temperature, Mitchell *et al.* (50, 51) proposed non-linear neuronal input-output relationships (Fig. 2). A change in neuronal sensitivity would not shift this relationship as in Hammel's model but would alter the sensitivity such that a new function is described. In figure 2 the warm input-output response curve is decreased from a to b. The cold sensory system would remain unaffected by pyrogens and the apparent threshold for heat loss responses would be shifted from A to B. However, the operational set-point remains unaffected at X.

Other examples of models in which sensitivity or gain of the thermoregulatory systems would not be constant are described when more than one population of thermodetectors interact to produce the responses. In guinea pigs the control of shivering thermogenesis is a function of the interaction of signals from thermodetectors of the skin and spinal cord and non-shivering thermogenesis a function of skin and hypothalamic temperatures (Fig. 3). The mathematical expression which shows the onset of either shivering or non-shivering thermogenesis to skin and spinal cord or skin and POH is hyperbolic in both cases (Fig. 3, upper left). For example, Brück and Schwennicke (12) were able to describe NST (Fig. 3) by the equation

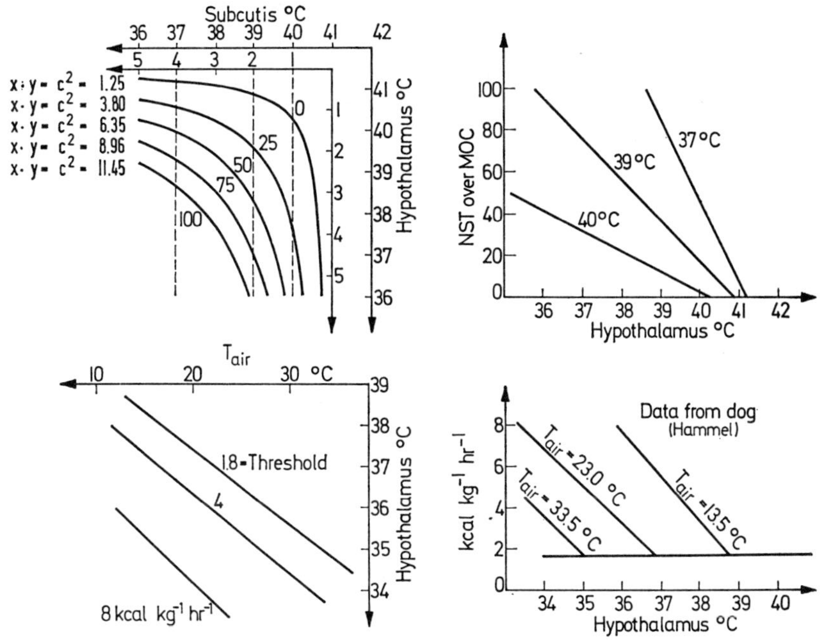

Fig. 3. *Comparison of "multiplicative" with an "additional" model. Upper left: The family of hyperbolas is derived from equation 2 and this in turn is based on studies in the guinea pig in which hypothalamic and body surface temperature were changed independently of each other and related to non-shivering thermogenesis (NST). The figures beside each hyperbola give NST in percent. Upper right: The three curves are obtained by plotting the data from the upper left diagram following the vertical dashed lines at skin temperatures of 37°, 39° and 40°C. MOC = minimal O_2 consumption. Lower right: Shivering heat production related to air temperature and hypothalamus temperature in the dog, as published by Hammel (19). Note absence of change in the slope of the curves. Lower left: Diagram derived from the lower right diagram. From Brück and Schwennicke (12).*

first proposed by Stolwijk and Hardy (61):

$$NST = K\ (T_{H_0} - T_H)(T_{S_0} - T_S)$$

where,

T_{H_0} and T_{S_0} = Respective reference temperatures for the hypothalamus and skin of the back.

T_H and T_S = Respective temperatures of the hypothalamus and skin of the back.

Because the response results from the error generated by two sets of thermodetectors, the processing of the signals is multiplicative. A theoretical limit of response to a forcing is represented by the asymptotes for the hyperbolic functions.

This equation represents a system in which T_{set} is a function of various reference temperatures which may be the product of activity in many CNS and peripheral nervous system areas. Nonthermodetectors may also influence the reference and the gain of such a system (65). If the data on non-shivering metabolism of Brück and Schwennicke (12) are replotted so that NST is a function of T_{POAH} (Fig. 3), then it becomes apparent, in the sense of Hammel's model, that the proportionality constant, or gain of the system (α), depends upon the thermal condition of two very diverse regions, the PO/AH and skin. Note that both the slopes, α, and the intercepts, T_{set}, are affected by the subcutaneous temperature and by no means constant. For comparison, see the parallelism of the replotted data of Hammel (20) on dogs (Fig. 3, lower left). It was suggested by Brück and Schwennicke (12) that if the data from Hammel's dog were replotted to a function of ambient temperature and hypothalamic temperature the straight lines might represent the center of relatively flat hyperbolas (Fig. 3, lower left), which would make it a multiplicative model.

These observations were not unprecedented. Jacobson and Squires (39) found that the relationship between skin and preoptic temperatures were multiplicative rather than additive insofar as metabolic and vasomotor responses of the cat are concerned. This means that both T_{set} and the sensitivity (α) of the system must be affected by peripheral information. Stolwijk and Hardy (61) had come to similar conclusions in their studies of human temperature regulation.

More recently, the careful studies of Stitt *et al.* (60) have shown that lowering skin temperature may alter thermosensitivity of the preoptic area but that only the reference point, not sensitivity, is altered during a febrile response to PGE_1 injected into the preoptic area.

Their data support the contention that the relationship between peripheral and preoptic temperatures is multiplicative rather than additive and may be fitted by equation 2. On the other hand, the shift in T_{set} accompanying a fever would indicate that the notion of a proportional control with an adjustable set-point may be compatible with the multiplicative model. By way of explaining some of their observations, they note that, "It is not unreasonable to suppose that the PO/AH area contains a population of neurons which exert an inhibition on heat production and excite heat losses when their temperature is raised."

The idea that additive and multiplicative control systems may coexist has been supported by the very recent work of Hammel *et al.* (23) on the harbor seal. It was found that the slopes of the regression lines of metabolic rate on preoptic temperature was a function of both ambient and rectal temperatures. They suggest that the central reference and thermal-sensitive neurons are both inhibited, and equally so, by the peripheral cold transducing elements.

The demonstration by Nakayama *et al.* (52) that some PO/AH neurons may alter their firing rates in response to local temperature laid the ground for more direct analyses of how peripheral and central information may be processed in order to achieve temperature regulation. Although one would expect to find units in the PO/AH that increased their rates of firing with rises in local temperature, observations of increased activity with the lowering of the local temperature created even more interest (see 19 and 37 for summaries of responses and species). Of obvious importance here are the findings of Hellon (38) that the curves relating brain temperature and firing rate of hypothalamic neurons were affected in two different ways when the ambient temperature was lowered. In one, the curve was shifted with no change in slope (sensitivity or α), which would be consistent with the model of Hammel. In the other, a change of slope, or sensitivity, was observed--a result that did little to resolve the issue as to whether the additive model of Hammel (20, 22) or the multiplicative hypothesis of Stowijk and Hardy (61) was obtained. As Hellon concluded, it well may be that both mechanisms are involved, as was suggested by Jacobson and Squires (39).

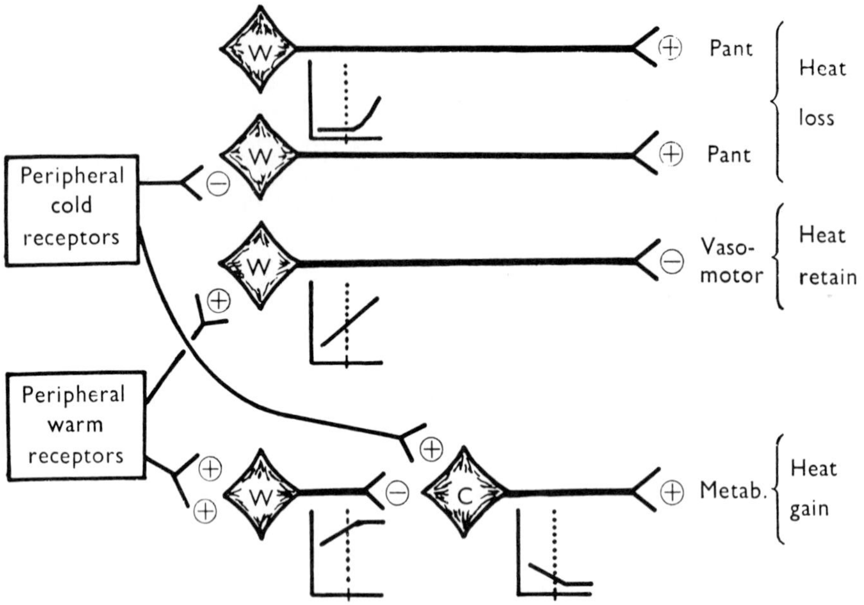

Fig. 4. *A neuronal model for thermoregulation. W indicates warm-sensitive or positive thermal coefficient unit. C indicates cold-sensitive or negative thermal coefficient unit. The graphs to the right of each neuron describe the thermoresponse curves (firing rate versus temperature) for that neuron. The + indicates an excitatory synaptic input and — indicates an inhibitory synaptic input. From Boulant (6).*

The observations of Hellon were confirmed by Boulant and Bignall (9), extended (11), and analyzed (6, 7, 10). A neuronal model (Fig. 4) was proposed by Boulant (6) which incorporated the salient information from the experiments of a number of investigators:

(i) To date it appears that the number of warm-sensitive neurons exceed the cold-sensitive ones by a factor of three. About a third of all neurons that could be recorded from for an extended period were temperature-insensitive (8).

(ii) The warm-sensitive neurons could be divided into four groups, depending on their level of firing rate at 38°C, as represented by the vertical dotted line in figure 4.

(iii) Based on the data of Boulant and Hardy (11), it was suggested that the low-firing group of warm-sensitive neurons would function in terms of heat loss and would be regulated primarily by central temperature; only 17% of them receiving thermal afferents from the skin or cord. In contrast, the higher firing rate neurons may be 80% innervated from the periphery and would function in terms of control of both heat loss and retention.

(iv) The cold-sensitivity (increase in firing with a fall in local temperature) is considered as actually being due to removal of synaptic inhibition by nearby warm-sensitive neurons as their firing rates diminished with a drop in local temperature. It was suggested that since similar proportions of the cold-sensitive and the very high firing rate neurons receive thermal afferents from the skin and spinal cord the two are likely to be functionally related, the one inhibiting the other to achieve a wide range of control. If this is so, it would be expected that there would occur an increase in preoptic thermosensitivity consequent to a change in peripheral temperature and an augmented rate of fire by the cold-sensitive units, the peripheral cold sensors facilitating the cold interneurons and the warm-sensitive ones remaining either unaffected or inhibited.

The attraction of Boulant's model lies in its flexibility and the fact that it appears to provide some insights which should be of considerable value in reconciling divergent experimental results and interpretations. Of particular importance is the suggestion that the quantity and type of afferent signals (whether from peripheral warm or cold receptors) a warm-sensitive neuron receives will determine its specific thermoregulatory role to a large extent. As Boulant (7) notes, this may explain the observations that changing ambient temperatures tend to have a relatively larger effect on heat production than on loss.

It would be well to emphasize that what Boulant labels as "C", or cold-sensitive, neurons are actually interneurons that are differentially inhibited or facilitated by other central high-firing rate neurons and peripheral cold receptors. By themselves, isolated from all other inputs, they would show a positive temperature characteristic.

In constructing models, especially neuronal, one frequently is constrained to ignore the variability and flexibility of the nervous system and to treat entire pools or populations of neurons as separate entities. In considering the applicability of Boulant's model, or any other one for that matter, it might be well to recall that each neuron that is indicated actually represents a large number of excitable cells. Each of them has characteristics which differ slightly from its functional fellow as determined by any number of physico-chemical factors ranging from size and distribution of dendrites and synapses, local O_2 tension, neurohumoral environment, etc. What is depicted then is an idealized, simplified presentation of what *might* be the situation under certain circumstances. Although there is not sufficient information available at this time, it would appear that the model of Boulant contains the elements which make it possible to observe both additive and multiplicative thermoregulatory responses. Do the units which show no change in thermosensitivity during stimulation of peripheral afferents mediate specific responses that would be seen as proportional control with adjustable set-point? A similar question may be asked about the units in which the sensitivity does change relative to their roles in multiplicative processes or signals originating from peripheral receptors. Perhaps the most important question would relate to how these different units may interact to achieve a common thermoregulatory end. Is the multiplicative model valid only when environmental extremes are encountered or is it actually the resultant of the activity of a single neuron pool subserving a given response? It would appear that it would be worth considering that whether or not preoptic thermosensitivity may be represented by a constant value of α for a specific response might depend on the number and probabilistic distribution of activities of the neural units involved in the response. This well might be determined, in turn, by the presence of biogenic amines, state of arousal, and so on. In short, far more work is required and it is much too early for doctrines to be promulgated.

III. EUTHERMIC HIBERNATORS

For the most part there is general agreement that during euthermia hibernators do not differ qualitatively from other mammals in terms of temperature control mechanisms. This is not to say that there are no points of controversy that would present modeling problems. In addition there are a few recent findings that should be reckoned with.

The thermoregulatory control system of hibernators must possess some particular properties in order for it to be able to operate over a wide thermal spectrum. To begin with, one may assume that central sensitivity to thermal stimuli does not differ markedly between hibernators and other mammals. From that assumption, it follows that "normal" PO/AH neurons would be operationally silent or, at the very best, there would be too small a driving force to activate the appropriate thermoregulatory responses at brain temperatures of 5°C (21, 36). This would follow as the consequence of the simple temperature effect on neuronal activity and the resultant disappearance of differential activity between the high- and low-Q_{10} neurons (Fig. 5). If euthermic hibernators should have very high values for α then it would be possible to maintain some effective driving force during deep hibernation.

In subsequent experiments, Hammel *et al.* (24), Heller and Colliver (32), Heller *et al.* (33), and Florant and Heller (17) did report exceptionally high values for preoptic thermosensitivity (Table 1). Heller *et al.* (33) suggested that these high levels of central thermosensitivity at euthermic temperatures would be sufficient to maintain an adequate driving force for temperature control at hibernating levels even an α 1/16th its upper level. As they point out, given Hammel's model, as brain temperature falls the firing rate/temperature curves of the two groups of sensory neurons approach one another and the thermosensitivity of the system would decline precipitously (Fig. 5).

The same authors (33) note that in two of their animals significant declines in preoptic thermosensitivity (α) did take place at higher ambient temperatures. They feel that this was related to an interaction between the

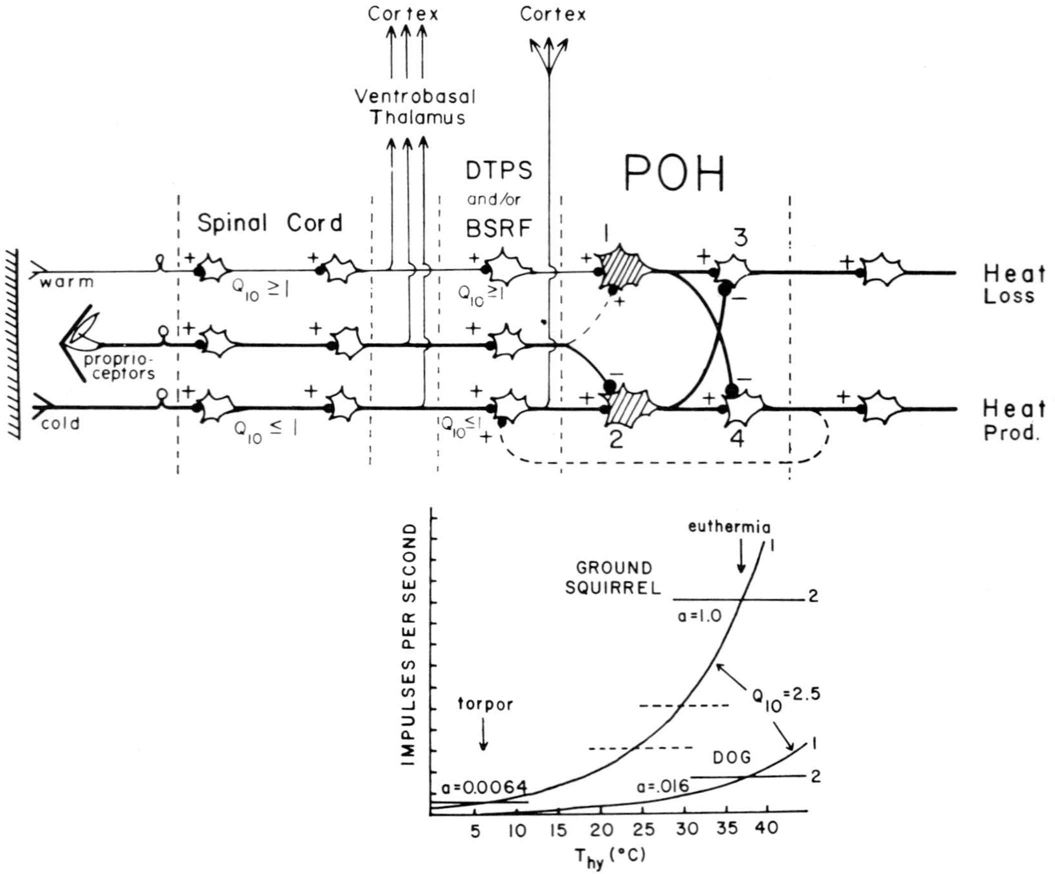
Cortex
Cortex
Ventrobasal Thalamus
DTPS and/or BSRF
POH
Spinal Cord
warm
proprio-ceptors
cold
$Q_{10} \geq 1$
$Q_{10} \leq 1$
1
2
3
4
Heat Loss
Heat Prod.
IMPULSES PER SECOND
euthermia
GROUND SQUIRREL
a=1.0
$Q_{10}=2.5$
torpor
DOG
a=0.0064
a=.016
5 10 15 20 25 30 35 40
T_{hy} (°C)

Fig. 5. "Hammel's neuronal model for CNS regulator of body temperature and predictions based on this model for how characteristics of regulator would change if T_b could be regulated at any level between 0 and 40°C. Heat loss and heat production represent target organs for affecting rates of heat loss from the body and rates of heat production by thermogenic tissue. Single neurons in model represent cell populations; 4 basic types are indicated in the POH. Neurons 1 and 2 differ in their degree of temperature dependence and both facilitate and inhibit neurons like 3 and 4 which have no spontaneous activity or temperature dependence. Neurons 1 and 2 receive facilitative and inhibitory inputs from a variety of sources. Graph at bottom of figure shows hypothetical firing rates of neurons 1 and 2 as a function of POH temperature for the dog and the ground squirrel. In both animals when firing rates of neurons 1 and 2 are equal, neurons 3 and 4 are inhibited more than facilitated and no net driving force is generated for heat loss or heat production mechanisms. If POH is heated, neuron 3 will be more facilitated than inhibited and will drive heat loss. Thus, threshold temperature (T_{set}) for a thermoregulatory response depends on relationship between neurons 1 and 2, and this can be changed by inputs to regulator mentioned above. For example, entrance into hibernation may be achieved by decreased input from BSRF (brainstem reticular formation) to neuron 2. As activity of neuron 2 decreased (dashed lines on ground squirrel curve), T_{set} would decrease. In both curves for ground squirrel and for dog, Q_{10} of neuron 2 is 1 and the Q_{10} of neuron 1 is 2.5, but ground squirrel curves are shifted to left. At any T_{set} proportionality constant for a thermoregulatory response (α) is a function of difference in slopes of firing rate versus temperature curves of neurons 1 and 2; hence at any T_{set} regulator of ground squirrel would produce a greater α than regulator of the dog. This may be an adaptation for thermoregulation during hibernation. Numbers labeled "a" represent measured α values in cal g^{-1} min^{-1} $°C^{-1}$. Dashed line indicating a feed back loop from neuron 4 to BSRF suggests existence of a positive feedback mechanism for elevating activity of neuron 2 and hence T_{set} during an arousal." From Heller and Colliver (32).

TABLE 1. Preoptic thermosensitivity of various mammals

Species	α ($W\ kg^{-1}\ {}^{0}C^{-1}$)	Source
		cited by
Dog	.81 - 1.5	Heath *et al.* (31)
Cat	.74 - 1.16	"
Rabbit	1.28	"
Sparrow	.90	"
C. tridecemlineatus	.47 - .99	"
C. lateralis	11.6	Hammel *et al.* (24)
C. lateralis	5.4, 5.9, 6.2	Heller *et al.* (33)
C. beldingi	2.4 - 3.2	"
M. flaviventris	1.1	Florant & Heller (17)
M. flaviventris	.57 - .87	Mills & South (49)
		Mills, Miller & South (48)

lower T_{set} and the exponential decrease of the Q_{10} curves for neuronal firing rates. This decrease in α was viewed by them as being too small to be accounted for by the model of Jacobson and Squires (39).

Important to the question of how different species may process thermal information is the relative degree to which they may be dependent upon peripherally derived signals from the skin, spinal cord, and internal receptors. By way of analyzing the problem, Heath *et al.* (31) compared preoptic thermosensitivity and other thermal control parameters in a variety of animals of different sizes, including euthermic hibernators (Table 1). According to the data they presented, weight specific preoptic thermosensitivity as a function of body size is quite constant. Since the rates of heat production and loss are inversely related to size (because of the weight specific heat transfer coefficient) it follows that smaller mammals would be more responsive to changes in skin (or ambient) temperature than larger ones. That is, there would occur a greater shift in the value of T_{set} for a given change in T_{skin}. However, these data and the interpretation have not gone unchallenged. As mentioned above, Heller *et al.* (33) have reported (Table 1) α values for small hibernators that are larger, by as much as an order of magnitude, than those of Heath *et al.* (31) (compare *Citridecemlineatus* with *C. lateralis*, Table 1). On the basis of their results on two species of ground squirrels relative to the dog (Table 1) Heller *et al.* suggest a contrary interpretation. They propose that preoptic thermosensitivity itself is inversely related to body size, especially in reference to hibernators--an interpretation which would be consistent with the retention of thermoregulatory control during hibernation in a manner that would be in accord with the model of Hammel. Thus they also predict that the large hibernator, marmot, would have an α value that was lower than that of the smaller ground squirrels. According to their data, the open-loop gain (OLG)[4] for the system of animals of different sizes remains fairly constant. This constancy of the OLG also is irreconcilable with the reports and interpretations of Heath *et al.* (31). As may be seen in Table 1, the data on marmots both supports their expectations and lies well within the range of values that Heath *et al.* (31) would expect--a rather conclusive result. Thus the question of relative preoptic and peripheral sensitivity in euthermic hibernators remains an open and important consideration.

In a series of experiments on euthermic marmots, Miller and South (46, 47) found ambient (skin) temperature altered metabolic responses to spinal cord temperature manipulations but did not alter the metabolic responses to hypothalamic temperature manipulations (48, 49). Very possibly the temperature range selected was too limited in the latter case. It was also shown that the hypothalamic T_{set} for shivering was lowered by heating the cord (47). Correlating nicely with these studies of periphal-central thermoregulatory interactions in the whole animal are the single unit recordings of Boulant and Hardy (11). They found that extrahypothalamic (i.e., spinal cord and peripheral) input to preoptic neurons markedly affected the firing rate of these neurons. In cases where extrahypothalamic thermal input was high, the hypothalamic unit's response to local temperature change diminished. Conversely under conditions where preoptic neuronal firing rate was high, the unit would exhibit decreased sensitivity to extrahypothalamic input. These interactions between neuronal inputs would manifest themselves in metabolic responses to hypothalamic and extrahypothalamic temperature manipulation that would be of differing gains and thresholds.

There also exists a body of evidence which implies that the interaction from other brain centers affects directly preoptic and extrahypothalamic thermosensitivity, more specifically these other brain centers are those involved with states of arousal, i.e., sleep and wakefulness. In kangaroo rats and ground squirrels (18, 35) the responsiveness to hypothalamic cooling was found to decrease as the animals proceeded from an awake state to slow wave sleep to paradoxical sleep. It has been interpreted that the thermoregulatory system is inhibited during sleep states. In the marmot, metabolic responses to spinal cord cooling and heating were studied relative to sleep states at differing ambient temperatures (46, 47). At an ambient temperature of 15°C the responses during an awake state were greater than during sleep states. However, at an air temperature of 5°C the response of the animal, as it proceeded from a sleep to an awake state, was greater than during either the sleep or awake state. These results are not incompatible with the idea of thermoregulatory inhibition during sleep. Also the increased response occurring during a change in arousal state suggests a compensation for a reduced body temperature which would occur with temperature regulation inhibition during sleep. An extension to arousal from hibernation can be drawn, in that similar inhibitory mechanisms alter thermosensitivity during hibernation, as might be suggested by the model of Luecke and South (45), and which would be released during arousal.

At least during the late summer and fall it would appear that the euthermic marmot is very dependent upon metabolic heat production rather than vasomotor control during exposure to ambient cold or cooling of the preoptic area (49, 57). An analogous situation obtains for higher temperatures in that the species will use drinking, postural adjustment, and, to a minor extent, some degree of panting under such circumstances--but no nuchal vasomotion is apparent. It is also of interest that, in the euthermic marmot, we never were able to induce a soporific state by warming the preoptic area nor to induce an arousal from sleep by cooling it. Quite the contrary, placing our animals in the cold or cooling the preoptic area either had no effect on the state of arousal or it tended to increase the frequency and duration of paradoxical sleep (57). It would be foolhardy to imply that these observations may be typical of all hibernating species; it will be of interest and importance to learn if they are confirmed in marmots and found to apply to other animals. The peculiarities of physical thermoregulation by the marmot well may be related to its relatively large body size. On the other hand, they could be the consequence of the organization of the central thermoregulatory control system and, hence, more characteristic of the ability to hibernate.

One point of agreement on euthermic hibernators is that these animals probably utilize the same general type of control system for regulating body temperature as do other mammals. This also means that many of the same controversies apply, as well as a couple of additional ones. Data are beginning to appear that would indicate that preoptic thermosensitivity is not necessarily representable by a constant value (33, 47). Whether this may be best interpreted by the model of Jacobson and Squires (39), of Mitchell *et al.* (50,51), or an adjustment to the Hammel model (23, 32) (Fig. 5) based on the data of Boulant and Hardy (11) awaits further information.

There still remains the vexing problem of comparative PO/AH thermosensitivity. Until the fundamental discrepancy between the data of Heath and his colleagues and those of Heller and his coworkers is resolved it will be most difficult to construct valid interspecies thermoregulatory control models. Obviously additional direct information is needed in this area.

The ability to retain neural control at low temperatures certainly implies that basic differences between hibernators and nonhibernators must exist at the neuronal level. Such a difference might consist of i) relative proportions of warm and cold-sensitive neurons, ii) the presence of a number of units with an unusually high firing rate which would retain thermosensitivity at very low temperatures, or iii) possession of a distinct population of neurons whose activity range would be limited to hibernating temperature levels. Indeed, Boulant and Bignall (8) did undertake such a study on the preoptic-hypothalamic area of ground squirrels (*C. lateralis*) and rats. From our standpoint, as it turned out, it was unfortunate that they limited their work to between 32° and 38°C; no differences of significance in proportions or characteristics of thermosensitive neurons of the two species were found. However, they did find several important characteristics of thermosensitive neurons in general which provided the basis for future work. Fortunately, the question was reopened by Wünnenburg *et al.* (68) with far different results from those of Boulant and Bignall, as discussed below.

IV. HIBERNATION

Although temperature regulation by hibernating animals has received its share of attention over the years, only recently have attempts been made to model it as a control system. Prior to the challenge of Hammel (21) there was no really proper paradigm for ordering what information did exist. Although, at the time, he felt that a true control system for body temperature regulation was unlikely to be present during deep hibernation, his model and general experimental approach provided a powerful impetus for further work by his and other research groups. Subsequently there has emerged solid evidence that true thermoregulatory control is a characteristic of the hibernating state--although species probably differ insofar as the details of the physiological mechanisms are concerned (see discussions in 31, 32, 43, 57). The first attempts at modeling a neuronal control system (24, 45), though no longer completely applicable, possessed features which warrant retention. In order to focus the discussion we will limit our consideration to the current form of the Hammel model as it relates to hibernation (32) and to a possible alternative we ourselves propose.

Rather than paraphrase the description of Hammel's model, the modification and legend of Heller and Colliver (32) is quoted in figure 5. Its operation depends upon: i) the retention of the driving force for low temperature control through a difference in the activity levels between the high-Q_{10} (warm) and low-Q_{10} neurons. ii) this is brought about by the existence of a very high basic (euthermic) firing rate of the high-Q_{10} neurons, i.e., an augmentation of thermosensitivity or α value. iii) it assumes, therefore, that identical pools of neurons control body temperature during both euthermia and hibernation. iv) the increase in T_{set} for arousal is brought about by a positive feedback loop which could extend from the thermogenic output of neuron 4 to the reticular formation. As suggested by the data of Beckman (2) and Beckman and Stanton (1), acetylcholine, or the responsiveness of the mesencephalic reticular formation to it, may also be involved in the positive feedback loop.

A general, but not necessarily crucial, problem of the first Hammel model (21) was that it could neither account for the various types of neuronal firing rate/temperature curves observed for the brain stem nor for the effectiveness of afferent input on thermosensitivity. Of course, such applies to the latter version of it as well (Fig. 5). A specific difficulty with this application of the model to hibernation is that it depends on an extraordinarily high value for preoptic thermosensitivity (α) and open loop gain at euthermic levels. Although supporting data (24, 32) looks very good, others have obtained data that would be incompatible with the hypothesis (1, 2, 43, 45-49, 56-59, 68). Finally, it is implied that responses (e.g., vascular) should occur as a consequence of slight warming, but such has not been observed. The most attractive feature is that, given its premises, it does not require a separate mechanism which must be brought into play for hibernation.

V. AN ALTERNATE MODEL

We tend to accept the model of Boulant (Fig. 4) as being the one most consistent with existing information on euthermic mammals. However, we propose a model for the situation during hibernation which is, perhaps, even simpler (Fig. 6). The neurons involved during actual hibernation may be either a separate pool brought into action as hibernation is entered, or portions of the euthermic pool that retain activity over an unusually wide range of temperature. One primary premise for this model, as in the one for euthermia, is that there exists a pool of warm receptor neurons (W) which exert an inhibitory influence on the cold neuron pool (C). Each receives excitatory peripheral input. The C units, of and by themselves, do not possess negative temperature coefficients, their firing rates being influenced or controlled by the relative inhibition received from W or excitation from the periphery. It is possible, but not necessary, that the output from C could influence thermolytic behavior (e.g., posture) through a negative feedback as indicated in the diagram. On the other hand, this type of behavioral regulation could be controlled directly through the warm pathway.

The second primary premise is that there is, at least, one pool of C neurons in the preoptic region that is functional only at low temperatures, being totally inhibited by W neurons at higher levels as indicated by the lower firing rate-temperature curve of figure 6. It is possible, but not required, that there also is a pool of W neurons specialized for hibernation. It seems more likely that the control of the C neurons during hibernation is carried out by that portion, of whatever size, of the W pool that retains differential activity at low temperatures. The remarkable ability of many tissues of hibernating species to be able to function over such a broad temperature range is well known (see 53 for examples).

Most exciting to us has been the recent demonstration by Wünnenburg *et al.* (68) of types of neuronal activity in golden Syrian hamsters that would fill the requirements of such a model handsomely. Two general types of preoptic unit responses relative to brain temperature were discovered. The first group evidenced an exponential increase in spontaneous firing rates at temperatures between 10° and 42°C. At 36-37°C they had very high rates: 29-59/s. According to the criteria of Boulant (6) and Boulant and Hardy (11), such neurons might be good candidates for the W units. The second group of units described by Wünnenburg and his coworkers (68) is even more interesting. These have bell-shaped temperature firing rate characteristics with positive temperature coefficients from 20° or 30°C to below 10°C and negative coefficients at higher temperatures, from 20° or 30°C to over 40°C preoptic temperature. It would seem reasonable to expect that even more dramatic differences might be obtained from preparations taken from animals in hibernation. It would also be of fundamental importance to study the effects of alterations in skin and spinal cord temperature on the activity of the preoptic units which would settle the problem of how important peripheral temperature reception is during hibernation. The data from hamsters were compared to those obtained from eight preoptic units of guinea pigs (68). In the latter, essentially all activity was lost below 30°C. One unit did have a bell-shaped firing rate-temperature curve, peaking at about 34°C.

In order to fit the results of Wünnenburg *et al.* (68) into our proposed model, three points should be made: 1) The hamsters that were studied had not been hibernating at the time of the experiments. 2) Enough information is available to suggest that many neural elements, both central (55) and peripheral (54), undergo additional modifications for low temperature function which are associated with the actual hibernating state. 3) Therefore one might expect that even though the units described by Wünnenburg and his colleagues are active at low temperatures their temperature-activity curves, or those of associated units, would be shifted to the left during hibernation--enhancing their ability to function at reduced temperatures. This would mean that they would exhibit firing rate-temperature relationships such as depicted in figure 6. If the exponentially high firing rate (>36°C) units are representable by those such as W in figure 6, then those with a bell-shaped temperature characteristic would be represented by C. At higher temperatures the C units would be strongly and progressively inhibited. At lower temperatures the C units undergo a disinhibition as the W become cold-inactivated. These effects should be even more pronounced if the C and W neuron pools were receiving appropriate excitatory afferent signals from the periphery. Under optimal circumstances an inverse exponential relationship between output of the system and the preoptic might be obtained (lower plot, Fig. 6).

It would be well to again emphasize a point that is very much a condition of this model. Although the discussion is in terms of W and C neurons, as if each were composed of homogeneous units, they actually are pools composed of large numbers of nerve cells. Each cell has its own excitability characteristics and individual intercellular synaptic associations. Each is subject to variations in afferent influences. Each has its own efferent responsibilities. Therefore, the output of the system in a hibernating animal will depend on the interaction of a huge number of factors including not only thermal information from peripheral and central sensors but wide variations in the temperatures of the controlling structures, with attendant physico-chemical alterations as well. Inevitably, this means that curves relating output to temperature would be represented by probabilistic or Gaussian distributions.

As a first approximation, we suggest that the control equation for a hibernating animal may be fitted by a logarithmic exponential equation in the form:

$$R = R_0 \, e^{-\beta(T_H - T_{set})} \qquad \text{(eqn. 3)}$$

where,

the notations have the same meaning as in equation 1 except that β is not regarded as a constant in the same sense as is α. A particular value for β would depend on the

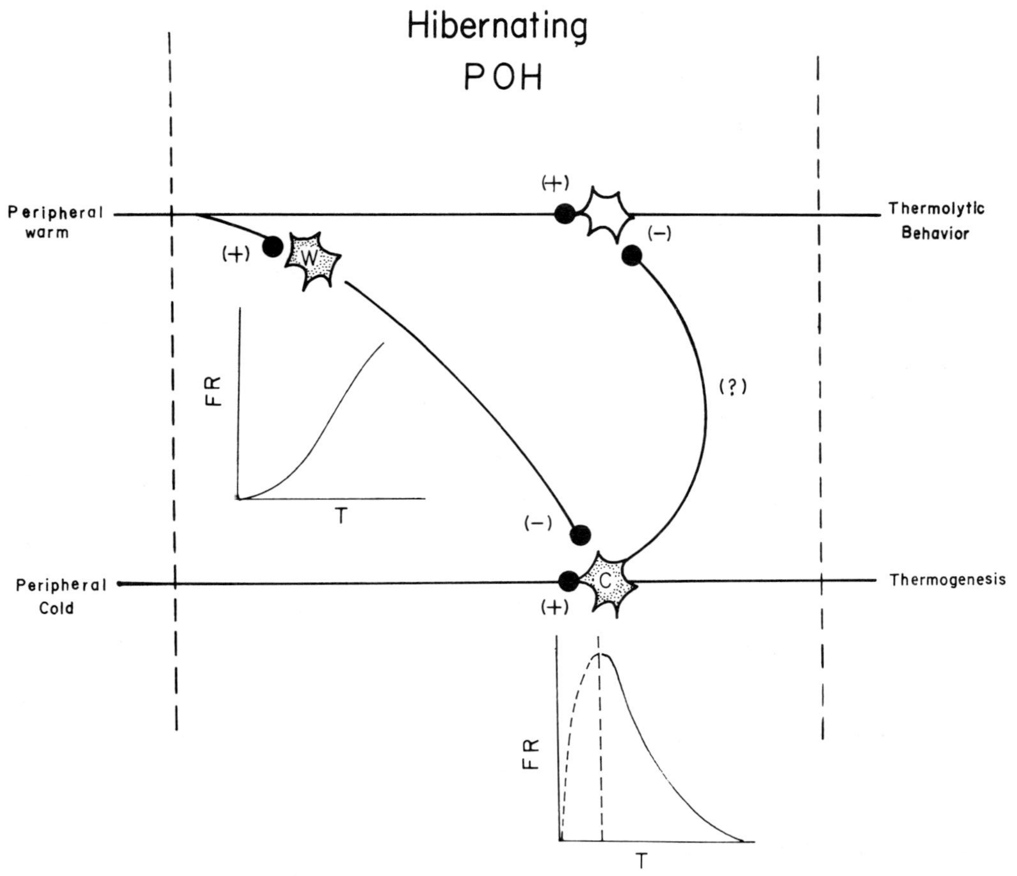

Fig. 6. *Proposed neuronal model for control of body temperature during hibernation. Extrahypothalamic inputs remain as in Fig. 5. Other symbols as in Fig. 4. The dashed line in the lower firing rate (FR)-temperature (T) plot indicates probable drop in activity between 0°C (vertical dashed line) and formation of ice-crystals (included for the sake of completeness).*

steady state conditions at a particular hibernating temperature, peripheral inputs and possible humoral factors that may affect the gain of the system. The quantity (T_H-T_{set}) would apply for thermogenic responses and (T_{set}-T_H) for thermolytic, e.g. panting.

T_{set} undoubtedly varies widely. Its value will depend on the particular physiological response being considered, ambient temperature, the complex interaction between minimal metabolic rate and weight-specific heat transfer coefficient and other factors which we are only beginning to learn about, such as possible interactions between stage of the hibernating cycle and species difference (17). It is commonly observed that most, if not all, of these animals will undergo hibernation at almost any ambient temperature, especially below 20°C. Recent direct and indirect evidence (17, 57) leaves no doubt that a T_{set} is adopted. The time required for this to occur may vary with temperature, species and environmental conditions. Our experiments (57) indicated that this required about three days of hibernation in the case of the marmot. On the other hand, the equally careful experiments of Florant and Heller (17) would seem to indicate that the T_{set} is adopted as the marmot reaches the steady state of deep hibernation.

VI. HIBERNATION, STEADY STATE BODY TEMPERATURE

As do other thermoregulatory control models, this model describes how a steady state situation may be retained. It addresses itself specifically to the hibernating state. Therefore, first to be considered will be the question of how it might function during deep hibernation. Entry and arousal will be discussed subsequently.

(i) Stimulation of peripheral cold receptors would excite the C pool, the thermogenic response would act to offset the perturbation at the control level and, we believe, the periphery as well. It is to be noted particularly that at no time has there been an indication of vasomotor activity in response to thermal stimuli during the course of our experiments with hibernating marmots (30, 45, 49, 56, 57) nor do we know of this being reported. In all cases when thermogenesis was elicited, surface and subcutaneous temperatures rose, in parallel, with deep body temperature. Therefore no vasomotor activity was incorporated into the model. The thermogenic response consists of cardioacceleration [a primary source of heat in the hibernating animal (41)], shivering and, presumably, non-shivering thermogenesis.

(ii) Stimulation of the central cold receptors would yield a response which, according to this model, would be proportional to the logarithm of the gain (ln OLG) of the system. This would be the consequence of the disinhibition brought about by cold inactivation of the strongly inhibitory W neuron pool. The exponential shape of the response curve would be a consequence of a large pool consisting of neurons having widely distributed sensitivities. An even more pronounced effect would be expected if the peripheral cold receptors were stimulated at the same time, facilitating the C pool. An exponential response curve was found in the marmot (57). In these experiments it was found that the ln OLG of the system, in terms of change in brain temperature, was inversely proportional to the prestimulus brain temperature. The regression line fit to the data indicated that the gain of the system would rise from a value of 7.4 at a brain temperature of 5°C to 149 at one of 0°C. Thus the greater the cooling, the greater will be the thermogenic response--exponentially (lower plot, Fig. 6). As the metabolic heat warms and activates the potent W neurons thermogenesis is minimized and the system returns to equilibrium. Should the stimulus be augmented--for example, such that peripheral and central receptors respond--then it is probable that the β value would increase markedly due to peripheral facilitation with a much greater increase in the gain. The consequence of this would be a reaction that would be indistinguishable from a partial or complete arousal. That is, this well might be the explanation for the "Alarm Reaction" of Heller and Hammel (36); we believe it is.

(iii) Warm stimuli applied peripherally or centrally will inhibit thermogenesis and cardiac activity (48, 49, 56, 57). As mentioned above, no vasomotor response will be obtained. On several occasions what might be taken as behavioral responses to preoptic warming have occurred (45, 50, 51). These range from slight postural adjustments within the nest to, on one occasion, an actual leaving of the nest and assuming a spread-out position (49). This latter animal returned to the nest some time later without coming to arousal, the brain temperature remaining at hibernating levels. The validity of these experiments may be moot since the stimuli themselves may have activated irrelevant neuronal mechanisms as discussed in the original paper. This doubt is indicated by a question mark in the feedback from the C to the "thermolytic behavior" neuron pool (Fig. 6), in the figure of the model.

It would be unreasonable to expect that every hibernator would utilize this proposed mechanism in the same way quantitatively. The relative dependence upon peripheral or central thermosensitivity undoubtedly is affected by such factors as species (including size, etc.) and all the factors involved in its natural history, as discussed by Lyman and O'Brien (42, 43).

A. Entry

Prior to actual entry a vast array of factors, ranging from nutritional, endocrine and humoral states to the influences of biorhythms, may interact to set the physiological stage for entry into hibernation. Of course, the models do not, cannot, address themselves to this preparatory, or organizational, phase. Regardless of what initiates it, entry appears to begin from a stage of sleep--specifically, from slow wave sleep (55) and metabolic heat production declines prior to any perceptible drop in T_b (40).

First, we suggest that the euthermic thermoregulatory control system is actively inhibited--presumably through the reticular formation of the brain stem--through a rather loosely defined "hibernation entry function" (45). Vasomotor regulation of heat loss seems to be inactive and the decline in T_b somewhat faster than would be the case with a dead animal, fitting a model of minimal heat production with circulatory forced convection (44, 45). Thermogenesis (shivering) is inactivated or, in light of the data of Florant and Heller (17) and Wünnenburg *et al.* (68), there is a progressive drop in T_{set} brought about, according to our model, by a gradual moderation of the inhibition of the C pool by the W.

Secondly, as hibernating temperature levels are attained, only the neuronal pools indicated in figure 6 remain active. In an operational sense, these neuronal associations would be parallel to those controlling T_b during euthermia. Whether or not some or all of those depicted in figure 6 are anatomically or physiologically the same as those in figures 4 or 5 is not relevant to our model. The question is relevant to adaptation of neuronal physiology for low temperature function.

B. Behavioral Thermoregulation of Entry

From time to time note is made of intense shivering or thermogenic bouts during entry into hibernation. This has been related to "test drops" or brakes on too rapid a drop in T_b on the part of *C. beechyi* (62) or to the decline of T_H below T_{set} in the case of *M. flaviventris* (17). It is very possible that both interpretations are far from the mark. Being equally puzzled by these bursts of activity when a marmot is well on its way into hibernation, on a number of occasions we carefully watched the animals on remote video throughout entire cycles. As will be reported in more detail in another paper, we found that on almost every occasion the activity was much more than mere shivering. The animals were improving their nests! During early entry (ca. T_{brain}=30°C) this frequently consisted of extraordinarily complex behavior in which the animal might even emerge from the nest and spend considerable time arranging it. This was always purposeful in that the nesting material was brought higher along the sides and closer to the animal. Frequently these operations were carried out from the nest itself. This was always preceded and accompanied by what could only be interpreted as an arousal type of EEG activity. Following the activity the animal would resume entry. As body temperature fell, this "behavior" became progressively stereotyped and ineffectual, and progressed from simple dorsiflexion of the head to very slight nonspecific movements below a T_{brain} of 15°C or so. Repeated observations and continuous recordings reaffirmed these observations. The effect of the home improvement activity was to increase the heat capacitance of the nest--thus making thermoregulation during hibernation more economical from an energetic standpoint.

In view of the above discussions one might construct the following general hypotheses: i) prior to actual entry into hibernation, the physiological groundwork is laid which permits the maintenance of integrated function at low body temperatures; ii) as the T_b drops during entry, temperature itself acts as a controller in that most, or all, of the neural mechanisms that are no longer relevant at a particular stage are shut off through cold inactivation; iii) as this occurs, other mechanisms--which are now very relevant--are released from inhibition and now become active. Still other mechanisms, which may be somewhat less cold insensitive, emerge as important because changes in their absolute activity now have a relatively profound effect. (An example of the latter instance is heart rate; during euthermia an increase in rate of 12 b/m has a trivial effect on body heat; during hibernation such a change represents a 200 or 300% increase in rate and an enormous increase in relative heat production because of the limited sources of body heat.) Thus, during entry, the decline in T_b itself and the previously laid groundwork interact to allow the emergence of the "behavioral thermoregulation of entry". As T_b declines further, the complex behavior pattern decreases in intensity and undergoes gradual cold inactivation itself. At about the same time the entire thermoregulatory control system of hibernation is operational through a similar process. In essence, temperature works on the nervous system in such a way that a series of "flip-flop" control switches are thrown as brain temperature changes. Another example in which we would interpret something of this sort as operating is provided by the brown bullhead (*I. nebulosis*) (15). The fish would select warmer portions of a temperature gradient box when its rostral brainstem was cooled slightly. However, when the brainstem was cooled below 10°C, another mechanism supervened and the fish moved to the *lowest* possible temperature in the gradient and became torpid.

C. Arousal

Regardless of what initiates arousal the effect of it would be that the C pool of neurons would be released from the thermoregulatory control exerted by the W inhibition and/or strongly facilitated and the full-blown thermogenic response of arousal initiated. Data has been found that would suggest that prearousal alterations in T_{set} do occur (17). As higher T_b levels are reached the euthermic control system would obtain as the full armory of regulatory neuronal associations becomes reactivated.

Regardless of which type of model or models are found to best represent thermoregulatory control during hibernation and/or euthermia, if they provoke controversy and stimulate experimental challenges they will have served their purpose. They have no other.

FOOTNOTES

[1]Office of the Director and Physiology Section, School of Life & Health Sciences, University of Delaware, Newark, DE 19711.

[2]Department of Psychology, University of Virginia, Charlottesville, VA 22904.

[3]Department of Biology, Northeastern University, Boston, MA 02115.

[4]Open loop gain (OLG) is defined as output/input in a feedback control system in which the control loop has been "opened" (66). For the experiments considered here the opening of the loop is accomplished by perturbing the temperature of the thermosensitive elements of the hypothalamus itself. Thus, if lowering the T_H by 2° increased the body temperature by 4°, the OLG = 2. A more complex expression and experimental procedures are disccussed in references 22 and 31.

REFERENCES

1. Beckman, A. L., and T. L. Stanton. Changes in CNS responsiveness during hibernation. Am. J. Physiol. 231: 810-816, 1976.
2. Beckman, A. L. Hypothalamic and midbrain function during hibernation. In: Current Studies of Hypothalamic Function. Lederis, K. and W. L. Veale (editors). Basel: Karger, 1978. (In press).
3. Bligh, J. Neuronal models of mammalian temperature regulation. In: Essays on Temperature Regulation. Bligh, J. and R. Moore (editors). New York: American Elsevier, 1972, pp. 105-120.
4. Bligh, J., and R. E. Moore (editors). Essays on Temperature Regulation. New York: American Elsevier, 1972.
5. Bligh, J. Temperature Regulation in Mammals and Other Vertebrates. New York: American Elsevier, 1973.
6. Boulant, J. A. The effects of firing rate on preoptic neuronal thermosensitivity. J. Physiol. 240: 661-669, 1974.
7. Boulant, J. A. A hypothalamic model for thermoregulation. In: Selected Topics in Environmental Biology. Bhatia, B., G. S. Chhina and B. Singh (editors). New Delhi: Interprint Publ., 1976, pp. 41-44.
8. Boulant, J. A., and K. E. Bignall. Determinants of hypothalamic thermosensitivity in ground squirrels and rats. Am. J. Physiol. 225: 306-310, 1973.
9. Boulant, J. A., and K. E. Bignall. Hypothalamic neuronal responses to peripheral and deep body temperatures. Am. J. Physiol. 225: 1371-1374, 1973.
10. Boulant, J. A., and R. R. Gonzalez. The effect of skin temperature on the hypothalamic control of heat loss and heat production. Brain Res. 120: 367-372, 1977.
11. Boulant, J. A., and J. D. Hardy. The effect of spinal and skin temperature on the hypothalamic control of heat loss and heat production. J.

Physiol. 240(3): 639-660, 1974.

12. Brück, K., and H. P. Schwennicke. Interaction of superficial and hypothalamic thermosensitive structures in the control of non-shivering thermogenesis. Int. J. Biometeor. 15: 156-161, 1971.
13. Brück, K., and W. Wünnenburg. Meshed control of two effector systems: nonshivering and shivering thermogenesis. In: Physiological and Behavioral Temperature Regulation. Hardy, J. D., A. P. Gagge and J. A. J. Stolwijk (editors). Springfield, Ill.: Thomas, 1970, pp. 562-580.
14. Cabanac, M., J. A. J. Stolwijk, and J. D. Hardy. Effect of temperature and pyrogens on single unit activity in rabbit's brain stem. J. App. Physiol. 24: 645-652, 1968.
15. Crawshaw, L. I., and H. T. Hammel. Behavioral regulation of internal temperature in the brown bullhead, *Ictalaurus nebulosus*. Comp. Biochem. Physiol. 47A: 51-60, 1974.
16. Eisenman, J. S. Pyrogen-induced changes in the thermosensitivity of septal and preoptic neurons. Am. J. Physiol. 216: 330-334, 1969.
17. Florant, G. L., and H. C. Heller. CNS regulation of body temperature in euthermic and hibernating marmots (*Marmota flaviventris*). Am. J. Physiol. 232: R203-R208, 1977.
18. Glotzbach, S. R., and H. C. Heller. Central nervous regulation of body temperature during sleep. Science 194: 537-539, 1976.
19. Guieu, J. D., and J. D. Hardy. Integrative activity of preoptic units. I: Responses to local and peripheral temperature changes. J. Physiol., Paris 63: 253-256, 1971.
20. Hammel, H. T. Neurons and temperature regulation. In: Physiological Controls and Regulations. Yamamoto, W. S. and J. R. Brobeck (editors). Philadelphia: Saunders, 1965, pp. 71-97.
21. Hammel, H. T. Temperature regulation and hibernation. In: Mammalian Hibernation. III. Fisher, K. C., A. R. Dawe, C. P. Lyman, E. Schönbaum and F. E. South (editors). Edinburgh & London: Oliver & Boyd, 1967, pp. 86-109.

22. Hammel, H. T. Regulation of internal body temperature. Ann. Rev. Physiol. 30: 641-710, 1968.
23. Hammel, H. T., R. W. Elsner, H. C. Heller, J. A. Maggert, and C. R. Bainton. Thermoregulatory responses to altering hypothalamic temperature in the harbor seal. Am. J. Physiol. 232: R18-R26, 1977.
24. Hammel, H. T., H. C. Heller, and F. R. Sharp. Probing the rostral brainstem of anesthetized, unanesthetized and exercising dogs and of hibernating and euthermic ground squirrels. Fed. Proc. 32: 1588-1597, 1973.
25. Hammel, H. T., D. C. Jackson, J. A. Stolwijk, J. D. Hardy, and S. B. Stromme. Temperature regulation by hypothalamic proportional control with adjustable set temperature. J. Appl. Physiol. 18: 1146-1154, 1963.
26. Hardy, J. D. Physiology of temperature regulation. Physiol. Rev. 41: 521-606, 1961.
27. Hardy, J. D. The "set-point" concept in physiological temperature regulation. In: Physiological Controls and Regulations. Yamamoto, W. S. and J. R. Brobeck (editors). Philadelphia: Saunders, 1965, pp. 98-116.
28. Hardy, J. D. Models of temperature regulation -- A review. In: Essays on Temperature Regulation. Bligh, J. and R. E. Moore (editors). New York: American Elsevier, 1972, pp. 163-186.
29. Hardy, J. D., A. P. Gagge, and J. A. J. Stolwijk (editors). Physiological and Behavioral Temperature Regulations. Springfield: Thomas, 1970.
30. Hartner, W. C., F. E. South, H. K. Jacobs, and R. H. Luecke. Preoptic thermal stimulation and temperature regulation in the marmot (*M. flaviventris*). Cryobiology 8: 312-313, 1971.
31. Heath, J. E., B. A. Williams, S. H. Mills, and M. J. Kluger. The responsiveness of the preoptic-anterior hypothalamus to temperatures in vertebrates. In: Hibernation and Hypothermia, Perspectives and Challenges. South, F. E., J. P. Hannon, J. S. Willis, E. T. Pengelly and N. R. Alpert (editors). New York: American Elsevier, 1973, pp. 605-627.

32. Heller, H. C., and G. W. Colliver. CNS regulation of body temperature during hibernation. Am. J. Physiol. 227: 583-587, 1974.
33. Heller, H. C., G. W. Colliver, and P. Anand. CNS regulation of body temperature in euthermic hibernators. Am. J. Physiol. 227: 576-582, 1974.
34. Heller, H. C., G. W. Colliver, and J. Beard. Thermoregulation during entrance into hibernation. Pflügers Arch. 369: 55-59, 1977.
35. Heller, H. C., and S. F. Glotzbach. Thermoregulation during sleep and hibernation. In: International Review of Science, Environmental Physiology II, vol. 15. Robertshaw, D. (editor). Baltimore: University Park Press, 1977, pp. 147-188.
36. Heller, H. C., and H. T. Hammell. CNS control of body temperature during hibernation. Comp. Biochem. Physiol. 41A: 349-359, 1972.
37. Hellon, R. F. Thermal stimulation of hypothalamic neurons in unanesthetized rabbits. J. Physiol., Lond. 193: 381-395, 1967.
38. Hellon, R. F. Temperature-sensitive neurons in the brain stem: their responses to brain temperature at different ambient temperatures. Pflügers Arch. 335: 323-334, 1972.
39. Jacobson, F. H., and R. D. Squires. Thermoregulatory responses of the cat to preoptic and environmental temperatures. Am. J. Physiol. 218: 1575-1582, 1970.
40. Lyman, C. P. Oxygen consumption, body temperature and heat rate of woodchucks entering hibernation. Am. J. Physiol. 194: 83-91, 1958.
41. Lyman, C. P. Circulation in mammalian hibernation. In: Handbook of Physiology, Circulation. Hamilton, W. F. and P. Dow (editors). Washington, D.C.: Am. Physiol. Soc., 1965, Sect. 2, Vol. III, pp. 1967-1989.
42. Lyman, C. P., and R. C. O'Brien. Sensitivity to low temperature in hibernating rodents. Am. J. Physiol. 222: 864-869, 1972.
43. Lyman, C. P., and R. C. O'Brien. A comparison of temperature regulation in hibernating rodents. Am. J. Physiol. 227: 218-223, 1974.

44. Luecke, R. H., E. W. Gray, and F. E. South. Simulation of passive behavior of a cooling biological system. Entry into hibernation. Pflügers Arch. 327: 37-52, 1971.
45. Luecke, R. H., and F. E. South. A possible model for thermoregulation during deep hibernation. In: Hibernation-Hypothermia Perspectives and Challenges. South, F. E., H. P. Hannon, J. S. Willis, E. T. Pengelley and N. R. Alpert (editors). New York: Elsevier, 1972, pp. 577-604.
46. Miller, V. M., and F. E. South. Effect of arousal state and ambient temperature on spinal cord thermosensitivity in the marmot. Fed. Proc. 35: 2395, 1976.
47. Miller, V. M., and F. E. South. Cord, hypothalamic and air temperature. Interaction with arousal states in the marmot. Am. J. Physiol. (submitted).
48. Mills, S. H., V. A. Miller, and F. E. South. Thermoregulatory responses of a hibernator to hypothalamic and ambient temperatures. Cryobiology 11: 465-472, 1974.
49. Mills, S. H., and F. E. South. Central regulation of temperature in hibernation and hypothermia. Cryobiology 9: 393-403, 1972.
50. Mitchell, D., J. W. Snellen, and A. R. Atkins. Thermoregulation during fever: change in set-point or change in gain. Pflügers Arch. 321: 293-302, 1970.
51. Mitchell, D., A. R. Atkins, and C. H. Wyndham. Mathematical and physical models of thermoregulation. In: Essays on Temperature Regulation. Bligh, J. and R. E. Moore (editors). New York: American Elsevier, 1972, pp. 37-54.
52. Nakayama, T., J. S. Eisenman, and J. D. Hardy. Single unit activity of anterior hypothalamus during local heating. Science 134: 560-561, 1961.
53. Precht, H., J. Christopherson, H. Hensel, and W. Larcher. Temperature and Life. New York: Springer-Verlag, 1973.
54. South, F. E. Phrenic nerve-diaphragm preparations in relation to temperature and hibernation.

Am. J. Physiol. 200: 565-571, 1961.

55. South, F. E., J. E. Breazile, H. D. Dellmann, and A. D. Epperly. Sleep, hibernation and hypothermia in the yellow-bellied marmot (*M. flaviventris*). In: Depressed Metabolism. Musacchia, X. J. and J. F. Saunders (editors). New York: Elsevier, 1969, pp. 277-312.

56. South, F. E., and W. C. Hartner. Hibernation and active thermoregulation: responses to manipulation of the temperature of the preoptic region of the marmot. Cryobiology 8: 389, 1971.

57. South, F. E., W. C. Hartner, and R. H. Luecke. Responses to preoptic temperature manipulation in the awake and hibernating marmot. Am. J. Physiol. 229: 150-160, 1975.

58. South, F. E., J. E. Heath, R. H. Luecke, L. T. Mihailovic, R. D. Myers, J. A. Panuska, B. A. Williams, W. C. Hartner, and H. K. Jacobs. Status of the role of the central nervous system and thermoregulation during hibernation and hypothermia. In: Hibernation and Hypothermia, Perspectives and Challenges. South, F. E., J. P. Hannon, J. S. Willis, E. T. Pengelley and N. K. Alperts (editors). New York: Elsevier, 1972, pp. 629-633.

59. South, F. E., and W. A. House. Energy metabolism in hibernation. In: Mammalian Hibernation III. Fisher, K. C., A. R. Dawe, C. P. Lyman, E. Schönbaum and F. E. South (editors). Edinburgh: Oliver & Boyd, 1967, pp. 305-324.

60. Stitt, J. T., J. D. Hardy, and J. A. J. Stolwijk. PGE fever: its effect on thermoregulation at different low ambient temperatures. Am. J. Physiol. 227: 622-629, 1966.

61. Stolwijk, J. A. J., and J. D. Hardy. Temperature regulation in man -- a theoretical study. Pflügers Arch. 291: 129-162, 1966.

62. Strumwasser, F. Thermoregulatory, brain and behavioral mechanisms during entrance into hibernation in the squirrel, *Citellus beecheyi*. Am. J. Physiol. 196: 15-22, 1959.

63. Whittow, G. C. (editor). Comparative Physiology of Thermoregulation. Vol. II. New York: Academic Press, 1973.

64. Whittow, G. C. (editor). Comparative Physiology of Thermoregulation. Vol. III. New York: Academic Press, 1973.

65. von Euler, K. Physiology and pharmacology of temperature regulation. Pharmacol. Rev. 13: 361-398, 1961.

66. von Euler, K. The gain of the hypothalamic temperature regulating mechanisms. Prog. Brain Res. 5: 127-131, 1964.

67. Williams, B. A., and J. E. Heath. Responses to preoptic heating and cooling in a hibernator, *Citellus tridecemlineatus*. Am. J. Physiol. 218: 1654-1660, 1970.

68. Wünnenberg, W., G. Merker, and E. Speulda. Thermosensitivity of preoptic neurons in a hibernator (golden hamster) and a non-hibernator (guinea pig). Pflügers Arch. 363: 119-123, 1976.

SLEEP AND HIBERNATION: ELECTROPHYSIOLOGICAL AND THERMOREGULATORY HOMOLOGIES

H. C. Heller[1], J. M. Walker[2], G. L. Florant[1], S. F. Glotzbach[1], and R. J. Berger[2]

ABSTRACT

Recent electrophysiological and thermoregulatory investigations have greatly enhanced our knowledge of sleep and hibernation in mammals. Studies on marmots and ground squirrels entering hibernation and shallow torpor, have shown that they are predominantly in slow wave sleep (SWS). In ground squirrels, total sleep time (TST) increases greatly as an animal enters torpor. A change in the distribution of arousal states occurs during entrance into hibernation such that ground squirrels spend progressively less time in REM sleep. Arousal states could not be identified by conventional criteria below a body temperature (T_b) of 25°C, however, at this T_b, TST approached 95% with less than 5% being REM. This is an interesting finding in light of thermoregulatory homologies which exist between SWS and hibernation. In SWS, T_b is regulated at a lower level than during wakefulness as indicated by lowered hypothalamic temperature (T_{hy}) sensitivity. Both threshold T_{hy} and the proportionality constant for the metabolic heat production (MHP) response to hypothalamic cooling are lower during SWS than during wakefulness. During REM sleep, the thermoregulatory system appears to be inactivated or severely inhibited. Changes in the characteristics of the thermoregulatory system during hibernation are qualitatively similar to the changes occurring during SWS. The entrance into hibernation involves a progressive decrease in the threshold T_{hy} and proportionality constant for the MHP response. These results lead us to hypothesize that the evolution of hibernation proceeded via a gradual extension of thermoregulatory adjustments occurring in all mammals during SWS.

ISBN 0-12-734550-7

I. HIBERNATION AS AN EXTENSION OF SLEEP

Hibernation is characterized by three behavioural criteria generally considered necessary to distinguish sleep from other behavioral states: immobility, stereotypic posture, and decreased responsiveness to stimuli (75). A fourth behavioral criteria of sleep, rapid reversibility of state, is not as evident in hibernators because of the influence of low body temperature on muscular movement. These behavioral similarities between sleep and hibernation have led some biologists and naturalists to call hibernation winter sleep, and of course, the German word for hibernation is Winterschlaf. Some physiologists have proposed that hibernation is an extension of sleep (26, 58, 62), but other authorities have considered hibernation and sleep distinct phenomena (17). Until recently, data did not exist which could clarify the relationship between these two physiological states (8).

Recordings of the electroencephalogram (EEG), electromyogram (EMG) and the electrooculogram (EOG) are the most prevalent criteria for the definition of arousal states, i.e. wakefulness, rapid eye movement (REM) sleep, and non-rapid eye movement (NREM) sleep which is largely synonomous with slow wave sleep (SWS) in the studies discussed in this paper (12, 17). Early studies of these parameters as well as activity in specific brain loci as recorded by depth electrodes were not directed toward comparisons between hibernation and euthermic sleep; but instead they focused on the influences of temperature on the activity of different brain areas and on the activity of specific brain areas as a function of events in the hibernation cycle (4, 41, 47, 61, 65, 67, for an excellent review, see 50).

The first publication of information enabling a comparison of sleep and hibernation was that of South *et al.* (62). These investigators made chronic recordings of cortical EEG, EMG, and activity in the mesencephalic reticular formation and in either the anterior or posterior hypothalamus in eight yellow-bellied marmots (*Marmota flaviventris*) throughout entire bouts of hibernation. Their results showed that during euthermic sleep the marmots spent about 20% of total sleep time (TST) in REM sleep and about 80% of TST in SWS, a distribution typical of other mammals. In their words, during entry into hibernation "...as brain temperature began to drop from 37°C, the immediate impression was that slow wave activity almost totally displaced low voltage-fast activity (i.e. REM sleep and wakefulness). However, this was more apparent than real since at 33°C the percentage of time occupied by the latter remained at approximately 20%." It became difficult to identify arousal states by conventional criteria as brain temperature (T_{br}) approached 20°C. The authors concluded that the initial entry into hibernation "...is not only initiated from an apparent condition of normal sleep but may well be a physiologic extension of it. Down to brain temperatures of approximately 25°C, the distribution of time in what would appear to be SS (SWS) remained close to 80%." The authors did not clearly state how they interpreted the low voltage fast activity which constituted the remaining 20% of the recording time; how much of it was REM and how much of it was wakefulness?

Preliminary results of EEG studies of ground squirrels entering hibernation published by Satinoff (58) led her to conclude, contrary to South *et al.* (62), that ground squirrels enter hibernation from wakefulness or a drowsy state resembling light SWS. Satinoff suggested, however, that hibernation might be a "deeper" stage of sleep. Other workers have also noted the occurrence of EEG sleep spindles in animals entering hibernation, but have not drawn conclusions about homology between sleep and hibernation (61, 66).

We have recently completed an extensive electrophysiological study of ground squirrels which was undertaken to determine the changes in arousal states which occur during entrance into hibernation. The detailed results of this study are published elsewhere (76); here we shall briefly describe the study and summarize the results. Electrodes for chronically recording cortical and hippocampal EEG, EOG, and neck EMG were implanted in ten male ground squirrels (2 *Citellus beldingi* and 8 *C. Lateralis*) along with a thermocouple re-entrant tube in the cortex, for measurements of brain temperature (T_{br}). All recordings were made from the animal in its home cage.

Continuous 24 hour recordings were made on five of the animals during July and August at an ambient temperature of 22°C. The other 5 animals were studied during 14 entrances into hibernation at an ambient temperature of 6°C from September through March. Five to ten hr recordings were obtained prior to entrance and following arousal on each animal on at least one occasion. Other 5 to 10 hr periods were sampled during deep hibernation. For three animals we were able to gradually raise T_a and therefore T_{br} to 24 to 26°C without initiating arousal and then to continue recording from these animals hibernating at a high T_{br} for up to 24 hr before they aroused to euthermia.

The electrophysiological recordings from euthermic ground squirrels (Fig. 1) were entirely typical of similar recordings during sleep and wakefulness in other rodents (72) and other mammals in general (12, 17). The distribution of arousal states in 24 hr recordings made during July and August were also quantitatively similar to values reported for other rodents (72). Total sleep time accounted for about 65% of the 24 hr period, and the % TST spent in REM was about 20%. The squirrels exhibited a distinct diurnal rhythm sleeping more during the 12 hr dark period than during the 12 hr light period.

The distributions of arousal states just before a bout of hibernation, during the early entrance into hibernation, and immediately following the return to euthermia are shown in Table I (all data) and in figure 2 (1 bout from each subject). Just prior to a bout of hibernation the animals spent an average of 53% of total recording time (TRT) asleep of which 22% consisted of REM.

During the early entrance (from a T_{br} of 35° to a T_{br} of 25°C) TST increased to 88%, but REM accounted for only 10% of TST. Moreover, it can be seen in figure 2 that the amount of REM expressed either as % TST progressively declined during entrance. In only 2 instances was REM sleep observed below a T_{br} of 27°C. It was not possible to score the EEG by conventional criteria below a T_{br} of 25°C because of the diminished amplitude and frequency (Fig. 3). However, computer analysis of the EEG (19) revealed that even below a T_a of 25°C the greatest proportion of time was spent in the

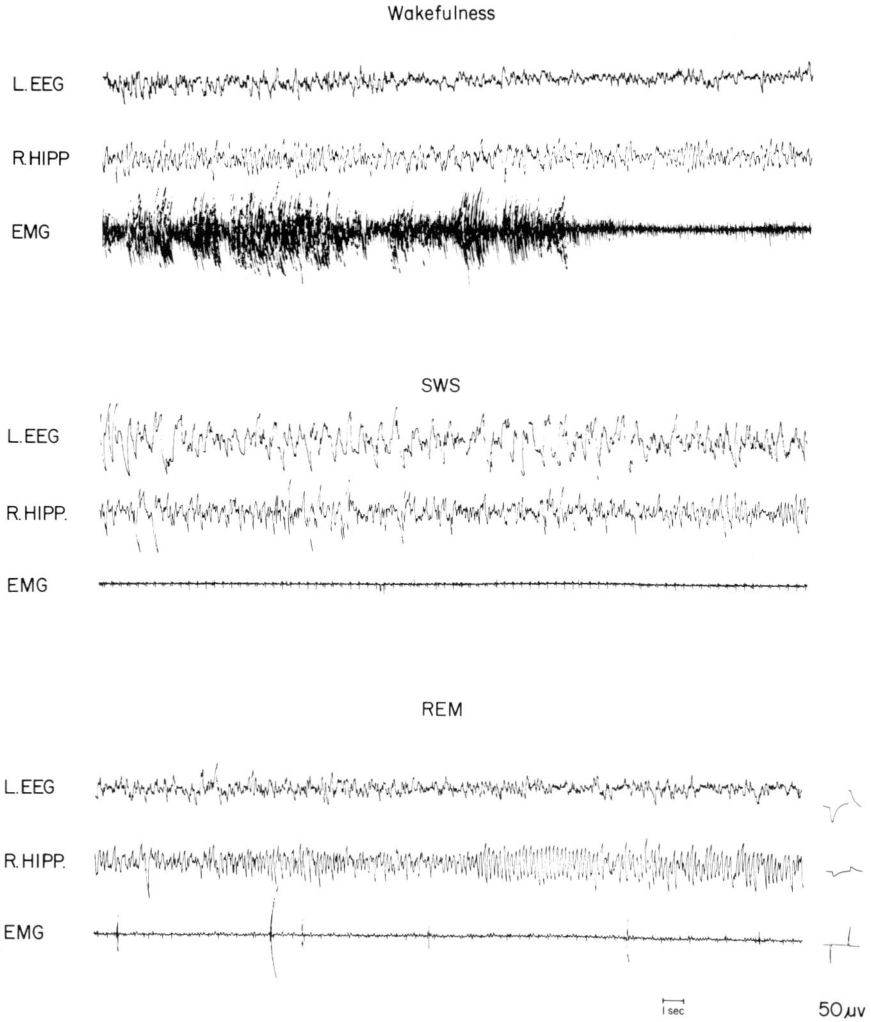

Fig. 1. *Electrophysiological correlates of wakefulness, slow-wave sleep (SWS), and rapid eye movement (REM) sleep in the golden-mantled ground squirrel, C. lateralis. Note the intermittent hippocampal (Hipp) theta rhythm (3 - 7 CPS) during wakefulness and the sustained theta activity during REM sleep. EEG was derived between occipital and parietal leads on the right or on the left side of the cortex. (From Walker et al., 1977, with permission.)*

Table I. Percent time spent in wakefulness, SWS, and REM sleep before, during entry into, and after bouts of deep hibernation

	5 hrs immediately preceding entrance			Entrance down to a T_{br} of 25°C			5 hrs immediately following return to euthermia		
	W (%TRT)	SWS (%TST)	REM (%TST)	W (%TRT)	SWS (%TST)	REM (%TST)	W (%TRT)	SWS (%TST)	REM (%TST)
A. *C. lateralis*									
3 animals, 10 bouts recorded between Oct. and March at a T_a of 6°C.	40	77	23	16	90	10	29	81	19
B. *C. belding*									
2 animals, 4 bouts recorded between Oct. and March at a T_a of 6°C.	57	80	20	7	90	10	31	76	24

04 CPS band which is more typical of SWS than either REM sleep or wakefulness.

The absence of REM at low T_{br}'s during hibernation was confirmed in recordings from hibernating ground squirrels gradually warmed to 26°C. At this T_{br} the cortical EEG consisted of a mixed pattern of spindles (75-150 μv, 3-5 CPS), intermittent slow waves (75-125 μv, <3 CPS), and low voltage activity (<25 μv, 5 to 8 CPS) (Fig. 4a). These recordings resembled SWS more than any other euthermic state. There was no evidence of REM, but brief periods of wakefulness were indicated by simultaneous EEG desynchronization and increased EMG activity.

The active arousal or rewarming from deep hibernation was characterized by a massive increase in EMG

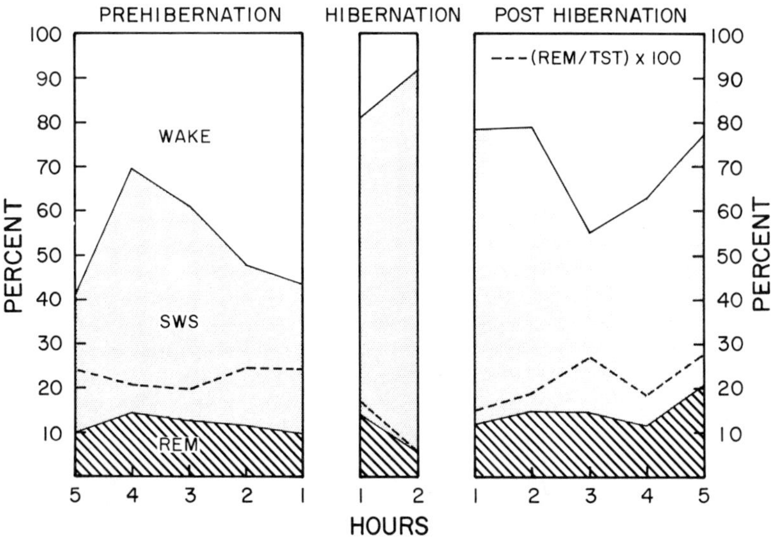

Fig. 2. *Mean hourly percentages of wakefulness, SWS, and REM sleep expressed as total recording time (TRT) for five ground squirrels for recording periods prior to, during early entrance into and immediately following arousal from hibernation. Only one recording for each squirrel for each condition is included. Dashed line is REM sleep expressed as percentage of total sleep time (TST) (from Walker et al., 1977 with permission).*

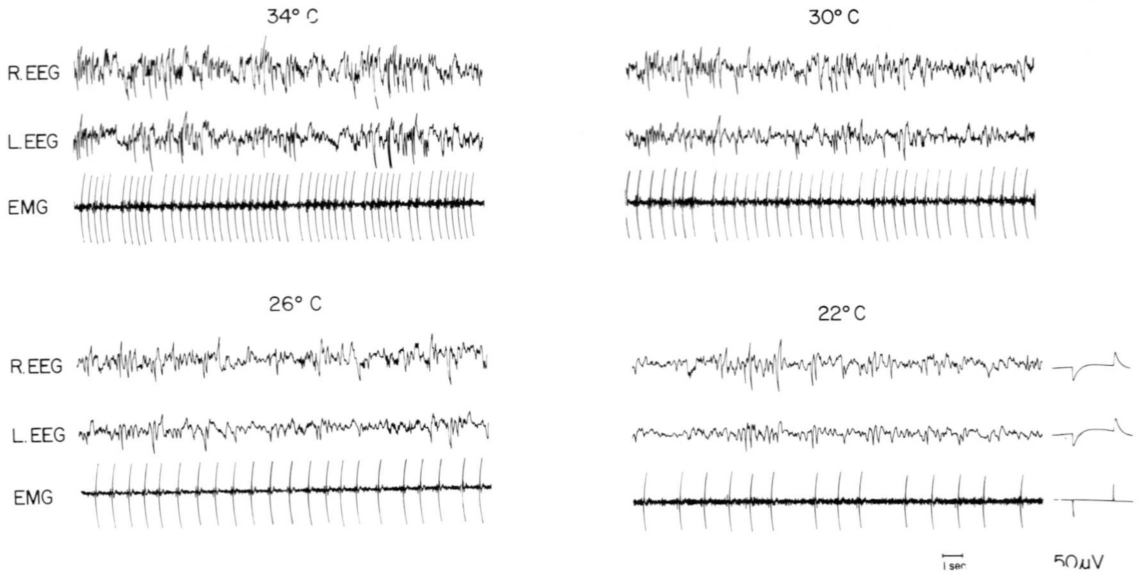

Fig. 3. *Examples of slow wave EEG activity and EMG from one ground squirrel during one entry into hibernation at brain temperatures of 34, 30, 26, and 22°C.*

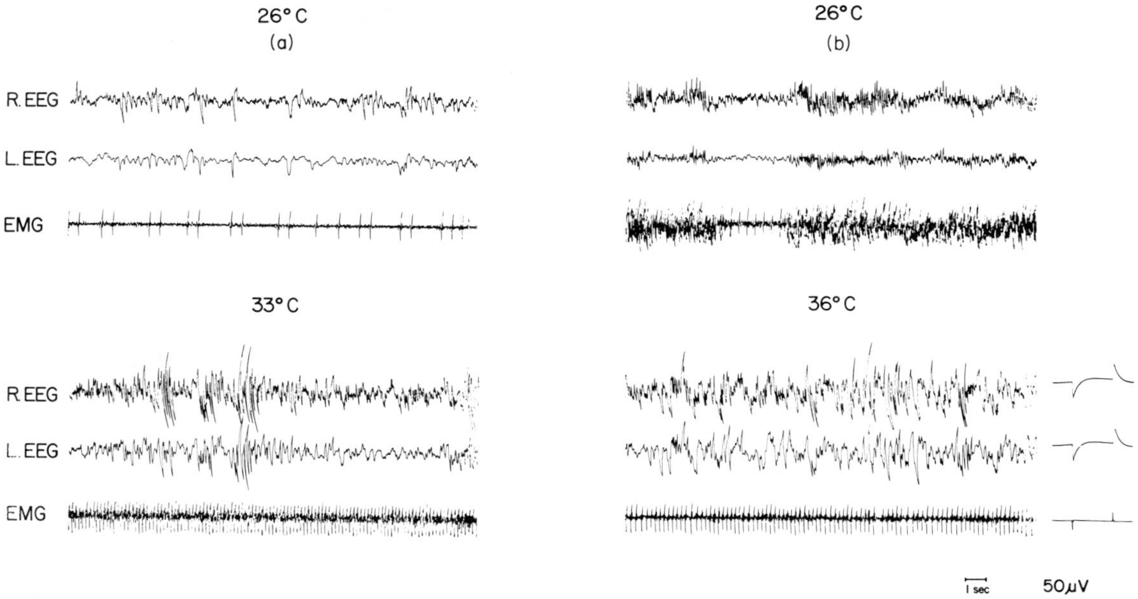

Fig. 4. *Examples of electrophysiological recordings obtained from a hibernating ground squirrel which has been gradually rewarmed by slowly raising T_a. In record "a" both T_a and T_b were 26°C and the animal showed no signs of arousal. Record "b" was obtained shortly after "a" at the same T_{br}, but the active arousal process had begun. Recordings at T_{br}'s of 33 and 36°C are from later in the same arousal. Activated EEG and EMG began to decrease at a T_{br} of approximately 33°C after which SWS dominated the record.*

activity and by a low voltage, moderately fast EEG (<80 μv, 5 to 8 cps) (Fig. 4) which resembled that of euthermic wakefulness. Occasionally there were brief periods (<1 min) of reduced EMG activity, and at T_{br}'s above 27°C slow wave EEG activity characterized these periods of quiescence. When T_{br} reached approximately 33°C, there was a considerable reduction in EMG activity and the EEG was largely that of SWS and intermittent wakefulness. These findings are similar to those observed by Kayser (40) who wrote "...a hibernator almost awakened from his hibernation shows a nervous activity comparable to that of a normothermic animal during sleep".

In subsequent studies during the winter months, 7 to 10 day continuous recordings from 2 golden-mantled ground squirrels showed some 1 to 3 day bouts of torpor in which T_{br} dropped below euthermic levels but remained 3 to 5°C above a T_a of 22°C. The decrease in T_{br} was characterized by similar distributions of arousal states as were seen during entrance into hibernation at lower T_a's (Figs. 5 and 6; Table II). Total sleep time increased to 88% of which 92% was SWS. The REM diminished as T_{br} fell and occupied about 3% of TST below a T_{br} of 27°C. When T_{br} stabilized at 25 to 27°C, arousal states could still be defined by standard electrophysiological criteria. SWS characterized by slow wave cortical EEG (<3 cps, 75-150 μv) and spindling activity, constituted the majority of the recordings, and the remainder consisted of brief episodes of wakefulness. The cortical EEG during sleep was similar in appearance to that observed when deeply hibernating subjects were warmed up to similar T_{br}'s by raising T_a; but there was a greater density of slow wave activity in these recordings than in those from rewarmed animals. These animals in shallow torpor could be awakened by sensory stimuli, and if the wakefulness persisted, T_{br} returned to euthermic levels. Spontaneous return to euthermia was more gradual and consisted of SWS alternating with wakefulness.

Following a bout of torpor the TST was significantly higher than just before the entrance as was also reported by Allison and Van Twyver for the echidna (4). However, once again REM constituted about 20% of TST (Table I, Fig 2). In spite of the apparent absence of REM

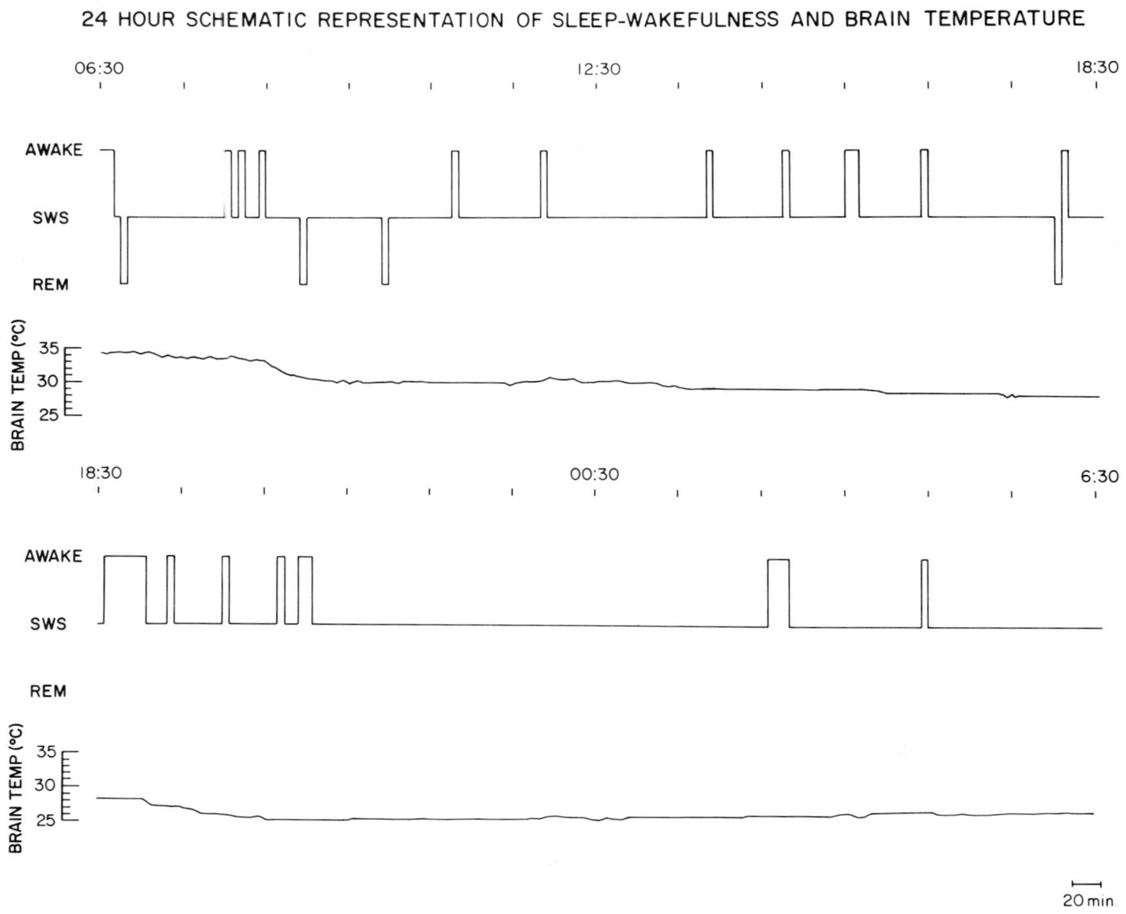

Fig. 5. *A 24 hr record of T_{br} and arousal states as a golden-mantled ground squirrel enters shallow torpor at a T_a of 22°C. The last episode of REM occurs at a T_{br} of about 27 °C. This entire bout of torpor is shown in figure 6.*

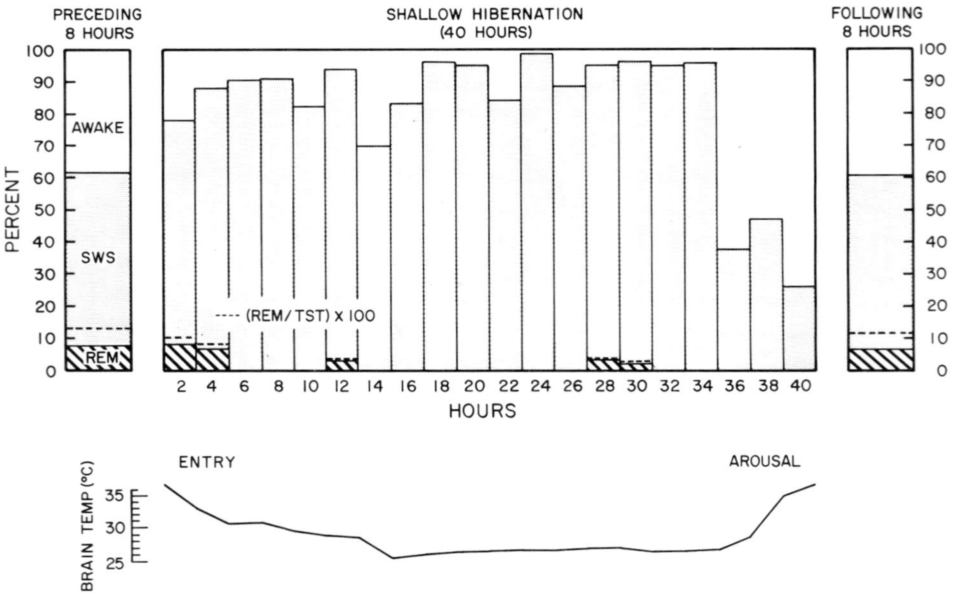

Fig. 6. *Arousal state distributions and T_{br} of a golden-mantled ground squirrel undergoing an entire bout of torpor at a T_a of 22°C plus summaries of arousal state distributions for the 8 hr preceding and the 8 hr following the bout.*

Table II. Percent time spent in wakefulness, SWS, and REM sleep before, during entry into, throughout, during exit from, and following bouts of shallow ($T_{br} > 25°C$) torpor.

	5 hrs immediately preceding entrance			Entry			Torpor			Exit from Torpor			5 hrs immediately following return to euthermia		
	W %TRT	SWS %TST	REM %TST	W %TRT	SWS %TST	REM %TST	W %TRT	SWS %TST	REM %TST	W %TRT	SWS %TST	REM %TST	W %TRT	SWS %TST	REM %TST
A. *C. lateralis* n=2 Nov. - Feb. Ta=22°C	37	83	17	12	92	8	11	97	3	31	100	0	39	89	11
B. *C. tereticaudus* n=4 June - July T_a=25°C	35	89	11	8	94	6	8	100	0	38	100	0	42	87	13

during hibernation, we saw no transient increase similar to the "REM rebound" observed after REM deprivation in euthermic animals.

Recently Walker, Garber and Berger (pers. comm) have extended electrophysiological investigations to a species undergoing daily bouts of shallow torpor. Four round-tailed ground squirrels (*C. tereticaudus*) received chronic implants as described above for *C. lateralis* and *C. beldingi*. Torpor was induced at a T_a of 25°C by food deprivation during June and July. In one animal torpor occurred after 2 days of deprivation whereas another animal required 14 days of deprivation before torpor was observed. After the first bout of torpor, subsequent bouts were entered at 24 and 48 hr intervals. Entrance typically occurred during the night and bouts lasted an average of 10 hrs. In each case the return to euthermia was spontaneous. Electrophysiological and T_{br} recordings were obtained during 20 episodes of torpor as well as during euthermic periods between bouts of torpor. Preliminary results derived from scoring one complete sequence of recordings for each of the 4 animals are presented in Table II.

The distribution of arousal states during entrance into shallow torpor was similar to those of *C. lateralis* and *C. beldingi* entering hibernation. Total sleep time increased to 90-95% while REM progressively decreased and constituted only 5 to 8% of the TST during entrance (Table II). At a T_a of 25°C it was possible to identify arousal states according to conventional electrophysiological criteria during entire bouts of torpor. At a stable T_{br} of 26-27°C, the electrophysiological recordings revealed brief episodes of wakefulness and longer periods of SWS during which the cortical EEG consisted of slow waves (<3 cps, >125μv) and spindles which were especially prevalent just following wakefulness. SWS comprised approximately 92% of the period of torpor over which T_{br} was stable. The other 8% consisted of wakefulness, and REM was completely absent.

The distribution of arousal states during the return to euthermia differed from that observed over the same range of T_{br}'s in ground squirrels arousing from deep hibernation. Whereas *C. lateralis* and *C. beldingi* showed a predominantly activated EEG and EMG up to a T_{br} of about 33°C, the return to euthermia in

C. tereticaudus consisted of only 38% wakefulness and 62% SWS. Again, REM was absent. In light of these results, the use of the word "arousal" to refer to the return to euthermia from a torpid condition should be abandoned as it may bear inaccurate connotations about the arousal state of the animal. Perhaps a better phrase would be "exit from hibernation" since the phrase "entrance into hibernation" is used almost universally to refer to the reciprocal phenomenon.

If hibernation is an extension of sleep, we should expect there to be a circannian rhythm of propensity to sleep in euthermic animals which would correspond with the circannian rhythm of hibernation. Preliminary findings from an ongoing study confirm this expectation. Twenty-four hr recordings of EEG, EOG, EKG, EMG, and T_{br} were obtained from euthermic ground squirrels during the months of July-August, November-December, February-March, and May-June. The animals were maintained at a constant T_a of 22°C and on a 12L:12D photoperiod. Recordings were scored "blind" (i.e., without knowledge of subject identity or date of recording) for arousal states. Figure 7 shows the progressive seasonal changes in percent time spent in sleep for 2 euthermic ground squirrels. Although there is some variation between animals in the percent TST, a consistent trend was evident. TST increased in the fall, remained high during winter, decreased in the spring and remained at the lowest level during the summer. Some animals exhibited higher amounts of sleep during late summer than during early summer. The distribution of TST into SWS and REM remained relatively constant throughout the year (Fig. 8) although mean duration of episodes of wakefulness as well as the % total recording time spent awake were significantly longer in summer than in winter (Fig. 8 and 9).

Our conclusion must be that where electrophysiological criteria can be applied, hibernation appears to be homologous with sleep, but more importantly it seems to consist predominantly of SWS with REM progressively decreasing as T_b falls and then disappearing completely below a T_b of 27°C. It would be extremely interesting to develop or make use of additional electrophysiological correlates of arousal states which would enable us to extend our resolution of arousal state homologies to lower T_b's. Later on we shall describe a different kind of physiological parameter which indicates continuity between euthermic SWS and deep hibernation.

II. AROUSAL-STATE-RELATED CHANGES IN THE THERMOREGULATORY SYSTEM

The hallmark of hibernation and shallow torpor is a decline in body temperature. Given the electrophysiological evidence for homology between these phenomena and sleep, it is of interest to also look for thermoregulatory homologies between them. In this section we shall review studies of sleep-related changes in body temperature and the thermoregulatory system.

A daily fluctuation in T_b is a well known phenomenon in man and other endotherms. Over the years a variety of hypotheses have been advanced to account for this fluctuation and many of those hypotheses have been effectively discounted (6). There remain two demonstrated underlying causes: one is a circadian rhythm in T_b

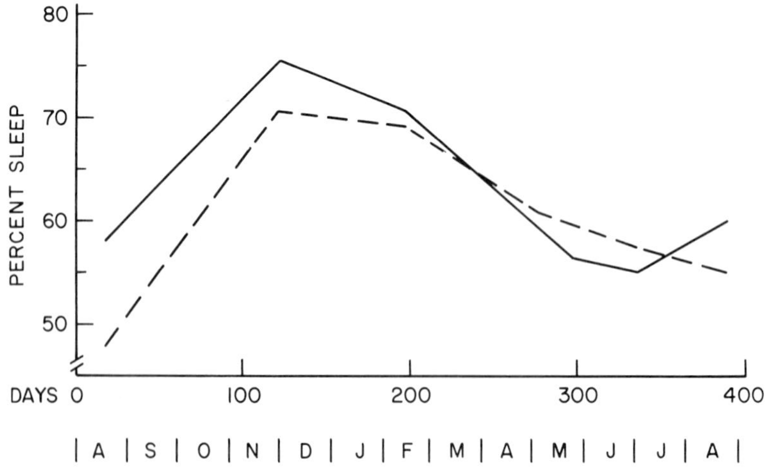

Fig. 7. *Annual changes in percent of 24 hr period spent asleep for two golden-mantled ground squirrels (solid and dashed lines) held at a constant temperature of 22° ± 2°C and a photoperiod of 12L:12D. Animals were euthermic for the duration of all recording periods.*

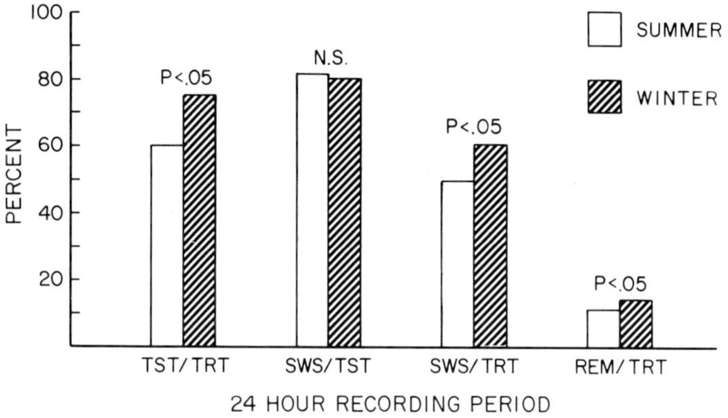

Fig. 8. *Distribution of arousal states in 4 golden-mantled ground squirrels in summer (August) and winter (November). The increase in TST of the euthermic squirrel does not involve a change in the distribution of sleep states.*

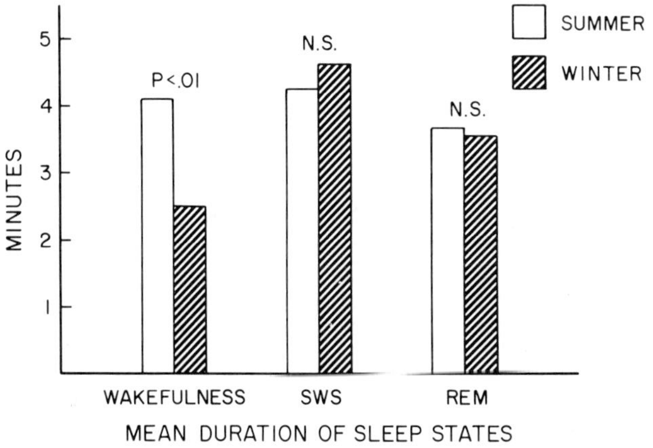

Fig. 9. *Mean duration of individual episodes of wakefulness, SWS, and REM in summer (August) and winter (November) in 4 golden-mantled ground squirrels. Episodes of wakefulness average significantly longer in summer than in winter, but sleep states show no changes.*

(6), and the other is sleep-related changes in T_b (37, 54). The independence of these two influences on T_b is revealed by the fact that in the absence of time cues the daily rhythms of sleep-wakefulness and T_b free run with different periodicities (7, 45, 77). Normally, these two influences on T_b are in synchrony with each other and we can think of sleep-related influences on T_b as being superimposed on an underlying circadian rhythm of T_b. In this paper we are concerned only with the sleep-related influences and their underlying causes.

Earlier studies of changes in T_b thermoregulatory responses during sleep did not include electrophysiological recordings and, therefore, cannot be interpreted in terms of electrophysiologically defined arousal states. Nevertheless, their results are quite interesting. In a pioneering study (16), Day measured rectal and skin temperatures and evaporative water loss during afternoon naps in nine children between the ages of 5 months and 4 years. He consistently observed a decline in rectal temperature averaging 0.55°C beginning at the onset of sleep and coincident with an increased evaporative water loss and a rise in skin temperature. The end of the decline in rectal temperature was preceded by decreases in evaporative water loss and skin temperatures. These results clearly demonstrated a regulated decline in T_b associated with sleep which occurred at a time of the day when the circadian rhythm of T_b should still be in the rising phase.

Thermoregulatory adjustments during the onset of sleep have also been seen in adult humans. In neutral or cool environments, the onset of sleep is accompanied by a rise in skin temperature and a fall in rectal temperature (30, 42, 43, 44). In neutral or warm environments sweating increases at the onset of sleep (23, 59, 68).

Animal studies have also indicated the existence of thermoregulatory adjustments associated with sleep. Even brief episodes of behavioral sleep were accompanied by declines in T_{br} in cats (2), rats (1), and rhesus monkeys (30). In the study of the rhesus monkey sleep was defined by eye closure, and sleep onset was seen to be accompanied by ear pinnae vasodilation and a subsequent fall in T_{br} (30).

Even though these studies demonstrated that T_b was regulated at a lower level during sleep than during wakefulness, the picture of thermoregulation during sleep became more complex when electrophysiological recordings were used to distinguish between thermoregulatory events occurring during REM and SWS (see 37 for a review). Transitions from wakefulness to SWS were associated with a decrease in the regulated T_b consistent with the earlier behavioural studies. In a neutral or cold environment this transition was accompanied by peripheral vasodilation and decreased shivering, whereas in a neutral-to-warm environment it was accompanied by an increase in evaporative water loss (3, 5, 9, 18, 31, 53, 57, 59). In contrast, measurements of thermoregulatory responses during REM revealed no consistent shift in the regulated T_b, but instead indicated a lack of regulation. At a warm T_a a cessation of panting or sweating and even vasoconstriction were seen during transitions from SWS to REM (9, 39, 53, 56, 57, 59, 60, 68, 74). In cool environments a cessation of shivering and even vasodilation may occur at the onset of REM (3, 9, 56, 57, 74).

Direct measurements of the characteristics of the central nervous thermoregulatory system are essential for an understanding of the causes of the various arousal-state-related changes in T_b's and thermoregulatory responses. The hypothalamus is a major integrative center in the thermoregulatory system (14, 27, 29). In addition, a major feedback loop in this regulatory system is the temperature of the anterior hypothalamus (T_{hy}). Results from a large number of species have shown that, if the T_{hy} of a mammal in a thermoneutral environment is gradually lowered, a threshold T_{hy} (T_{set}) will be reached below which the rate of metabolic heat production (MHP) is inversely proportional to T_{hy}. Similarly, if the T_{hy} of this mammal is gradually raised, a different T_{set} will be reached above which the activity of heat loss effectors is directly proportional to T_{hy}. The characteristics of this central nervous regulator of T_b can be expressed by the following equation:

$$R_1 - R_0 = \alpha\ (T_{hy} - T_{set})$$

where R_0 is the rate of a thermoregulatory response when $T_{hy} = T_{set}$ and R_1 is the rate when T_{hy} is above the T_{set} for a heat loss response or below the T_{set} for a heat production response. Other factors that influence the thermoregulatory system, such as skin temperature, spinal cord temperature, exercise, pyrogens, as well as different states of arousal can be formulated as inputs that modulate T_{set} and/or the proportionality constant α.

The first explorations of hypothalamic thermosensitivity during sleep were by Parmeggiani and colleagues (55), who used diathermic heating to elicit thermal polypnea and panting in cats. Results were not reported for wakefulness, but comparisons were made between REM and SWS. Marked thermal polypnea and even panting were elicited by hypothalamic heating during SWS, but the same alterations of T_{hy}'s had no effect on respiratory rate during REM, showing that the central thermoregulatory system is functioning during SWS but is severely inhibited during REM, as was suggested earlier by Parmeggiani and Rabini (56) (Fig. 10).

The first quantitative comparisons of hypothalamic thermosensitivity as a function of arousal states were obtained from kangaroo rats (*Dipodomys sp.*) by Glotzbach and Heller (24). In these experiments EEG, EMG, T_{hy}, and rate of MHP were continuously recorded while T_{hy} was manipulated. Rate of MHP was plotted as a function of T_{hy} for wakefulness, SWS, and REM. The results demonstrated proportional regulation of the MHP response during SWS but, in comparison with wakefulness, the T_{set} and the α for the response were lower, thereby accounting for the regulated lowering of T_b during SWS. Increases in MHP could not be elicited in kangaroo rats by lowering T_{hy} during REM, indicating once again, that the thermoregulatory system is severely inhibited during REM. In figure 11 data on hypothalamic thermosensitivity as a function of arousal state are presented for a kangaroo rat not included in our earlier study.

Hypothalamic thermosensitivity as a function of arousal state has recently been measured in a hibernator, the marmot (*M. flaviventris*) (22). These experiments were performed on three individuals, and the results were similar to those obtained from kangaroo rats. The plot of rate of MHP versus T_{hy} exhibited a lower threshold and a lower slope during SWS than during wakefulness (Fig. 11). No hypothalamic thermosensitivity was apparent

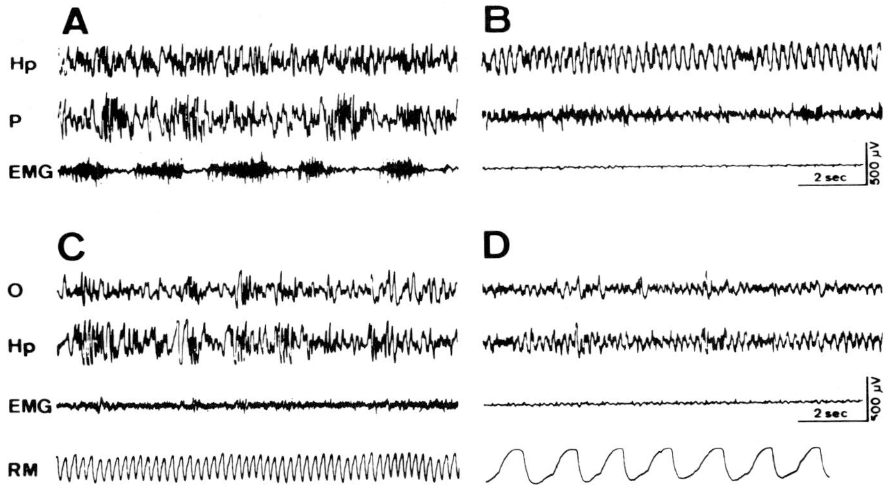

Fig. 10. *Recordings of EEG, EMG, and respiratory movements (RM) in a cat during sleep. Traces labelled A and B were recorded at a T_a of 6°C during SWS and REM respectively. Bursts of shivering are evident in the neck muscle EMG during SWS but not during PS. Traces labelled C and D were obtained at a T_a of 36.5°C during SWS and REM respectively. Panting during SWS is evidenced by the high frequency of respiratory movements (255/min) in C. During the bout of REM shown in D there is no panting as evidenced by the slower, deeper respiratory movements. Abbreviations are as follows: HP = hippocampus dorsalis; P = parietal; EMG = neck electromyogram; o = occipital, (reproduced from Parmeggiani and Rabini, 1967, with permission).*

during REM. We are presently confirming the generality of these findings on another species of hibernator, the golden-mantled ground squirrel.

Since these results demonstrate that during SWS, the regulation of T_b is qualitatively the same as during wakefulness, but reset to a lower level, and since the entrance into hibernation is an extension of SWS as evidenced by electrophysiological recordings, might not hibernation represent an extension of the thermoregulatory adjustments which occur during SWS?

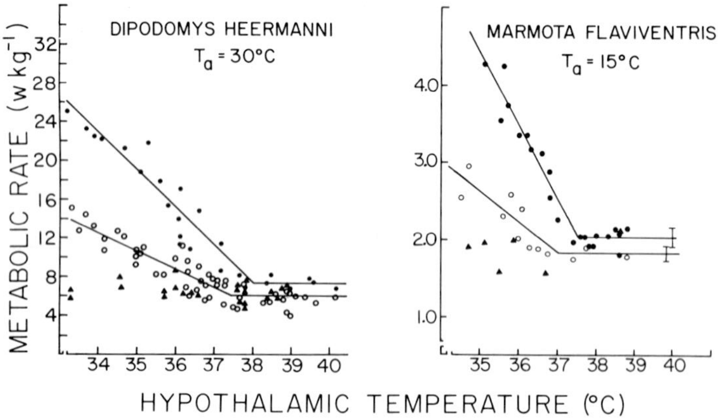

Fig. 11. *Characteristics of hypothalamic thermosensitivity as a function of arousal states in a kangaroo rat, a non-hibernator, and in a marmot, a hibernator. Data points obtained during wakefulness are the solid circles, those obtained during SWS are the open circles, and those obtained during REM are the closed triangles. The* α_{MHP} *is -3.9 W/kg·°C for the wakeful kangaroo rat and -1.9 WKG*$^{-1}$*°C*$^{-1}$ *for the wakeful kangaroo rat in SWS, values similar to those published for other kangaroo rats (24). For the marmot the wakeful* α_{MHP} *is -1.10 W/Kg·°C and the SWS* α_{MHP} *is -0.4 W/kg·°C.*

III. THE REGULATION OF BODY TEMPERATURE DURING TORPOR

The evidence is now conclusive that lowered T_b's during torpor and deep hibernation are not due to an absence of or an inadequacy in thermoregulatory abilities; on the contrary, the torpid animal can regulate its T_b with considerable precision (for a review see 37). The questions we shall address in this section are: what are the characteristics of the central nervous regulator of T_b during hibernation? Are they continuous with values observed during euthermia? If so, how is the regulatory system reset to achieve the entrance into hibernation? Is it continuously operative during hibernation? How is it reset to achieve the exit from hibernation?

At the time of the Third Hibernation Symposium it seemed very unlikely that the same central thermoregulatory mechanism could be operative at both euthermic and hibernating T_b's (26). This opinion was based on a neuronal model of the hypothalamic thermoregulatory mechanism and quantitative measures of hypothalamic thermosensitivity in the dog, as hypothalamic thermosensitivity had not been measured in any small mammal let alone any hibernator. The model predicted that the proportionality constant (α) for a thermoregulatory response should decrease with a biologically reasonable temperature coefficient (Q_{10}) as the T_{set} for the response decreased. When a Q_{10} of 2.5 was applied to the value for hypothalamic thermosensitivity measured on the dog, it was apparent that the driving force for the metabolic heat production response would become insignificant long before a T_{set} low enough for deep hibernation could be reached.

Initial experiments on golden mantled ground squirrels seemed to corroborate this view that the central thermoregulatory system operative in euthermia was inoperative during deep hibernation. Calorimetric measurements on hibernating ground squirrels over a range of T_a's revealed that metabolic rate was a passive function of T_b (28). Broad manipulations of T_{hy} of hibernating ground squirrels seemed to have no proportional influence on metabolic rate as it did during euthermia (38). However, in both of these studies it was noted that the hibernating squirrel remained sensitive to very low temperatures; T_{hy}'s somewhat below 3.0°C could initiate arousals from hibernation. Since these arousal responses could be reversed only by rapidly heating T_{hy} above 10-12°C it was assumed that they were alarm responses involving a reactivation of the central thermoregulatory system. That this alarm temperature (T_{alarm}) was sensed in the brain was indicated by the fact that T_a and therefore T_b could be lowered below T_{alarm} without causing an increase in MHP as long as T_{hy} was held above T_{alarm}.

Other workers studying different species also reported a low temperature sensitivity of the brains of hibernating animals. Using spoon-shaped external thermodes Lyman and O'Brien (48, 49) showed that two Citellids (*C. lateralis* and *C. tridecemlineatus*), hamsters (*Mesocricetus sp.*), and the dormouse (*Glis glis*) responded to cooling of the head by increasing heart rate, respiratory rate, and skeletal muscle action potentials. In the Citellids it was possible to keep T_{br} between 4°C and 10°C while cooling T_a as low as -7.8° (and, therefore, T_{skin} to as low as 0.5°C) without eliciting responses in heart rate, respiratory rate, and EMG. South and colleagues (51, 63) using single hypothalamic thermodes to manipulate T_{hy} of hibernating marmots demonstrated a T_{set} below which heart rate, EMG, and various T_b's increased. Conversely they showed that warming the hypothalamus of the hibernating marmot could result in a decrease of EMG and heart rate. These results combined with many reports of animals hibernating for long periods while holding T_b considerably above T_a (37) led to the supposition that the hibernator's thermoregulatory repertoire consisted of more than on/off responses triggered by a low T_a or T_b.

Subsequent experiments on hibernating golden-mantled ground squirrels employing finer manipulations of T_{hy} at levels close to the "alarm level" revealed a range of T_{hy}'s over which the characteristics of hypothalamic thermosensitivity of the hibernating ground squirrel were qualitatively similar to those measured in the euthermic animal (33, 34) (Fig. 12). A T_{set} for the metabolic heat production response was continuously present, and when T_{hy} was cooled below this T_{set} the increase in rate of MHP was proportional to (T_{hy} - T_{set}). Similar results were recently obtained from the marmot (21). Of course, both the T_{set} and the proportionality constant for the metabolic heat production response were much lower in hibernation than in euthermia.

Quantitative comparisons of α_{MHP} as a function of T_{set} in euthermia and in hibernation yielded interesting results. First, α_{MHP} expressed in weight specific terms was 6 to 10 times higher in the ground squirrel than in the dog (34). This eliminated the earlier objection to the concept of the same hypothalamic regulatory mechanism functioning during both euthermia and hibernation. It also led to the hypothesis that high hypothalamic thermosensitivity might be a specific adaptation of hibernators. Hypothalamic thermosensitivity has subsequently been shown to be inversely related to body size for hibernators and non-hibernators alike (32). Second, if we express the relationship between α_{MHP} and T_{set} as a temperature coefficient, it is of the order of 2.5 (21, 33), a value compatible with the concept that the same regulatory mechanism is continuous between euthermia and hibernation.

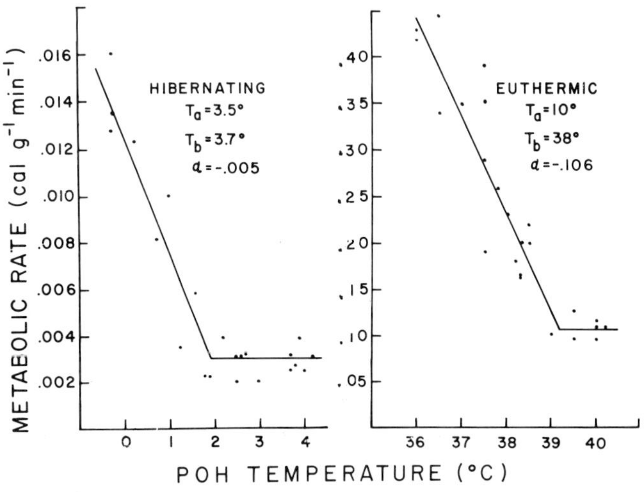

Fig. 12. *Characteristics of hypothalamic thermosensitivity for a golden-mantled ground squirrel during euthermia and hibernation.*

Experiments were undertaken to follow the change in the T_{set} for the MHP response in golden mantled ground squirrels during entrance into hibernation. Animals with chronic thermode and thermocouple reentrant tube implants were adapted to a metabolism chamber. As they entered hibernation T_{hy} was manipulated while rate of MHP was continously measured. Because the characteristics of the thermoregulatory system were continually changing, manipulations were grouped into series lasting 10 to 30 minutes. Within these 10 to 30 minute periods attempts were made to find the smallest decrease in T_{hy} which would stimulate an increase in MHP. Typically we would find one level of cooling which would not elicit a response and a lower level of cooling which would elicit a response and assumed that the actual T_{set} was between those two T_{hy}'s at that time. Data obtained from multiple entries of three ground squirrels are shown in figure 13. Clearly, the T_{set} for the MHP response is gradually lowered during the entrance into hibernation and the central thermoregulatory system appears to be continuously operative over the entire range of T_b's experienced by the hibernator. These results also led us to postulate an explanation for observations of "braking" or slowing the entrance into hibernation by bursts of MHP (46, 66). If T_{set} is suppressed more rapidly than T_b can fall, then the entrance appears passive. If, on the other hand, T_b tends to fall more rapidly than T_{set}, a MHP response will occur thus slowing or braking the entrance into hibernation. A prolonged plateauing of T_{set} would result in T_b being regulated at some level between euthermia and deep hibernation. In similar experiments on marmots, cases were found in which T_{set} was above actual T_{hy} for large portions of the entrance (21).

Recent experiments on marmots have added convincing evidence for the continuity of the central thermoregulatory mechanism over the entire range of T_b's experienced by the hibernator (22). Since the marmot enters hibernation more slowly than the ground squirrel, we decided to attempt to measure α_{MHP} at various T_{sets} between euthermia and deep hibernation. The characteristics of the system are still continuously changing as the marmot enters hibernation albeit more slowly than in

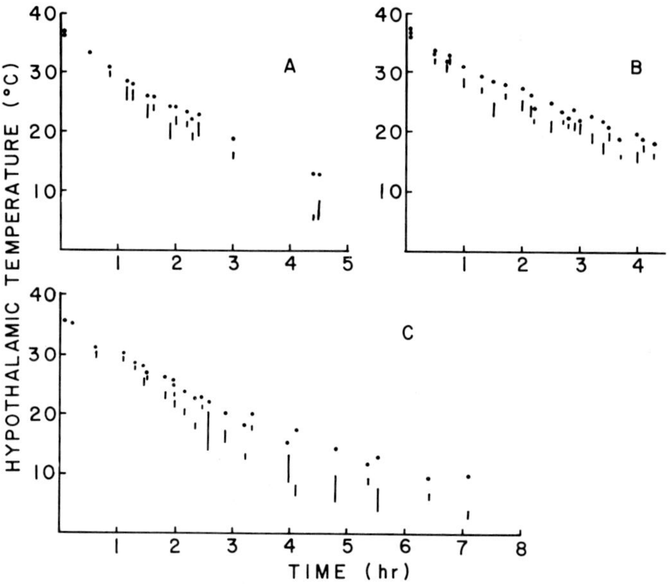

Fig. 13. Demonstration of the continuous decline in the threshold hypothalamic temperature (T_{set}) for the metabolic heat production response during entrances into hibernation. Each graph consists of data obtained on one ground squirrel. Data were obtained from A during two entrances, from B during 3 entrances, and from C during 4 entrances. Each dot indicates the average T_b over a 15 to 30 minute period and the vertical line under that dot connects the lowest T_{hy} which did not elicit an increase in MHP and the highest T_{hy} which did elicit an increase in MHP over that period of time. Therefore, T_{set} at the time each T_b was recorded falls on the vertical line beneath the dot. From (35).

the ground squirrel, so many manipulations of T_{hy} at each T_{set} are not possible. We, therefore, adopted a two point method for estimating α_{MHP} during entrance. At each T_{set} enough manipulations of T_{hy} were made to be able to measure two response levels, and α_{MHP} was calculated from these two points. Many such estimates of α_{MHP} were made during multiple entries of 4 marmots. The results are plotted in figure 14 together with measurements of α_{MHP} made during euthermic wakefulness and sleep, during deep hibernation, and at times when T_b was regulated at intermediate levels between euthermia and deep hibernation for long enough periods to construct the usual hypothalamic thermosensitivity curves from the responses to many manipulations of T_{hy}. Curves obtained during two plateaus in an entrance into hibernation are shown in figure 15. The values of α_{MHP} obtained from plateaus confirmed the results of the two point method. Clearly α_{MHP} is a continuous function of T_{set} from euthermia to deep hibernation.

The extrapolation of the α_{MHP} vs T_{set} relationship

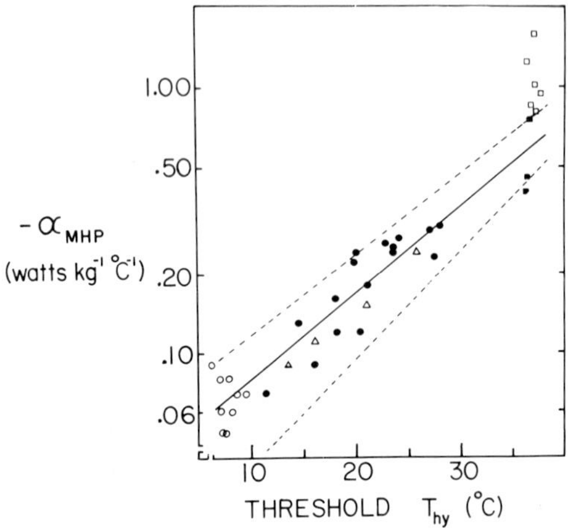

Fig. 14. *Hypothalamic thermosensitivity of 3 marmots plotted as a function of the threshold hypothalamic temperature for the metabolic heat production response. Data points obtained during deep hibernation are open circles, points obtained during plateaus in entrances are open triangles, points obtained by the two point method during entrance are closed circles, euthermic SWS values are closed squares, and euthermic wakeful values are open squares. The regression line was fitted to the hibernating values and then extrapolated to the euthermic range. The dotted lines delineate the 99% confidence limits of the regression line.*

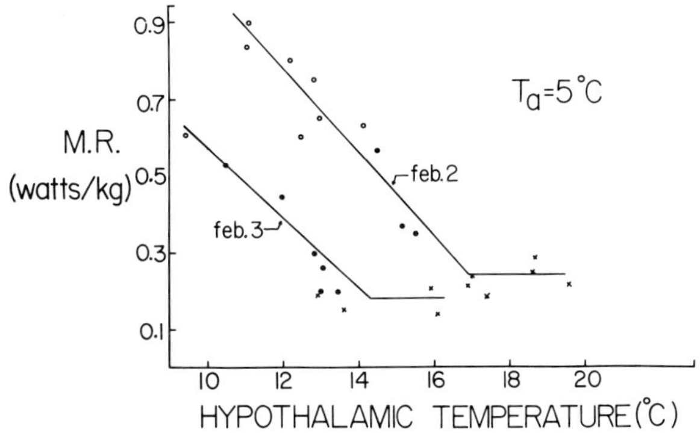

Fig. 15. *Characteristics of hypothalamic thermosensitivity of a marmot during two plateaus in an entrance into hibernation.*

calculated from hibernating values to euthermic levels yields a very interesting result . The relationship appears to be continuous with SWS α_{MHP} values and not with wakeful α_{MHP} values (Fig. 14). The T_{sets} over which α_{MHP} values were measured during hibernation were mostly below the T_b's at which conventional electrophysiological criteria can be used to define arousal states. The continuity of this relationship with euthermic SWS, however, lends indirect support to the proposal we advanced in section I that hibernation is an extension of SWS.

The concept of a unitary central nervous regulatory mechanism continuously operative over such a wide range of temperatures is also supported by recent studies of Wünnenberg and colleagues on the temperature sensitivity of pre-optic anterior hypothalamic neurons (64, 75). They have demonstrated temperature sensitive pre-optic units in anesthetized euthermic hibernators which have continuous firing rate curves over a 30°C range of temperatures. Such units were not found in a non-hibernator. Hence, the neuronal substrate for a very broad range thermoregulatory mechanism exists in the hibernator, apparently constituting a specialized adaptation.

Is the thermoregulatory system continuously functioning throughout entire bouts of hibernation? We believe that it is. Marmots remained in the metabolism chamber for entire bouts of hibernation which usually lasted 7 days and ended with a spontaneous arousal. Manipulations of T_{hy} were made throughout each bout and at all times it was possible to locate a T_{set} for the MHP response. Moreover, T_{set} showed a regular pattern of change in all subjects and in all bouts studied. Even after T_b reached equilibrium with T_a, T_{set} continued to decline for the first two thirds of the bout. Data for the last third of bouts were not extensive, but they indicated that one or two days before the active arousal occurs, T_{set} begins to rise (21). Spontaneous arousals usually begin with a slow, almost imperceptible increase in metabolic rate and T_b gradually giving way to exponential increases. We think it likely that the initial increases in metabolism occur when the slowly rising T_{set} exceeds T_{hy} thereby generating an error signal and a thermoregulatory drive. As this error signal increases it may trigger a positive feedback mechanism which increases the rate at which T_{set} is rising and therefore the error signal itself. Such a positive feedback mechanism would explain the very rapid rise in T_{set} seen very soon after an arousal is initiated (29, 38).

Any non-thermal environmental stimulus which induces a metabolic response from an animal in deep hibernation most probably acts via an increase in T_{set}. Since an observable metabolic response would only occur when T_{set} rose above actual T_{hy}, the farther T_{set} was below T_{hy} the stronger the stimulus needed to induce a metabolic response. Perhaps, therefore, spontaneous changes in T_{set} during bouts of hibernation as observed in the marmot cause the progressive changes in irritability during bouts of hibernation (71, 11). Seasonal differences in the pattern of T_{set} change may also be the underlying cause of the seasonal progression of hibernation bout lengths (70, 15). Of course, this hypothesis merely pushes the question one step backwards. Now we may ask what input to the thermoregulatory mechanism causes these fluctuations in the setting of the regulator?

We have no answer, but we can draw attention to still another possible relationship between hibernation and euthermic sleep. Twente and Twente (69) have shown that hibernation bout length is inversely related to T_b. Hence the input controlling the cycle of change in the setting of the temperature regulator during a bout of hibernation is temperature sensitive. When this relationship was extrapolated to euthermic T_b it coincided with the daily TST of the euthermic animal.

IV. SUMMARY AND CONCLUSIONS

Several conclusions can be drawn from the evidence reviewed in this paper:

1. Hibernation and shallow torpor in mammals are extensions of sleep, predominantly slow wave sleep. Rapid eye movement sleep probably does not occur during deep hibernation.

2. A unitary central nervous regulator of T_b is continuously operative over the entire range of T_b's experienced by the hibernator and seems to be functional at all times during a bout of hibernation.

3. The resetting of the central nervous regulator of T_b which occurs during hibernation appears to be homologous with that which occurs during the transition from wakefulness to SWS in the euthermic mammal.

It is very likely that torpor or hibernation as we see it expressed in various species has had polyphyletic origins. We have studied only a handful of these species to date, and therefore we must be cautious in assuming the generality of these conclusions. For example, Batholomew and Cade (10) reported that the pocket mouse, *Perognathus longimembris*, may engage in intense and coordinated behaviors as its T_b falls during entrance into hibernation. Certainly we would not claim it impossible for a means of resetting the T_b regulator to lower levels during wakefulness to have evolved. Additional comparative studies will be very interesting, especially on birds since their central nervous thermoregulatory system is considerably different from that of mammals.

The functional homologies between sleep, shallow torpor, and hibernation prompt us to propose that they may have had a common evolutionary history and may all serve at least one common adaptive function -- energy conservation. The energy conserving function of hibernation is quite obvious. Based on amount of body fat or on amount of food reserves determined from excavated food caches, and in consideration of average winter conditions, it is possible to show that a hibernator in nature could not survive without lowering T_b (36, 52). The energy conserving function of sleep may be less obvious, but various lines of evidence leading to this conclusion have been discussed by Berger (13).

The muscular relaxation of a sleeping or torpid animal conserves energy as compared with a quiet, alert animal, but a fall in T_b adds considerably to that energy savings. This conservation of energy may be of enormous selective advantage especially to small endotherms during periods of obligate inactivity or food scarcity. As body size decreases, thermal conductance and therefore the rate of weight-specific heat production of an endotherm increases. Basal metabolic rate is a power function of body weight (W) with a factor of approximately 2/3, but the potential energy reserves an animal can carry in the form of body fat is directly proportional to W. Over the entire size range of mammals, 50% of total body weight seems to be the maximum level of fat stores. Since basal metabolism is proportional to $W^{2/3}$ and energy reserves are proportional to W, it is clear that maximum fasting time decreases as body size decreases ($\alpha W^{1/3}$) (52). These calculations underestimate the relationship between body size and fasting time because they are based on basal metabolic rates. The smaller the animal, the higher will be the lower limit of its thermoneutral zone. Therefore, as T_a decreases a small mammal must incur an elevation in its rate of MHP before a larger mammal.

The influences of T_a, T_b, and W on the maximum time mammals can survive without food are illustrated in figure 16. These calculations are based on the assumptions that each animal begins the fast with a fat reserve equal to 20% of its lean body weight, that it does not engage in activity during the fast, that the Q_{10} of its metabolic rate is 2.5, and that its thermal conductance equals 3.57 $W^{.49}$ (a value given by Bartholomew 25). Based on these assumptions a 100 g mammal can survive 9 days without food at thermoneutrality, but

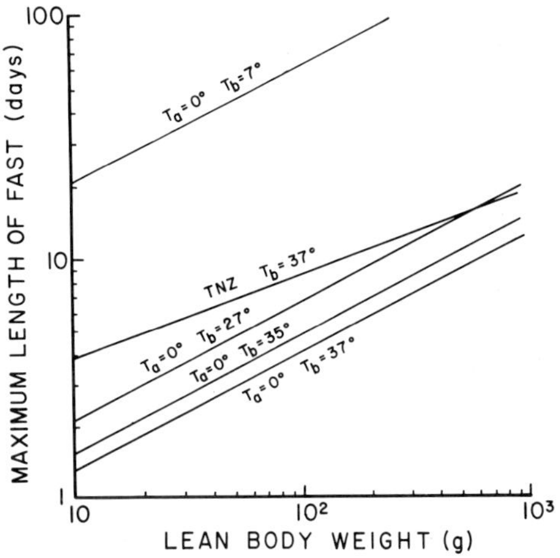

Fig. 16. *Estimations of the maximum time mammals of different body sizes can survive without food at thermoneutrality (TNZ) and at 0°C with body temperature regulated at different levels. It is assumed that stored fat at the beginning of the fast is equal to 20% of lean body weight.*

only 4 days at a T_a of 0°C. If, however, the animal sleeps during the fast allowing T_b to fall 2°C, it can survive for an additional day at T_a = 0°C. If it undergoes shallow torpor and lowers its T_b 10°C below euthermic levels, it can fast for almost 7 days, and if it hibernates and regulates its T_b at 7°C, it can survive a 65 day fast. These calculations are gross approximations, but they point out the adaptive advantage of lowered T_b during periods of food unavailability.

Our hypothesis for the evolution of hibernation is based on the physiological homologies between sleep and hibernation and on the supposition that they serve in part the common adaptive function of energy conservation. Since SWS is ubiquitous in birds and mammals, we assume that it is a more primitive phenomenon than hibernation. We think it is reasonable to suspect that SWS evolved early in the history of the endotherms in response to the energetic costs of the endothermic strategy. In the past it would have been selectively advantageous, especially for small endotherms, to decrease energy expenditure during periods of obligate inactivity and food scarcity. Even though terrestrial habitats were quite warm in the Jurassic and Cretaceous periods, nevertheless, valuable energy savings could have been achieved by lowering T_b during inactive periods. Perhaps the downward resetting of the regulator of T_b during SWS evolved for this purpose. Substantial cooling of the earth's surface did not occur until the mid to late Tertiary period (20), long after the major adaptive radiation of the mammals and birds had occurred. At this time endotherms occupying the more northerly and southerly latitudes would have experienced increasing environmental seasonality with extended periods of cold and food unavailability. Those species smallest in body size and those occupying the harshest habitats would have been subjected to increasing selective pressures for energy conservation. We believe that the progenitors of today's hibernators responded to that selective pressure by extending the downward resetting of the thermoregulatory system already present during SWS to lower and lower levels. This, of course, would have been associated with secondary selective pressures for the host of low temperature adaptations which we now subsume under the phenomenon of hibernation.

ACKNOWLEDGEMENTS

We are very grateful to Dr. Lawrence Wang and to Dr. Jack Hudson for arranging this symposium.

Our own research presented in this paper was supported by N.I.H. grants NS 10367 and GM 26395 to H.C.H. and GM 26394 to R.J.B.

FOOTNOTES

1. Department of Biological Sciences
Stanford University
Stanford, California 94305

2. Thimann Laboratories,
University of California
Santa Cruz, California 95064

REFERENCES

1. Abrams, R., and H. T. Hammel. Hypothalamic temperature in unanesthetized albino rats during feeding and sleeping. Am. J. Physiol. 206: 641-646, 1964.
2. Adams, T. Hypothalamic temperature in the cat during feeding and sleep. Science 139: 609-610, 1963.
3. Affani, J. M., E. Lisogorsky, and A. M. Scaravilli. Sleep in the giant South American armadillo, *Priodontes giganteus* (Edentata, Mammalia). Experientia 28: 1046-1047, 1972.
4. Allison, T., and H. Van Twyver. Electrophysiological studies of the echidna, *Tachyglossus aculeatus* II. Dormancy and hibernation. Arch. Ital. Biol. 110: 185-194, 1972.
5. Allison, T., H. Van Twyver, and W. R. Goff. Electrophysiological studies of the echidna, *Tachyglossus aculeatus*. I. Waking and sleep. Arch. Ital. Biol. 110: 145-184, 1972.
6. Aschoff, J. Circadian rhythm of activity and of body temperature. In: Physiological and Behavioral Temperature Regulation. Hardy, J. D., A. P. Gagge and J. A. J. Stolwijk (editors). Springfield: Charles C. Thomas, 1970, pp. 905-920.
7. Aschoff, J., U. Gerecke, and R. Wever. Phasenbeziehungan zwischen den circadianen perioden der Aktivität und der Kerntemperatur beim Menschen. Pflügers Arch. 295: 173-183, 1967.
8. Bach, L. M. N. Analysis of correlates between levels of consciousness and activity of the central nervous system. In: Hibernation-Hypothermia Perspectives and Challenges. South, F., J. Hannon, J. Willis, E. Pengelley, and N. Alpert (editors). New York: Elsevier, 1972, pp. 535-550.
9. Baker, M. A., and J. N. Hayward. Autonomic basis for the rise in brain temperature during paradoxical sleep. Science 157: 1586-1588, 1967.

10. Bartholemew, G. A., and T. J. Cade. Temperature regulation, hibernation, and aestivation in the little pocket mouse, *Perognathus longimembris*. J. Mammal. 38: 60-72, 1957.
11. Beckman, A. L., and T. L. Stanton. Changes in CNS responsiveness during hibernation. Am. J. Physiol. 231: 810-816, 1976.
12. Berger, R. J. Oculomotor control: A possible function of REM sleep. Psycholog. Rev. 76: 144-164, 1969.
13. Berger, R. J. Bioenergetic functions of sleep and activity rhythms and their possible relevance to aging. Federation Proc. 34: 97-102, 1975.
14. Bligh, J. The thermosensitivity of the hypothalamus and thermoregulation in mammals. Biol. Rev. 41: 317-367, 1966.
15. Daan, S. Periodicity of heterothermy in the garden dormouse, *Eliomys quercinus* (L.). Netherlands J. of Zool. 23: 237-265, 1973.
16. Day, R. Regulation of body temperature during sleep. Am. J. Dis. Child. 61: 734-746, 1941.
17. Dement, W. Sleep in Mammalian Hibernation III. Fisher, K., A. Dawe, C. Lyman, E. Schönbaum, and F. South (editors). New York: American Elsevier, 1967, pp. 175-195.
18. Euler, C. V., and U. Söderberg. Co-ordinated changes in temperature thresholds for thermo-regulatory reflexes. Acta. Physiol. Scand. 42: 112-129, 1958.
19. Feinberg, I., J. D. March, G. Fein, T. C. Floyd, J. M. Walker, and L. Price. Period and amplitude analysis of 0.5 - 3c/sec activity in NREM sleep of young adults. Electronenceph. Clin. Neurophysiol. (In press).
20. Flohn, H. Ein geophysicalisches Eiszeit-Modell. Eiszeitalter Ggw. 20: 204-231, 1969.
21. Florant, G. L., and H. C. Heller. CNS regulation of body temperature in euthermic and hibernating marmots (*Marmota flaviventris*). Am. J. Physiol. 232(5): R203-208, 1977.
22. Florant, G. L., B. M. Turner and H. C. Heller. Temperature regulation during wakefulness, sleep, and hibernation in marmots (*Marmota*

flaviventris). Am. J. Physiol. (In press).
23. Geschickter, E. H., P. A. Andrews, and R. W. Bullard. Nocturnal body temperature regulations in man: a rationale for sweating in sleep. J. Appl. Physiol. 21: 623-630, 1966.
24. Glotzbach, S. F., and H. C. Heller. Central nervous regulation of body temperature during sleep. Science 194: 537-539, 1976.
25. Gordon, M. S. Animal Physiology, 3rd. ed. New York: Macmillan, 1977, 699 pp.
26. Hammel, H. T. Temperature regulation and hibernation. In: Mammalian Hibernation III. Fisher, K. C., A. R. Dawe, C. P. Lyman, E. Schönbaum, and F. E. South (editors). New York: American Elsevier, 1967, pp. 86-96.
27. Hammel, H. T. Regulation of internal body temperature. Ann. Rev. Physiol. 30: 641-710, 1968.
28. Hammel, H. T., T. V. Dawson, R. M. Abrams, and H. C. Anderson. Total calorimetric measurements on *Citellus lateralis* in hibernation. Physiol. Zool. 41: 341-357, 1968.
29. Hammel, H. T., H. C. Heller, and F. R. Sharp. Probing the rostral brainstem of anesthetized, unanesthetized, and exercising dogs and of hibernating and euthermic ground squirrels. Fed. Proc. 32: 1588-1597, 1973.
30. Hammel, H. T., D. C. Jackson, J. A. J. Stolwijk, J. D. Hardy, and S. B. Strømme. Temperature regulation by hypothalamic proportional control with an adjustable set point. J. Appl. Physiol. 18: 1146-1154, 1963.
31. Hayward, J. N., and M. A. Baker. A comparative study of the role of the cerebral arterial blood in the regulation of brain temperature in five mammals. Brain Res. 16: 417-440, 1969.
32. Heller, H. C. Hypothalamic thermosensitivity in mammals. Experientia (In press).
33. Heller, H. C., and G. W. Colliver. CNS regulation of body temperature during hibernation. Am. J. Physiol. 227: 583-589, 1974.

34. Heller, H. C., G. W. Colliver, and P. Anand. CNS regulation of body temperature in euthermic hibernators. Am. J. Physiol. 227: 576-582, 1974.
35. Heller, H. C., G. W. Colliver, and J. Beard. Thermoregulation during entrance into hibernation. Pflügers Arch. 369: 55-59, 1977.
36. Heller, H. C., and D. M. Gates. Altitudinal zonation of chipmunks (Eutamias): energy budgets. Ecology 52: 424-433, 1971.
37. Heller, H. C., and S. F. Glotzbach. Thermoregulation during sleep and hibernation. In: Environmental Physiology II, Int. Rev. Physiol. 15: 147-188, 1977.
38. Heller, H. C., and H. T. Hammel. CNS control of body temperature during hibernation. Comp. Biochem. Physiol. 41A: 349-359, 1972.
39. Henane, R., A. Buquiet, B. Roussel, and J. Bittel. Variations in evaporation and body temperature during sleep in man. J. Appl. Physiol. 42: 50-55, 1977.
40. Kayser, C. The Physiology of Natural Hibernation. Oxford: Pergamon Press, 1961, 325 pp.
41. Kayser, C., F. Rohmer, and C. Hiebel. L'E.C.G. de l'hibernant. Lethargie et reveil spontane du spermophile. Rev. Neurol. 84: 570-578, 1951.
42. Kirk, E. Untersuchunger über den Einfluss des normalen Schlafes suf die Temperatur der Füsse. Scand. Arch. Physiol. 61: 71-78, 1931.
43. Kreider, M. B., E. R. Buskirk, and D. E. Bass. Oxygen consumption and body temperatures during the night. J. Appl. Physiol. 12: 361-366, 1958.
44. Kreider, M. B., and P. F. Iampietro. Oxygen consumption and body temperature during sleep in cold environments. J. Appl. Physiol. 14: 765-767, 1959.
45. Lund, R. Circadian periodicity of physiological and psychological variables in seven blind subjects with and without time cues (Ph.D. thesis). Technische Universität München, 1974.

46. Lyman, C. P. Oxygen consumption, body temperature and heart rate of woodchucks entering hibernation. Am. J. Physiol. 194: 83-91, 1958.
47. Lyman, C. P., and P. O. Chatfield. Hibernation and cortical electrical activity in the woodchuck. Science 117: 533-534, 1953.
48. Lyman, C. P., and R. C. O'Brien. A comparison of temperature regulation in hibernating rodents. Am. J. Physiol. 227: 218-223, 1974.
49. Lyman, C. P. and R. C. O'Brien. Sensitivity to low temperature in hibernating rodents. Am. J. Physiol. 222: 864-869, 1972.
50. Mihailović, L. J. T. Cortical and subcortical electrical activity in hibernation and hypothermia. In: Hibernation-Hypothermia Perspectives and Challenges. South, F., J. Hannon, J. Willis, E. Pengelley, and N. Alpert (editors). New York: Elsevier, 1972, pp. 487-534.
51. Mills, S. H., and F. E. South. Central regulation of temperature in hibernation and normothermia. Cryobiology 9: 393-403, 1972.
52. Morrison, P. Some interrelations between weight and hibernation function. Bull. Mus. Comp. Zool. 124: 75-90, 1960.
53. Ogawa, T., T. Satoh, and K. Takagi. Sweating during night sleep. Jpn. J. Physiol. 17: 135-148, 1967.
54. Parmeggiani, P. L. Interaction between sleep and thermoregulation. Waking and Sleeping 1: 123-132, 1977.
55. Parmeggiani, P. L., C. Franzini, P. Lenzi, and G. Zamboni. Threshold of respiratory responses to preoptic heating during sleep in freely moving cats. Brain Res. 52: 189-201, 1973.
56. Parmeggiani, P. L., and C. Rabini. Shivering and panting during sleep. Brain Res. 6: 789-791, 1967.
57. Parmeggiani, P. L., and L. Sabattini. Electromyographic aspects of postural respiratory and thermoregulatory mechanisms in sleeping cats. Electroencephalogr. Clin. Neurophysiol. 33: 1-13, 1972.
58. Satinoff, E. Hibernation and the central nervous system. In: Progress in Physiological

Psychology, Vol. 3. Stellar, E. and J. M. Sprague (editors). New York: Academic Press, 1970, pp. 201-236.

59. Satoh, T., T. Ogawa, and K. Takagi. Sweating during daytime sleep. Jpn. J. Physiol. 15: 523-531, 1965.

60. Shapiro, C. M., A. T. Moore, D. Mitchell, and M. L. Yodaiken. How well does man thermoregulate during sleep? Experientia 30: 1279-1280, 1974.

61. Shtark, M. B. The Brain of Hibernating Animals. Novosibirsk, Nauka Press, 1970 (Eng. Trans. NASA TTF-619).

62. South, F. E., J. E. Breazile, H. D. Dellman, and A. D. Epperly. Sleep, hibernation and hypothermia in the yellow-bellied marmot (*M. flaviventris*). In: Depressed Metabolism. Mussacchia, X. J. and J. F. Saunders (editors). New York: Elsevier, 1969, pp. 277-312.

63. South, F. E., W. C. Hartner, and R. H. Luecke. Responses to preoptic temperature manipulation in the awake and hibernating marmots. Am. J. Physiol. 229: 150-160, .

64. Speulda, E., and W. Wünnenberg. Thermosensitivity of preoptic neurones in a hibernator at high and low ambient temperatures. Pflügers Arch. 370: 107-109, 1977.

65. Strumwasser, F. Regulatory mechanisms, brain activity and behavior during deep hibernation in the squirrel, *Citellus beecheyi*. Am. J. Physiol. 196: 23-30, 1959.

66. Strumwasser, F. Thermoregulatory, brain and behavioral mechanisms during entrance into hibernation in the squirrel, *Citellus beecheyi*. Am. J. Physiol. 196: 15-22, 1959.

67. Strumwasser, F., J. L. Smith, J. J. Gilliam, and F. R. Schlechte. Identification of active brain regions involved in the process of hibernation. XVI International Congress of Zoology 2, 53, 1963.

68. Takagi, K. Sweating during sleep. In: Physiologica and Behavioral Temperature Regulation. Hardy, J. D., A. P. Gagge, and J. A. J. Stolwijk (editors). Springfield: C. C. Thomas,

1970, pp. 669-675.
69. Twente, J. W. and J. A. Twente. Regulation of hibernating periods by temperature. Proc. Natl. Acad. Sci., USA 54: 1058-1061, 1965.
70. Twente, J. W., and J. A. Twente. Seasonal variation in the hibernating behavior of *Citellus lateralis*. In: Mammalian Hibernation III. Fisher, K. C., A. R. Dawe, C. P. Lyman, E. Schönbaum, and F. W. South (editors). New York: Elsevier Press, 1967, pp. 47-63.
71. Twente, J. W., and J. A. Twente. Progressive irritability of hibernating *C. lateralis*. Comp. Biochem. Physiol. 25: 467-474, 1968.
72. Van Twyver, H. Sleep patterns of five rodent species. Physiol. and Behav. 4: 901-905, 1969.
73. Van Twyver, H., and T. Allison. Sleep in the opossum *Didelphis marsupialis*. Electroencephalogr. Clin. Neurophysiol. 29: 181-189, 1970.
74. Van Twyver, H., and T. Allison. Sleep in the armadillo *Dasypus novemcinctus* at moderate and low ambient temperatures. Brain, Behav. and Evol. 9: 107-120, 1974.
75. Walker, J. M., and R. J. Berger. A polygraphic study of the tortoise (*Testudo denticulata*). Brain, Behav. and Evol. 8: 453-467, 1973.
76. Walker, J. M., S. F. Glotzbach, R. J. Berger, and H. C. Heller. Sleep and hibernation in ground squirrels (*Citellus*): Electrophysiological observations. Am. J. Physiol. (In press).
77. Wever, R. The circadian multi-oscillator system of man. Int. J. Chronobiol. 3: 20-55, 1974.
78. Wünnenberg, W., G. Merker, and E. Speulda. Thermosensitivity of preoptic neurones in a hibernator (golden hamster) and a non-hibernator (guinea pig). Pflügers Arch. 363: 119-123, 1976.

THERMOSENSITIVITY OF PREOPTIC NEURONS AND HYPOTHALAMIC INTEGRATIVE FUNCTION IN HIBERNATORS AND NONHIBERNATORS

W. Wünnenberg
Institute of Zoology
University of Kiel
West Germany

G. Merker, E. Speulda
Institute of Physiology
University of Giessen
West Germany

ABSTRACT

Studies on central nervous mechanisms underlying temperature regulation in hibernators were summarized and compared with results of corresponding experiments in non-hibernators. In golden hamsters thermoreceptive neurones of the preoptic region were still responsive to hypothalamic temperatures below 10°C, while in a non-hibernator (guinea pig) preoptic thermoreceptive neurones became insensitive below 30°C. The thermosensitive preoptic units were localized and studied histologically. In further experiments the electrical activity of the central nervous system of animals in hibernation has been studied by recording the spontaneous electroencephalogram of the diencephalon. At the entry into hibernation, small-amplitude EEG-waves became predominant. This transition into a small-amplitude EEG which seems to be typical for the hibernating state was not caused by the temperature decline but preceded the decrease of body temperatures. This result is consistent with the view that entry into hibernation is caused by active inhibitory mechanisms in which the midbrain reticular formation or the limbic system might be involved. In contrast to the preoptic area the medial and dorsal part of the hypothalamus is not thermosensitive. Structures of this part of the diencephalon, however, play an important role in the integration of input signals from thermoreceptive areas and in the control of thermoregulatory effector systems.

ISBN 0-12-734550-7

I. INTRODUCTION

During the last few years a number of reviews have appeared dealing with temperature regulation in hibernators and non-hibernators (7, 22, 35, 42). As a result of these comprehensive publications it became clear that in principle there is no difference in temperature regulation of euthermic hibernators and non-hibernators and furthermore, that the temperature control system is in function during hibernation.

Relatively little is known of the central nervous mechanisms underlying hibernation and temperature control during hibernation. It is the aim of this paper to summarize studies on central nervous structures involved in thermoregulation in hibernators and to compare these data with results of corresponding experiments in non-hibernators.

II. THERMOSENSITIVITY OF THE PREOPTIC AREA (POA) IN HIBERNATORS AND NON-HIBERNATORS

In euthermic hibernators, local thermal stimulation of the POA elicited thermoregulatory responses which did not show qualitative differences to corresponding effects in non-hibernators (23, 24, 26, 27, 42, 48, 56, 61, 62). This finding generated the desire to find a measure for the temperature sensitivity of the POA which would allow a quantitative comparison of preoptic thermosensitivity in different species. In a number of studies (13, 19, 24, 26, 28, 31, 36, 38, 48, 62), hypothalamic thermosensitivity was measured by determining the proportionality-constant for the heat production response to hypothalamic cooling, according to the following equation:

$$R - Ro = \alpha \ (T_{hy} - T_{set}) \quad (22)$$

in which R - Ro is the metabolic response, T_{hy} is the hypothalamic temperature and T_{set} a reference temperature.

The proportionality-constant α (W/kg · °C) expresses the weight-related metabolic response to preoptic thermal stimulation. In some species investigated so far (26, 31, 36, 48, 62), the α-value was dependent on the ambient temperature (T_a); an increase of T_a led to a decrease of α.

The α-values of some species of hibernators and non-hibernators are shown in Table 1. Though the number of data is limited, it becomes clear that there is no significant difference in the preoptic thermosensitivity of euthermic hibernators and non-hibernators provided that animals of nearly the same body size are compared.

The α-values, namely, depend on the body size, inasmuch as in small species relatively high proportionality constants were determined. This finding does not necessarily mean that preoptic thermosensitivity increases with decreasing body weight. In the small rodents, listed in Table 1, basal metabolic rate (BMR) is approximately 2- to 3-times higher than in rabbits and cats. This follows; the metabolic response to local cooling does not show significant dependance on the body size when the effect of preoptic cooling is related to BMR.

Thermosensitivity of the POA could also be demonstrated in animals during hibernation (23, 24, 25, 42, 43, 44, 55, 56). However, as there are only a few quantitative data on the thermoregulatory response in hibernating animals, it is difficult to find out whether preoptic thermosensitivity is reduced during hibernation. In the hibernating *Citellus lateralis* the weight-related metabolic response to preoptic cooling (Table 1) is 10 - 15 times lower than during the non-hibernating state (25). On the other hand, metabolism during hibernation is only 1/30 of the summer BMR (39, 40). Thus, the relative metabolic response to preoptic cooling in the hibernating animal is high.

III. THERMORESPONSIVE PREOPTIC NEURONES IN HIBERNATORS AND NON-HIBERNATORS

The finding of a metabolic response to hypothalamic cooling suggested the existence of hypothalamic neurones that are sensitive to the local temperature. The earliest successful attempt to study with neurophysiological methods, the activity of temperature-sensitive units in the POA was made by Nakayama *et al.* (50). The results of this investigation were confirmed

Table I

	SPECIES	WEIGHT (kg)	α (W/kg·°C)	Ref. No.
Hibernators	Marmota flaviventris	4.0	-0.48 to 0.6	48
(euthermic)	Citellus tridecemlineatus	0.22	-0.59	62
	Citellus lateralis	0.180	-4.8 to -6.9	26
	Citellus beldingi	0.300	-2.3 to -3.6	26
Hibernator	Citellus lateralis	0.180	-0.44	25
(during hibernation)				
	Dipodomys ingens	0.130	-2.4 to -3.4	19
	Ammospermophilus nelsoni	0.115	-5.1 to -7.6	28
	Rabbit	3.0	-1.28	13
Non-hibernators	Cat	2.4	-0.74	36
	Dog	21.0	-0.81	38
	Dog	11.0	-1.1 to -1.7	31

Legend: Preoptic thermosensitivity (α) of some species of hibernators and non-hibernators.

in further studies in which thermosensitive preoptic neurones were found in several species (7, 15, 16, 65).

As preoptic thermosensitivity has been demonstrated even in animals during deep hibernation (23, 24, 25, 42, 43, 44, 55, 66), it could be expected that in hibernators, preoptic structures involved in temperature regulation are specialized to maintain their function over the wide range of temperatures which occur during the different phases of hibernation. This would imply that in hibernators, preoptic thermoresponsive units are qualitatively different from those of non-hibernators. Any significant differences were not found in a study by Boulant and Bignell (8) who compared temperature-responsive preoptic neurones in ground squirrels and rats. In their study, however, only moderate thermal stimuli were applied to the preoptic region; single unit activity at hypothalamic temperatures below 32°C was apparently not studied.

In previously reported experiments on euthermic hibernators (golden hamster) and non-hibernators (guinea pig), we studied the activity of thermoresponsive preoptic neurones at hypothalamic temperatures ranging from 5-43°C (57, 68).

In a first series of studies on 23 euthermic golden hamsters (weighing 105 - 145 g) and 13 guinea pigs (weighing 650 - 750 g) only T_{hy} was manipulated by a fine thermode device (68). Skin temperatures (T_{sk}) were maintained at 36-37°C by immersing the animals (up to the head) in a temperature-regulated, well-stirred water bath (57).

In golden hamsters, 54 single units of the POA could be kept under observation long enough to complete several heating and cooling cycles. Of these, 37 were temperature-insensitive, and 17 proved to be sensitive to local thermal stimulation with a negative or positive thermal coefficient of more than 1.0 imp/sec · °C.

An example of a temperature-responsive preoptic unit in the golden hamster is given in figure 1. A section of the course of an experiment shows the changes in firing rate of the unit due to step-wise variations of preoptic temperatures (left part, Fig. 1). Local heating of the hypothalamus led to an increase of the firing rate up to 90 imp/sec; the unit became

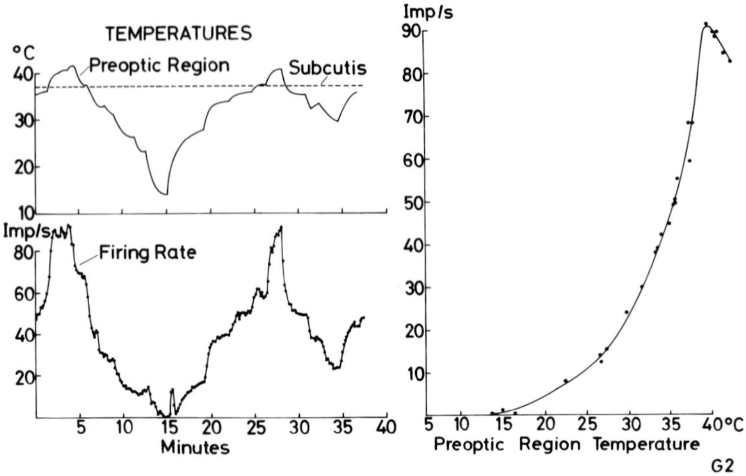

Fig. 1. *Left side: response of a preoptic neurone to thermal stimulation of the POA in a golden hamster in the non-hibernating state. Right side: impulse frequencies of this unit as a function of various hypothalamic temperatures under steady state conditions (From 68).*

inactivated when the hypothalamus was cooled to temperatures below 15°C. This procedure could be repeated several times; the results of all heating and cooling cycles are summarized in the diagram on the right side of figure 1.

The temperature-impulse rate characteristic of this unit shows an exponential course ranging from 40 - 14°C.

A second type of preoptic neurone that responded to hypothalamic temperatures far below those of euthermia is shown in figure 2. This unit had a negative temperature coefficient at hypothalamic temperatures between 40 - 30°C and a positive temperature coefficient in the temperature range between 10 - 20°C. This resulted in a bell-shaped temperature-impulse rate characteristic with a maximal firing rate at T_{hy} between 20 - 30°C.

The temperature-impulse rate characteristics of the 19 thermosensitive units recorded from the POA in the golden hamsters are illustrated in figure 3 (upper

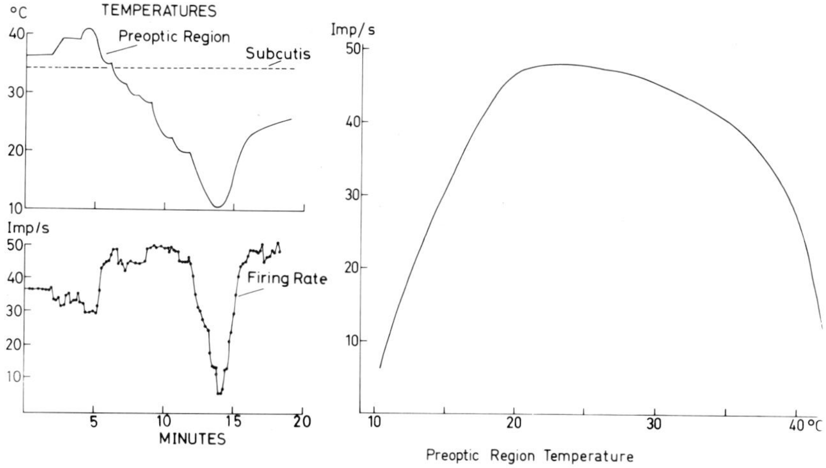

Fig. 2. *Left side: response of a preoptic neurone to thermal stimulation of the POA in a golden hamster in the non-hibernating state. Right side: impulse frequencies of this unit as a function of various hypothalamic temperatures under steady state conditions. Data were obtained from three heating and cooling cycles.*

part). Of the total of 54 neurones which were studied, 13 (i.e. 24.1%) showed an exponential temperature-firing rate characteristic. On the average these units had a thermal coefficient of + 3.7 (S.E. ± 0.3) imp/sec · °C in the temperature range from 40 - 30°C, + 1.5 (S.E. ± 0.1) imp/sec · °C at T_{hy} 30 - 20°C and + 0.6 (S.E. ± 0.1) imp/sec · °C at T_{hy} between 15 and 20°C. Four units (i.e. 7.4%) had a negative thermal coefficient of -1.1 (S.E. ± 0.2) imp/sec · °C on the average at T_{hy} between 40 and 30°C. These units had a positive thermal coefficient of + 1.9 (S.E. ± 0.7) imp/sec · °C on the average at T_{hy} 10 - 20°C.

In control experiments on a nonhibernator (guinea pigs), thermosensitivity of preoptic neurones was studied under experimental conditions that were comparable to those performed in the golden hamsters. From a total of 35 units that were recorded in the

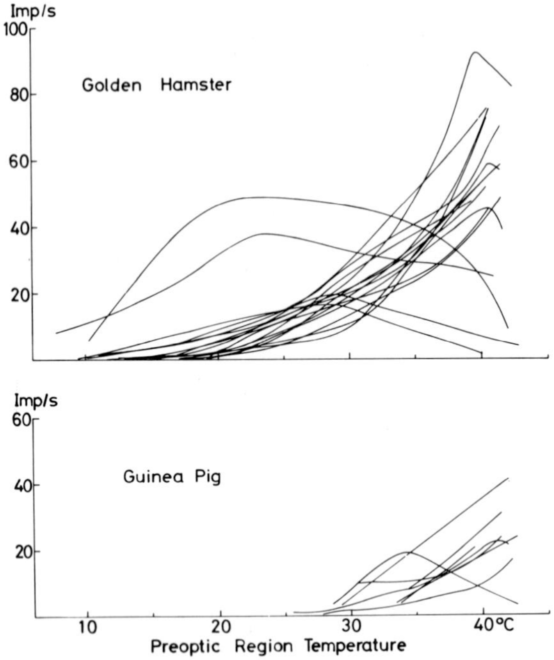

Fig. 3. *Responses of preoptic single units to thermal stimulation of the POA in euthermic golden hamsters (upper part) and guinea pigs (lower part). (Adapted from ref. 68).*

guinea pigs 27 were temperature-insensitive and 8 (i.e. 23%) proved to be responsive to thermal stimulation of the POA with a mean thermal coefficient of + 2.3 (S.E. ± 0.3) imp/sec • °C at T_{hy} 40 - 35°C. However, in all experiments performed in this species thermoresponsive units of the preoptic region became inactive or insensitive to thermal stimulation as soon as brain temperatures fell below 28 - 30°C (Fig. 3, lower part).

A comparison of the results obtained in golden hamsters and guinea pigs (Fig 3) shows that the range of hypothalamic temperatures in which the thermoresponsive preoptic units are active is much wider in the hibernator than in the non-hibernator. Thermosensitivity over a temperature range of more than 30°C is based on firing rates that are much higher than discharge activities of preoptic units in non-hibernators. The spontaneous activity of the units with exponential characteristics, averaged 40.0 (S.E. ± 2.5) imp/sec at T_{hy} 36.5°C, while in guinea pigs the 6 neurones with positive thermal coefficients had a mean spontaneous activity of 15.8 (S.E. ± 3.2) imp/sec at T_{hy} of 38.5°C.

Though these studies show that preoptic single units in golden hamsters responded to hypothalamic temperatures far below euthermia, they do not prove that preoptic neurones are capable of sensing temperatures during deep hibernation. Most single units - especially those with the exponential temperature-firing rate characteristic - became inactive at hypothalamic temperatures between 10 and 15°C (Fig. 3 upper part). Hypothalamic temperatures of 5 to 6°C, however, could be recorded in the golden hamsters used in these experiments when hibernating at T_a of 3 - 4°C. On the other hand the experimental procedures of these studies did not simulate temperature conditions during hibernation, as only the diencephalon was cooled while temperatures of other parts of the body were kept at the normothermic level. As temperatures of various thermoreceptive areas, e.g. spinal cord and skin can influence the activity of POA units (9, 10, 20, 29, 30, 63) it was speculated that the temperature-impulse rate characteristics of the preoptic units could be shifted to lower temperatures or modified in their temperature sensitivities by a high activity of the cutaneous cold-receptors occurring during hibernation (53).

To investigate the factors which could modify the temperature response curves of preoptic neurones in such a way that they are sensitive to T_{hy} 0 - 10°C, further experiments were undertaken in which T_{hy} and T_{sk} were varied independent of each other (57).

In these experiments on 25 euthermic golden hamsters, 20 of a total of 73 preoptic single units proved to be sensitive to local thermal stimulation. Figure 4 shows portions of a record obtained from an experiment in which the response of a preoptic single unit to local thermal stimulation was tested at T_{sk} = 21°C (left side, figure 4) and T_{sk} = 36°C (right side, figure 4). At both skin temperatures this unit was sensitive to heating of the POA with a positive thermal coefficient; the thermosensitivity of this unit, however, was evidently higher when the skin was warmed.

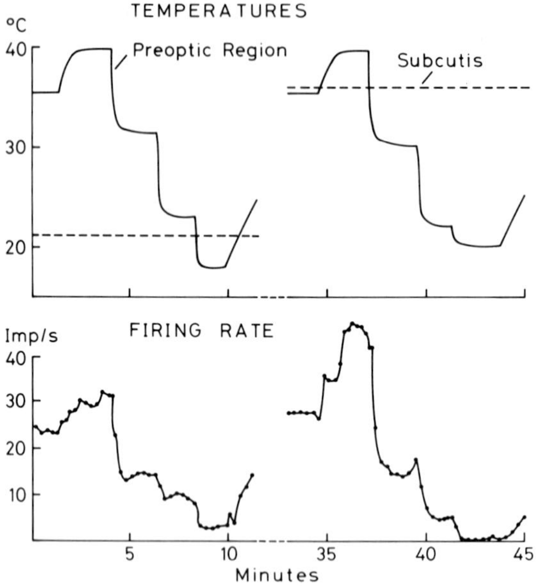

Fig. 4. Responses of a preoptic neurone in a euthermic golden hamster to local thermal stimulation at T_{sk} = 21°C and T_{sk} = 36°C (from 57).

The response to low T_{hy} was also influenced by the temperatures of the body surface. At T_{sk} = 36°C the unit was inactivated at T_{hy} = 20°C, while at T_{sk} = 21°C this neurone showed a discharge frequency of 3 imp/sec even at T_{hy} = 18°C.

Figure 5 shows the average static responses to hypothalamic thermal stimulation of 20 preoptic units recorded at T_{sk} = 36°C (solid line). Of these, only 7 could be additionally tested at T_{sk} = 20°C (broken line), as in most cases single unit recordings were disturbed by vigorous shivering when skin temperature was lowered. A comparison of both response curves demonstrates that the temperature range in which the neurones are responsive is shifted more than 5°C to lower values by skin cooling. This effect is concurrent with a reduced thermal sensitivity of the units, i.e. the slope of the temperature-impulse rate characteristic of the hypothalamic neurones is significantly reduced at low T_{sk}.

These experiments indicate that in hibernators preoptic thermoreceptive structures are capable of sensing temperatures even in deep hibernation, when they are influenced by a high activity of cutaneous

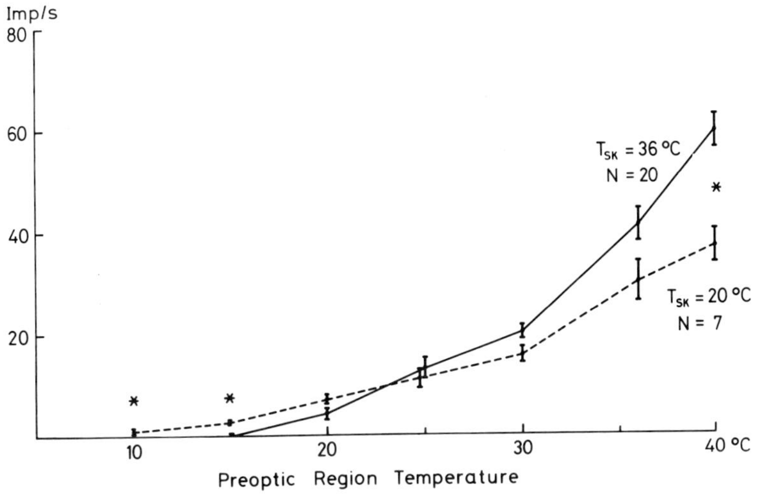

Fig. 5. *Average temperature response of preoptic neurones at T_{sk} = 36°C (solid line) and T_{sk} = 20°C (broken line). Vertical bars indicate S.E. At preoptic region temperatures marked by an asterisk, the differences in the firing rates are significant ($P < 0.001$). (From ref. 57).*

cold receptors. It seems to be possible that not only cutaneous cold receptors modify the temperature response curves of preoptic units but also other extrahypothalamic thermoreceptive structures, e.g. the spinal cord. Furthermore, seasonal variations of the thermosensitivity of preoptic units should be taken into consideration.

The thermosensitive units were localized by iron markings and Prussian-blue reaction in the preoptic area of the golden hamster and of the guinea pig. In the POA of the golden hamster, a dorsal and ventral accumulation of iron markings can be seen in the areas of the *Nucleus praeopticus medialis* and *N. praeopticus lateralis* (Fig. 6 a,b). In the region between the marked dorsal and ventral area no thermosensitive units could be traced. Thermosensitive units of the guinea pig POA are scattered in a medio-dorsal area of the *N. praeopticus periventricularis* and in the adjacent part of the *N. praeopticus medialis* (Fig. 6 c). The lateral extension of the thermosensitive region has not

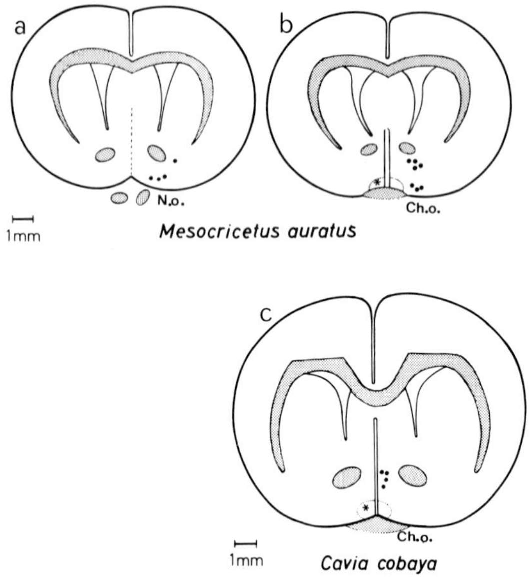

Fig. 6. *Localization of thermosensitive structures (•) in the preoptic area of the golden hamster (Mesocricetus auratus, a, b) and of the guinea pig (Cavia cobaya, C.). Frontal sections schematized. Ch.O.: Chiasma opticum; N.O.: Nervus opticus; asterisk: Nuc. suprachiasmaticus.*

yet been defined in the guinea pig; furthermore it cannot be excluded that a more ventral area might also contain thermosensitive structures.

As shown by their distribution, thermosensitive preoptic structures are not restricted to one of the classic parvocellular preoptic nuclei. No cytologically circumscribed "thermosensitive center" can be emphasized, as the findings in the golden hamster POA rather indicate a reticular distribution of thermosensitive structures.

It can be concluded after comparison with Golgi-impregnations of the corresponding preoptic regions, that the marked thermosensitive structures are localized within circumscribed nerve cell clusters. The neurones of these cell groups are manifoldly connected, the perikarya are often in close contact with each other (Fig. 7). Nerve cell clusters in thermosensitive

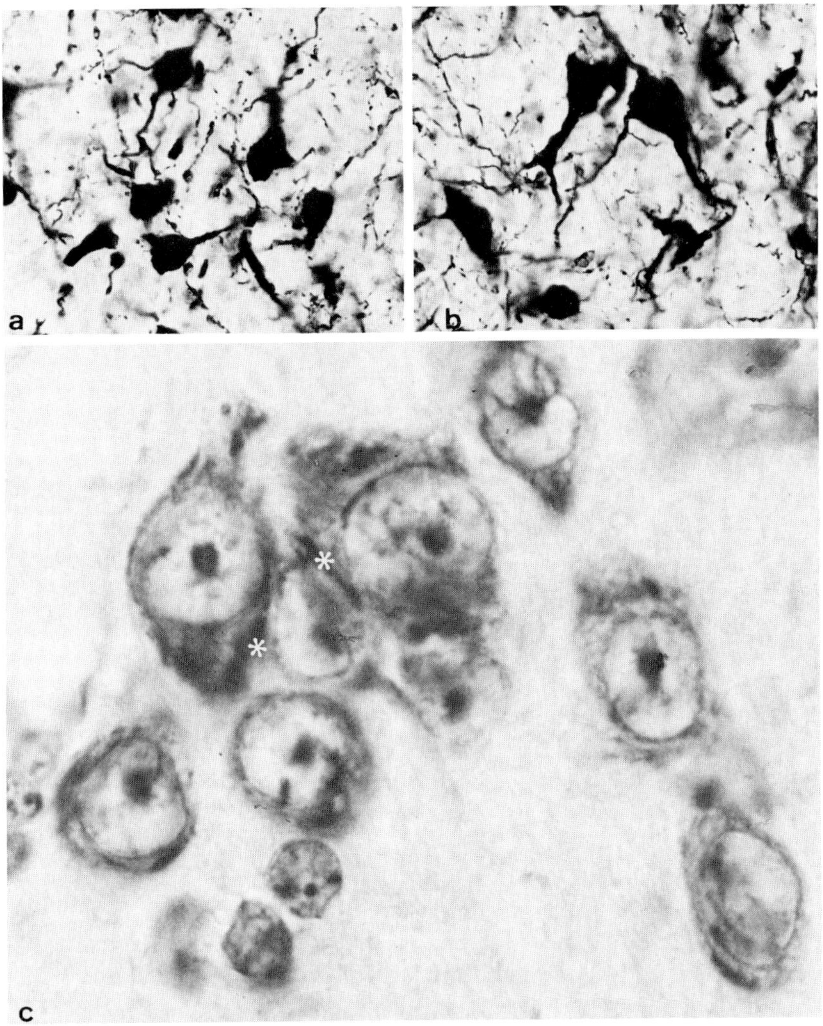

Fig. 7. *Nerve cell clusters in the thermosensitive areas of the golden hamster (a) and of the guinea pig (b, c). a: Golden hamster, ventral thermosensitive area; arrows: Beaded, spin bearing dendrites. Golgi-impregnation, 1:400. b: Guinea pig, thermosensitive area of the Nucleus praeopticus medialis. Golgi-impregnation, 1:400. c: Guinea pig, groups of nerve cells in the thermosensitive area of the Nucleus praeopticus medialis. Note the different size of the cells and the close contact between some perikarya (asterisk). Stain: Cresyl violet; 1:1600.*

areas of the golden hamster and of the guinea pig receive an input of ascending aminergic fibres (46). As revealed by the microspectralfluorometric technique, the aminergic terminals and preterminals contain noradrenaline or dopamine. The noradrenaline terminals dominate in this area. Terminals containing 5-hydroxitryptamine could not be observed.

The circumscribed cell aggregations might represent nuclear subunits; in this case the pattern of synaptic connections must be essential for the functional interaction or isolation of certain cell groups or cellular nets (46, 52).

IV. ELECTRICAL ACTIVITY OF THE CENTRAL NERVOUS SYSTEM DURING HIBERNATION

The electrical activity of the central nervous system of animals in hibernation has been studied by recording the spontaneous electroencephalogram (47, 54). When an animal enters hibernation, EEG-activity decreases in a step-wise sequence from high to low levels in the brain (54). Suppression of electrical activity first occurs in the neocortex, later in the midbrain reticular formation (MRF) and finally in areas of the limbic system. Strumwasser (58) showed that even in deep hibernation (T_{hy} = 6.1°C) the medial preoptic area, septum, and the ventromedial nucleus of the hypothalamus remain active with small-amplitude EEG-waves.

With the aim to find evidence for the existence of active inhibitory mechanisms inducing the onset of hibernation, EEG-recordings from the ventromedial region of the hypothalamus of hedgehogs were performed. An example of this type of experiments is given in figure 8. Here, portions of EEG-recordings are shown (lower part, Fig. 8) which were predominant in the periods marked by the corresponding capital leters (upper part of the Fig.). A comparison of period D and E shows that the transition into the small-amplitude EEG, which seems to be typical for the hibernating state is not caused by the temperature decline but precedes the decrease of body temperatures. This result is consistent with the view that entry into hibernation is caused by active inhibitory mechanisms in which the MRF or the limbic system might be involved (3, 22, 25, 42).

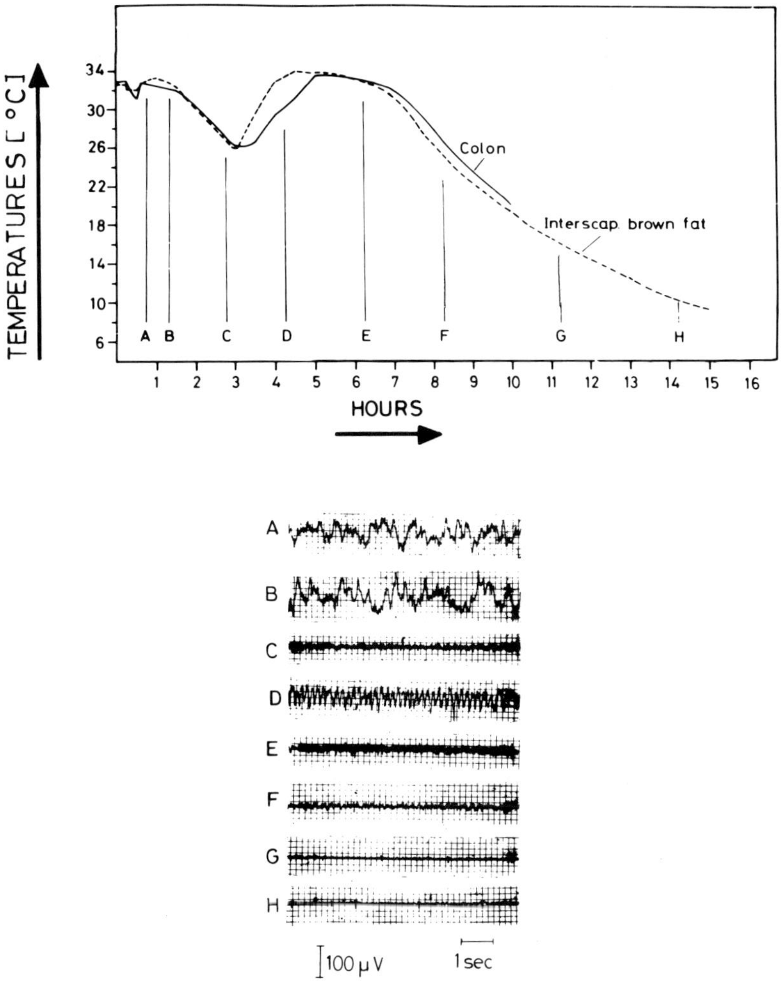

Fig. 8. *EEG-recordings from the ventromedial region of the hypothalamus of a hedgehog entering hibernation. The portions of the EEG-recordings (lower part) were predominant in the periods marked by the corresponding capital letters (upper part).*

V. INTEGRATIVE FUNCTION OF HYPOTHALAMIC STRUCTURES

Most current theories on temperature regulation are based on the concept that information arising from different thermoreceptive regions, e.g. - subcutis, spinal cord, preoptic area - is integrated by hypothalamic structures. This idea of a dual function of hypothalamic nervous tissue involved in temperature regulation i.e. both as a sensor and as a signal integrator (4, 34), is based on numerous experiments in which the effects of local stimulation or lesions were studied (7, 35).

The experiments of Bard *et al.* (2), of Keller and McClaskey (41) pointed to the posterior hypothalamus as a site important for the integration and transmission of neural activity related to thermoregulation. Ablation of this area made the animals incapable of maintaining body temperatures when exposed to cold. This effect cannot be explained by a destruction of thermoreceptive neurones, as posterior hypothalamic tissue shows no or only little sensitivity to changes of its own temperature (12, 18, 21, 33, 37, 45).

The view that input signals from different thermoreceptive regions are integrated by hypothalamic structures, has been supported by single-unit studies in which neurones were found to respond to thermal stimulation of more than one thermosensitive area. Units of this type were found in the POA (17, 20, 29, 63), the posterior hypothalamus (14, 51, 67), and the midbrain (49). The results of these studies show in principle that there is a convergence of input signals. However, it does not prove that a particular unit - responding to thermal stimulation of various thermoreceptive areas - has an integrative function, as convergence of inputs may have occurred in other brain areas and then projected to the neurone whose activity was recorded.

To gain further insight into the process of integration of input signals related to thermoregulation we made an attempt to study the electrical activity of larger complexes of diencephalic tissue. In these experiments, which were performed in guinea pigs, the local electroencephalogram (EEG) of different parts of the hypothalamus was recorded during thermal stimulation of the thermosensitive part of the spinal cord (64).

Modification of the local EEG due to spinal cord heating was observed in all experiments in which the recording electrodes were placed in the anterior or posterior part of the diencephalon. In the anterior hypothalamus, mean EEG frequency increased when the spinal cord was heated, while a reverse response to spinal cord heating, i.e. a decrease of the mean frequency, could be observed in all animals with electrodes in the posterior hypothalamus. No change of the local EEG activity during thermal stimulation of the spinal cord was recorded in animals with electrode positions in the rostral part of the preoptic region or in the midbrain area caudal to the mammillary corpora.

The finding that the EEG-response to spinal cord heating was different depending on the electrode positions suggested that the involved structures of the anterior hypothalamus and those of the posterior region are part of two diverse circuits within the thermoregulatory system. This hypothesis could be confirmed by further experiments in which the thermoregulatory responses to spinal cord heating were studied prior to and after electrocoagulation of hypothalamic tissue (60, 65).

Figure 9 shows, as an example of this type of study, an experiment in the course of which a bilateral lesion in the anterior hypothalamus (upper part of Fig. 9) was carried out. Before electrocoagulation, respiratory rate could be increased by heating of the thermosensitive spinal region. After electrocoagulation of the anterior hypothalamus, however, this excitatory effect of spinal cord heating on heat polypnea was abolished. This result was confirmed by a series of corresponding studies which are summarized in figure 10. In all experiments in which the anterior part of the hypothalamus was coagulated (animal no. 895, 02, 01, 988, 926, 868), only a poor respiratory response to spinal cord heating could be obtained after electrocoagulation. Lesions outside the anterior hypothalamus did not abolish the effect of spinal cord heating on the respiratory rate (e.g. animal no. 1, 04, 06, 05). The results of these studies lead to the conclusion

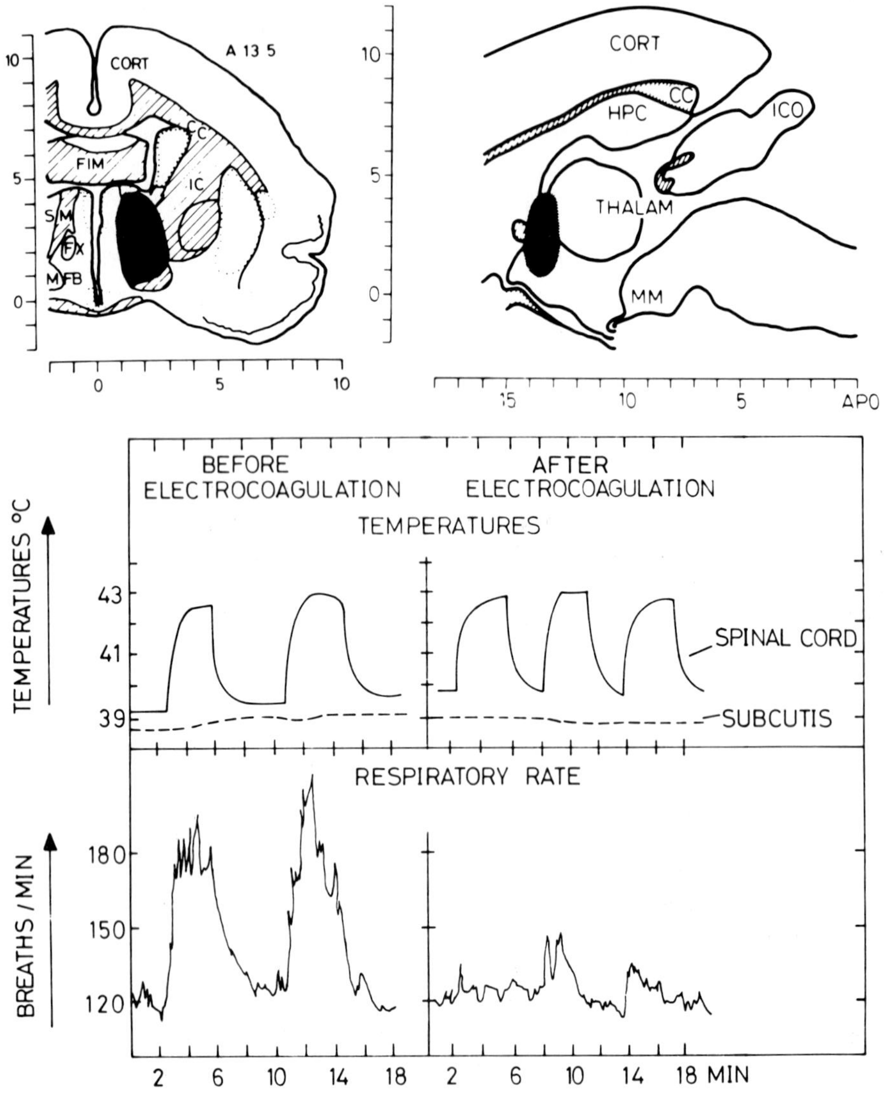

Fig. 9. *Effect of spinal cord heating on respiratory rate before and after electrocoagulation in the anterior hypothalamus. To represent the localizations of the lesions, a sterotaxic atlas of the forebrain of the guinea pig was used (60). (From ref. 65, courtesy of Charles C. Thomas, Publisher, Springfield, Ill.).*

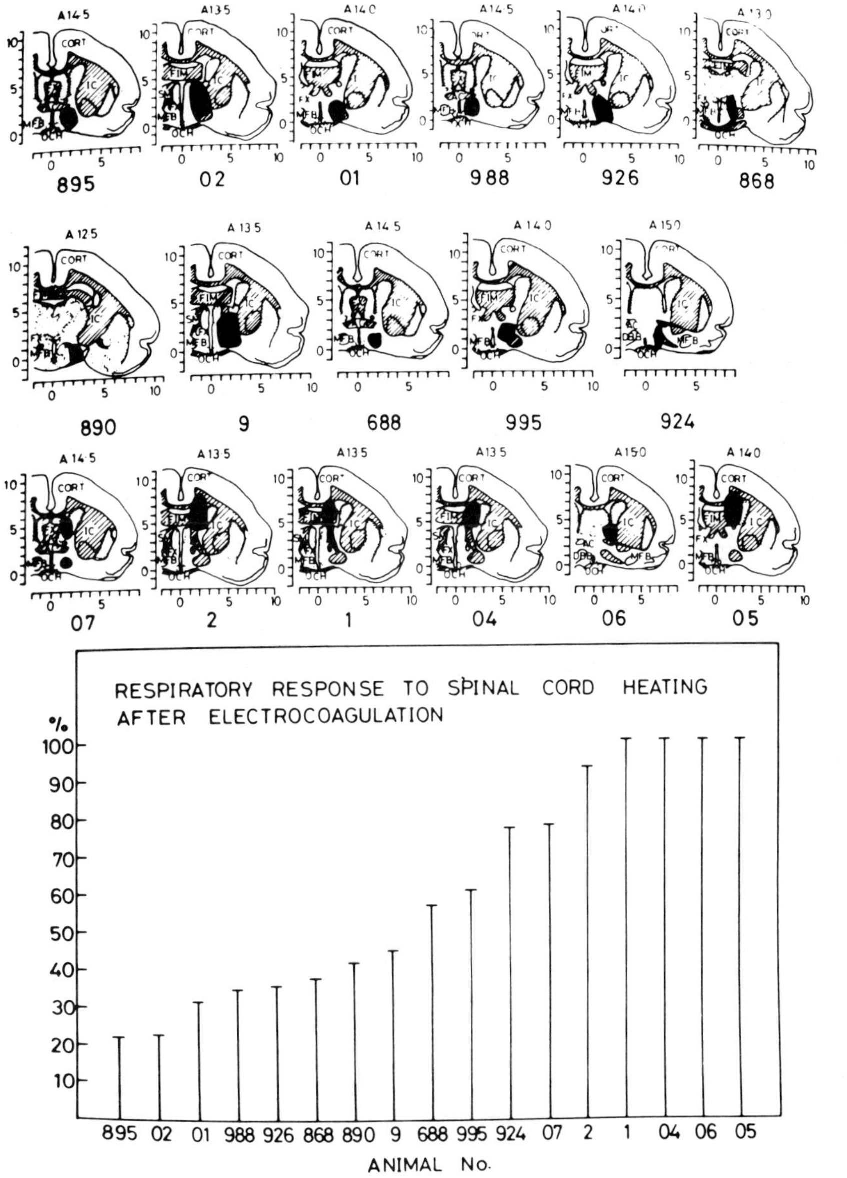

Fig. 10. *Responses of respiratory rate to spinal cord heating in seventeen guinea pigs shown in lower section. Results after coagulation are related to values recorded before the lesions were made. Site and size of the lesions are represented in upper section. To represent the lesions, a sterotaxic atlas of the forebrain of the guinea pig was used (60). (From 65, courtesy of Charles C. Thomas, Publisher, Springfield, Ill.).*

that pathways ascending from the thermosensitive spinal region are switched over to efferent systems controlling heat polypnea in the anterior part of the hypothalamus.

In further studies (66), the inhibitory action of spinal cord heating on shivering was examined before and after electrocoagulation (Fig. 11). Localization and size of the lesions were varied intentionally. Depending on the effect of local heating of the spinal cord on the intensity of shivering, the results could be divided into three groups. When a lesion in the medio-dorsal part of the hypothalamus had been made, it was still possible to cause shivering by external cooling. However, inhibition of shivering by heating the thermosensitive spinal structures was no longer possible after electrocoagulation. Results of this type (group I) were obtained in all experiments in which the region limited by fornix, mammillo-thalamic tract, reuniens thalamic nucleus and medio-dorsal thalamic nucleus had been coagulated. In this group the ascending fibres from the spinal thermosensitive elements were evidently interrupted. In the experiments summarized in group II and III the effect of local heating of the spinal cord on shivering was not abolished by electrocoagulation. In two animals (group III), in which the area above the mammillary medial nucleus had been destroyed, cold-induced shivering intensity was reduced to 50 - 80% of the amount before the electrocoagulation. It is possible that in these animals part of the integrative tissue controlling shivering had been destroyed or that the descending shivering pathway had been damaged by the lesion.

The lesions in the medio-dorsal hypothalamus (group I) only cancelled the inhibitory effect of spinal cord heating on shivering but did not abolish shivering response to external cooling. Thus, the localizations of the lesions (group I) are not identical with the sites of thermointegrative structures controlling shivering. They mark the area where the ascending pathway from the spinal thermosensitive elements reach this integrative region. Thermointegrative interneurones that control shivering are supposed to lie more caudal from the area delimited by the lesions (group I), i.e. in a region between the

<table>
<tr><th>Group</th><th>Localization and size of the lesion</th><th>Animal-No</th><th>Occurrence of shivering</th><th>Shivering suppressed by local heating at $C_5 - T_2$</th></tr>
<tr><td rowspan="5">I</td><td rowspan="5">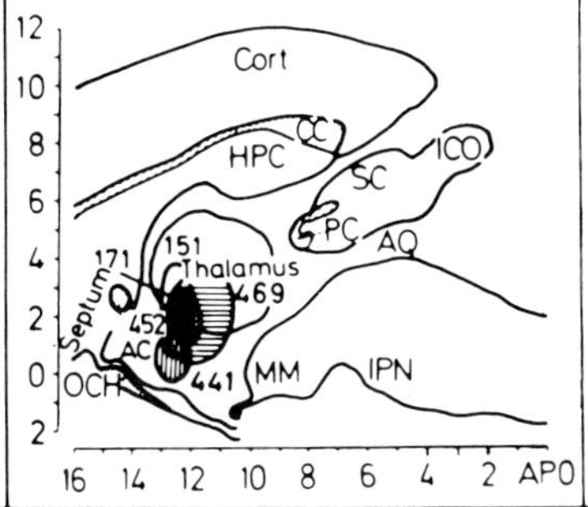</td><td>151</td><td>yes</td><td>no</td></tr>
<tr><td>469</td><td>yes</td><td>no</td></tr>
<tr><td>452</td><td>yes</td><td>no</td></tr>
<tr><td>441</td><td>yes</td><td>no</td></tr>
<tr><td>171</td><td>yes</td><td>no</td></tr>
<tr><td rowspan="8">II</td><td rowspan="8">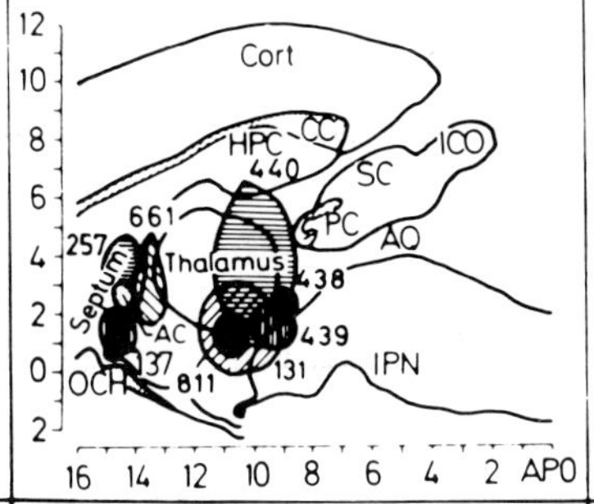</td><td>131</td><td>yes</td><td>yes</td></tr>
<tr><td>440</td><td>yes</td><td>yes</td></tr>
<tr><td>438</td><td>yes</td><td>yes</td></tr>
<tr><td>439</td><td>yes</td><td>yes</td></tr>
<tr><td>257</td><td>yes</td><td>yes</td></tr>
<tr><td>137</td><td>yes</td><td>yes</td></tr>
<tr><td>661</td><td>yes</td><td>yes</td></tr>
<tr><td>811</td><td>yes</td><td>yes</td></tr>
<tr><td rowspan="2">III</td><td rowspan="2">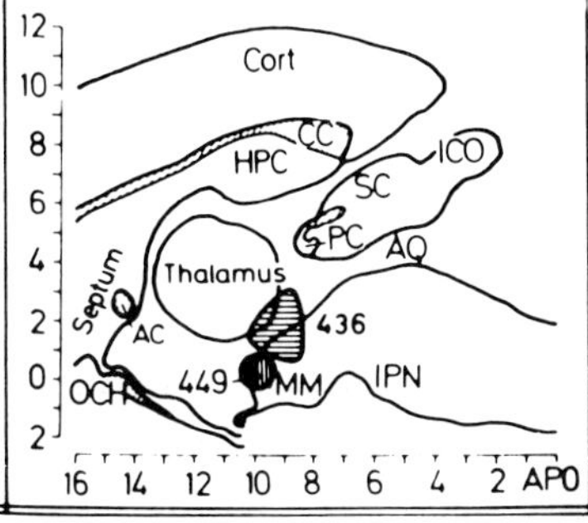</td><td>436</td><td>el. Activity reduced</td><td>yes</td></tr>
<tr><td>449</td><td>el. Activity reduced</td><td>yes</td></tr>
</table>

Fig. 11. *Influence of local RF-heating of the spinal segments C_5-T_2 on shivering, following circumscribed lesions made by electrocoagulation at various sites of the diencephalon. In Groups II and III, shivering could be reduced by at least 70% when the temperature of the cervical spinal cord was increased to an average of 40°C. In Group I, shivering remained unchanged even with an average temperature increase to 40.5°C. To represent the localization of the lesions (shaded areas), a stereotaxic atlas was used (60). (From 66).*

ascending pathways and the central portion of the efferent shivering pathway as described by Birzis and Hemingway (5).

The experiments summarized in the figures 10 and 11 indicate that hypothalamic structures, which are obviously not sensitive to their own temperatures, play an important role in the integration of input signals from thermoreceptive areas and in the control of heat loss mechanisms. As shown in figure 12 the field of neurones involved in the control of heat polypnea extends to the POA. This is in accordance with the observation of Andersson *et al.* (1) who showed in goats that entire destruction of the rostral part of the hypothalamus by proton irradiation led to the loss of open mouth panting during heat exposure.

The importance of posterior hypothalamic tissue for the control of shivering has been emphasized by several authors. Stuart *et al.* (59) concluded that "the primary centre for shivering" is in the posterior hypothalamus. The central portion of the "shivering pathway" was described by Birzis and Hemingway (5) and Hemingway (32).

VI. POSSIBLE MODE OF INTEGRATION OF THERMAL INFORMATION

Possible modes of integration of temperature signals have been discussed in several neuronal models (7, 11, 12, 22, 42). Some of these models also include thermoregulation in hibernators and discuss proportional control during deep hibernation rather than a passive

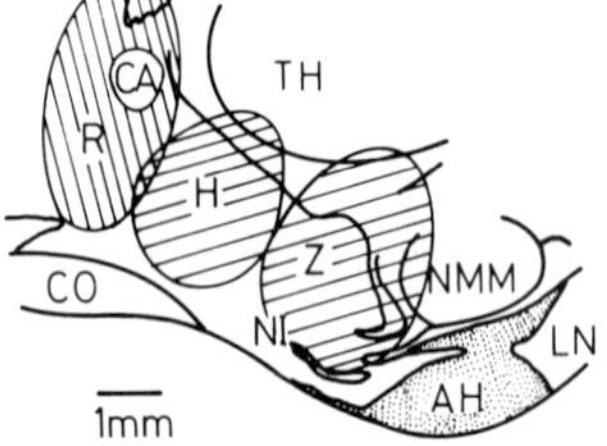

Fig. 12. *Thermoreceptive region (R) and site of thermointegrative structures involved in the control of heat polypnea (H) and shivering (Z). Data obtained from studies in guinea pigs (65, 66).*

turning off of the control system. There also seems to exist unanimity of opinion that the onset of hibernation bases on an active inhibitory mechanism that probably proceeds from the brain stem reticular formation (3, 22, 25, 42).

In the well-known model proposed by Hammel *et al.* (22), the setpoint for body temperature is created by the intersection of the temperature-impulse rate characteristics of hypothalamic neurones with a high and a low temperature sensitivity, respectively. The single unit studies on golden hamsters (Fig. 1, 2, 3) seem to present a well-fitting neurophysiological supplement to this model. The units found in these studies show a striking conformity with the high Q_{10} units proposed by Hammel *et al.* (22). Furthermore, there is also neurophysiological evidence for the existence of "reference units". In the POA of golden hamster, units were found with a constant discharge frequency in the temperature range between 10 - 43°C.

There is some discrepancy between the different neuronal models (e.g. 11, 12, 22) with respect to the interaction of temperature signals from the spinal cord and the skin. In the model by Hammel *et al.* (22) cutaneous, spinal cord, and hypothalamic temperature sensors are placed in series while Brück and Wünnenberg (12) proposed a parallel input of temperature signals.

Both hypothesis are partly supported by results reported in this paper. The results summarized in Fig. 11 (group II), f. i., show that destruction of the POA does not abolish the thermoregulatory response to thermal stimulation of the spinal cord and the skin. This finding indicates that the afferent pathways from different thermoreceptive areas are independent of each other. On the other hand there is evidence from neurophysiological studies (Fig. 4, 5) that thermoresponsive preoptic units are influenced by the activity of other thermoreceptive sites. These results support the model proposed by Hammel *et al.* (22).

Obviously both models are suitable to illustrate hypothesis revealed from a number of experimental findings - they are, however, not complex enough to explain all results found in experiments on thermoregulation.

ACKNOWLEDGEMENT

Supported in part by the Deutsche Forschungsgemeinschaft Wu 63/2 and Wu 63/4.

REFERENCES

1. Andersson, C., G. Gale, B. Hökfelt, and B. Larsson. Acute and chronic effects of preoptic lesions. Acta Physiol. Scand. 65: 45-60, 1965.
2. Bard, P., J. W. Woods, and R. Blier. The effects of cooling, heating and pyrogen on chronically decerebrate cats. In: Physiological and Behavioral Temperature Regulation. Springfield, Ill.: Thomas 1970, 519-545.
3. Beckman, A. L., and T. L. Stanton. Changes in CNS responsiveness during hibernation. Am. J. Physiol. 231: 810-816, 1976.
4. Benzinger, T. H. Heat Regulation: Homeostasis of central temperature in man. Physiol. Rev. 49: 671-759, 1969.
5. Birzis, L., and A. Hemingway. Descending brain stem connections controlling shivering in cat. J. Neurophysiol. 19: 37-43, 1956.
6. Bligh, J. The thermosensitivity of the hypothalamus and thermoregulation in mammals. Biol. Rev. 41: 317-367, 1966.
7. Bligh, J. Temperature Regulation in Mammals and other Vertebrates. Amsterdam and London: North Holland Publishing Company, 1973.
8. Boulant, J. A., and K. E. Bignall. Determinants of hypothalamic neuronal thermosensitivity in ground squirrels and rats. Am. J. Physiol. 225: 306-310, 1973.
9. Boulant, J. A., and K. E. Bignall. Hypothalamic neuronal responses to peripheral and deep-body temperatures. Am. J. Physiol. 225: 1371-1374, 1973.
10. Boulant, J. A., and J. D. Hardy. The effect of spinal and skin temperatures on the firing rate and thermosensitivity of preoptic neurones. J. Physiol. (Lond.) 240: 639-660, 1974.
11. Brück, K. Temperature regulation and catecholamines. Isr. J. Med. Sci. 12: 924-933, 1976.
12. Brück, K. and W. Wünnenberg. "Meshed" control of two effector systems: non-shivering and shivering thermogenesis. In: Physiological

and Behavioral Temperature Regulation. Hardy, J. D., A. P. Gagge and J. A. Stolwijk (editors). Springfield, Ill.: Thomas, 1970, 562-580.

13. Downey, J. A., R. F. Mottram, and G. W. Pickering. The location by regional cooling of the central temperature receptors in the conscious rabbit. J. Physiol. (London) 170: 415-441, 1964.

14. Edinger, H. M., and J. S. Eisenman. Thermosensitive neurons in tuberal and posterior hypothalamus of cats. Am. J. Physiol. 219: 1098-1103, 1970.

15. Eisenman, J. S. Unit activity studies of thermoresponsive neurons. In: Essays on Temperature Regulation. Bligh, J. and R. E. Moore (editors). Amsterdam and London: North Holland Publishing Company, 1972.

16. Eisenman, J. S. Sensory organs and thermogenesis. Isr. J. Med. Sci. 12: 916-923, 1976.

17. Eisenman, J. S., and D. C. Jackson. Thermal response patterns of septal and preoptic neurons. Exp. Neurol. 19: 33-45, 1967.

18. Folkow, B., G. Ström, and B. Unväs. Cutaneous vasodilatation elicited by local heating of the anterior hypothalamus. Acta Physiol. Scand. 17: 317-326, 1949.

19. Glotzbach, S. F., and H. C. Heller. CNS regulation of metabolic rate in the kangaroo rat *Dipodomys ingens*. Am. J. Physiol. 228: 1880-1886, 1975.

20. Guieu, J. D., and J. D. Hardy. Effects of heating and cooling of the spinal cord on preoptic unit activity. J. Appl. Physiol. 29: 675-683, 1970.

21. Hammel, H. T., J. D. Hardy, and M. M. Fusco. Thermoregulatory response to hypothalamic cooling in unanaesthetized dogs. Am. J. Physiol. 198: 481-486, 1960.

22. Hammel, H. T., H. C. Heller, and F. R. Sharp. Probing the rostral brain stem of anaesthetized, unanaesthetized, and exercising dogs and of hibernating and euthermic ground squirrels. Fed. Proc. 32: 1588-1597, 1973.

23. Hartner, W. C., F. E. South, H. K. Jacobs, and R. H. Luecke. Preoptic thermal stimulation and temperature regulation in the marmot (*M. flaviventris*). Cryobiology 8: 312-313, 1971.
24. Heath, J. E., B. A. Williams, S. H. Mills and M. J. Kluger. The responsiveness of the preoptic-anterior hypothalamus to temperature in vertebrates. In: Hibernation and Hypothermia, Perspectives and Challenges. South, F. E., J. P. Hannon, J. S. Willis, E. T. Pengelley and N. R. Alpert (editors). Amsterdam: Elsevier, 1972, pp. 605-627.
25. Heller, H. C., and G. W. Colliver. CNS regulation of body temperature during hibernation. Am. J. Physiol. 227: 582-289, 1974.
26. Heller, H. C., G. W. Colliver and P. Anand. CNS regulation of body temperature in euthermic hibernators. Am. J. Physiol. 227: 576-582, 1974.
27. Heller, H. C., and H. T. Hammel. CNS control of body temperature during hibernation. Comp. Biochem. Physiol. 41A: 349-359, 1972.
28. Heller, H. C., and J. A. Henderson. Hypothalamic thermosensitivity and regulation of heat storage in a day-active desert rodent *Amnospermophilus nelsoni*. J. Comp. Physiol. 108: 255-270, 1976.
29. Hellon, R. F. The stimulation of hypothalamic neurones by changes in ambient temperature. Pflügers Arch. 321: 56-66, 1970.
30. Hellon, R. F. Temperature-sensitive neurons in the brain stem: their responses to brain temperatures at different ambient temperatures. Pflügers Arch. 335: 323-334, 1972.
31. Hellstrøm, B., and H. T. Hammel. Some characteristics of temperature regulation in the unanaesthetized dog. Am. J. Physiol. 213: 547-566, 1967.
32. Hemingway, A. Shivering. Physiol. Rev. 43: 397-442, 1963.
33. Hemingway, A., T. Rasmussen, H. Wikoff and A. T. Rasmussen. Effect of heating hypothalamus of dogs by diathermy. J. Neurophysiol. 3: 329-338, 1940.

34. Hensel, H. Physiologie der Thermorezeptoren. Ergebn. Physiol. 47: 166-368, 1952.
35. Hensel, H., K. Brück and P. Raths. Homeothermic organisms. In: Temperature and Life. Precht, H., J. Christophersen, H. Hensel and W. Larcher (editors). Berlin, Heidelberg and New York: Springer Verlag, 1973.
36. Jacobsen, F. H., and R. D. Squires. Thermoregulatory responses of the cat to preoptic and environmental temperatures. Am. J. Physiol. 218: 1575-1582, 1970.
37. Jessen, C. Two-dimensional determination of thermosensitive sites within goat's hypothalamus. J. Appl. Physiol. 40: 514-520, 1976.
38. Jessen, C., and E. T. Mayer. Spinal cord and hypothalamus as core sensors of temperature in the conscious dog. I. Equivalence of responses. Pflügers Arch. 324: 180-204, 1971.
39. Kayser, C. Stoffwechsel und Winterschlaf. Helgoländer Wiss. Meeresunters. 9: 158-186, 1964.
40. Kayser, C. La dépense d'energie des mammifères en hibernation. Arch. Sci. Physiol. (Paris) 18: 137-150, 1964.
41. Keller, A. D., and E. B. McClaskey. Localization by the brain slicing method, of the level or levels of the brain stem upon which effective heat dissipation is dependent. Am. J. Physiol. 43: 181-213, 1964.
42. Luecke, R. H., and F. E. South. A possible model for thermoregulation during deep hibernation. In: Hibernation and Hypothermia, Perspectives and Challenges. South, F. E., J. P. Hannon, J. S. Willis, E. T. Pengelly, and N. R. Alpert (editors). Amsterdam: Elsevier, 1972, pp. 577-604.
43. Lyman, C. P., and R. C. O'Brien. Sensitivity to low temperature in hibernating rodents. Am. J. Physiol. 222: 864-869, 1972.
44. Lyman, C. P., and R. C. O'Brien. A comparison of temperature regulation in hibernating rodents. Am. J. Physiol. 227: 218-223, 1974.
45. Magoun, H. W., F. Harrison, J. R. Brobeck, and S. W. Ranson. Activation of heat loss mechanisms by local heating of the brain. J. Neurophysiol. 1: 101-144, 1938.

46. Merker, G., and H. G. Hartwig. Thermosensitive und thermointegrative areale des hypothalamus: Struktur und Funktion. Anat. Anz. (In press).
47. Mihailović, L. T. Cortical and subcortical electrical activity in hibernation and hypothermia. In: Hibernation and Hypothermia. South, F. E., J. P. Hannon, J. S. Willis, E. T. Pengelly, and N. R. Alpert (editors). Amsterdam: Elsevier, 1972, pp. 487-534.
48. Mills, S. H., V. M. Miller, and F. E. South. Thermoregulatory responses of a hibernator to hypothalamic and ambient temperatures. Cryobiology 11: 465-472, 1974.
49. Nakayama, T., and J. D. Hardy. Unit responses in the rabbit's brain stem to changes in brain and cutaneous temperature. J. Appl. Physiol. 27: 848-857, 1969.
50. Nakayama, T., H. T. Hammel, J. D. Hardy and J. S. Eisenman. Thermal stimulation of electrical activity of single units of the preoptic region. Am. J. Physiol. 204: 1122-1126, 1963.
51. Nutik, S. L. Effect of temperature change of the preoptic region and skin on posterior hypothalamic neurons. J. Physiol. (Paris) 63: 368-370, 1971.
52. Oksche, A. The neuroanatomical basis of comparative neuroendocrinology. Gen. Comp. Endocrinol. 29: 225-239, 1976.
53. Raths, P., und H. Hensel. Cutane Thermorezeptoren bei Winterschläfern. Pflügers Arch. 293: 281-302, 1967.
54. Shtark, M. B. The Brain of Hibernating Animals. Washington D.C.: NASA Technical Translation TTF-619, 1972 (Translation of Mozg Zimnespyashchekt. Novosibirsk: Nauka Press, 1970).
55. South, F. E., and W. C. Hartner. Hibernation and active thermoregulation: responses to manipulation of the temperature of the preoptic region in the marmot. Cryobiology 8: 389, 1971.
56. South, F. E., W. C. Hartner, and R. H. Luecke. Responses to preoptic temperature manipulation in the awake and hibernating marmot. Am. J. Physiol. 229: 150-160, 1975.

57. Speulda, E., and W. Wünnenberg. Thermosensitivity of preoptic neurones in a hibernator at high and low ambient temperatures. Pflügers Arch. 370: 107-109, 1977.
58. Strumwasser, F. Regulatory mechanisms, brain activity and behavior during deep hibernation in the squirrel, *Citellus beecheyi*. Am. J. Physiol. 196: 23-30, 1959.
59. Stuart, D. G., Y. Kawamura, A. Hemingway and W. M. Price. Effects of septal and hypothalamic lesions. Exp. Neurol. 5: 335-347, 1962.
60. Tindal, J. S. The forebrain of the guinea pig in sterotaxic coordinates. J. Comp. Neurol. 124: 259-266, 1965.
61. Williams, B. A., and J. E. Heath. Responses to preoptic heating and cooling in a hibernator, *Citellus tridecemlineatus*. Am. J. Physiol. 218: 1654-1660, 1970.
62. Williams, B. A., and J. E. Heath. Thermoregulatory responses of a hibernator to preoptic and environmental temperatures. Am. J. Physiol. 221: 1134-1138, 1971.
63. Wit, A., and S. C. Wang. Temperature-sensitive neurons in preoptic anterior hypothalamic region: effects of increasing ambient temperature. Am. J. Physiol. 215: 1151-1159, 1968.
64. Wünnenberg, W. Alteration of EEG activity of the hypothalamus by thermal stimulation of the spinal cord. Pflügers Arch. 344: 75-82, 1973.
65. Wünnenberg, W. Thermointegrative function of the hypothalamus. In: Regulation of Depressed Metabolism and Thermogenesis. Janský, L. and X. J. Musacchia (editors). Springfield, Ill.: Charles C. Thomas Publishing, 1976, pp. 26-41.
66. Wünnenberg, W. and K. Brück. Studies on the ascending pathways from the thermosensitive region of the spinal cord. Pflügers Arch. 321: 233-241, 1970.
67. Wünnenberg, W. and J. D. Hardy. Responses of single units of the posterior hypothalamus to thermal stimulation. J. Appl. Physiol. 33: 547-552, 1972.

68. Wünnenberg, W., G. Merker, and E. Speulda. Thermosensitivity of preoptic neurones in a hibernator (golden hamster) and a non-hibernator (guinea pig). Pflügers Arch. 363: 113-123, 1976.

TIME SEQUENCE OF PHYSIOLOGICAL CHANGES DURING HIBERNATION: THE SIGNIFICANCE OF SEROTONERGIC PATHWAYS

L. Janský
Department of Comparative Physiology
Faculty of Sciences
Charles University
Prague, Czechoslovakia

ABSTRACT

Mechanisms controlling serotonin metabolism in the brain and the significance of serotonergic pathways for physiological processes of hibernating and awake golden hamsters were studied. The important role of 5-HT in hibernation becomes evident from experiments showing that feeding a tryptophan rich diet facilitates the occurrence of hibernation and that the 5-HT turnover is markedly increased during prehibernation and hibernation periods. On the other hand, in awake animals infusions of 5-HT into the brain ventricle induce only small changes in body temperature and fail to evoke hibernation. The 5-HT turnover is influenced temporarily by changes in environmental temperatures, the changes being unspecific to temperature stimulus. We suggest, that unknown seasonal mechanisms induce changes in 5-HT metabolism. Since the availability of the substrate (tryptophan), rather than the 5-HT destruction rates, are responsible for the increase in 5-HT turnover, an enhanced food consumption observed during prehibernation period and, thus, an increased substrate availability may increase the activity of 5-HT pathways in the brain. The lowered activity of liver tryptophan pyrrolase observed in hibernation contribute to the increased 5-HT turnover. Activation of 5-HT pathways may influence the production of hypothalamic releasing factors and consequently the hormonal status of the whole organism. Multiple biochemical changes in tissues occurring during preparation to hibernation might be induced by such mechanisms.

ISBN 0-12-734550-7

I. INTRODUCTION

Hibernation is a remarkable and unique physiological phenomenon, attracting the attention of physiologists for almost a century. There have been thousands of interesting papers published on this subject [for bibliography until 1964 see Mayer (34)]. A great deal of physiological and biochemical changes have already been described in hibernating animals and yet there is little understanding of the basic mechanisms enabling the gradual and controlled entering of hibernation and survival of this hypothermic state. Studies on hibernation are hampered by the fact that it is difficult to distinguish between primary and secondary changes and to establish causal relationships between individual phenomena. In other words, it is difficult to establish the causes and the consequences of observed changes.

There is no doubt that survival of lowered body temperatures is made possible by multiple biochemical adaptations of cell intermediary metabolism (for reviews see 3, 5, 18). Furthermore, there exists a distinct possibility that the biochemical changes occurring in various tissues of hibernating animals may be the consequence of some regulatory (nervous and humoral) mechanisms.

Literary data indicate that neuronal mechanisms controlling hibernation originate in the limbic system (57). The endogenous circannual cycle is probably responsible for the timing of entrance into hibernation in many species (42). In addition, functional changes in the ventromedial nucleus of the hypothalamus, enabling a substantial increase in food intake prior to hibernation have been observed (35, 36). On the other hand, hypothalamic mechanisms regulating body temperature in awake hibernators are found to be essentially the same as in non hibernators (63). During hibernation the normal hypothalamic regulation of body temperature is first inactivated. After some days, however, thermoregulation is demonstrable in response to both lowering or raising temperature in the preoptic area (55). The lowering of the "set point" for temperature regulation by mechanisms so far unknown, is believed to be responsible for the controlled entrance into hibernation (17).

Furthermore, hibernation can be characterized by

striking changes in activity of some endocrine glands. The subject has been recently reviewed (22, 36, 46). In short, it can be stated that the activity of the thyroid, adrenal cortex and gonads is lowered, while the neurosecretion of the hypothalamus and the pancreas activity persist during hibernation (1, 58). Mechanisms inducing the involution of endocrine glands and the reasons for the lowered hormonal activity are not known. It can be expected, however, that it is the changed hormonal status in hibernating organism, which creates the necessary condition for adaptations of cellular metabolism in target organs, thus enabling survival under hypothermic conditions.

In spite of a rather detailed knowledge of neuronal and humoral regulatory mechanisms, the relative importance and the time sequence of individual hibernation phenomena are not known.

In an attempt to trace the most fundamental mechanism controlling hibernation we focused our attention on serotonergic pathways in the brain, which are believed to play a considerable role in regulation of body temperature, sleep phenomena, sexual behavior and other centrally mediated physiological functions (for review see 62). Since hibernation is considered to be a specialized thermoregulatory model and since it has been found that it is realized from a state of sleep (54), these studies have been of specialized interest. Finally, the aim of this paper was to test our earlier hypothesis, that serotonergic pathways in the brain may influence the release of hypothalamic hormonal factors controlling the adenohypophysis and consequently the activity of related hormonal glands.

II. SEROTONIN METABOLISM IN THE BRAIN OF HIBERNATING ANIMALS

A. Serotonin Content and Turnover in the Brain

Our earlier experiments (41) have shown that substantial changes in serotonin metabolism occur during a hibernation cycle. These experiments were performed on male golden hamsters kept under natural conditions, where rhythms of illumination and temperature fluctuated according to seasons of the year. In the winter season animals which entered hibernation were analysed, while in other seasons experiments were performed on awake animals. Concentrations of 5-HT and 5-HIAA were measured according to the method by Johnson and Lewander (25), the 5-HT turnover was studied by inhibition of 5-HT degrading enzyme, the monoamine oxidase (MAO), with pargyline and by estimating the rate of 5-HT accumulation (59).

In agreement with other data (31, 51, 56, 61), it was shown that the 5-HT content in the brain was increased in hibernating animals (Fig. 1). In addition to that, the 5-HT turnover was also increased markedly. The activation of 5-HT pathways was already apparent during the prehibernation period. Immediately after arousal from hibernation the 5-HT turnover rapidly decreased again to values similar to that measured in the summer (Fig. 1). No distinct changes in 5-HT turnover were observed in other groups of golden hamsters kept under constant conditions at 30°C throughout the year (41).

These data indicated a certain relationship between hibernation and the activity of serotonergic pathways in the brain. They also prompted a series of questions, however, the most pertinent being the control of the enormously increased 5-HT turnover in hibernating animals.

B. <u>Monoamine Oxidase Activity in the Brain</u>

To contribute to the solution of this problem the activity of the monoamine oxidase (MAO) in the brain stem in hibernating and awake golden hamsters was measured using the Wurtman and Axelrod method (64). In this case, the activity of MAO was examined in hibernating animals kept under outdoor conditions and in wakeful animals kept indoor at 25°C. Both groups were examined in the month of February. No difference in MAO activity was found between these two groups of animals (Fig. 2), indicating that the increased 5-HT turnover was not related to 5-HT destruction rate (41). This was in accordance with data which showed that the rate of 5-HT synthesis is rather dependent on substrate concentrations (24, 43). These data indicate that the concentrations of the precursor (tryptophan) normally present in the mammalian brain are less than Michaelis constant (Km) of the key enzyme, tryptophan hydroxylase, which regulates the synthesis of 5-HT. The

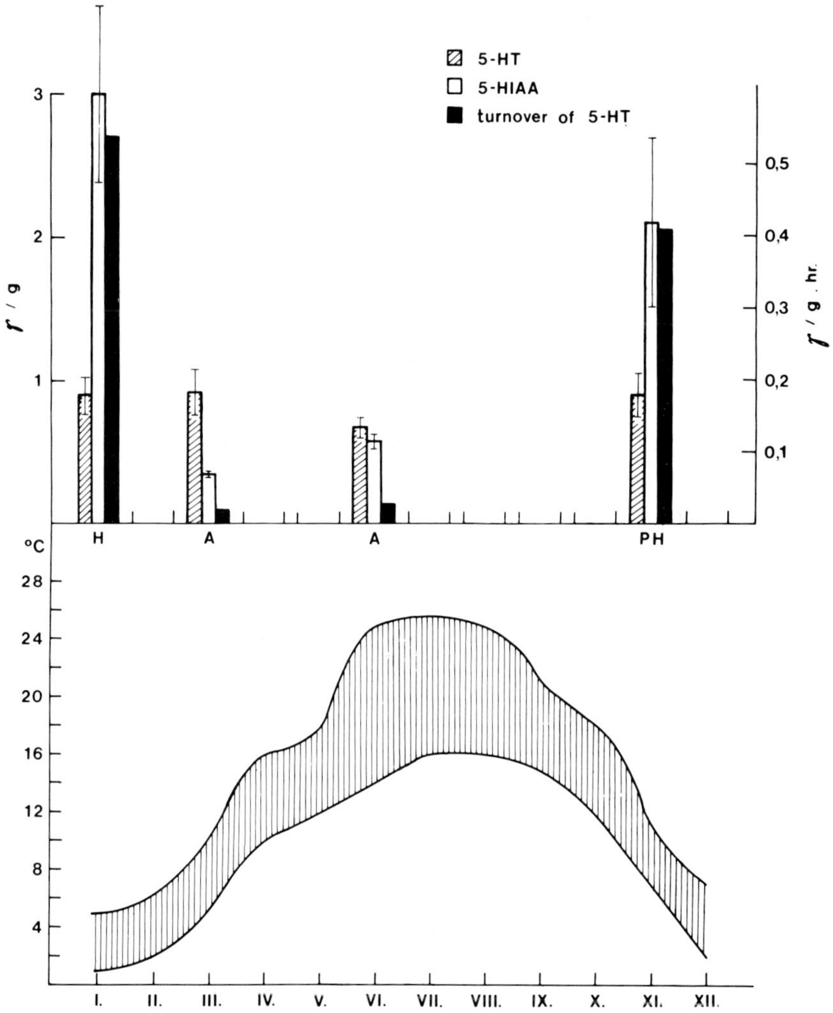

Fig. 1. *Seasonal changes in 5-HT turnover (black columns), 5-HT and 5-HIAA concentrations (hatched and white columns) in the brain-stem of golden hamsters kept under natural conditions. A= active, PH= prior to hibernation, H= hibernating. Lower part of the figure shows seasonal variations of environmental temperatures in animal rooms (From 41).*

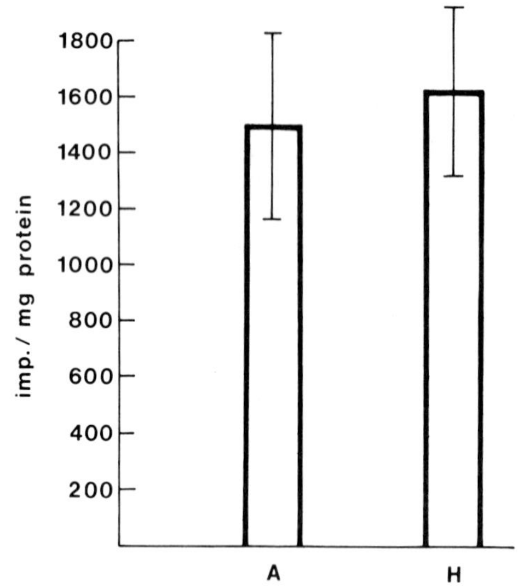

Fig. 2. *Monoamine oxidase activity in the brain-stem of non-hibernating (A) and hibernating (H) golden hamsters (From 41).*

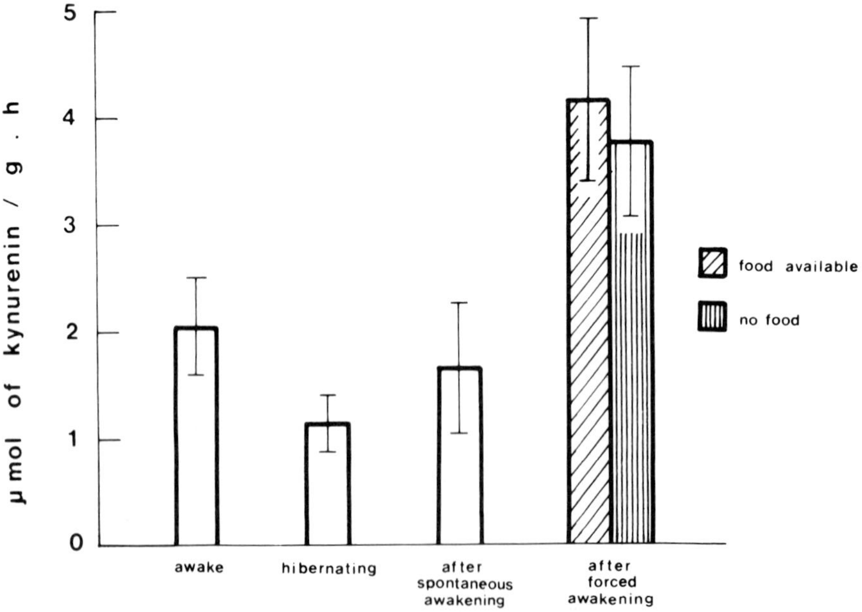

Fig. 3. *Changes in liver tryptophan pyrolase activity (measured at 37°C) in non-hibernating and hibernating golden hamsters (From 39).*

intraneural concentrations of tryptophan may not be sufficient to saturate the enzyme and small variations in the availability of tryptophan may, thus, rapidly affect the rate of 5-HT synthesis (10, 14, 33). In addition, tryptophan hydroxylase does not seem to be inhibited by high concentrations of 5-hydroxytryptophan, 5-HT or 5-hydroxyindolacetic acid (5-HIAA); this indicates that there is no product feedback inhibition (32, 38).

Brain tryptophan is derived from plasma and a proportionality exists between brain and plasma tryptophan concentrations (15). All factors controlling the level of plasma tryptophan should be taken into consideration therefore, when explaining mechanisms controlling 5-HT turnover in the brain.

C. Liver Tryptophan Pyrrolase Activity

Most of the tryptophan from the blood is metabolized by liver tryptophan pyrrolase (LTP). We therefore, measured the LTP activity in the awake and hibernating hamsters, following the method of Knox and Auerbach (30). Experiments were performed in the winter season. Active animals were kept under constant conditions at 25°C; hibernating individuals were kept in a dark, quiet room at 4 - 6°C.

We found that the activity of this enzyme measured at 37°C decreased by more than 40% during hibernation (Fig. 3) (8, 39). Furthermore, we have found that the activity of LTP was strongly temperature dependent and decreased to minimal values when measured at 10°C (8, 39). Thus, during hibernation, at very low body temperatures, practically no tryptophan can be metabolized and it becomes fully available for the 5-HT synthesis in the brain. This forms a necessary prerequisite for an increased 5-HT turnover in the brain of hibernating animals because there is no dietary supply of tryptophan.

Immediately after arousal, the liver tryptophan pyrrolase activity increased (Fig. 3). This increase was not induced by the substrate, since it occurred also in individuals without access to food. It was presumed, therefore, that some hormonal factors may be responsible for this increase. As it is known that LTP activity can be increased by adrenal steroids (30, 49), the effect of hydrocortisone on the acitvity of this enzyme was estimated. It was found that hydrocortisone (1.5 mg/kg i.p.) increases LTP activity by 97%, 6 hours after i.p. injection (Novotná, unpublished observation). Since it is also known (12) that during awakening from hibernation the level of corticosterone in the blood increases, it is possible that the increase in the LTP activity after awakening is induced by adrenal steroids.

These data may also contribute to explanation of the obvious discrepancy observed in hibernating individuals between the increased 5-HT content and an increased 5-HT turnover in the brain. The lowered activity of LTP, contributing to an increased 5-HT synthesis, on the one hand, and the unchanged MAO activity, limiting the rate of 5-HT destruction, on the other hand, may induce the temporary accumulation of 5-HT.

D. Effect of Tryptophan Feeding on Induction and Frequency of Hibernation

In the prehibernation period, the increase of 5-HT turnover could be controlled by changes in tryptophan concentration in the blood, arising from the food. It is well known that during the prehibernation period substantial changes in feeding pattern and food consumption occur (35, 36). In addition, an increased metabolism in the cold must also contribute to an increase in food consumption.

As a result of the increased food consumption the concentration of amino acids in the blood may increase substantially. It is tempting, therefore, to speculate that the feeding behaviour might induce changes in brain 5-HT turnover and that this may influence eventually, the pattern of hibernation.

Following this line of reasoning the effect of a tryptophan-rich diet on the induction and frequency of hibernation was studied. Golden hamsters were placed in a quiet dark room, where the temperature ranged between 4-6°C. Control animals were fed corn diet (containing 9 mg of tryptophan/g protein) *ad libitum*, while experimental groups received 5 mg or 15 mg of d,l-tryptophan/g of corn food, respectively. This corresponds to 58 or 174 mg of tryptophan/g protein. Patterns of hibernation were studied using the sawdust method.

Animals of all groups entered hibernation after about 8 weeks in the cold. Then the occurrence of hibernation increased in the group receiving the tryptophan-rich diet (Fig. 4). The effect seemed to be dose dependent. The experiment was repeated in the spring (Fig. 4). In this season very few animals entered hibernation, but again, in the group receiving tryptophan-rich diet there was an increased tendency for hibernation.

It can be concluded, therefore, that the presence of tryptophan in the food may facilitate the occurrence of hibernation. On the other hand, it must be stated, that neither the seasonal trends for hibernation nor the length of induction period prior to onset of hibernation were influenced by tryptophan feeding. In addition to that, no significant changes in the length of individual hibernating periods were observed (Fig. 5). The increased frequency of hibernation was mostly due to shortening of awakening periods between individual hibernation bouts (Table 1).

On the basis of the above findings it seems reasonable to conclude that an increase in food consumption and a lowering of the LTP activity contribute to an increase in concentration of tryptophan in the blood. This forms the prerequisite for increasing the 5-HT turnover. Increased activity of 5-HT pathways facilitates the occurrence of hibernation. Processes, which are influenced by the increased activity of 5-HT pathways during hibernation are not identifiable from our experiments, however.

III. SEROTONIN METABOLISM IN THE BRAIN OF AWAKE HIBERNATORS

In order to explain the role of 5-HT pathways in hibernation, studies on non-hibernating individuals were first conducted. These experiments were designed to clarify the role of 5-HT in thermoregulation and in influencing the activity of some hormonal glands.

The significance of various neurotransmitters for thermoregulation of non-hibernators has been studied extensively in recent years (for review see 20). Even though great interspecies differences have been found, two models which specified the role of various neurotransmitters in thermoregulation can be formulated (Fig. 6).

In the monkey, it is postulated (37) that two anatomically distinct neurochemical pathways originate in

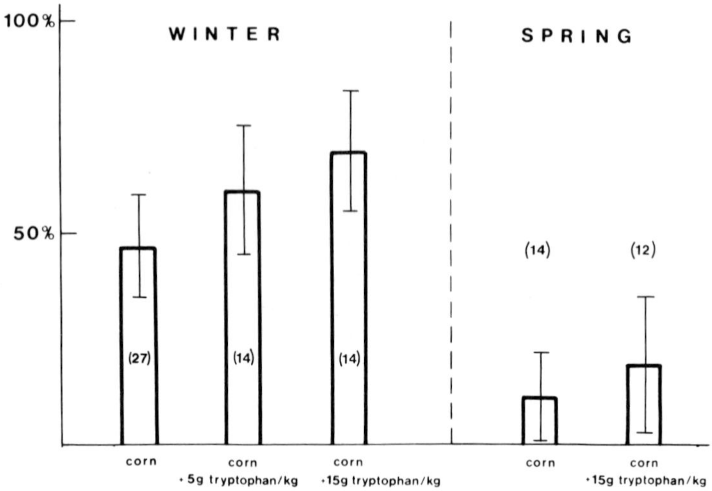

Fig. 4. *Effects of tryptophan feeding (5g, or 15g/kg corn) on occurrence of hibernation in golden hamsters expressed as percentage of all animals used. Numbers in parenthesis denote number of animals. Vertical bars indicate ± S.D.*

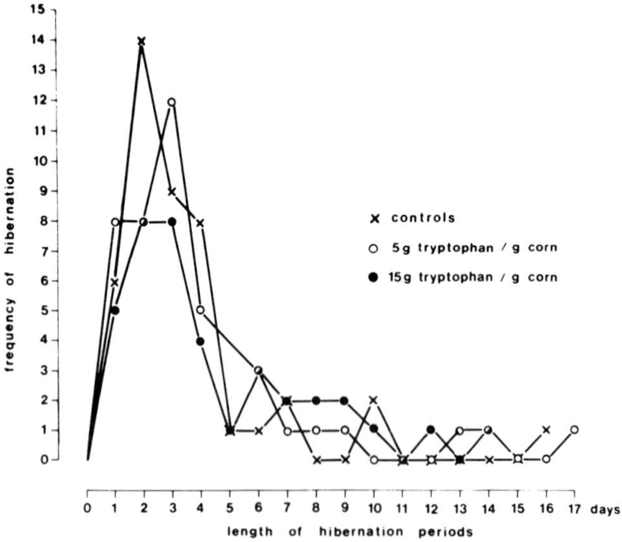

Fig. 5. *Frequency of hibernation periods of different length in golden hamsters fed by corn diet with different concentrations of tryptophan. Data calculated for the month of January.*

TABLE 1

Total time of awakening during hibernation periods of golden hamsters fed by tryptophan-rich diet (related to an individual)

controls	5 mg of tryptophan/g of corn	15 mg of tryptophan/g of corn
15.6 days	12.6 days	10.6 days

Data calculated for the month of January

the anterior preoptic region of the hypothalamus and project to the posterior hypothalamus via cholinergic structures. While the activation of 5-HT pathways is considered necessary for maintenance of heat production, the noradrenaline (NA) pathways seem to either inhibit the cholinergic part of the heat production pathway or to antagonize the hypothalamic action of 5-HT. Furthermore, it has been suggested that the suppression of the heat production system permits the cholinergic system in the posterior hypothalamus to activate the efferent heat loss pathway, which could also be influenced by extrahypothalamic impulses.

The above hypothesis is not applicable to other groups of animals, namely to ungulates and rodents. For these animals a neuronal model with reciprocal inhibition between heat production and heat loss was suggested by Bligh *et al.* (4). According to this model, 5-HT acts to increase the heat loss and to reduce heat production, whereas NA is thought to cause inhibition of heat loss and heat production systems at synapses posterior to those for 5-HT. According to this model the serotonergic pathways would come into action during an animal's exposure to heat, while NA pathways would be stimulated in cold exposure. This concept is supported in biochemical studies which show that in rats the 5-HT turnover increases with heat exposure, while the NA turnover is enhanced with cold exposure (52, 53).

No substantial attempt has been made thus far, in

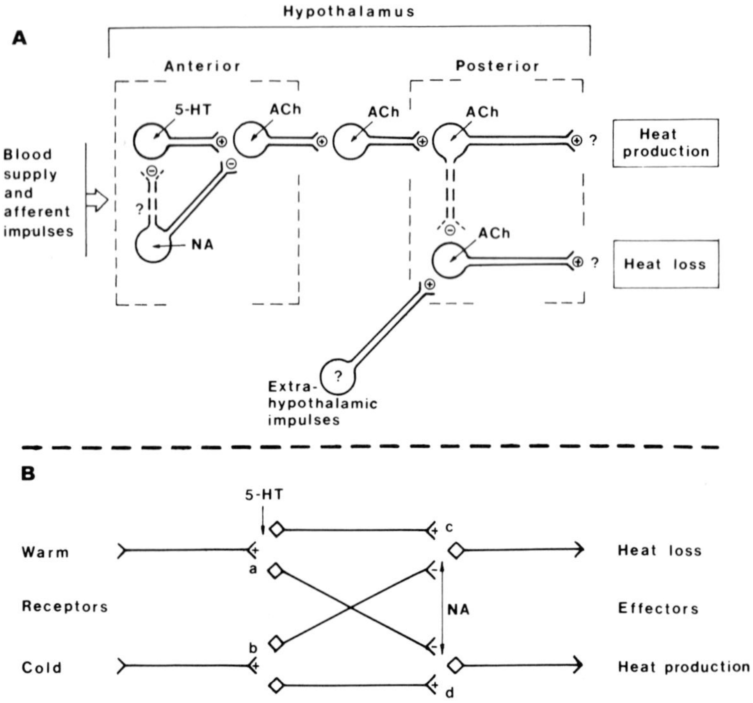

Fig. 6. *Two neuronal models of hypothalamic systems regulating body temperature.*

A. In the monkey, the serotonin (5-HT) containing cells in the anterior hypothalamus seem to be activated by cooling which stimulates heat production mechanisms via cholinergic structures in the posterior hypothalamus. Noradrenaline (NA) containing cells may be activated by warming and may inhibit either the 5-HT or the cholinergic pathways in the anterior hypothalamus. Due to suppression of heat production pathways the second cholinergic system in the posterior hypothalamus may activate efferent heat loss pathways (From 37). B. In the sheep, the stimulation of 5-HT containing cells by warming activates the heat loss pathway. Activation of (NA) containing cells either reduces heat production at high temperatures (synapse d) or decreases heat loss at low temperatures (synapse c) (From 4).

elucidating the role of central aminergic transmission of hibernators. Only Beckman and Satinoff (2) have shown that the intraventricular injections of amines induce arousal from hibernation.

A. Intraventricular Infusions of Biogenic Amines at Different Temperatures

In our experiments the effect of intraventricular infusions of 5-HT (5μg/kg, at a rate of 5μl/min) was studied in awake golden hamsters exposed to 30 or 15°C in different seasons of the year (21). Animals were kept all year round in a constant environment (25°C, 12L:12D cycle of illumination). A short-lasting (about 1 hr) decrease in body temperature, depending on the temperature of the environment (0.95°C at 30°C and 1.74 at 15°C, respectively) was observed, indicating the activation of heat loss mechanisms (Fig. 7). Heat production remained essentially unchanged, although there was a considerable variation in individual data during any one experiment. No differences were observed in different seasons of the year and no signs of entering torpidity could be detected. These data did not reveal any striking differences between golden hamsters and other rodents, as observed by other authors. It was concluded, therefore, that the mere surplus of 5-HT in the brain was not sufficient for inducing hibernation. An additional possibility could be that the serotonergic pathways responsible for hibernation are not located in the vicinity of the third ventricle and that the infused 5-HT becomes metabolized before reaching the active serotonergic area. This is not probable, however, since the concentration of 5-HT used in our experiments was sufficiently large. On the other hand, it must also be taken into consideration that it has been found that the increased concentration or turnover of 5-HT in the brain may not necessarily lead to an increase in functional activity of the serotonergic pathways (16).

In this connection our experiments studying the effect of NA and acetycholine (ACh) infusions into the third brain ventricle should also be mentioned (Fig. 8 and 9). In accordance with published data (20) on non-hibernators, an inhibition of the increased heat production in the cold was observed after NA (5μg/kg, at a rate of 5μl/min). No effect was visible at the thermo-

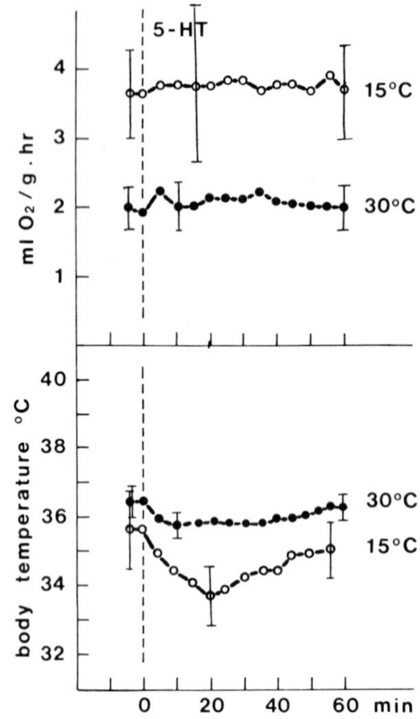

Fig. 7. *Effects of intraventricular infusions of serotonin (5-HT) (5 μg/kg) on body temperature and oxygen consumption at different environmental temperatures in non-hibernating golden hamsters. Vertical bars indicate ± S.D. (From 21).*

neutral zone. Acetylcholine (5μg/kg, at a rate of 5μl/min) stimulated heat production in the thermoneutral environment, while in the cold it had an inhibitory effect on heat production initially, followed by a compensatory increase in heat production. All changes observed were short-lasting and, again, no signs of entering torpodity were observed. Similarly as for the infusion of 5-HT, no seasonal differences in response could be detected.

B. Serotonin Turnover in the Brain after Exposure to Different Temperatures

The 5-HT turnover was studied in non-hibernating golden hamsters during exposure to different temperatures (40). Animals were kept at 25°C and then were exposed to

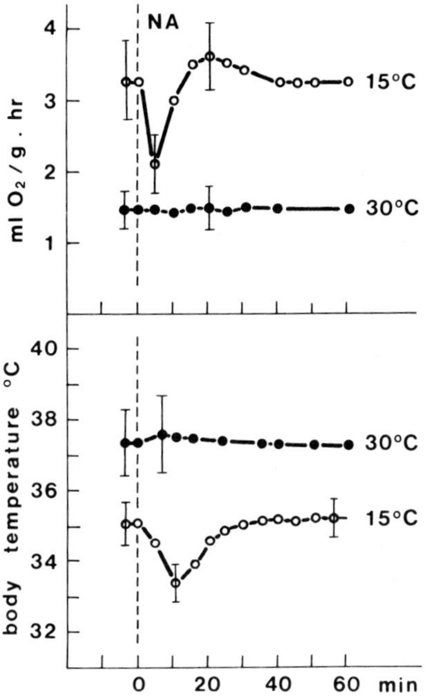

Fig. 8. *Effects of intraventricular infusions of noradrenaline (NA) (5 µg/kg) on body temperature and oxygen consumption at different environmental temperatures in non-hibernating golden hamsters. Vertical bars indicate ± S.D. (From 21).*

high (36°C), or low (6°C) temperatures for different periods of time.

Exposure to 6°C increased the 5-HT turnover, the greatest increase (400%) being observed after 12 hours. Later on, the 5-HT turnover decreased and after 7 days in the cold it returned to and remained at the original level for 21 days. Essentially the same picture could be traced in warm-exposed individuals, indicating that the activity of 5-HT pathways is not related specifically to the temperature stimulus (40; Novotná, unpublished observation).

These data are only partially consistent with those obtained from non-hibernating species, such as white rat and mouse. In these animals short-term exposure to

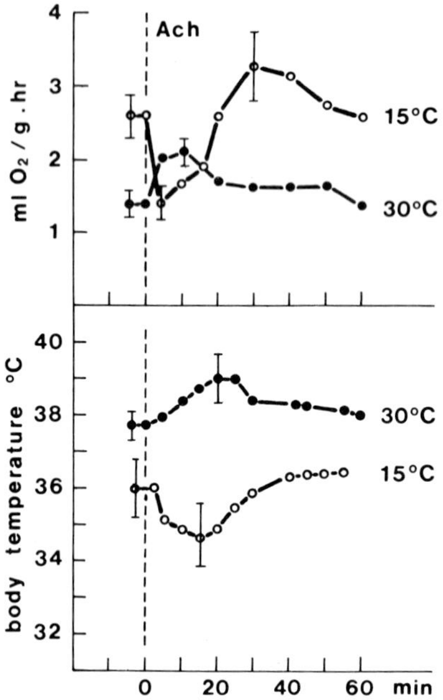

Fig. 9. *Effects of intraventricular infusions of acetylcholine (Ach) (5 µg/kg) on body temperature and oxygen consumption at various environmental temperatures in non-hibernating golden hamsters. Vertical bars indicate ± S.D. (From 21).*

warm raises the 5-HT turnover, whereas cold exposure decreases it (19, 45, 53). No time course of changes was measured in earlier studies, however.

C. Liver Tryptophan Pyrrolase Activity after Exposure to Different Temperatures

Since a certain reciprocity between 5-HT turnover and LTP activity was observed during the hibernation cycle in our previous experiments, the LTP activity was also measured in non-hibernating golden hamsters kept at 25°C. Experiments were performed in the spring. The aim of these experiments was to relate the time course of LTP activity to the 5-HT turnover. Experiments were performed on non-hibernating golden hamsters kept at 25°C and then exposed to 6°C or 36°C for different periods of time (39). A lowering of LTP activity was observed 3 - 6 hr after exposure either to warm (by 41%) or to cold (by 45%). Within 24 hr the activity returned to the original level. After 7 days of cold exposure the activity was markedly increased. Although an inverse relationship between LTP activity and 5-HT turnover could be traced in the early phase (up to 6 hr after temperature exposure), further analysis did not reveal any relationship between 5-HT turnover and LTP activity in the later phase. Similarity of the responses both in warm- and cold-exposed animals, however, seemed to indicate that changes in 5-HT turnover and LTP activity are not related to temperature stimulus *per se*, but rather a reaction common to exposure of stressful conditions.

D. Steroid Production after Exposure to Different Temperatures

As it has been suggested by Selye (50) that in non-hibernating mammals during the stress reaction there is an increased production of steroids from the adrenal cortex, it was of interest to study the release of steroids in golden hamsters in response to temperature stress. The plasma cortisol concentration of non-hibernating golden hamsters exposed to 6 or 36°C for different periods of time was measured by the method of Purwes and Sirett (44). Cortisol is the main corticoid fraction in this species (13), although corticosterone in smaller quantities is also present. It was found that neither the exposure to cold, nor the exposure to warm increased the cortisol level in the blood (Table 2). These observations agree with earlier findings (11, 12, 47). Thus, it was confirmed, that golden hamsters differ from other homeotherms in that they do not show an alarm reaction typical for other homeotherms when exposed to cold. Furthermore, these experiments indicated that in golden hamsters exposed to various temperatures in the non-hibernating state the changes in the LTP activity and the 5-HT turnover are not related to the production of steroids. This contrasts with the situation during arousal from hibernation in the golden hamsters and with our recent data on rats which show similar patterns of changes in LTP activity and in production of steroids after cold exposure. In rats, corticosterone level in the

Table 2

Cortisol concentration in the blood (μg/100 ml plasma) of golden hamsters exposed to different temperatures

	COLD			WARM	
controls	(9)	11.71 ± 4.17*	controls	(9)	11.71 ± 4.17
3 hours in 6°C	(3)	12.95 ± 7.94	3 hours in 36°C	(4)	10.57 ± 7.14
6 hours in 6°C	(8)	12.81 ± 4.05	6 hours in 36°C	(8)	11.10 ± 8.99
24 hours in 6°C	(8)	10.86 ± 4.36	24 hours in 36°C	(9)	6.66 ± 3.71
			7 days in 36°C	(5)	8.00 ± 1.32

*± S.D. Numbers in parenthesis are individuals used.

blood increases 4 times after 3 hours in 5°C, then decreases gradually, and reaches the original level after several days. LTP activity increases by 50% after 6 hours and returns to the original level within the first day of cold exposure (Table 3).

IV. THE POSSIBLE ROLE OF 5-HT PATHWAYS IN HIBERNATION

Data presented above suggest that the control of body temperature in awake animals is not under the strong influence of serotonergic pathways. To answer the question concerning physiological processes related to the increased 5-HT turnover in hibernation we presented a hypothesis in our earlier paper (23), that the activation of serotonergic pathways is responsible for the release of hormonal factors from the hypothalamus controlling the adenohypophysis. Our recent experiments showing that in the non-hibernating golden hamsters there is no relationship between steroid release from adrenals and 5-HT turnover in the brain seem to exclude the possibility that the pituitary adrenal axis is under control of serotonergic pathways in the non-hibernating individuals. This, however, does not mean that other hormones released from the adenohypophysis, namely gonadotrophic hormones and TSH, may not be under the control of serotonergic pathways. On the contrary, there is ample evidence in the literature showing that the release of these hormones is suppressed by the increased activity of serotonergic pathways in non-hibernating species (6, 9, 27, 28, 29, 48, 60).

The evidence presented above allows us to suggest a model describing the time sequence of phenomena leading to the onset of hibernation. It should be emphasized, however, that the following model is highly speculative. Our purpose is to stimulate further discussion on the subject, rather than to present a definitive and clear picture of how hibernation is initiated.

It is assumed that serotonergic pathways play an important role in controlling processes leading to hibernation. The crucial point in this model is that there exists an annual clock triggering hibernation mechanism which changes the responses of the animal to increase supply of tryptophan in the diet. This clock prevents hibernation in the spring and facilitates it in

Table 3

Corticosterone concentration in the blood and liver tryptophan pyrrolase activity of white rats exposed to 5°C for different periods of time

		corticosterone µg/100 ml of plasma		LTP activity mmol of kinurenin/g.h
controls	(10)	4.08 ±* 2.10	(7)	10.1 ± 1.40
3 hours			(5)	13.0 ± 0.55
6 hours	(15)	19.17 ± 2.25	(5)	17.2 ± 2.10
24 hours	(10)	9.62 ± 3.38	(9)	10.4 ± 2.44
4 days			(5)	11.1 ± 1.10
7 days	(3)	7.80 ± 1.80		
10 days	(5)	6.20 ± 1.74		

*± S.D. Numbers in parentheses are individuals used.

the fall. The clock is associated with a food intake center in the ventromedial hypothalamus. When winter approaches, an increased amount of food is consumed. As a consequence, the concentration of tryptophan in the blood increases and it becomes available in a greater extent for the increased activity of 5-HT pathways. These pathways influence the release of some hypothalamic factors controlling the adenohypophysis and related endocrine glands. In this manner the production of gonadotrophic and thyrotrophic hormones may be lowered. The lowered concentrations of hormones in the blood in turn may lower the activity of LTP which results in an increased supply of 5-HT to the brain and potentiates the further lowering of activity of the endocrine glands. Due to changes in production of some hormones, changes in peripheral tissues occur, enabling the survival of the animal under hypothermic conditions.

REFERENCES

1. Aron, C., and C. Kayser. Sommeil hivernal et pancreas endocrine. C.R. Soc. Biol. 150: 410-413, 1956.
2. Beckman, A. L. and E. Satinoff. Arousal from hibernation by intrahypothalamic injections of biogenic amines in ground squirrels. Am. J. Physiol. 222: 875-879, 1972.
3. Beyer, R.F. Effects of low temperature on cold sensitive enzymes from mammalian tissues. In: Hibernation-Hypothermia. Perspectives and Challenges. South, F. E., J.P. Hannon, J. R. Willis, E. T. Pengelley, and N. R. Alpert (editors). Amsterdam: Elsevier Publ. Co., 1972, pp. 17-54.
4. Bligh, J., W. H. Cottle, and M. Maskrey. Influence of ambient temperature on the thermoregulatory response to 5-hydroxytryptamine, noradrenaline and acetylcholine injected into the lateral cerebral ventricles of sheep, goats and rabbits. J. Physiol. 212: 377-392, 1971.
5. Burlington, R. F. Recent advances in intermediary metabolism of hibernating mammals. In: Hibernation-Hypothermia. Perspectives and Challenges. South, F. E., J. P. Hannon, J. R. Willis, E. T. Pengelley, and N. R. Alpert (editors). Amsterdam: Elsevier Publ. Co., 1972, pp. 3-15.
6. Califaris, L., and S. Taleisnik. Involvement of neurones containing 5-hydroxytryptamine in the mechanism of prolactin release induced by oestrogen. J. Endocrinol. 62: 25-33, 1974.
7. Cannon, B., and C. F. Polnaszek. The role of mitochondrial membrane lipids in cold adaptation and hibernation. In: Regulation of Depressed Metabolism and Thermogenesis. Janský, L., and X. J. Musacchia (editors). Springfield, Illinois: Charles C. Thomas Publ., 1976, pp. 93-116.
8. Civín, J. Effect of hibernation on liver tryptophan pyrrolase activity. Ph. D. Thesis, Prague, 1977.

9. Chen, H. J., and J. Meites. Effects of biogenic amines and TRH on release of prolactin and TSH in the rat. Endocrinol. 96: 10-14, 1975.
10. Costa, E., and J. L. Meek. Regulation of biosynthesis of catecholamines and serotonin in the CNS. Ann. Rev. Pharmacol. 14: 491-511, 1974.
11. Deane, H. W., and C. P. Lyman. Body temperature, thyroid and adrenal cortex of hamsters during cold exposure and hibernation, with comparisons to rats. Endocrinol. 55: 300-315, 1954.
12. Denyes, A., and R. H. Horwood. A comparison of free adrenal cortical steroids in the blood of a hibernating and non-hibernating mammal. Can. J. Biochem. Physiol. 38: 1479-1487, 1960.
13. Frenkel, J. K., K. Cook, H. J. Grady, and S. K. Pendleton. Effects of hormones on adrenocortical secretion of golden hamsters. Lab. Invest. 14: 142-156, 1965.
14. Glowinski, J., M. Hamon, and F. Héry. Regulation of 5-HT synthesis in central serotonergic neurons. In: New Concepts in Neurotransmitter Regulation. Mandell, A. J. (editor). New York: Plenum, 1973, pp. 239-257.
15. Grahame-Smith, D. G. Studies *in vivo* on the relationship between brain tryptophan, brain 5-HT synthesis and hyperactivity in rats, treated with a monoamine oxidase inhibitor and L-tryptophan. J. Neurochem. 18: 1053-1066, 1971.
16. Grahame-Smith, D. G. How important is the synthesis of brain 5-hydroxytryptamine in the physiological control of its central function? Adv. Biochem. Psychopharmacol. 10: 83-91, 1974.
17. Hammel, H. T., H. C. Heller, and F. R. Sharp. Probing the rostral brainstem of anesthetized, unanesthetized, and exercising dogs and of hibernating and euthermic ground squirrels. Fed. Proc. 32: 1588-1597, 1973.
18. Hannon, J. P., R. R. Beyer, R. F. Burlington, G. N. Somero, and J. H. Bland. A guide for future studies of low temperature metabolic function. In: Hibernation-Hypothermia Per-

spectives and Challenges. South, F. E., J. P. Hannon, J. R. Willis, E. T. Pengelley, and N. R. Alpers (editors). Amsterdam: Elsevier Publ. Co., 1972, pp. 99-120.

19. Harri, M., and R. Tirri. Brain monoamines in the temperature acclimation of mice. Acta Physiol. Scand. 75: 631-635, 1969.

20. Hellon, R. Central transmitters and thermoregulation. In: Essays on Temperature Regulation. Bligh, J., and R. E. Moore (editors). Amsterdam: North-Holland, 1972, pp. 71-85.

21. Hlavička, P. The role of aminergic transmission in thermoregulation of golden hamsters. Ph. D. Thesis, Prague, 1977.

22. Hoffman, R. A. Terrestrial animals in cold: hibernators. In: Handbook of Physiology, Sect. 4., Dill, D. B., E. F. Adolph, and C. G. Wolber (editors). Baltimore, Maryland: Waverly Press, Inc., 1964, pp. 379-403.

23. Janský, L., and R. Novotná. The role of central aminergic transmission in thermoregulation and hibernation. In: Regulation of Depressed Metabolism and Thermogenesis. Janský, L., and X. J. Musacchia (editors). Springfield, Illinois: Charles C. Thomas Publ. 1976, pp. 64-80.

24. Jéquier, E., D. S. Robinson, W. Lovernberg, and A. Sjoerdsma. Further studies on tryptophan hydroxylase in rat brainstem and beef pineal. Biochem. Pharmacol. 18: 1071-1081, 1969.

25. Jonsson, J. and T. Lewander. A method for the simultaneous determination of 5-hydroxy-3-indoleacetic acid(5-HIAA) and 5-hydroxytryptamine (5-HT) in brain tissue and cerebrospinal fluid. Acta Physiol. Scand. 78: 43-45, 1970.

26. Jouvet, M. Biogenic amines and the state of sleep. Science 163, 25-31, 1969.

27. Kamberi, I. A., R. S. Mical, and J. C. Porter. Effect of anterior pituitary perfusion and intraventricular injection of catecholamines and indoleamines on LH release. Endocrinol. 87: 1-12, 1970.

28. Kamberi, I. A., R. S. Mical, and J. C. Porter. Effects of melatonin and serotonin on the release of FSH and prolactin. Endocrinol. 88: 1288-1293, 1971.
29. Kordon, C., C. A. Blake, J. Terkel, and C. H. Sawyer. Participation of serotonin-containing neurons in the suckling - induced rise in plasma prolaction levels in lactating rats. Neuroendocrinol. 13: 213-223, 1973/74.
30. Knox, W. E. and V. H. Auerbach. The hormonal control of tryptophan peroxidase in the rat. J. Biol. Chem. 214: 307-313, 1955.
31 Kudryavtseva, N. N., and N. K. Popova. Serotonin content in various portions of the brain during hibernation and awakening. Bull. Exp. Biol. Med. 75: 44-47, 1973.
32. Lin, R. C., N. H. Neff, S. H. Ngai, and E. Costa. Turnover rates of serotonin and norepinephrine in brain of normal and pargyline-treated rats. Life Sci. 8: 1077-1084, 1969.
33. Mandell, A. J., S. Knapp, and L. L. Hsu. Some factors in the regulation of central serotonergic synapses. Life Sci. 14: 1-17, 1974.
34. Mayer, W. V. Natural mammalian hibernation, a bibliography. Arctic Aeromedical Laboratory, Fort Wainwright, Alaska, 1964.
35. Mrosovsky, N. The adjustable brain of hibernators. Scientific Amer. 218: 110-118, 1968.
36. Mrosovsky, N. Hibernation and the Hypothalamus. Towe, A. (editor). New York: ACC Meredith Corporation, 1971, pp. 1-287.
37. Myers, R.D. and T. L. Yaksh. Control of body temperature in the unanesthetized monkey by cholinergic and aminergic systems in the hypothalamus. J. Physiol. 202: 483-500, 1969.
38. Nagatsu, T., K. Mizutani, I. Nagatsu, S. Matsura, and T. Sugimoto. Pteridines as cofactor or inhibitor of tyrosine hydroxylase. Biochem. Pharmacol. 21: 1945-1953, 1972.
39. Novotná, R., J. Civín, and L. Janský. The relation between tryptophan pyrrolase activity 5-HT turnover and induction of hibernation. Proc. 27. Int. Congr. Physiol. Sci., Paris, 555. 1977.

40. Novotná, R., and L. Janský. Effect of different environmental temperatures on the serotonin concentration and turnover in the brain stem of a hibernator. Physiol. bohemoslov. 25: 37-42, 1976.
41. Novotná, R., L. Janský, and Z. Drahota. Effect of hibernation on serotonin metabolism in the brain stem of the golden hamster (*Mesocricetus auratus*). Gen. Pharmac. 6: 23-26, 1975.
42. Pengelley, E.T., and K. H. Kelly. A "circannian" rhythm in hibernating species of the genus Citellus with observations on their physiological evolution. Comp. Biochem. Physiol. 19: 603-617, 1966.
43. Peters, D. A. V., P. L. McGeer, and E. G. McGeer. The distribution of tryptophan hydroxylase in cat brain. J. Neurochem 15: 1431-1435, 1968.
44. Purwes, H. D., and X. E. Sirett: The fluorimetric estimation of cortisol in human plasma. Aust. J. Exp. Biol. Med. Sci. 47: 589-599, 1969.
45. Reid, W. D., L. Volicer, H. Smookler, M. A. Beaven, and B. B. Brodie. Brain amines and temperature regulation. Pharmacol. 1: 329-344, 1968.
46. Reiter, R. J., and R. A. Hoffman. The endocrine system. In: The Golden Hamster its Biology and Use in Medical Research. Hoffman, R.A., P. F. Robinson, and H. Magalhaes (editors). Ames, Iowa: The Iowa State Univ. Press, 1968, pp. 139-155.
47. Schindler, W. J., and K. M. Knigge. Adrenal cortical secretion by the golden hamster. Endocrinol. 65: 739-747, 1959.
48. Schneider, H. P., and S. M. McCann. Mono- and indolamines and control of LH secretion. Endocrinol. 86: 1127-1133, 1970.
49. Seifert, J. Tryptophan pyrrolase in rat liver after phenobarbital adminstration. Toxicol. 1: 179-186, 1973.
50. Selye, H. The physiology and pathology of exposure to stress. Acta: Montreal, 1950, p. 203.

51. Shaskan, E. G. Hypothalamic and whole brain levels of 5-HT in bats: Some aspects of central control of thermoregulation. Ph.D. Dissertation, Univ. of Arizona, 1969.
52. Simmonds, M. A. Effect of environmental temperature on the turnover of 5-hydroxytryptamine in various areas of rat brain. J. Physiol. 211: 93-108, 1971.
53. Simmonds, M. A. Effect of environmental temperature on the turnover of 5-hydroxytryptamine in various areas of rat brain. J. Physiol 211: 93-108, 1970.
54. South, F. E. J. D. Heath, R. H. Luecke, L. T. Mihailovic, R. D. Myers, J. A. Panuska, B. A. Williams, W. C. Hartner and H. K. Jacobs. Status of the role of the central nervous system and thermoregulation during hibernation and hypothermia. In: Hibernation-Hypothermia Perspectives and Challenges. South, F.E., J. P. Hannon, J. R. Willis, E. T. Pengelley, and N. R. Alpers (editors). Amsterdam: Elsevier Publ. Co., 1972, 629-633.
55. South, F. E., W. C. Hartner, and R. H. Luecke. Response to preoptic temperature manipulation in the awake and hibernating marmot. Am. J. Physiol. 229: 150-160, 1975.
56. Spafford, D. C., and E. T. Pengelley. The influence of the neurohumor serotonin on hibernation in the golden-mantled ground squirrel, *Citellus lateralis*. Comp. Biochem. Physiol. 38A: 239-250, 1971.
57. Strumwasser, F. Thermoregulatory, brain and behavioral mechanism during entrance into hibernation in the squirrel, *Citellus beecheyi*. Am. J. Physiol. 196: 15-22, 1959.
58. Suomalainen, P. Stress and neurosecretion in the hibernating hedgehog. Bull. Mus. Comp. Zool. 124: 271-282, 1960.
59. Tozer, T. N., N. H. Neff, and B. B. Brodie. Application of steady-state kinetics to the synthesis rate and turnover time of serotonin in the brain of normal and reserpine-treated rats. J. Pharmacol. Exp. Ther. 153: 177-182, 1966.

60. Tuomisto, J., T. Ranta, P. Männistö, A. Saarinen, and J. Lappäluoto. Neurotransmitter control of thyrotropin secretion in the rat. Eur. J. Pharmacol. 30: 221-229, 1975.
61. Uuspaä, V. J. The 5-hydroxytryptamine content of the brain and some other organs of the hedgehog (*Erinaceus europeus*) during activity and hibernation. Experientia 19: 156-159, 1963.
62. Vogt, M. Tryptaminergic neurotransmission. In: Proc. 6th Intern. Congr. Pharmacol. Tuomisto, J., and M. K. Paasonen (editors). Finland: 3-16, 1975.
63. Williams, B. A. and J. E. Heath. Responses to preoptic heating and cooling in a hibernator *Citellus tridecemlineatus*. Am. J. Physiol. 218: 1654-1660, 1970.
64. Wurtman, R. J., and J. Axelrod. A sensitive and specific assay for the estimation of monoamine oxidase. Biochem. Pharmac. 12: 1439,1964.

AUTONOMIC REGULATION OF HIBERNATION BY *CITELLUS* AND *EPTESICUS*

John W. Twente and Janet Twente
Dalton Research Center and
Division of Biological Sciences
University of Missouri-Columbia
Columbia, Missouri, U.S.A. 65201

ABSTRACT

Autonomic control of natural hibernation was demonstrated in *Citellus lateralis*, *C. columbianus* and *Eptesicus fuscus*. Hibernators are difficult to arouse early in the period of hibernation; later, less stimuli are required to provoke premature arousal. Arousal is stimulated through a sympathetic reflex. This reflex was mimicked by a β-adrenergic, dose-response relationship of injected catecholamines, potency: (isoproterenol > epinephrine ≥ norepinephrine > phenylephrine); progressively smaller dosages provoked premature arousal as the period of hibernation proceeded. Termination of the period of hibernation is metabolically determined and expressed through progressive irritability. Hibernation is maintained, actively, through parasympathetic suppression of sympathetic stimuli resulting in non-thermogenic, inhibited arousal. This autonomic hibernation maintenance response (AHMR) was described as the regulator of hibernation which operated until it was incapable of suppressing the sympathetic. Failure of AHMR to inhibit arousal was considered the result of changes (either ±) in thresholds of central neurones which determine progressive irritability. A model of autonomic control is presented. Diving responses (DR) elicited during hibernation suggested a possible origin of AHMR; evolutionary parallelism for DR utilization would explain species similarity.

ISBN 0-12-734550-7

I. INTRODUCTION

Most of our efforts these last fifteen years have been devoted to the investigation of the factors which cause provoked and spontaneous arousal from hibernation. Our goal has been to try to understand the bases of the mechanisms which govern the initiation of arousal and, thereby, to determine the physiological regulators of the period of hibernation itself. Concurrently, we have explored the endocrinological features which determine the seasonal pattern of hibernating behavior.

Routinely, we monitor, continuously, the body temperature (T_b), respiratory patterns and electrocardiograms (ECG's) of several hibernating mammals in varying physiological states. In the course of these long-term observations, we have gained insight into certain aspects of the regulation of hibernation by the nervous system.

Current emphasis concerning the mechanisms of the initiation of arousal from hibernation and the maintenance of hibernation involves the regulation of thermogenesis and the consequent active thermoregulation throughout the period of hibernation (2,8,31,34). We feel that not only is this approach to the problem misleading, it is basically incorrect. Instead, we are becoming increasingly convinced that the maintenance of the state of hibernation is independent of thermoregulation.

There seems to be no question that the central nervous system is active during hibernation (4,5,17,26, 32,33,37,40). Controversies seem to center around matters of where in the brain, how much and when does activity occur (4,5,8,9,17,26,32,33,34,38,40).

The autonomic nervous system, however, definitely has been indicated as a regulator of some aspects of hibernation (13). Lyman and O'Brien (28,29) demonstrated that cardiac arrythmias during induction could be abolished by atropine and concluded that the parasympathetic was functional although not essential for induction. Chatfield and Lyman (3) and Lyman and O'Brien (27, 28, 29) showed that the arousal process is under sympathetic control. Lyman and O'Brien (28,

29) considered the heart to be under sympathetic control but that the effect of the parasympathetic during hibernation was minimal and could not be evidenced unless the heart was in a state of cardio-acceleration. Finally they concluded (28) "...the parasympathetic system has a regulatory, but not essential effect on the heart rate as the animal enters hibernation, and has minimal effect during hibernation and arousal."

Lyman and O'Brien (30) also demonstrated that muscle action potentials (MAP's) precede cardioacceleration. Because MAP's could be abolished by curarization or by spinal section, they considered cardioacceleration to be a sympathetic reflex from the MAP's. Johansen, Krog and Reite (12) reported that vagal function persisted in the hypothermic hedgehog at T_b 5 - 7° C.

Cardiac arrythmias (suggesting to us parasympathetic activity) during hibernation or induction have been shown to be present in the ECG's of hedgehogs, *Erinaceus europaeus* (6,16,19); birchmice, *Sicista betulina* (11); European hamsters, *Cricetus cricetus* (16); garden dormice, *Eliomys quercinus* (16,35); marmots, *Marmota monax* (24) and *M. flaviventris* (39) and ground squirrels, *Citellus citellus* (16); *C. beecheyi* (40); *C. franklini* (6); *C. lateralis* (46); *C. parryi* (6); and *C. tridecemlineatus* (23,27).

In this report we will demonstrate that hibernation is maintained and regulated actively through a parasympathetic response and that arousal proceeds when this parasympathetic response can no longer overcome sympathetic drive. We will try to demonstrate that progressive changes of levels in central nervous thresholds to sympathetic stimuli will determine whether arousal occurs or whether it will be inhibited through parasympathetic influences.

II. MATERIALS AND METHODS

Golden-mantled ground squirrels (*Citellus* = *Spermophilus lateralis*) undergo seasonal changes in hibernating behavior consisting of an "adaptive" autumn pattern, a constant and predictable winter pattern and a "deadaptive" spring pattern of hibernating behavior (45). If animals are improperly fed and maintained, or if handled or otherwise disturbed when they are aroused between periods of hibernation, they undergo physiological changes which result in erratic patterns of hibernating behavior resembling the seasonal patterns indicated above. We have considered these responses to be indicators of hibernation stress (45,50).

Table 1 summarizes some of the responses that relate to these phenomena.

Similar seasonal and stress patterns have been measured for the Columbian ground squirrel (*C. columbianus*) (Moy and Twente, unpublished). We have preliminary indications that hibernating behavior of the big brown bat, (*Eptesicus fuscus*) basically is not unlike that of the ground squirrels in respect to seasonal patterns and stress (unpublished).

Although some of our methods for the care and maintenance of ground squirrels have been published (45) we should elaborate upon them. A repetition of our methods which allows for the attainment and discernment of unstressed animals would enhance a repetition of results and at the same time help to explain certain discrepancies between our findings and those of others. Our studies have been made possible through the following procedures which result in predictable hibernation which probably resembles that which occurs under natural conditions.

Our colony of *C. lateralis* is added to yearly with wild-trapped, adult animals to replace those lost through attrition. Therefore, we are studying animals of ages ranging from one to 12 years, the majority of which, of course, are in the younger bracket. The essence of occasional gerontological excess shown by some of these ground squirrels seems to lie in good and gentle care, proper food and comfortable housing.

A. Diet

A simple, relatively natural diet of sunflower seeds and oranges will carry a ground squirrel in a healthy, non-obese state through to old age with successful, and natural, yearly hibernation. This diet has none of the complicating factors of many laboratory chows which often cause obesity and have been shown to contain animal hormones which may deleteriously influence experimental results (10). After the season of hibernation when the animals have been returned to the animal colony, they are given drops of vitamin D-3 and sunlamp treatments at weekly intervals for a period of approximately one month.

B. Housing

Animals are housed individually in roomy cages 30 x 37 x 23 cm, at 21 to 22°C. Cedar shavings are furnished in abundance and cotton batting is supplied for nesting material. When cages are changed, the nests are transferred with the animal. Animals are weighed, lower incisors cut and cages changed at biweekly intervals. When summer weight patterns cease escalating and gonadal regression is completed, animals are considered ready for hibernation and are placed in steel hibernation cages 30 x 25 x 15 cm with their nests, shavings and food, and placed in environmental rooms at 4° - 5°C. Hibernation usually begins within 24 hr (if it had not already been initated in the animal quarters). During the season of hibernation, the ground squirrels never are disturbed excepting for experiments; cages never are changed (they do not get dirty) and manipulations such as weighing or electrode placement take place only when the animals are in the hibernating state. Effort is made not to enter the cold rooms when an animal is aroused. Active and noisy individuals are not allowed to remain in the environmental chamber because their disturbance may cause stressful reactions to other aroused hibernators. Several hibernating species frequently are kept in the same environmental room with no adverse effects being observed upon hibernating behavior.

C. Records

A thermocouple placed in the nest measures the duration of the segments of the cycle of hibernation. Durations of the periods of hibernation, arousal,

Table 1. Typical pattern of torpidity and its modification by stress in the Golden-mantled ground squirrel *Spermophilus lateralis* under laboratory conditions.

	Autumn Pattern	Winter Pattern	Spring Pattern	Stress (♀)	Stress (♂)
Duration of period of hibernation	Initially brief becoming longer each successive period first 4 - 6 periods (45)	Constant ($\pm$ 10%) for an animal at a level of temperature (45)	Becomes shorter each successive period Last 2-4 periods before cessation (45)	Shorter than winter periods Mimics autumn pattern (45)	Shorter than winter periods Mimics spring pattern. Hibernation ceases prematurely (45)
Duration of period of activity	Initially long averaging 20 hours becoming shorter each successive period (45)	Constant $\overline{X}$ = 10.5 h. never more than 15 h. (41)	Becomes longer each successive period (45)	Becomes longer as in autumn pattern (45)	Becomes longer as in spring pattern (45)
Body Temperature (Tb)	Initially 4 - 5°C above $T_{\mu e}$ becoming lower each successive period (45)	Within 0.5 $T_{\mu e}$ (45)	Usually > 1°C above $T_{\mu e}$ (45)	Usually > 1°C above $T_{\mu e}$ (45)	Usually > 1°C above $T_{\mu e}$ (46)

Heart Rate at 4 - 7°C	>7 b/m	<6.5 b/m (50)	>7 b/m (50)	>7 b/m (50)	>7 b/m (50)
Partial arousal to stimuli	Frequently	Never	Frequently	Frequently	Frequently
Response to 0°C $T_{\mu e}$	Unknown	None if unstressed (48,50)	Slight increase in Tb or complete arousal (48,50)	Varying, depending upon degree of stress. May show slight increase in Tb (50)	As females
Arousing response to i.p. control	Unpredictable may arouse at any time	Never arouse at any time before 50% of average duration of period of hibernation Typically arousal not provoked before 80 - 90% (46)	Unpredictable may arouse at any time	Unpredictable may arouse at any time	As females

activity and induction are determined to the nearest hour from recordings printed at 6 min intervals on time-printed chart paper by L & N 16-point recording potentiometers. Each animal is designated by number and individual records, charted daily, reveal the status of that animal in respect to its cycle of hibernation, seasonal pattern and degree of hibernation stress shown, if any.

D. Environmental Rooms

Remote, dark, quiet cold rooms (Forma Scientific) with little air circulation have produced an environment conducive to natural hibernation. Each room is equipped with two coils and two compressors. Each compressor is equipped with a hand capacity valve to regulate the temperature of the coil. This allows for the maintenance of relatively warm coils with the avoidance of icing. In addition, defrost consists of a compressor shutting down at four hr intervals for a few minutes to allow for the evaporation of water that has condensed on the coils. This system allows for non-cyclical temperature records and a constancy of room temperature varying less than ± 0.5°C. The microenvironment ($T_{\mu e}$) within the cage is virtually invariable. Such precision is desirable although by no means essential to good hibernation research, incidentally.

E. Bats

Hibernating big brown bats, *Eptesicus fuscus* were captured in local caves, taken to the laboratory, weighed and placed singly in specially-designed recording chiropteran hibernacula. These tin cans with small holes punched in the lids for ventilation were lined with sponge rubber which was kept constantly moist. This simple technique allows for hibernation as it occurs in the wild state. A small dish of water was kept filled for drinking. Three thermocouples one each at high, middle and low levels of the container recorded the position of the bat whenever it aroused and wherever it moved when awake. The durations of periods of activity, induction, hibernation and arousal could be estimated to within

5 min. In late spring, the bats were returned unharmed to the wild without having exhibited excessive weight loss. The ability to fly did not seem to be impeded by the winter's confinement. Data obtained from several years, using these methods, indicate that patterns of hibernation are similar to those of ground squirrels.

F. Electrocardiograms

ECG's were measured by a Grass Model 7 Polygraph. Recordings were made continuously, usually at chart speeds of 25 mm/min. ECG's of ground squirrels were obtained from electrodes made from #27 stainless steel hypodermic needles. An important modification of technique of that which we have reported (46) involved pushing the needle under the skin and out the other side. The tip was then secured with a small piece of cork. This procedure eliminated needle prick stimulation which frequently occurs when needles merely are placed under the skin with the tip exposed since under these conditions any movement causes a stimulus from the needle tip.

ECG's of bats were obtained from electrodes made from #33 stainless steel orthodontic wire. These were twisted into a form inspired by a safety pin. Approximately 5-7 mm of electrode was pushed through the skin and out the other side where it was secured to itself by a bend in the wire. ECG's were obtained from a quasi Lead I positioning in that the negative pole was placed at the level of the right shoulder and the positive on the chest near the proximal part of the left humerus.

G. Respiration

Respiratory patterns of ground squirrels usually were recorded simultaneously with the ECG's. These were obtained from pulmonary belts manufactured from a strip of a 6 mm rubber surgical tubing. Pieces of Velcro were sewed to the ends of the rubber strip so that they could be secured around the chest of the ground squirrels. A thread from the pulmonary belt was tied to a strain gauge transducer. Any movement registered by the animal was recorded from this by the Polygraph. This system is so sensitive that pulse pressure from the heart may be clearly indicated on the records as well as respiratory and other movements.

H. Drugs

Catecholamine dosages were administered i.p. on a µg base/animal basis. Using a Hamilton microliter syringe, appropriate quantities were withdrawn from the vials or bottles and mixed in a 1.0 ml syringe with 0.5 ml .015M pH 7.65 TES buffer (N-tris [Hydroxymethyl] methyl 1-2-aminoethane sulfonic acid). Prior experience has shown that the absolute weight of the animal is not critical for the determination of the arousing effects of catecholamines (47). Should a µg/kg dosage transformation be desired, the *C. lateralis* used in these experiments averaged 240 ± 50 g (s.d.). The compounds administered included isoproterenol (I) (Isuprel[R], Winthrop); epinephrine (E) (Adrenaline[R], Parke, Davis); norepinephrine (NE) (Levophed[R], Winthrop); and phenylephrine (P) (Neo-synephrine[R], Winthrop).

I.p. injection was administered dorsolaterally without otherwise touching the animal. The distinction between the abdominal and the pleuropericardial cavity may be observed by the presence of a concavity on the side of the hibernating ground squirrel immediately anterior to the thigh. At this point, a 9 mm #26 hypodermic needle is pushed straight in to its extremity. This site of injection insures the best success avoiding the pleural cavity and minimizes the possibilities of ineffective injections because of injection directly into the intestine or mesenteric fat. Success of catecholamine injections was determined by ECG patterns after latent periods ranging from 5 to 45 min. Latent periods of thermogenic effects (47), as measured by thermocouple implants (44) or thermocouple belts (49), appeared within 30 to 45 min after the first cardiac effects were evident.

Atropine sulfate (Wellcome) was administered i.p. as follows: 200 µg to ground squirrels, *C. lateralis* and *C. columbianus*, and 20 µg to the bat, *E. fuscus*.

I. Diving tests

Hibernating *C. lateralis* with ECG electrodes attached previously were removed from their hibernation cages and placed in small, wire baskets which allowed for the continuation of hibernation in the characteristic, curled posture. Water (the temperature of which was found to be irrelevant since equal responses were obtained with water ranging from 1° to 44°C) was then slowly squirted on the muzzle from beneath the animal with a needleless 35 ml syringe. The stimulus, applied for 15 - 20 sec, was indicated simultaneously on the Polygraph via a signal marker. A control stimulus for the water involved gentle stroking of the muzzle with cotton-tipped applicators.

III. RESULTS

A. Arrythmias

Variations in the duration of diastole universally distinguish the cardiac patterns of *C. lateralis*, *C. columbianus* and *E. fuscus* throughout the period of hibernation as well as during induction and early arousal. Electrocardiographic patterns (Fig. 1) illustrate these sinus arrythmias which originate at the S - A pacemaker which is under central sympathetic and parasympathetic control. As will be shown, diastolic arrythmias characterized by long R - R intervals are the result of parasympathetic stimulation of the heart. Shortened R - R intervals indicate sympathetic stimulation. Frequency histograms of R - R intervals (usually n = 50) under each of the experimental situations show the relative activities of central sympathetic and parasympathetic action as it is expressed peripherally by the heart.

Parasympathetic determination of these arrythmias was confirmed by i.p. injection of atropine to hibernating animals. Atropinization in these hibernators was achieved after a latent period of a minimum of two hr after injection. In most instances, the maximum development of the atropine effect was apparent between 3 to 4 hrs. Usually the effects of atropine were no longer apparent (nor statistically detectable) at 20 to 30 hrs after injection. Criteria

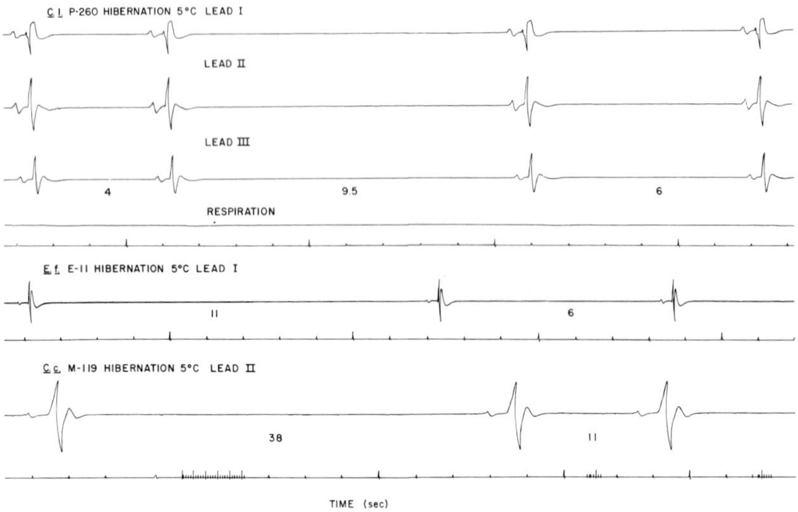

Fig. 1. *ECG's of hibernating Citellus lateralis (with 3 leads taken simultaneously), Eptesicus fuscus and C. columbianus illustrate diastolic arrythmias characteristic of hibernating mammals. The duration of the depolarization of the heart of E. fuscus (0.8 sec), C. lateralis (1.4 sec) and C. columbianus (2.0 sec) is consistent with body weights of 15 - 20 g, 200 - 300 g, and 400 - 700 g respectively and therefore the relatively increased cardiac masses.*

for atropinization included partial or complete abolition of diastolic arrythmias; abbreviation of the duration of the R - R intervals; and the resultant tachycardia. The effect of atropine plainly can be determined through inspection of the ECG's themselves, the sequential plots of the durations of the R - R intervals and the frequency histograms of the durations of the 50 R - R intervals (Fig. 2).

Table 2 reports a statistical comparison of the results before and after atropinization. For each individual of each species, the average R - R interval was reduced (Student *t*-test P > 0.001). Based on 50 consecutive R - R intervals for hibernation and

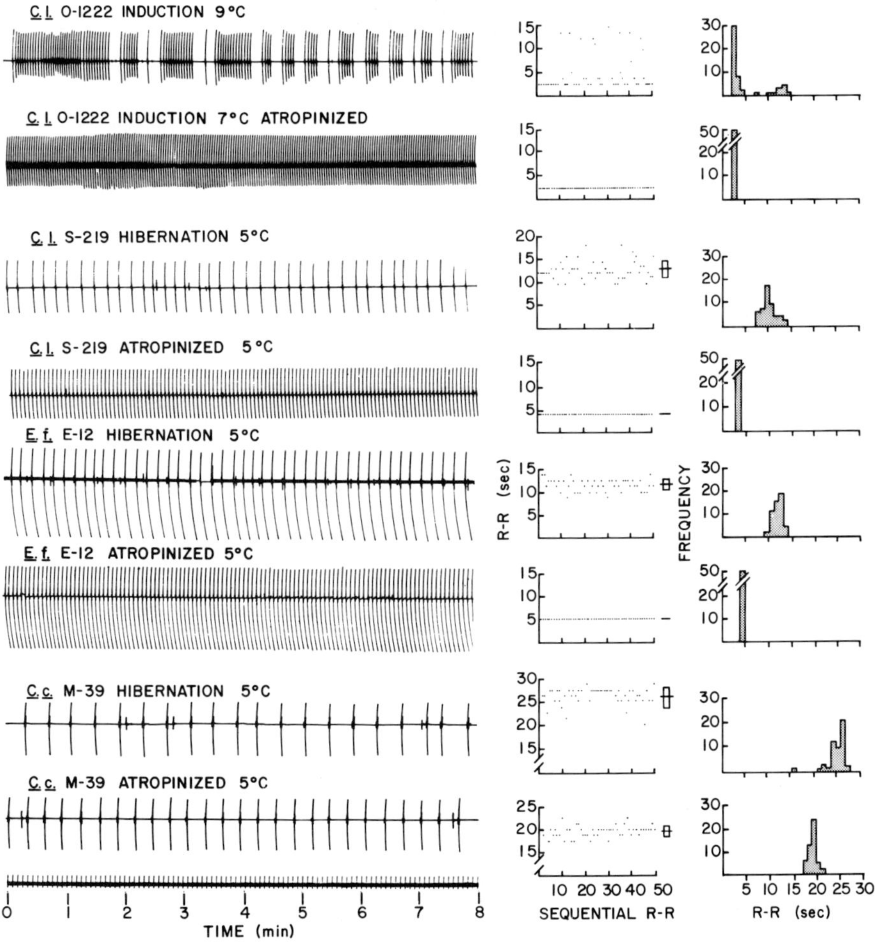

Fig. 2. *ECG's of hibernating Citellus lateralis, Eptesicus fuscus and C. columbianus before and after injection of atropine illustrate the reduction in the duration of the R - R interval. The ECG's figured represent the order of the sequential R - R intervals (up to 50) plotted in the middle column. These same intervals are plotted nonsequentially in the frequency histograms.*

Table 2. Effects of atropine on the heart rate and duration of R - R intervals.

	During the 60 minutes before atropine $\bar{X} \pm$ s.d.	During 60 minutes of maximum atropinization $\bar{X} \pm$ s.d.	Change %
C. lateralis N = 11			
Average heart rate	5.30 ± 0.64	9.42 ± 2.35	+ 78
Average R - R (sec)	11.27 ± 1.32*	6.61 ± 1.80+	- 41
Minimum R - R (sec)	7.98 ± 1.30	6.08 ± 1.39	- 24
Maximum R - R (sec)	14.46 ± 2.32	7.77 ± 2.53	- 46
Max R - R/Min R - R	1.84 ± 0.36	1.26 ± 0.25	- 32
C. columbianus N = 3			
Average heart rate	2.5 ± 0.4	3.5 ± 0.9	+ 37
Average R - R (sec)	24.9 ± 3.1*	18.8 ± 4.5+	- 25
Minimum R - R (sec)	16.5 ± 1.0	16.7 ± 3.8	+ 1
Maximum R - R (sec)	27.1 ± 4.0	21.7 ± 3.8	- 20
Max R - R/Min R - R	1.6 ± 0.1	1.3 ± 0.1	- 20
E. fuscus N = 1			
Average heart rate	5.2	11.3	+117
Average R - R (sec)	11.5 ± 1.3*	5.3 ± 0+	- 54
Minimum R - R (sec)	8.8	5.3	- 40
Maximum R - R (sec)	13.8	5.3	- 62
Max R - R/Min R - R	1.6	1.0	- 38
C. lateralis Non-hibernating N = 4			
Average heart rate	355. ± 63	462. ± 22	+ 30
Minimum R - R (sec)	0.18 ± 0.03*	0.14 ± 0.05+	- 22
Maximum R - R (sec)	0.38 ± 0.01	0.14 ± 0.05	- 62
Max R - R/Min R - R	2.12 ± 0.35	1.05 ± 0.06	- 50

* Based on 50 R - R intervals immediately prior to injection.

\+ Based on 50 consecutive R - R.

for atropinized hibernation of *C. lateralis* (n = 11), the increase in average heart rate, the reductions in the durations of the average R - R interval and the maximum R - R interval are significant (Student *t*-test P > 0.001). A reduction in the minimum R - R interval associated with an increase in sympathetic activity did not always occur. A reduction in the maximum R - R interval (2 - 12 sec) indicating a direct effect of atropine upon the blocking of cardiac, parasympathetic receptors, appeared in all animals. The reduction of the ratio of the maximum R - R/minimum R - R was approximately the same for all animals of the three species.

Data from atropinized, non-hibernating *C. lateralis* (200 μg/animal with a 10 - 20 min latent period) also show an increase in average heart rate and reductions in maximum and minimum R - R intervals and a similar reduction in the ratio of these intervals.

B. Induction into Hibernation

Intervals of tachycardia followed by diastolic arrythmias characterize induction into hibernation of *C. lateralis* (Fig. 3). As T_b declines, the periods of extended diastole become increasingly frequent. A similar pattern of arrhythmia of slightly less magnitude occurs during induction of *E. fuscus* (Fig. 3). For *C. lateralis*, the maximum R - R/minimum R - R ratio is reduced from about 5 at 10°C to approximately 2 at 4.5°C. In *E. fuscus*, the maximum R - R/minimum R - R at 10°C is 2 and does not decline further with lowered body temperatures.

C. Hibernation

ECG's were obtained from unstressed animals hibernating in the established winter pattern of hibernating behavior at Tb = 5°C. The ECG's, sequential plots of durations of R - R intervals, and frequency histograms of the R - R durations for hibernating *C. lateralis*, *E. fuscus* and *C. columbianus* are illustrated in figure 4. At this level of temperature, the heart rates of *C. lateralis* and *E. fuscus* range from 4 to 7 b/min, whereas the heart rate of *C. columbianus* is 2 to 4 b/min. The maximum R - R/minimum R - R (approximately 1.5 to 2) is the

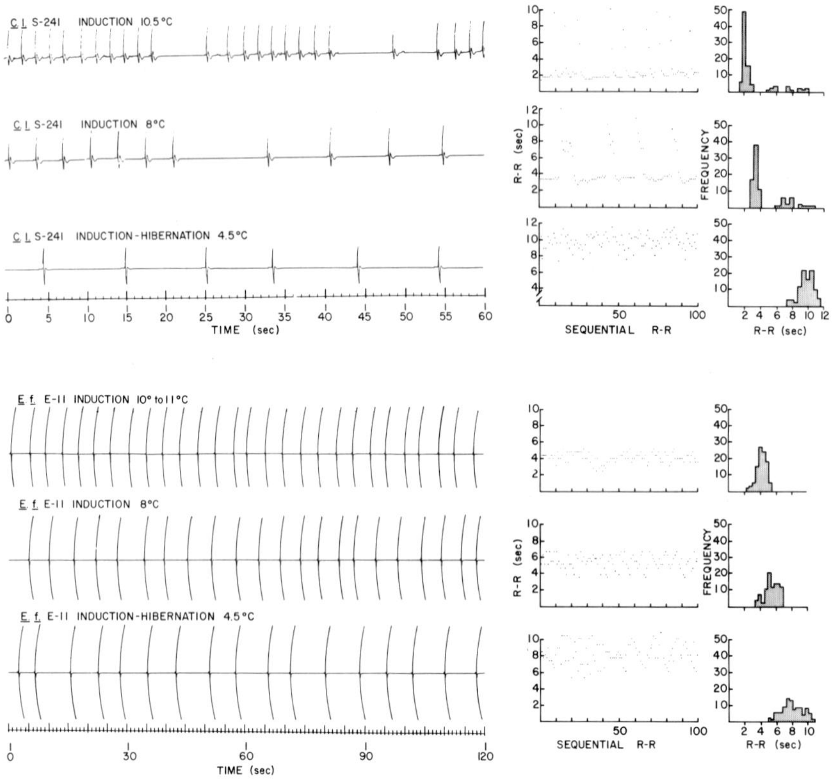

Fig. 3. *ECG's of Citellus lateralis and Eptesicus fuscus show the pronounced diastolic arrythmias characteristic of induction. The biphasic distribution of the R - R intervals indicate the variable parasympathetic/sympathetic dominance. The reduction in the frequency of the short duration R - R intervals and the increased frequency of the long duration R - R intervals reflects increased parasympathetic control. Microenvironmental temperatures of the ground squirrel's nest and the bat hibernaculum were 4.5°C. The body temperatures during the time the ECG's were obtained is indicated.*

same for all three species (Tables 2,3,4 and 5). The frequency histograms of the duration of the R - R intervals indicate a single homogeneous population.

The range of the durations of the R - R intervals during hibernation is less than that shown

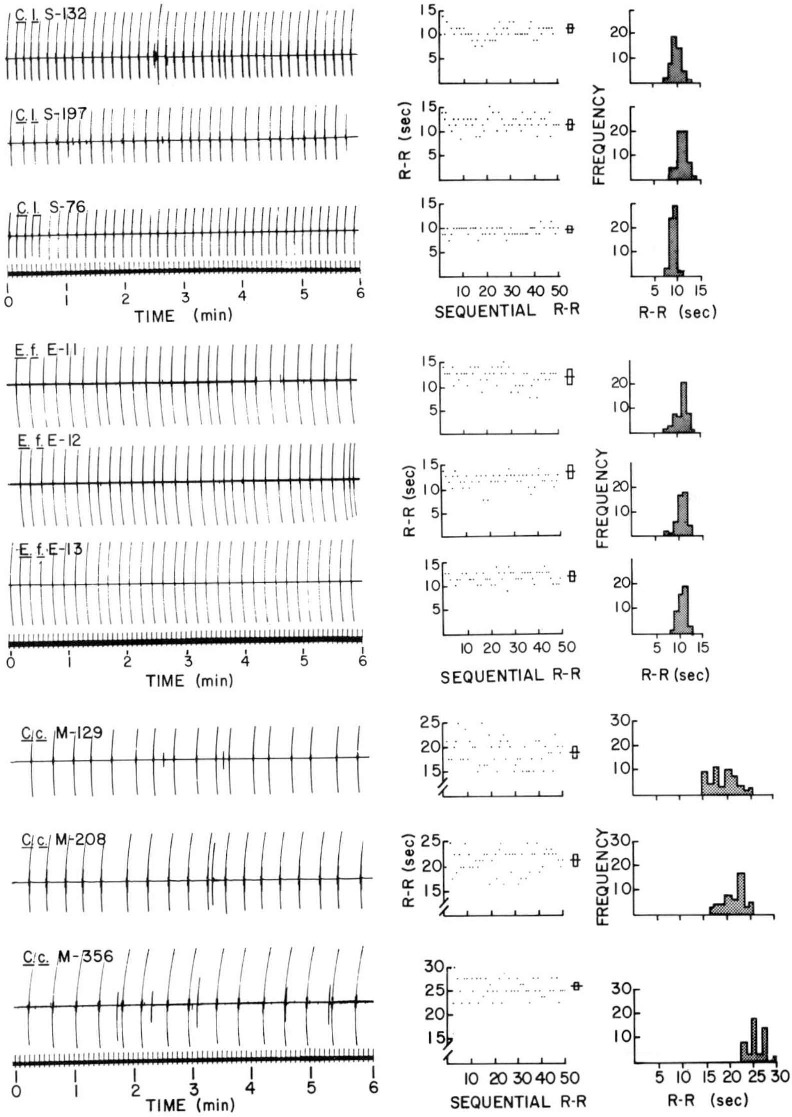

Fig. 4. *ECG's of three hibernating individuals (5°C) of each species show similar two-fold variation in the duration of the R - R intervals as well as the homogeneous frequency distribution. Both traits are characteristic of hibernation. Averages and standard deviations are indicated beside the sequential plot.*

Table 3. Changes in duration of R - R intervals during spontaneous inhibited arousals.

	During the 30 minutes before inhibited arousal $\bar{X} \pm$ s.d.	During inhibited arousal $\bar{X} \pm$ s.d.	Change %
C. lateralis N = 26			
Average heart rate	6.10 ± 1.08		
Minimum R - R (sec)	7.63 ± 1.60	4.98 ± 1.38	- 35
Maximum R - R (sec)	13.63 ± 3.49	16.30 ± 4.67	+ 19
Max R - R/Min R - R	1.83 ± 0.55	3.32 ± 0.77	+ 83
C. columbianus N = 11			
Average heart rate	2.88 ± 0.43		
Minimum R - R (sec)	18.48 ± 2.64	7.76 ± 1.25	- 58
Maximum R - R (sec)	26.0 ± 2.42	27.86 ± 6.42	+ 7
Max R - R/Min R - R	1.45 ± 0.14	3.70 ± 1.06	+155
E. fuscus N = 3			
Average heart rate	5.2 ± 0.1		
Minimum R - R (sec)	7.5 ± 0	2.9 ± 1.9	- 61
Maximum R - R (sec)	14.2 ± 0.7	17.1 ± 3.1	+ 21
Max R - R/Min R - R	1.9 ± 0.1	8.4 ± 6.4	+349

during induction and suggests that autonomic activity, therefore, may be reduced. In actuality, sympathetic activity is reduced, and thus, parasympathetic influences (as evidenced by the long R - R intervals) are not so obvious. There appears to be little question that the parasympathetic is dominant during hibernation.

D. Inhibited Arousal

Inhibited arousal consists of alternate sympathetic and parasympathetic bursts. Frequency histograms of the durations of the R - R intervals of *C. lateralis*, *C. columbianus* and *E. fuscus* indicate a trend towards a biphasic distribution representing the two arms of the autonomic nervous system (Fig. 5).

Table 4. Changes in duration of R - R intervals during the first 10 minutes of spontaneous arousal from hibernation.

	During the 60 minutes before arousal $\bar{X} \pm$ s.d.	During the first 10 minutes of arousal $\bar{X} \pm$ s.d.	Change %
C. lateralis N = 8			
Average heart rate	5.24 ± 0.86		
Minimum R - R (sec)	7.98 ± 0.94	4.79 ± 1.53	- 40
Maximum R - R (sec)	16.45 ± 3.70	11.15 ± 6.52	- 32
Max R - R/Min R - R	2.10 ± 0.63	2.38 ± 1.08	+ 13
C. columbianus N = 5			
Average heart rate	3.4 ± 0.9		
Minimum R - R (sec)	17.8 ± 4.5	6.3 ± 1.2	- 65
Maximum R - R (sec)	24.8 ± 3.6	17.4 ± 4.7	- 30
Max R - R/Min R - R	1.4 ± 0.2	2.8 ± 0.6	+ 93
E. fuscus N = 1			
Average heart rate	5.9		
Minimum R - R (sec)	5.0	0.7	-614
Maximum R - R (sec)	12.5	3.3	-278
Max R - R/Min R - R	2.5	4.7	+ 88

Sequential plots of the R - R intervals indicate that reestablishment of the pattern of hibernation is rapidly attained.

Inhibited arousals occur spontaneously or after disturbance. Extrinsic factors which at times may provoke inhibited arousal include handling, i.p. injection, electrode implacement, loud noise or cage bumping. Intrinsic factors which have been observed occasionally to initiate inhibited arousal include movement and deep inhalation. Tachycardia associated with breathing occurs only in some individuals and may appear during, before,or after respiration in *C. lateralis*.

Changes in ECG patterns, sequential R - R intervals and frequency histograms for the durations of R - R intervals are illustrated during spontaneous,

Table 5. Changes in duration of R - R intervals during diving tests.

	During the 60 minutes before diving test $\bar{X} \pm$ s.d.	During the 10 minutes before diving test $\bar{X} \pm$ s.d.	Max R - R diving response $\bar{X} \pm$ s.d.	Change from the maximum during the 60 minutes before diving test %	Change from the maximum during the 10 minutes before diving test %
C. lateralis non-hibernating 35 - 37°C N = 5					
Average heart rate	384. ± 46	423. ± 19			
Minimum R - R (sec)	0.17 ± 0.02	0.16 ± 0.02			
Maximum R - R (sec)	0.35 ± 0.60	0.26 ± 0.08	1.00 ± 0.38	+185	+285
Max R - R/Min R - R	2.04 ± 0.60	1.62 ± 0.43	6.40 ± 2.63	+215	+295
C. lateralis hibernating 5°C N = 15					
Average heart rate	5.93 ± 0.82	10.57 ± 4.5			
Minimum R - R (sec)	8.71 ± 1.03	4.83 ± 1.96			
Maximum R - R (sec)	12.91 ± 2.53	7.99 ± 3.42	13.67 ± 2.93	+166	+ 71
Max R - R/Min R - R	1.73 ± 0.59	1.71 ± 0.68	3.18 ± 1.19	+184	+ 85

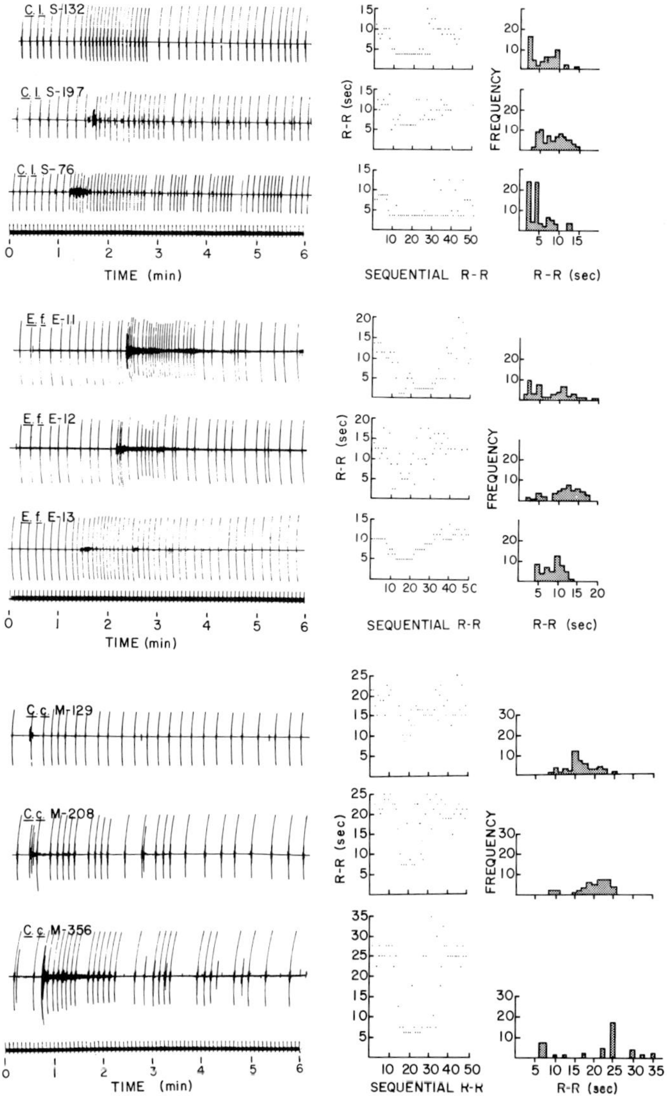

Fig. 5. *Comparison of ECG's of the three species during short inhibited arousal shows that the sympathetic activation followed by parasympathetic inhibition is similar. Reduction in the duration of the R - R interval indicates increased sympathetic activity.*

inhibited arousals of short duration for the three species in figure 5. Comparison of these data with those for the same animals recorded 30 min previously (Fig. 4) shows that an increase in heart rate with a consequent decrease in the R - R duration occurs prior to the long, diastolic arrythmia which signals the beginning of the inhibition of arousal. The ECG patterns of inhibition of arousal are similar for the three species under consideration.

Table 3 presents data obtained during short, inhibited arousals (Fig. 5) as well as longer inhibited arousals of 30 to 60 min duration. Individuals of the three species studied show several inhibited arousals during a single period of hibernation. The durations of the minimum R - R and maximum R - R intervals during inhibited arousal within a single period of hibernation were similar and were seen to occur at any time until the initiation of complete arousal. Their appearance was not a function of the elapsed fraction of the period of hibernation. Therefore, only one set of data from an animal per period of hibernation has been included in Table 3.

This table shows that the duration of the minimum R - R interval is decreased during inhibited arousals in all three species and indicates relative activation of the sympathetic nervous system.

The maximum duration of the R - R interval usually is lengthened in all species during inhibited arousal indicating an increase in parasympthetic activity. An estimate of the shift in dominance between the sympathetic and parasympathetic during inhibited arousal is given by the greater ratio of maximum R - R/minimum R - R.

E. <u>Arousal</u>

Complete arousal apparently begins when autonomic dominance shifts from the parasympathetic to the sympathetic. Unlike its role in inhibited arousal, the parasympathetic appears unable to reassert itself and it is thereby overridden by the sympathetic. Because of the physiological variability of the animals involved, the initiation of complete arousal in winter-pattern animals of each species is a continuum. Electrocardiographic records show variability ranging from patterns resembling long series of inhibited arousals to patterns of relatively gradual and progressively shortened R - R intervals.

This latter pattern actually takes place relatively quickly and is exhibited by winter-pattern animals that have hibernated to the predicted termination of their period of hibernation and are provoked to arouse through some intrinsic factor. The *C. lateralis* shown in figure 6 (212 hr at 5°C) illustrates this type of spontaneous arousal. Sequential R - R intervals and the frequency histograms for this animal reveal that the role of the parasympathetic becomes increasingly diminished. A single diastolic pause appears in the figured ECG.

An arousal illustrated for *E. fuscus* (Fig. 6) on the other hand shows series of several, apparently inhibited arousals, before sympathetic influences prevailed. This bat had hibernated over 400 hr at 5°C in the period of hibernation under consideration and was near the predicted termination of that period. A provoked, premature arousal for the same animal after 24 hr of hibernation (Fig. 7) reveals a long series of inhibited arousals before complete arousal suggesting a more effective parasympathetic early in the period of hibernation.

The terminal, spontaneous arousal of *C. columbianus* (350 hr at 6°C) (Fig. 6) indicates strong parasympathetic influences which are only gradually overcome. *C. columbianus* during provoked, premature arousal showed long series of inhibited arousals as did the provoked, premature arousal of *C. lateralis* (Fig. 7). These characteristic, inhibited arousals which occur in premature arousals before the sympathetic prevails illustrate the strength of the parasympathetic in its role in the maintenance of hibernation.

Data recorded one hr prior to arousal are compared to those during the first 10 min of arousal (Table 4). During early arousal, reduction of the duration of the minimum R - R interval in all species indicates sympathetic activation. The concomitant reduction in the duration of the maximum R - R interval reflects sympathetic override of the parasympathetic. The ratio of the maximum R - R/minimum R - R

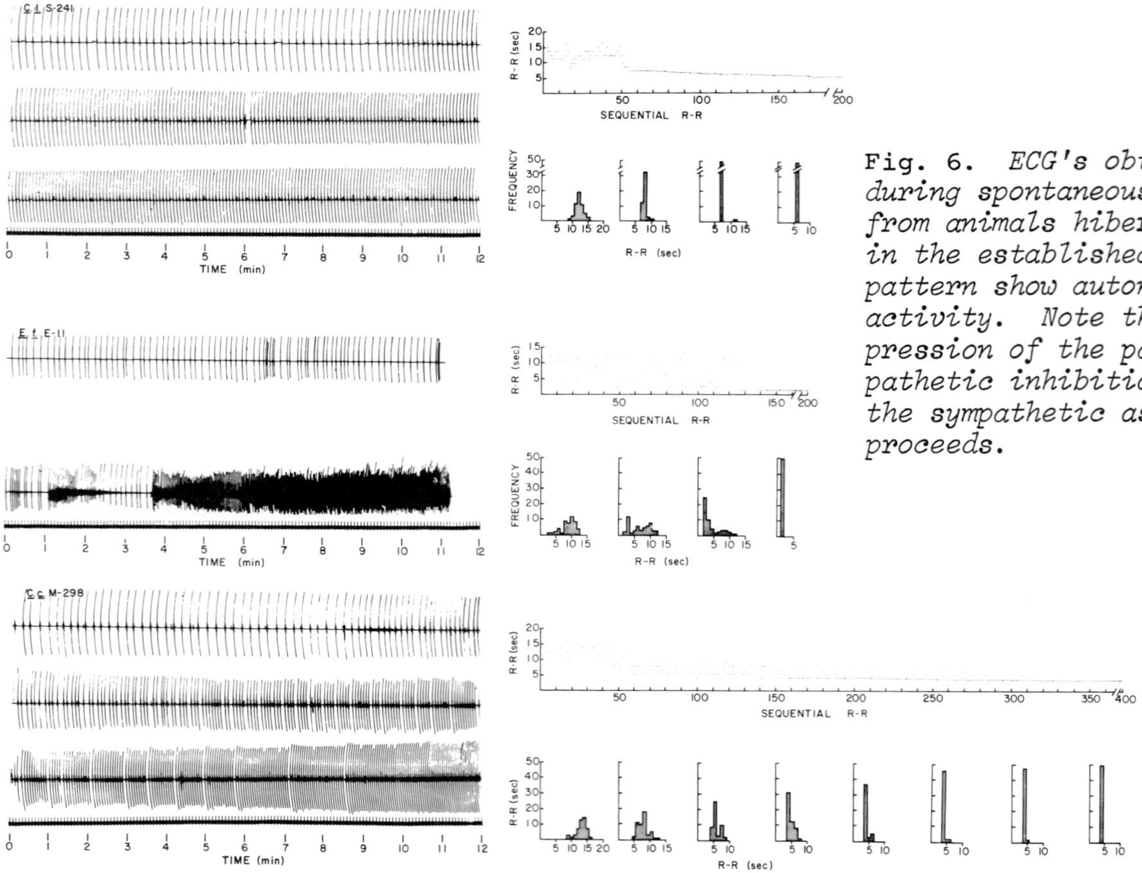

Fig. 6. *ECG's obtained during spontaneous arousal from animals hibernating in the established winter pattern show autonomic activity. Note the suppression of the parasympathetic inhibition by the sympathetic as arousal proceeds.*

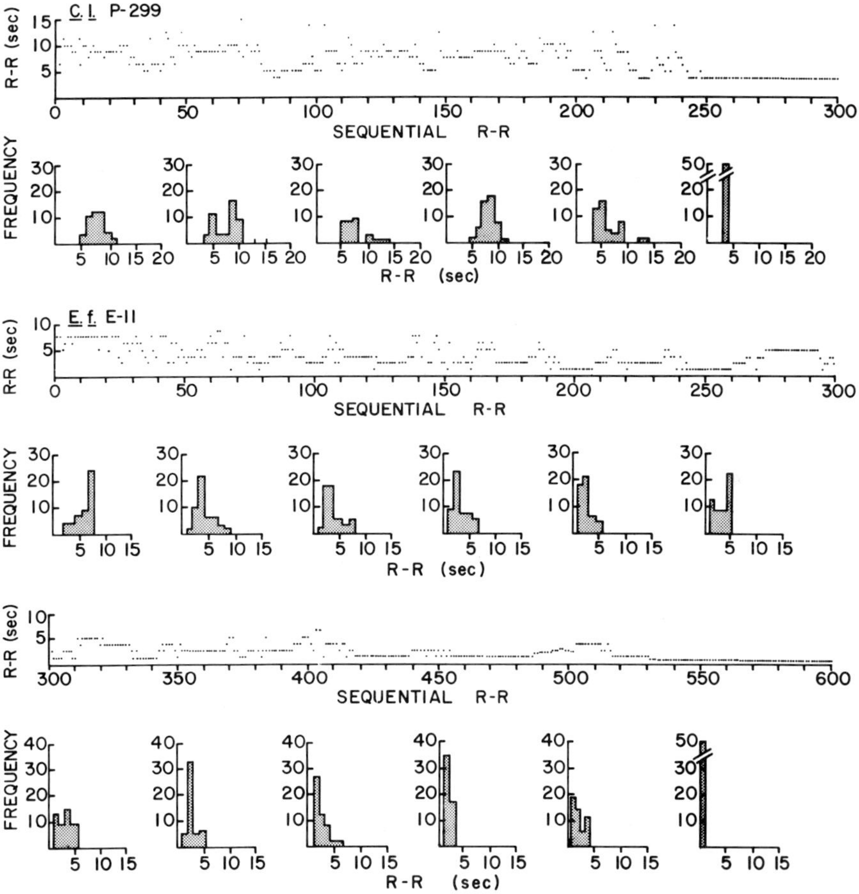

Fig. 7. *Plots of durations of R - R intervals during premature arousal demonstrate extensive intervals of inhibited arousal throughout until final suppression by the sympathetic. Compare the incidence of inhibition with that shown during terminal, spontaneous arousal (Fig. 6).* <u>*Citellus lateralis*</u> *(P-299) had hibernated approximately 60 percent and* <u>*Eptesicus fuscus*</u> *(E-11) had hibernated approximately 5 percent of their predicted periods of hibernation.*

indicates that during the first 10 min of arousal domination of the sympathetic over the parasympathetic is not complete and that the autonomic balance lies midway between conditions seen in inhibited arousal and those of atropinized animals.

F. Diving Responses

The similarity of the pronounced, diastolic arrythmias of inhibited arousal to arrythmias seen in the ECG during the diving response suggested that these phenomena could have the same origin. Squirting water (10 - 20 cc) on the nose of strapped-down, euthermic *C. lateralis* causes pronounced, diastolic arrythmias and associated bradycardia typical of the diving response (Fig. 8 and Table 5) indicating that this fossorial mammal also possesses this widespread reflex (1). Comparisons of ECG's, sequential R - R intervals and frequency distributions of the durations of the R - R intervals for the same animals after atropinization show complete abolition of the arrythmic and bradycardic aspect of diving response.

Administration of the same stimulus of squirts of 10 - 30 cc water on the nose of unrestrained, hibernating *C. lateralis* resulted in what could be considered the diving response in that diastolic arrythmias appeared several seconds after the stimulus. The strength of this response in the hibernating ground squirrel is remarkable in that some animals show the maximum duration of the R - R interval after this stimulus (Fig. 9 and Table 5). However, as with other arousal-provoking stimuli, not every animal responded to the stimulus of the diving test. Animals were further tested for the diving response by lightly stroking the sides of the muzzle. Bradycardia and diastolic arrythmias occasionally were elicited by stroking the muzzle although not as efficiently or frequently as with the water stimulus.

G. Comparison of Results

Cardiac patterns of *C. lateralis* represented by one individual in the same period of hibernation (Fig. 10) illustrates the responses described heretofore. ECG data revealed the typical arrythmic

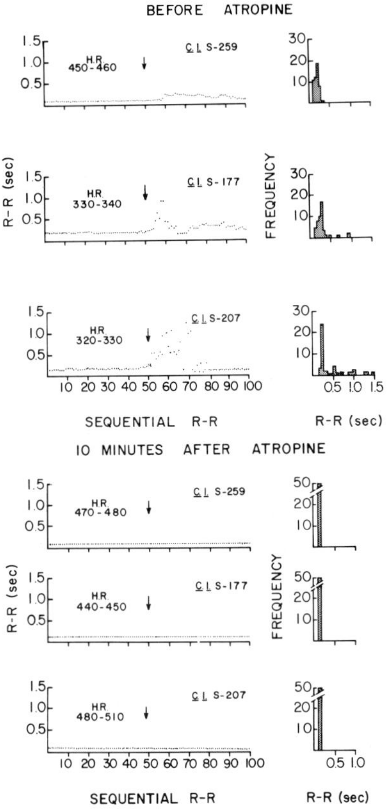

Fig. 8. *Euthermic (T_b= 37°C) Citellus lateralis show pronounced increases in the duration of the R - R interval during tests for the diving response (arrow). These arrhythmias were abolished by atropine. Sequential durations of R - R intervals are plotted 50 before and 50 after the stimulus of water. The frequency histogram is based upon the 50 R - R intervals after the initiation of the stimulus, denoted by the arrow.*

pattern of hibernation and the stronger, biphasic R - R distribution characteristic of inhibited arousal. Atropinization was evidenced by abolition of arrythmia. The strength of the parasympathetic and its role in maintaining hibernation is demonstrated by pronounced, inhibited arousal in the atropinized animal. Effects of atropine were reduced after 22 hr and by 26 hr appeared to be gone even though the results of a diving test were negative. At 28 hr after the administration of atropine, an arrythmia with an R - R interval of 23 sec resulted from the diving stimulus.

Atropinization in the dosages administered never caused premature arousal. Animals were no

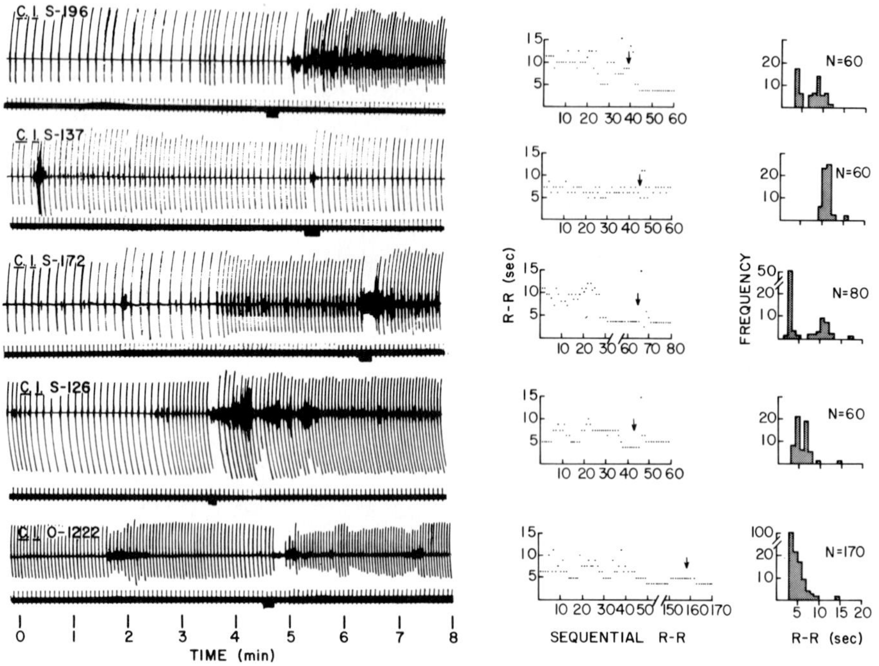

Fig. 9. *Hibernating C. lateralis were tested for the diving response by squirting water on the muzzle. The signal marker in the ECG's indicates the duration of the stimulus. The diving response is most apparent during sympathetic activity. The induced arrhythmia is greater than that seen during the inhibited arousal (S-196). Sequential R - R intervals during 5 to 10 min prior to the diving test (including the ECG segments shown) are plotted. Frequency histograms (variable number) show relative magnitude of R - R after the diving test compared to R - R intervals determined 5 to 10 min before the diving test.*

more irritable during atropinization than before. This was determined through control injections given at the peak of atropinization which failed to elicit arousing responses. These findings suggest that the parasympathetic blockade was acting primarily at the level of the cardiac receptors. Durations of the

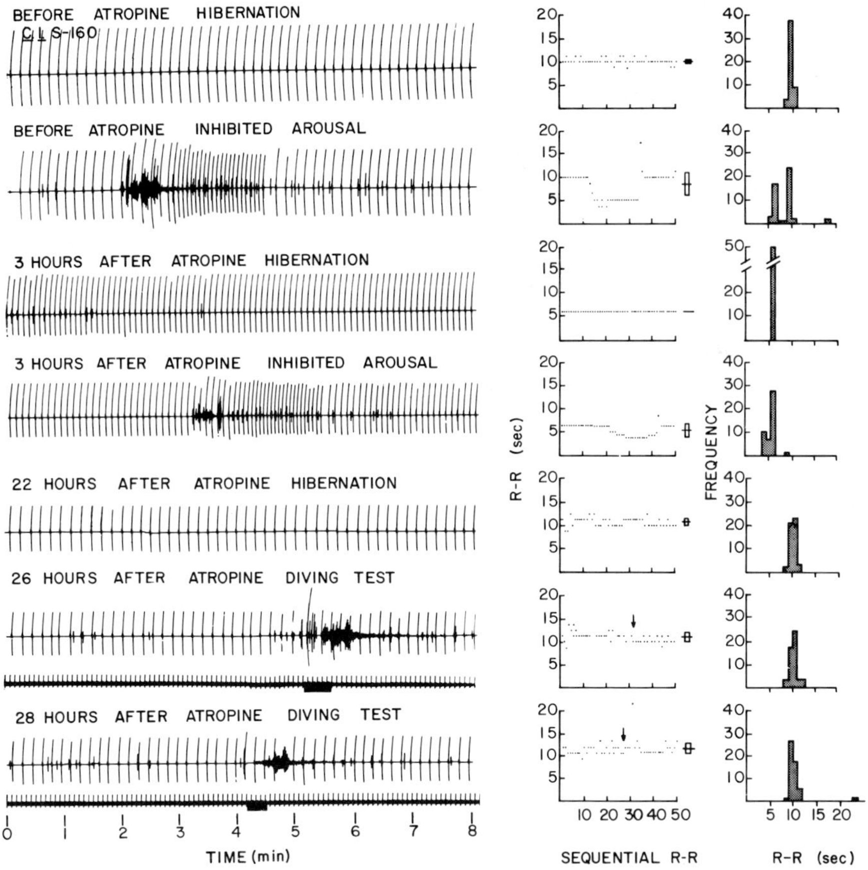

Fig. 10. *ECG's of Citellus lateralis (S-160) recorded during the same period of hibernation illustrate patterns of hibernation, inhibited arousal, effects of i.p. injected atropine up to 28 hr after injection, and response to diving tests. Note the negative response to diving test at 26 hr and the positive at 28 hr after atropine. Frequency histograms denote relative sympathetic/parasympathetic influences. Averages and standard deviations are noted on the sequential plots.*

period of hibernation in which atropinization occurred were no shorter than predicted.

The durations of the maximum and minimum R - R intervals of hibernating *C. lateralis* have been plotted with respect to heart rate in figure 11. Ground squirrels with heart rates between 6 - 7 b/min have relatively shorter maximum R - R intervals indicating less parasympathetic activity. Since the minimum duration of the R - R interval is the same for heart rates of 4 - 7 b/min, the major difference between slow (4 b/min) and fast (7 b/min) heart rates is accomplished by increasing the maximum duration of the R - R interval via the parasympathetic.

The comparison of the control R - R durations of *C. lateralis* (Fig. 11) with those seen during atropinization, inhibited arousals, spontaneous arousal, and diving tests is illustrated in figure 12.

In *C. lateralis* (Fig. 12), a reduction of the minimum R - R interval occurs in all test situations. This reduction denotes an increase in sympathetic activity. During atropinization and the first 10 min of spontaneous arousal it is obvious from reduction in the durations of the maximum R - R interval that parasympathetic influences are diminished. During inhibited arousal, and to a lesser extent during diving tests, the maximum R - R interval is increased designating the establishment of parasympathetic dominance.

H. Catecholamines

Arousal from hibernation seems to be initiated by a peripheral stimulus which, in turn, causes a sympathetic reflex. The incidence of the sympathetic episode, however subtle, first stimulates the parasympathetically-mediated inhibited arousal response. Arousal-inducing agents, whether mechanical or pharmacological, are not always effective. In the process of determining how a certain stimulus sometimes provokes premature arousal and sometimes does not, we demonstrated the concept of progressive irritability (46,47).

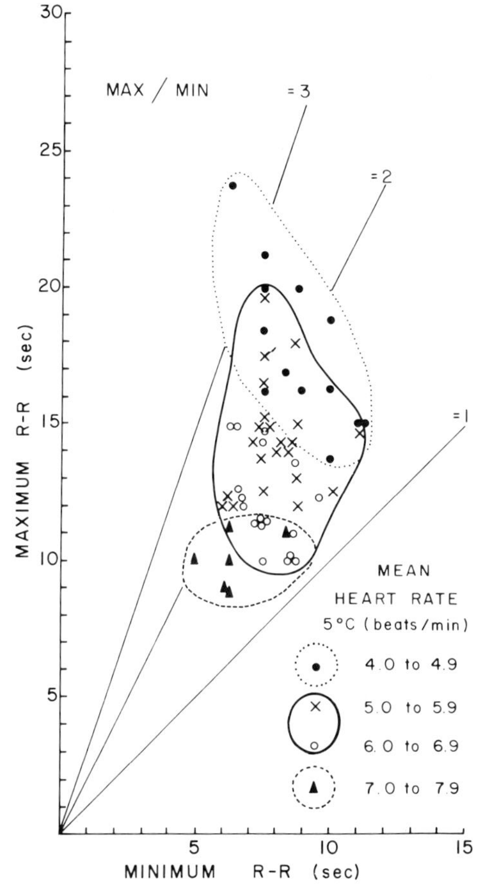

Fig. 11. *Range of durations of R - R intervals in a normal population of hibernating* *__Citellus lateralis__* *showing the continuum of hibernation stress from none to mild stress. All animals are hibernating at 5°C; variation in heart rates is not an effect of temperature. Maximum durations of R - R intervals represent parasympathetic activity whereas minimum duration of R - R intervals represent sympathetic activity. Reductions in the ratios between the two (diagonal lines) indicate reduced parasympathetic influences. Each datum is based upon the maximum and the one minimum duration of the R - R interval for an animal during one hr of continuous recording without an inhibited arousal.*

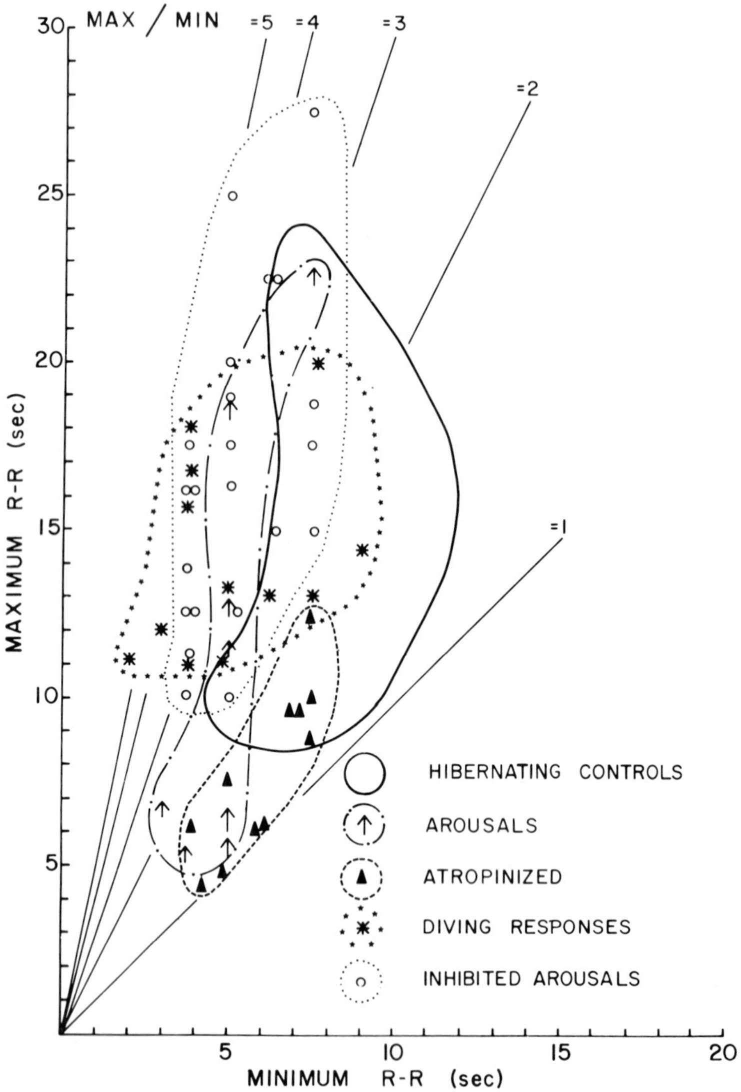

Fig. 12. *Changes in ratio of the durations of the R - R intervals of Citellus lateralis during test situations indicate changes in dominance of autonomic activity. The controls (Fig. 11) are here described by a solid line. Greatly increased parasympathetic activity is indicated in all situations except during atropinization.*

Generally, the hibernating animal tends to become increasingly sensitive to stimuli, whether extrinsic or intrinsic, as the period of hibernation proceeds. Pharmacologically, this may be proved in that if progressive irritability exists, then progressively smaller dosages of a compound would be required to elicit premature arousal as the period of hibernation proceeds--the arousing dosage would be a function of the elapsed fraction of the period of hibernation at the time of injection.

Therefore, in order to test progressive irritability, pharmacologically, and at the same time mimic the effects of the sympathetic events which naturally stimulate the proposed, central, arousal inhibiting response, the endogenous catecholamines, epinephrine (E) and norepinephrine (NE) were injected i.p. to hibernating *C. lateralis*. Isoproterenol (I) and phenylephrine (PE) strong β- and α-agonists, respectively, were compared with the former two catecholamines in an effort to determine the pharmacological basis for the stimulation of arousal.

The dosage of catecholamines necessary to provoke premature arousal and thereby to override the parasympathetic is a function of how long the animal has hibernated. Figure 13 illustrates this relationship of the arousing dosage to the elapsed fraction of the period of hibernation at the time of injection. An obvious line of demarcation exists between arousal and non-arousal for I, E, and NE. The results of the few arousing responses from PE are problematical. These may be secondary responses in that periods of latency for the initiation of thermogensis ranged from 3.5 - 4 hr whereas that for the other catecholamines ranged from 0.5 - 1.5 hr.

Most of the non-arousing dosages of I, E, and NE caused partial arousals of 2 - 18°C above the T_b of hibernation. In our opinion, these most likely are the result of the direct effect of circulating catecholamines upon the brown fat. Lines of regression calculated from the earliest arousal obtained at each dosage used are shown in figure 14. A relative potency is revealed I > E $\geq$ NE > PE which is typical of β-adrenergic responses. The intercepts of the lines give dosages (μg/animal) required to provoke

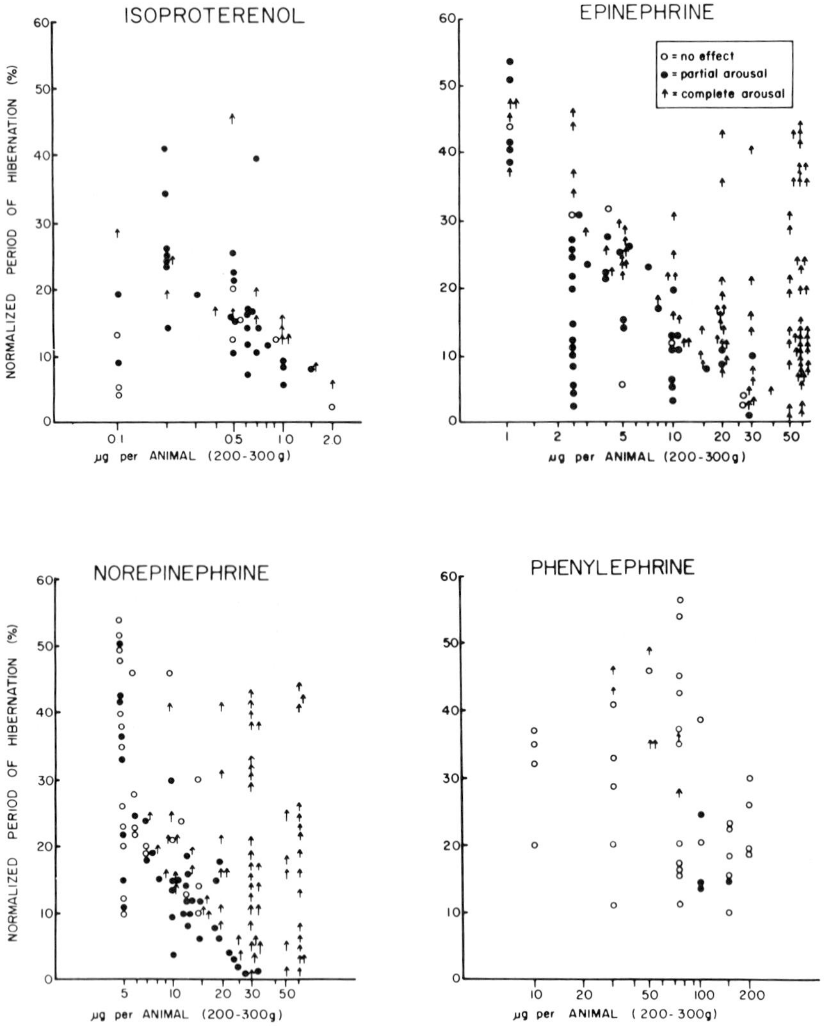

Fig. 13. *Arousing effects of catecholamines in respect to the elapsed fraction of the predicted winter-pattern period of hibernation of Citellus lateralis. Progressive irritability is demonstrated in that complete arousal is provoked by progressively smaller dosages as the period of hibernation proceeds. Sympathetic effects were considered to mimic those that naturally occur and act as stimuli to arousal.*

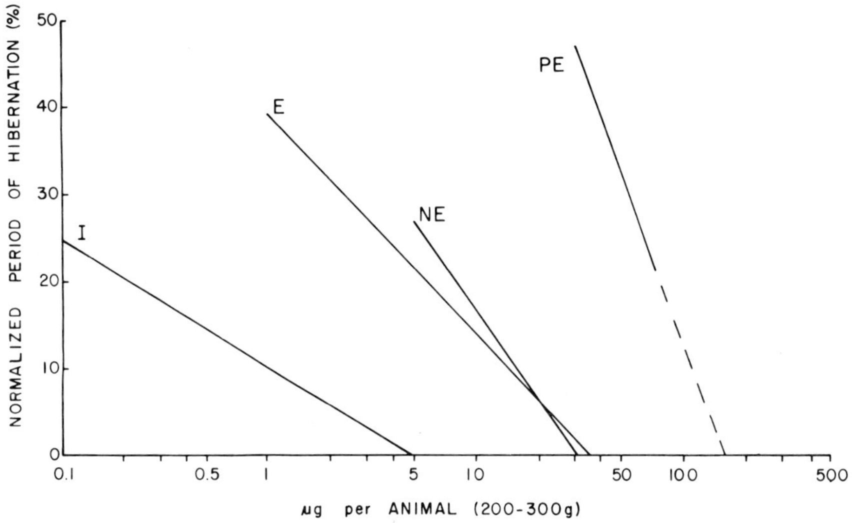

Fig. 14. *Lines of regression of arousing effects of catecholamines were derived from data shown in Fig. 13. The factors that stimulate arousal appear to act as β-adrenergic agonists. Data pertaining to the arousing effects of phenylephrine are questionable (see text) although the lack of arousing effects should be considered to be significant.*

premature arousal at the beginning of the period of hibernation of 5 μg, 35 μg, 30 μg and 125 μg respectively.

IV. DISCUSSION

Autonomic activity appears to regulate the maintenance of the state of hibernation. Occasionally, in unstressed, winter-pattern hibernators, suggestions of sympathetic activity signal the initiation of spontaneous arousal before the predicted termination of that period of hibernation. However, because the animal has not attained the physiological state which determines the normal conclusion of that period of hibernation, strong parasympathetic influences seem to override sympathetic drive and hibernation continues.

This shift in central domination from sympathetic to parasympathetic is signified peripherally by the pronounced, diastolic arrhythmias always present in the ECG's. Our data make it evident that it is through this mechanism of parasympathetically-mediated, inhibited arousal that the state of hibernation is actively maintained. We designate this mechanism as the autonomic hibernation maintenance response--the AHMR.

There seems to be little doubt that the parasympathetic is active throughout induction into hibernation, the period of hibernation and early arousal. Using the heart as an index of parasympathetic activity, it is clear that atropine tends to abolish arrhythmias and cause tachycardia during induction, the period of hibernation and inhibited arousal. It seems significant, at the effective dosages administered, that atropine does not cause arousal; irritability in atropinized animals does not increase; and the period of hibernation, therefore, proceeds to its normal conclusion. These observations suggest that the parasympathetic effects that assist in the maintenance of hibernation are not inhibited by atropine and that the heart, fortuitously, is most sensitive to atropine. The heart, thereby, serves as a relatively non-involved indicator of the parasympathetic mechanisms apparently active elsewhere.

Since we have concluded that hibernation is maintained actively via parasympathetic inhibition of arousal through the AHMR, we must try to explain why it does not keep the animal in a constant state of hibernation and how arousal eventually occurs. There comes a time in every period of hibernation when the AHMR can no longer function to suppress sympathetic impulses and arousal results. If this "area of time" is achieved when the animal is undisturbed, arousal is considered to be spontaneous. If arousal occurs as a result of extrinsic factors, arousal is said to be provoked. We do not mean to imply, however, that spontaneous arousal occurs without stimulation. Evidence from respiratory belts indicates that the hibernating animal at the time of spontaneous arousal stimulates itself, albeit involuntarily, probably by breathing or other movement thereby stimulating the sympathetic processes that lead to the beginning of arousal.

Stressed animals or those exhibiting the autumn or spring pattern of hibernating behavior may be stimulated to partial or complete arousal at any time during the period of hibernation (Table 1). Generally, these relatively erratic hibernators are irritable and unpredictable and do not maintain hibernation as well as do winter-pattern animals. As such, they are difficult subjects to study, especially for research involving the regulation of the period of hibernation.

A theoretical ceiling set temperature of about 12°C was proposed by Heller and Hammel (9) for hibernating *C. lateralis*. This concept was used by Beckman *et al.* (2) to explain inhibited, thermogenic partial arousals by the *C. lateralis* in their experiments. These partial arousals of 1.5 to 6°C above T_b of hibernation occurred spontaneously as well as in response to handling and after injection of ACh into the midbrain reticular formation. This physiological behavior suggests to us that their animals were exhibiting symptoms of hibernation stress (45, and Table 1).

Such a proposed ceiling set temperature which is defended by the inhibition of thermogenic responses through the thermoregulatory centers does not account for the observed, predictable hibernation at temperatures above 15°C. We have established that winter-pattern hibernation occurs for *C. lateralis* at T_b = 19°C and 25°C (43) and for *C. columbianus* at T_b = 19°C (49).

The inhibited arousals which are under consideration in this presentation are not accompanied by thermogenic responses. Even the spontaneous arousal responses that we are reporting are evident from cardiac patterns 30 to 60 min prior to any thermogenesis.

Unstressed hibernation of all species that we have studied, including bats, is independent of body and environmental temperature (between temperature levels of freezing and thermoneutrality) except as a passive regulator of metabolic processes.

Winter-pattern animals, actually are difficult to arouse (except towards the end of the period of hibernation at which time arousal is expected). It is these animals which have convinced us that hibernation is actively, strongly, and continuously being maintained by central influences.

Correlations of the concept of irritability with metabolism and the duration of the period of hibernation were first made by one of us from field observations of the bat *Myotis velifer* (42). Kristofferson and Soivio (18) noted that hedgehogs (*Erinaceus europaeus*) did not arouse in response to pricking with injection needles just after entering into hibernation, but that similar treatment shortly before the time of the estimated arousal always caused arousal.

We referred to the increasing sensitivity to stimuli as the period of hibernation proceeded as progressive irritability and quantified the response as follows (46).

As the winter-pattern period of hibernation proceeds, the stimulus required to provoke premature arousal becomes lessened. To state this another way, the same stimulus, applied at successive intervals throughout the period of hibernation, does not provoke premature arousal until a threshold of sensitivity to the stimulus is achieved. *C. lateralis*, early in their periods of hibernation (before 50% of the normalized period of hibernation had elapsed) never were provoked to arouse by i.p. injection of 0.5 ml normal saline. Provoked arousals to the stimuli of these injections occurred with increasing frequency as the normal termination of the period of hibernation was approached. Series of injections given at daily intervals similarly did not provoke arousal until late in the period of hibernation.

As a measure of progressive irritability, arousing effects of epinephrine were studied (47). A dose-response curve gave evidence that progressively smaller dosages were required to provoke premature arousal as the period of hibernation continued. At the time of that report we were still unaware of the possibilities of the sympathetically-stimulated AHMR response and considered the effect of epinephrine was probably central and direct. We also published that norepinephrine was not a reliable arousing agent. This latter observation was based upon miscalculations of dosage. On both accounts, therefore, we were wrong.

We now consider that it is the peripheral, rather than a direct, central effect of these injected catecholamines that are responsible for the sympathetically-mediated vascular changes and cardioacceleration and associated hypo- and hypertensive factors which we are postulating may constitute the intrinsic stimulus that initiates the arousal and which may be inhibited through the AHMR.

The thermogenesis of partial arousal effected by catecholamines (sometimes to T_b as high as 23°C) appears to be their direct effect, through the circulation, upon the brown fat. The data show that thermogenic effects of partial arousal are irrelevant to whether partial or complete arousal occurs. Therefore, the stimulation of arousal or inhibited arousal is unrelated to changes in body temperature as held by Beckman *et al.* (2).

An additional experiment involving the effects of catecholamines upon progressive irritability was performed recently and reported here. I, E, NE and PE were administered i.p. to determine whether changes in sensitivity by the centers which regulate the AHMR could be related to the pharmacological properties of the catecholamines. A classical β-adrenergic response was demonstrated by the potency of the dosages required to provoke premature arousal in the I > E ≥ NE > PE. The nature of the peripheral stimuli which result in the AHMR is suggested by the picture presented by these β-adrenergic agonists. The role of hypo- and hypertensive factors as reflex mediators for the maintenance of hibernation seems to constitute a problem that could yield significant results.

Lyman and O'Brien (29,30) demonstrated that stimuli from MAP's originated a sympathetic reflex that results in cardioacceleration. We postulate that sympathetic reflexes, in turn, activate the parasympathetically-mediated AHMR. The question may be posed as to whether it is the strength of the original stimulus which is responsible for initiation of arousal (29,30) or whether it is the strength of the secondary stimulus, the sympathetic reflex from cardioacceleration and associated vascular change. The results from the dose-response curves from the catecholamine injections suggest that the latter is most likely. Although the sympathetic stimulus for a given dosage may have been the same whenever administered, the central response was to override the sympathetic (via the AHMR) until it was no longer capable of so doing, and arousal was manifested.

The duration of the period of hibernation of *C. lateralis* is a semi-log function of body temperature over a range of 2° - 25°C thereby indicating that the frequency of arousal is regulated by temperature and therefore by some aspect of metabolism (43,44). That this relationship was not limited to a single species was later demonstrated by a comparative study which showed that the frequency of arousal was regulated by temperature by representatives of two additional subgenera of *Citellus*. *C. tridecemlineatus* and *C. columbianus* showed statistically similar slopes of regression to that shown for *C. lateralis* (49).

Temperature dependency of the duration of the period of hibernation has been suggested, although with too little quantification, for other hibernators as follows: dormice, *Glis glis* (14) and *Eliomys quercinus* (22,35); hedgehogs, *Erinaceus europaeus* (20); ground squirrels, *Citellus citellus* (14) and *C. lateralis* (36); golden hamsters, *Mesocricetus auratus* (21); greater horseshoe bats, *Rhinolophus ferroequinum* (37); and the cave Myotis, *Myotis velifer* (42). We have preliminary data which indicates that several genera of vespertilionid bats, including *Eptesicus*, show a similar temperature dependency of the duration of the period of hibernation (unpublished).

The temperature dependency of the duration of the period of hibernation leaves little question in our minds that some aspect of metabolism is intimately involved with the regulation of the frequency of arousal. These metabolic changes would appear to result in progressive irritability which, in turn, culminates in spontaneous arousal.

Virtually all of our data indicate that spontaneous arousal occurs as a culmination of the factors that produce the centrally-mediated phenomenon of progressive irritability.

When the AHMR becomes so weakened, or more likely inhibited, that it cannot suppress sympathetic stimuli, arousal occurs. We postulate that it is the metabolic change in the neurones associated with the AHMR that result in progressive irritability. We suspect that the nature of the change is through hypo- and hyperpolarization which results in gradual changes in thresholds of transmission thereby causing progressive irritability.

The simplest theory we can derive which best fits our findings can be best stated by a hypothetical model. This model is based upon a gradual change in the polarization of neurones associated with the mediating feedback components between autonomic, sympathetic (baro) receptors and parasympathetic effectors of the AHMR. The hyperpolarization of some components and hypopolarization of others would proceed to the point that the AHMR could only discharge weakly and the sympathetic events would be compounded until complete arousal becomes manifested.

Because we have no method of distinguishing the two phenemona, the significance of the apparent similarity between the AHMR and the diving response must remain conjectural. It is not inconceivable that the AHMR may have been derived as an evolutionary modification of the diving response. Should this widespread, parasympathetically-mediated response (1) prove to have been the forerunner of mammalian, hibernatory, regulatory mechanisms, then its parallel, selective, acceptance by diverse groups of mammals would constitute a rational explanation for the consequent similarity between unrelated species. The palatability of the concept of species similarity, instead of species differences, would seem to be greatly enhanced.

When *C. lateralis* die during the season of hibernation, it occurs without apparent cause, or pathology, while the animal is still in a state of deep hibernation. Although we have yet to record ECG's and respiratory patterns of ground squirrels experiencing such death, our observations of the beginnings of thousands of arousals make us suspect that it may result from overaccentuation of parasympathetic influences. Knowledge acquired from hibernating animals could well be extended to humans in that many features of sudden infant death syndrome are suspiciously like those of inhibited arousal and the diving response.

ACKNOWLEDGEMENT

This investigation was supported, in part, by the United States Health Service Grant AM-11320 and by the Dalton Research Center.

REFERENCES

1. Andersen, H. T. Physiological adaptations in diving vertebrates. Physiol. Rev. 46: 212-243, 1966.
2. Beckman, A. L., T. L. Stanton, and E. Satinoff. Inhibition of the CNS trigger process for arousal from hibernation. Am. J. Physiol. 230: 1018-1025, 1976.
3. Chatfield, P.O., and C. P. Lyman. Circulatory changes during the process of arousal in the hibernating hamster. Am. J. Physiol. 163: 566-574, 1950.
4. Chatfield, P. O., and C. P. Lyman. Subcortical electrical activity in the golden hamster during arousal from hibernation. Electroencephalog. Clin. Neurophysiol. 6: 403-408, 1954.
5. Chatfield, P. O., C. P. Lyman, and D. P. Purpura. The effects of temperature on the spontaneous and induced electrical activity in the cerebral cortex of the golden hamster. Electroencephalog. Clin. Neurophysiol. 3: 225-230, 1951.
6. Dawe, A. R., and P. R. Morrison. Characteristics of the hibernating heart. Am. Heart J. 49: 367-384, 1955.
7. Hammel, H. T., T. J. Dawson, R. M. Abrams and H. T. Andersen. Total calorimetric measurements on *Citellus lateralis* in hibernation. Physiol. Zool. 41: 341-357, 1968.
8. Heller, H. C., and G. W. Colliver. CNS regulation of body temperature during hibernation. Am. J. Physiol. 227: 583-589, 1974.
9. Heller, H. C., and H. T. Hammel. CNS control of body temperature during hibernation. Comp. Biochem. Physiol. 41: 349-359, 1972.
10. Herroux, O., and V. M. Petrovic. Effect of high- and low-bulk diets on the thyroxine turnover rate in rats with acute and chronic exposure to different temperatures. Can. J. Physiol. Pharmacol. 47: 963-968, 1969.
11. Johansen, K., and J. Krog. Diurnal body temperature variations and hibernation in the birch

mouse, *Sicista betulina*. Am. J. Physiol. 196: 1200-1204, 1959.

12. Johansen, K., J. Krog, and O. Reite. Autonomic nervous influence on the heart of the hypothermic hibernator. Ann. Acad. Sci. Fennicae Series A IV 71: 245-254, 1964.
13. Johansson, B. W. Heart and circulation in hibernators. In: Mammalian Hibernation III. Fisher, K.C., A.R. Dawe, C.P. Lyman, E. Schönbaum, and F.E. South (editors). Edinburgh: Oliver and Boyd, 1967, p. 200-218.
14. Kayser, Ch. Echanges respiratoires et production de chaleur en fonction de la température chez le Loir (*Myoxus glis*) a l'étai de veille et de sommeil hibernal. Compt. Rendus de. la Soc. de biol. 128: 1201-1204, 1938.
15. Kayser, Ch. La dépense d'énergie des mammifères hibernants pendant toute la duree de l'hibernation. Recherches fáites sur le spermophile (*Citellus citellus*). Arch. des Sci. Physiol. VI: 193-212, 1952.
16. Kayser, Ch. The heart frequency during hibernation. In: The Physiology of Natural Hibernation. Ch. Kayser. Oxford: Pergamon Press, 1961, p. 104-108.
17. Kayser, C., and A. Malan. Central nervous system and hibernation. Experientia 19: 441-451, 1963.
18. Kristofferson, R., and A. Soivio. Hibernation of the Hedgehog (*Erinaceus europaeus* L.) The periodicity of hibernation of undisturbed animals during the winter in a constant ambient temperature. Ann. Acad. Sci. Fennicae Series A IV. 80: 5-22, 1964.
19. Kristofferson, R., and A. Soivio. Hibernation in the Hedgehog (*Erinaceus europaeus* L.) Changes of respiratory pattern, heart rate and body temperature in response to gradually decreasing or increasing ambient temperature. Ann. Acad. Sci. Fennicae. Series A IV. 82: 1-17, 1964.
20. Kristofferson, R., and A. Soivio. Studies on the periodicity of hibernation in the hedgehog

(*Erinaceus europaeus* L.) I. A comparison of induced hypothermia in constant ambient temperatures of 4.5 and 10 C. Ann. Zool. Fenn. 1: 370-372, 1964.

21. Kristofferson, R., and A. Soivio. Duration of hypothermia periods and type of respiration in the hibernating golden hamster, *Mesocricetus auratus* Waterh. Ann. Zool. Fenn. 3: 66-67, 1966.
22. Lachiver, F., and R. Boulouard. Evolution du poids corporel chez un hibernant le Lérot (*Eliomys quercinus* L.) durant la periode hivernale et au cours de l'exposition au froid en été. Mammalia 29: 343-366, 1965.
23. Landau, Barbara R., and A. R. Dawe. Respiration in the hibernation of the 13-lined ground squirrel. Am. J. Physiol. 194: 75-82, 1958.
24. Lyman, C. P. Oxygen consumption, body temperature, and heart rate of woodchucks entering hibernation. Am. J. Physiol. 194: 83-91, 1958.
25. Lyman, C. P., and P. O. Chatfield. Mechanisms of arousal in the hibernating hamster. J. Exp. Zool. 114: 491-516, 1950.
26. Lyman, C. P., and P. O. Chatfield. Hibernation and cortical electrical activity in the woodchuck (*Marmota monax*). Science 117: 533-534, 1953.
27. Lyman, C. P., and R. C. O'Brien. Circulatory changes in the 13-lined ground squirrel during the hibernating cycle. Bull. Mus. Comp. Zool. 124: 353-372, 1960.
28. Lyman, C. P., and R. C. O'Brien. Autonomic control of circulation during the hibernating cycle in ground squirrels. J. Physiol. 168: 477-499, 1963.
29. Lyman, C. P., and R. C. O'Brien. The effect of some autonomic drugs on *Citellus tridecemlineatus* during the hibernating cycle. Ann. Acad. Sci. Fennicae. Series A IV. 71: 311-323, 1964.
30. Lyman, C. P., and R. C. O'Brien. Hyperresponsiveness in hibernation. Symp. Soc. Exp. Biol. 23: 489-509, 1969.

31. Lyman, C. P., and R. C. O'Brien. Sensitivity to low temperature in hibernating rodents. Am. J. Physiol. 222: 864-869, 1972.

32. Mihailovic, Lj. T. Cortical and subcortical electrical activity in hibernation and hypothermia. In: Hibernation and Hypothermia, Perspectives and Challenges. South, F.E., J.P. Hannon, J.S. Willis, E.T. Pengelley, and N.R. Alpert (editors). Amsterdam: Elsevier, 1972, p. 487-534.

33. Mihailovic, Lj., B. D. Beleslin, D. Ĉupić, and D. Popesković. Interaction between the diffuse thalamo-cortical system and the brain stem activating system during arousal from hibernation. Iugoslav. Physiol. Pharmacol. Acta 5: 289-294, 1969.

34. Mills, S. H., and F. E. South. Central regulation of temperature in hibernation and normothermia. Cryobiology 9: 393-403, 1972.

35. Pajunen, Irmeli. Body temperature, heart rate, breathing pattern, weight loss and periodicity of hibernation in the French garden dormouse, *Eliomys quercinus* L., at 4.2 ± 0.5°C. Ann. Zool. Fenn. 11: 107-119, 1974.

36. Pengelley, E. T., and K. C. Fisher. Rhythmical arousal from hibernation in the Golden-mantled ground squirrel, *Citellus lateralis tescorum*. Can. J. Zool. 39: 105-120, 1961.

37. Ransome, R. D. The effect of ambient temperature on the arousal frequency of the hibernating Greater horseshoe bat, *Rhinolophus ferroequinum*, in relation to site selection and the hibernation state. J. Zool., Lond. 164: 353-371, 1971.

38. Raths, P. Die bioelektrische Hirntätigkeit des Hamsters in Verlaufe des Erwachens aus Winterschlaf und Kältenarkose. Z. Biol. 110: 62-80, 1958.

39. South, F. E., W. C. Hartner and R. H. Luecke. Responses to preoptic temperature manipulation in the awake and hibernating marmot. Am. J. Physiol. 229: 150-160, 1975.

40. Strumwasser, F. Some physiological principles governing hibernation in *Citellus beecheyi*. Bull. Mus. Comp. Zool. 124: 285-318, 1960.

41. Torke, K. G., and J. W. Twente. Behavior of *Spermophilus lateralis* between periods of hibernation. J. Mamm. 58: 385-390, 1977.
42. Twente, J. W. Jr. Some aspects of habitat selection and other behavior of cavern-dwelling bats. Ecology 36: 706-732, 1955.
43. Twente, J. W., and J. A. Twente. Regulation of hibernating period by temperature. Proc. Nat'l Acad. Sci. USA 54; 1058-1061, 1965.
44. Twente, J. W., and J. A. Twente. Effects of core temperature upon duration of hibernation of *Citellus lateralis*. J. Appl. Physiol. 20: 411-416, 1965.
45. Twente, J. W., and J. A. Twente. Seasonal variation in the hibernating behavior of *Citellus lateralis*. In: Mammalian Hibernation III. Fisher, K.C., A.R. Dawe, C.P. Lyman, E. Schönbaum, and F.E. South (editors). Edinburgh: Oliver and Boyd, 1967, p. 47-63.
46. Twente, J. W., and J. A. Twente. Progressive irritability of hibernating *Citellus lateralis*. Comp. Biochem. Physiol. 25: 467-474, 1968.
47. Twente, J. W., and J. A. Twente. Effects of epinephrine upon irritability of hibernating *Citellus lateralis*. Comp. Biochem. Physiol. 25: 475-483, 1968.
48. Twente, J. W., and J. A. Twente. Responses of hibernating *Citellus lateralis* to lowered environmental temperatures. In: Adaptations to Arid Zone Conditions: Hibernation. Vinogradov, U.W. and S.G. Kolawa (editors). Novosibirsk: Academy of Sciences USSR, 1970, p. 181-187.
49. Twente, J. W., J. A. Twente, and R. M. Moy. Regulation of arousal from hibernation by temperature in three species of *Citellus*. J. Appl. Physiol. Respirat. Environ. Exercise Physiol. 42: 191-195, 1977.
50. Wit, L. C. Metabolic responses of hibernating *Citellus lateralis* to lowered temperatures. Ph.D. Dissertation, Univ. of Missouri-Columbia, 1975.

Session III

Cellular and Biochemical Adaptations in Natural Torpidity

COLD TOLERANCE OF MAMMALIAN CELLS: PREVALENCE AND PROPERTIES

J. S. Willis
Department of Physiology and Biophysics
University of Illinois
Urbana, Illinois 61801

ABSTRACT

Four possible determinants of cold tolerance of mammalian cells to low temperature are species, tissue, isolation from the organism, and prior experience of cold. Perfused hearts of deeply hibernating species possess unusual ability to beat at low temperature. Among hearts of other species a range of tolerance is found. Such variation may also be found within an organism (e.g. peripheral vs. core tissue) or even within a tissue (heterogeneity of cells). Characteristic tolerance of cells of deep hibernators can persist for at least a few generations in culture. Long established aneuploid cell lines are unusually tolerant of cold but so highly variable among themselves that tissue- and species-related differences have not been found. Prior experience of the organism to cold sometimes alters aspects of cell function in the cold. Whether a cell may detect and respond directly to a prior cold experience is not clear. Damage of cells by exposure to lowered temperature may be of two fundamental types, reproductive (eventual) and physiological (immediate). The former has hardly yet been studied in the context of the whole organism. The latter may operate through energy metabolism or ion regulation. Differences in temperature sensitivity of ion regulation have been shown in erythrocytes and cultured kidney cells to be due to differences both in passive permeability and in ion pumps. Use of isolated membrane preparations and Na-K ATPase for estimating temperature sensitivity of pump activity and membrane fluidity is subject to severe unforseen pitfalls.

ISBN 0-12-734550-7

Recognition and understanding of homeostatic regulation of factors of the internal environment is a cornerstone of physiology. A deeply engrained corollary of that concept is that when a factor is rigorously regulated and permitted to vary only within narrow limits in an organism the cells of that organism lose their tolerance to wide fluctuation in that factor. A prime example of this corollary is the general sensitivity of mammalian cells to lowered temperature. There are, however, numerous cases of mammalian cells exhibiting remarkable resistance to lowered temperatures, some of which confirm the cited corollary and others of which do not. In either case, familiarity with these exceptions and analysis of them is important for biology and medicine. My purpose, therefore, is to identify the situations in which cold tolerant mammalian cells are found and then to discuss mechanisms that may underlie that resistance.

I. CIRCUMSTANCES OF RESISTANCE

A useful focal point for the first task is the paper of Lyman and Blinks (39) in which a comparison was made of the ability of perfused hearts of four species of hibernating and four species of non-hibernating rodents to continue beating at low temperature. Heart was chosen because it was felt to be the most temperature-sensitive tissue, and among the points investigated were the effect of hibernation on the sensitivity of hamster hearts and the effect of cooling on beating of cultured heart cells from newborn rats and hamsters. Thus, within this now classic study were considered either implicitly or explicitly all of the variables which may be determinants of temperature sensitivity of cells: 1) species specificity (i.e., the question of whether resistance is phylogenetically determined or is related purely to the thermal habits of the species); 2) tissue specificity within an organism; 3) persistence in isolation from the organism (and the related issue of the influence of the degree of differentiation); and 4) dependence on acclimation or conditioning.

Before discussing each of these categories it is necessary to state more explicitly what is to be meant by "cold tolerance". Here, too, the paper of Lyman and Blinks is useful in laying the guidelines. They used simple EKG to detect the occurrence of heart beat, which was at the same time an indicator of the integrity of the constituent cells, a reflection of the capacity of the organ to function *in vivo*, and a variable which is a continuous and easily measured function of temperature. They then determined both persistence of activity at low temperature (by finding the lowest temperature at which beating could be observed, generally above 9°C) and the effect of cooling on subsequent function at high temperature (by comparing the beat rate of hearts as they were cooled and then rewarmed). Hence in discussing the sequels of this work I shall confine consideration to tolerance of temperatures near, but above, freezing and shall be concerned where possible with the survival of the cells both while they are cold and after they have been rewarmed.

A. Species Specificity

Lyman and Blinks of course were interested in the cold resistance of hibernators and a principal issue in their study was whether cold tolerance would be confined to the hibernating species or would exhibit some systematic phylogenetic pattern. To explore this they carefully selected the eight species of rodents they studied so that there were two subfamilies included, each of which contained species of hibernators and non-hibernators. Thus, in the sciurinae, hearts of grey squirrels could be compared with those of ground squirrels, chipmunks, and woodchucks, and in the circitinae, hearts of cotton rats could be compared with those of hamsters. They observed a wide difference in the tolerance of the hearts of hibernators (persistence of beating at temperatures from 3°C to below 0°C) compared with those of non-hibernating species (failure at temperatures ranging from 8°C to 12°C). There was no indication of a familial or phylogenetic tendency to cold tolerance with the minor exception of occasional results with hearts of grey squirrels (see below).

The correlation of cold resistance with thermal habits of the organism has been somewhat modified, however, by later comparisons. Lyman (38) found that the isolated hearts from antelope ground squirrels (*Ammospermophilus leucurus*, the erstwhile *Citellus leucurus*), a species which does not show any tendency for voluntary torpor, stopped beating at 2°C. Conversely, Hudson (29) reported that the isolated, perfused hearts from round-tailed ground squirrels (*Citellus tereticaudus*), a species which does undergo shallow torpor, stopped at about 4°C. (More precisely, they slowed to 2 beats/min at 5.7°C, compared with -0.7°C for 2 beats/min in hearts of *Citellus tridecemlineatus*.)

Similarly, Hudson (30) found that in isolated hearts of several small mice representing two families (the heteromyid *Perognathus* and the cricetids *Baiomys* and *Peromyscus*, four species) the "cut-out" temperature, which varied between 7.2°C and 1.5°C correlated only in general, and not in detail, with the tendency of the organism to lower its body temperature during torpor. None of the species tested, for example, allowed its body temperature to fall below about 17°C, and *Baiomys* which tolerated the least amount of body cooling (to 22°C) possessed hearts with the lowest "cut-out" temperature of the group (1.5°C). Thus, insofar as the ability of the heart to beat constitutes a limit for survival in forced hypothermia, all of the species possessed a considerable "margin of safety".

It appears that the species chosen by Lyman and Blinks represented an extreme of a continuum rather than a discontinuity. Deeply hibernating species require and possess highly cold-resistant hearts. Organisms less likely to face deep hypothermic periods may or may not have unusually cold-resistant hearts, and other factors, such as, for example, adaptations to other aspects of the organism's niche, may influence the degree of the resistance possessed.

B. Tissue Specificity

In discussing species differences the comparison has so far been limited, like that of Lyman and Blinks, to a single organ, the heart. How much greater variation, then, would be found amongst other tissues and is tolerance of low temperature within an organism correlated with the thermal experience and demands upon the tissue?

1. Periphery vs. core

Even in the most rigorously homeothermic mammals and birds the temperature of the peripheral tissues and organs may be allowed to decrease, in some cases to temperatures approaching 0°C. Consequently, in comparing cold tolerance among various tissues principal attention has been given to differences between "core" tissues (heart, brain, kidney) and peripheral tissues (skin, deep tissues of appendages).

The most astounding case of cold tolerance of a peripheral tissue was also the earliest described and is perhaps the best known. Chatfield, Lyman and Irving (14) found that the distal portion of the sciatic nerve of cold adapted herring gull which runs through the poorly insulated metatarsal region of the foot and leg was capable of conducting at 0°C whereas the proximal tibial portion was blocked at 8°C. The remarkable feature of this discovery is of course that the difference in adaptation existed within different portions of the same nerve cells. Chatfield and Lyman (13) confirmed the existence of peripheral cold resistance in a mammal by showing that the caudal nerve of the poorly insulated tail of rats was capable on the average of conducting at temperatures down to 5°C in comparison with the rat tibial nerve which is blocked at about 9°C. Miller (48) has reported other, more impressive, examples of this in caudal nerves of several arctic species. He found some correlation as well between the lowest temperature of conduction (-5°C in the beaver) and the degree of insulation and cooling permitted in the tail (block at 3.1°C in the caudal nerve of the furry fox tail).

2. Core tissues and cell heterogeneity

The species-related differences discussed above for hearts can also in general be found in other core tissues. Thus, phrenic nerves of hamsters conduct at lower temperatures than those of rats, muscles of hamsters directly stimulated with standard stimuli contract at lower temperatures than those of rats, and red blood cells and kidney cells of hamsters and ground squirrels are better able to retain K at low temperature than those of guinea pig or rats (34, for other references see 73). But as was the case for heart, paradoxical, or at least non-conforming, examples can also be cited: vagus nerve of rabbit conducts at 0°C and cells of rabbit kidney cortex retain K as well as those of hibernators and the auditory nerve of hamsters is blocked at low temperature (for references, see 73). Furthermore, the hearts of rabbits are not cold resistant as are the kidneys and vagus nerve, so diversity may be found among the core tissues of a single organism.

Indeed, such diversity may be apparent within a single tissue. Because of heterogeneity Lyman and Blinks were unable to find a difference between rats and hamsters with regard to temperature coefficient of the rate of beating of foci of disaggregated cells of ventricles from day-old animals. In one case three foci within one culture each had different temperature coefficients. Similarly, Andrus (3) could not find any correlation between cut-out temperature of cultured cells of hearts of new-born animals and the thermal tolerance of the adult organism in six species of cricetids. Although it was suggested from this that the cells of new-born and non-hibernators might possess cold resistance, Lyman and Jarrow (40) found that there was also no difference in average cut-out temperature of cultured heart cells of adult rats and ground squirrels. In the latter case there was a wide spectrum for each species; some rat heart cells continued down to 0°C whereas some ground squirrel cells blocked at about 8°C (total range 0-13°C for rat, 0-8.9°C for ground squirrel).

Although it has not been studied systematically, such cellular heterogeneity is probably present within the intact tissue and is not merely a quirk of isolation or culture. Thus, Lyman and Blinks found that some hearts of grey squirrels, if cooled slowly, could develop ectopic foci that would enable them to beat at several degrees below their typical cut-out temperature of 12°C. Marshall (43) found that although rabbit atria stopped beating at 17°C, the pacemaker region continued to fire down to 10°C. Conversely, Nardone (52) detected branch bundle block in hearts of hibernating arctic ground squirrels. In hamster papillary muscle Jacobs and South (31) found considerable decrease of the resting potential with cooling even at 12°C, and at 12°C the variation was much greater than at 37°C. Presumably, the less depolarized fibers would be the ones still capable of conducting action potentials, and this expectation was borne out at least in the hearts of hibernating hamsters, in which there was little change in height of action potential or of magnitude of overshoot down to 12°C.

In atria of thirteen-lined ground squirrels Marshall and I (44) found little decrease in resting membrane potential between 31°C and 6°C and no increase in variability at the lower temperature. In those studies of ground squirrel atria, however, we took our measurements from film records of action potentials (i.e., resting potential before firing, total height, duration, rate of rise, and overshoot of action potential). Hence only the resting potentials of actually conducting fibers were scored and reported, and it is not unlikely that in that tissue, as well, there were quiescent and possibly depolarized regions at low temperature.

C. Persistence in Isolation

A central issue of the study of Lyman and Blinks was whether cold resistance was intrinsic to the tissue or whether it depended upon factors residing in the organism. There have been similar demonstrations of acute retention of function following removal from the organism for peripheral nerve, skeletal muscle, and using K retention as a criterion, for kidney and red blood cells (34, for other references see 73).

[The one notable exception to this pattern may be brain. Slices of brain cortex of awake hamsters do not retain K *in vitro* at 5°C better than those of rat, though no K is lost from cerebral cortex *in vivo* during hibernation (23, 80). Slices of cerebral cortex from hibernating hamsters were found to be only marginally better at retaining K *in vitro*. On the other hand, brain slices from hamster show less decrease of respiration than those of guinea pig when rewarmed after a cold incubation of several hours (50)].

Beyond the freshly isolated tissue one can envision a progression of preparations, each more distantly related to the donor organism: stored tissues or organs, primary cultures grown from explants of tissues, passaged cells (secondaries), established (usually "aneuploid" that is, no longer diploid) cell lines, and malignantly transformed cells. The question then is, how far in this series does the intrinsic cold tolerance (or lack thereof) persist? This question is of importance not merely because it bears upon the degree to which the phenomenon of cold tolerance resides intrinsically in the cell freed of organismic influences, but also because each of the steps in the progression offers a wider scope of possibilities for exploring mechanisms of difference at the cellular level.

Lyman and Blinks took the first step along this route with their three or four day cultures of heart cells, but it is not clear whether these cells were growing. In any case, as already discussed, neither in their own nor in succeeding studies in which minimum temperature or autorhythmicity was scored, in growing and non-growing, adult and neonatal cultures of heart cells was any persistent, species-related pattern of cold resistance observed.

1. Stored tissues

With other criteria, however, such patterns have been detected both in stored tissues and in cell cultures. At 5°C excitability of diaphragm muscle persisted for 70 and for 160 hours for that of hamster and ground squirrel, respectively, whereas diaphragm of rat survived only about 50 hours (69). Both kidney slices (75, 78, 79) and red blood cells (34) from hibernators persist in showing greater retention of K for a week or more when kept constantly at 5°C, than do those of non-hibernators.

When survival *in vitro* of isolated organs or tissues for such long intervals is considered, questions arise concerned with stability of the tissue at the high or "control" temperature (usually near 37°C). For example, it is well known that slices of parenchymatous mammalian tissues at 37°C often, by several criteria, begin to deteriorate after about an hour of incubation. Even undamaged cells or organs (e.g., red blood cells, hearts) when stored in closed systems at higher temperature will quickly exhaust metabolites. In either case, cooling to 5°C may have the effect of preserving the tissue, and such tendency will be countervailing to, and will complicate the interpretation of, the deleterious effects of the low temperature. On the other hand, it is also possible that some kinds of repair or of acclimation to the incubation conditions may take place at permissive temperatures. In this case, holding a tissue or organ at the higher temperature *in vitro* before cooling will produce a result different from that of the more usual practice of immediate cooling (for examples of such an effect, see 79 and 64).

In this regard, if one could assume that species- or tissue-specific differences in cold tolerance persist in culture, use of cultured cells would present a further advantage. Cells which have not only survived but have managed to grow in a culture medium (normally at temperatures near 37°C) may be presumed to be in a stable condition, at least in terms of a time scale of weeks, rather than hours.

2. Primary and secondary cell cultures

Zeidler and Willis (84) grew primary cultures of kidneys from adult hamsters, ground squirrels and guinea pigs for three days (probably two or three cell generations) and then exposed them for up to 4 hours at 5°C. The cells cultured from guinea pig kidney lost K rapidly to about 60 per cent of that at 37°C. Cells of hamster lost only 10 percent of their K and ground squirrels about 20 per cent. Later studies showed the high retention in hamster primaries persists for at least 48 hours.

All the cells of primary cultures of fibroblasts grown from fetal hamsters survived continuous exposure at 5°C (judged by exclusion of eosine dye) for at least seven days and when returned to 37°C were able to proliferate and produce new colonies (5). Indeed, surviving hamster cells grew after transfer even after 10 days in the cold. In contrast, more than 20 per cent of the cells of fibroblastic cultures grown from human embryos were unable to exclude eosine dye after only three days in the cold and after seven days even the surviving (dye excluding) cells could not grow (5).

Thus, in primary cultures from both adult and embryonic sources, when criteria are used which are related to the cell's own survival rather than to its differentiated function, species-related differences in cold tolerance can be shown to persist. What about tissue-specific differences? Baudysova (5) compared survival and regrowth of primary cultures of skin with a core tissue (lung) from human embryonic fibroblasts and found that the skin cells survived better (i.e., significant cell death after seven days of cold exposure instead of after three days) and were able to regrow after longer periods of cold exposure.

Turning to the question of persistence of tolerance following transfer and subculturing of primaries, Baudysova and I measured dye exclusion and K retention in fibroblastic secondaries of whole hamster embryos and of human embryonic lung (76). The species difference observed with primaries was still apparent, but the hamster lines were more variable in their resistance than primaries had been, one or two showing extensive cell death even in three days. We also compared early with late passages of both hamster and human lines of fibroblasts and found no further changes related to "age" of the line (i.e., cold-sensitive hamster lines remained cold-sensitive, and cold resistant lines remained cold resistant).

So far as comparisons are possible, the cold tolerance of the three-day primaries of kidney cortex of hibernators seemed to resemble that of freshly excised tissue. The results with cells from embryonic sources suggests that the determination of cold tolerance occurs during the early differentiation of tissues. Further, more extensive and more detailed studies may fracture this happy, though fragile, picture, but for the moment at least these two findings would seem to dispel the bugaboo of "dedifferentiation" as a limitation of the application of the cell culture technique to the problem of analyzing cellular differences of cold tolerance. To be sure, in the kidney cortex, some dedifferentiation of epithelial membrane function does occur (the extent and nature of which we are currently investigating, ref. 6), but at least in primary cultures the mechanisms of cold tolerance are deeply engrained.

[A related, lingering issue is whether in either hibernators or non-hibernators the fetal or neonatal tissues are more cold-tolerant than the corresponding adult tissues. This question, arising from old observations of greater cold tolerance in neonatal whole organisms than in adults of the same species, has never been adequately explored. It is surprising, for example, that no one has yet attempted to repeat the main experiment of Lyman and Blinks with perfused hearts of newborn or even fetal animals. Such an experiment, or some modification thereof, such as with isolated atria or sinus venosus, would seem to be entirely feasible, though admittedly requiring some delicacy of touch! The only relevant study of which I am aware is that of van Rossum (67) who found that liver slices from fetal rats at between 17 and 19 days of gestation showed little loss of K or gain of Na or water at 1°C whereas liver slices of adult, newborn and fetal rats after more than 19 days of gestation lost K and gained Na and water when incubated at 1°C.]

3. Established cell lines

The survival of aneuploid cell lines in the cold has been studied extensively by Holeckova and her co-workers in the Tissue Culture Laboratory of the Czechoslovak Academy of Sciences. Both in terms of eosine exclusion and of regrowth after cold exposure as criteria for survival, they have found that a human established line (HeLa) survives much better in the cold (5°C) than a diploid line (lung epithelium) (27, 28, 46). The difference is apparently not due to the malignant transformation, since a culture of Detroit-6 cells (human bone marrow, not known to be malignantly transformed) survived one week of cold exposure better than HeLa. Mouse L cells also are very resistant to cold, one "wild" (i.e., stock, unselected) culture survived three weeks of cold exposure with only 20 per cent mortality (28).

It is apparent, however, that even though aneuploid cells may be generally more cold-resistant than cultured diploid cells, there is much variation among sublines of a given type. For example, in the HeLa cells compared with human lung epithelium above, HeLa mortality at 5 days was only about 10 per cent whereas in the culture compared with Detroit-6 cells it was 70 per cent in one week. Similarly, survival of "wild-type" L cells of various cultures is widely varying (81). Such variations render meaningless any attempt to compare cells of different species.

In my laboratory, for example, Foster once attempted to compare the cold resistance of BHK (an established, near-diploid fibroblastic cell line derived originally from hamster kidney) with that of a selection of other cell types from non-hibernating species. At first the results were very encouraging (for a hibernation-chauvinist), cells of BHK placed at 5°C for four hours lost virtually no K whereas all the others did. (For example, in Girardi human heart cells K after 4 hours in the cold was 70 ± 5 per cent of that at the start. In BHK the value after 4 hours in the cold was 98 ± 5). Several months later it was necessary for another purpose to check again the K retention in BHK, and to our surprise we found that it now varied enormously and in some cases was as low as 65 per cent (78 ± 3 per cent for 10 subcultures in calf serum). No amount of fiddling with serum source (fetal vs. calf, suppliers, etc.) or other details of our procedure could recover the original, high, consistent retention of K.

Such variability suggests that the degree of cold adaptation of established lines is the result of genetic variation and selection. Possibly the apparently greater survival observed by the Prague group is a result of the hundreds of transfers which any cell line has experienced and which require fluid changing, trypsinizing, centrifugation and other operations at room temperature.

Holeckova and her co-workers examined in some detail the effect of conditioning exposure to cold on aneuploid cell lines and in every case found that repeated cold treatment led to a large increase of survival during subsequent cold exposure (27, 28). For example, they found an increase from 25 per cent survival in an unconditioned strain of mouse L cells exposed to cold for one week, to 90 per cent survival in a conditioned strain, and a similar improvement for a control and conditioned subculture of HeLa. By contrast, no improvement with conditioning was observed in human diploid cells (27, 46).

While one cannot rule out some physiological adaptation playing a role in such results, it seems likely in view of the long conditioning intervals (multiple weeks) with consequent cell death, that selection is the major factor and that the result is another reflection of the heterogeneity of intrinsic resistance discussed previously.

4. Summary

Taken together, the experiences with cells in culture show that the earlier results with incubated heart cells notwithstanding, large species-related differences in cold-sensitivity do persist for at least several days in isolation from the organism even after growing for several generations in warm medium and without humoral factors from the donor. The technique of cell culture would therefore appear to be suitable for investigating, at the level of intact cells, mechanisms of resistance to acute cold exposure and also for testing either the direct effects of longer term or repeated cold exposure or the effects of humoral or extracellular agents which in the whole organism may impinge on cells and affect their resistance. Recent, as yet primitive, examples of such applications are provided below.

With multiple passages, other factors may come into play. In cultures of cells from hibernators less resistant cells may overgrow and dominate, whereas, at least in long established, aneuploid lines from cold-sensitive organisms and tissues, selection for cold resistant cells may also occur. On the other hand, neither age of diploid cultures nor malignant transformation of aneuploid ones seems to play a detectable role.

D. Acclimation

In the preceding section evidence is given that prior exposure to cold can select resistant cells from among a growing and heterogeneous population, but even in nongrowing populations, including those of the tissues of an organism, there is a question of the extent to which prior experience may affect cold sensitivity. This issue has in fact mainly been raised and studied within the context of experiences of the whole organism, and, again, we can find a good early example in the paper of Lyman and Blinks. They attempted to determine whether hibernation itself affected the cold sensitivity of hamster hearts, and they did so in two ways. They removed hearts from actually hibernating hamsters and from hamsters which were awake prior to use but which had recently hibernated. In neither case was there any significant difference from hearts of awake hamsters in minimum temperature at which beating was sustained.

Despite this negative result, numerous other cases of improvement of performance or maintenance of cellular function in hibernation have been reported. Indeed, in the hamster heart itself, changes in temperature dependence of membrane potentials and of metabolism of both intact tissue and isolated mitochondria have been described (37, 61, 62). Similarly, in preparations from hibernating hamsters, compared with those from awake hamsters, kidney cortex accumulates K more rapidly, the phrenic-nerve diaphragm neuromuscular junction function at a lower temperature, and cold-induced swelling in diaphragm muscle is less (for references see 73, 78).

What about acclimation in mammals that are not "deep" hibernators? Hudson (30) found that when *Peromyscus* were kept at 12°C their hearts, when removed and perfused had cut-out temperatures below those of deermice of the same species acclimated to 33°C. Miller and Irving (49) found that the temperature at which rats were housed appeared to affect the conduction properties of their caudal nerves *in vitro*, but there was no improvement of nerve function at low temperature.

Such results raise the question of whether the influence of prior experience of low temperature is direct or whether it is mediated through the organism. Actually, there are two separate issues contained within this question. The first is whether the effect of prior exposure to lowered temperature is exerted directly upon the tissue or whether it is mediated by neural or humoral signals. The other issue is whether the change in resistance of the cells is a result of the cell's own intrinsic activity (synthesis of new proteins, for example) or whether it is the result of factors imposed on it from the milieu (change in fatty acids available to it, for example). Thus, in short, is the cell the receptor of the conditioning change in temperature or is it the effector of the response?

Taking the latter issue first, results (see Behrisch, this symposium) with the pyruvate kinase of liver of ground squirrel and bats in which the regulatory kinetics of its molecular effectors are different in the summer and winter form and apparently involve synthesis of different peptides adequately demonstrate that seasonal changes can be part of a cell's genetic repertoire. On the other side of the coin, there are cases of exogenous nonessential factors causing changes in the resistance of cells in culture to low temperature. Michl and Svobodova (47) found that inclusion of uridine in the incubation medium of unadapted mouse L cells improved their survival in the cold to the level of that of cold adapted cells. Similarly, Matsumura *et al.* (45) found that serum in the medium improved the survival of cultured cells of a mouse fibroblast and of a rat liver parenchymal line at 4°C.

With regard to the detection of the conditioning temperature change there are two possible approaches; to find tissue changes in the absence of change in body temperature or to show that change of temperature *in vitro* can improve the subsequent performance of non-growing tissues. With regard to the first point, it is worth noting that several of the examples of acclimation were found in hamster, and this is a particularly favorable species because in captivity Syrian hamsters only hibernate when exposed to cold and characteristically wait for an interval of three to six weeks in the cold before beginning to hibernate. Hence one could study the time course of such changes and judge by whether they appear before or after the onset of hibernation whether they could result from direct effects of temperature on cells or are the result of the organism's response.

Unfortunately this has not been done for any of the cases mentioned, but as discussed in an earlier review (78), the changes in Na-K ATPase activity of brain and kidney, associated with hibernation in hamsters occur before the hamsters hibernate, do not progress further after the animals have started to hibernate, and do not occur in hamsters which have remained a long time in the cold without hibernating. So, unless the hamster is sneaking periods of low body temperature in the prehibernation preparatory period, which no one has noticed, these results permit only one conclusion, that in these cases at least the acclimation represents an organismic response to the cold.

With regard to the other approach, *in vitro* acclimation, I know of no clear case of this for a mammalian tissue, but at least one study bears consideration as a possibility. Exposure of embryonic hamster fibroblasts (secondaries) to cold for 3 days led to better survival during subsequent cold exposure, an effect not observed in human diploid cells (46). This result could of course be a result merely of cell selection as was previously proposed for similar results with aneuploid cells, but in this case there was only a single exposure for a short period, instead of the long and repeated exposures required to produce "adaptation" in the aneuploid cells. Since primary cultures do retain species-specific cold resistance and secondary cultures seem to lose this to a varying extent, it may be that in secondaries the expression of this property can still be elicited by cold treatment.

E. Summary of Part 1

It appears that tolerance of low temperature is part of the genetic heritage of a wide spectrum of mammalian cells. In part this arises from the expected correlation of tolerance with the necessity to experience widely varying temperatures, either because of the location of the tissue within the organism or because of the thermal habits of the organism as a whole. But in part it also occurs as unexpected resistance even in tissues not normally called upon to face the threat of cold. Heterogeneity of resistance can be found even within a central core tissue and continues to be expressed in isolated cells grown for generations in culture. The degree of intrinsic tolerance of any given cell seems to be part of its early program of differentiation, but the expression of that tolerance can be modified at least by organismic influences and in addition extrinsic factors may enhance tolerance. Hence the clear demarcations observed by Lyman and Blinks represent the extremes of perhaps several continua, and differences between hibernators and non-hibernators and between core and peripheral tissues may represent more a diversity in the expression, than a diversity in the possession of the properties underlying tolerance.

II. CAUSES OF COLD DEATH OF MAMMALIAN CELLS

Since, to such a large extent, characteristic intrinsic tolerance or sensitivity to low temperature persists in cells and tissues isolated from the organism, one can consider the mechanisms of cold death (or avoidance thereof) as a question separate from that of death of an intact mammal in hypothermia. To do so, however, requires recognition of a distinction not hitherto made in this discussion. "Survival of cells" has been judged, both in this review and by investigators, either by criteria of cell function or - in the case of cultured cells - by their ability to reproduce a viable cell population when returned to high temperature. But of course cells may themselves survive, and yet still lose reproductive capacity.

Death of cells in the cold involves the upsetting of a steady state, loss of reproductive capacity upon rewarming involves the interruption of a progression. Correspondingly, the candidates nominated as critical limiting factors in the cold are those of minimal maintenance of that steady state (metabolism, ion regulation), whereas the factors considered crucial for loss of reproductive viability are those concerned with the cell cycle. In pursuing mechanisms of cold injury in isolated cells, it is necessary to enquire not only whether the source of damage can be isolated to a single unitary effect within either of these two "realms" (death in the cold and reproductive death after rewarming), but also whether the dichotomy between these "realms" is real.

A. Reproductive Death after Cooling

The key events of interest for progression in the cell cycle are duplication of DNA, segregation of chromosomes and cell division, and the effect of low temperature upon each has been investigated. Cooling does indeed cause mitotic block and delay in division upon rewarming (15, 51, 65), non-disjunction of chromosomal breaks and other aberrations (25), perhaps arising from partial or imperfect DNA synthesis in the cold (15, 60, 70).

Are these acute effects of low temperature the source of later damage? The presence of chromosomal aberrations and a tendency toward higher ploidy in lines surviving repeated cold exposure (10, 27) suggest that such effects are not necessarily irreparable, but, on the other hand, loss of reproductive capacity has been correlated with cold-induced hyperploidy (51) and loss of chromosomes through non-disjunction can hardly be salutary.

Two further lines of evidence support the notion that these replicative effects are critical. First, the errors in chromosomal duplication lead to the prediction that serious damage is caused in late G_1 or early S (25, 53, 70), and, accordingly, with synchronized cultures Nelson and Kruuv (53) found that cells at these stages exposed to 5°C were the most vulnerable in terms of loss of colony forming ability. Secondly, the loss of colony forming ability is an *inverse* function of cooling (54). That is, to produce a given loss of viability requires less time at a high suboptimal temperature than at a low one. This observation is in accord with the fact that most of the replicative effects are severe even at moderate temperatures such as 25°C or 15°C (51, 57, 60, 65, see, however, 25).

With regard to whether the dichotomy between reproductive death and death in the cold is real, it would be difficult to reconcile the last observation (inverse effect of temperature) with an effect upon energy and mass balance of the cell. Although there is too little information on this point, it seems very unlikely that such factors as ATP levels and ion concentrations can be very far out of balance at such moderate temperatures. Provisionally, at least, therefore, we must conclude that loss of reproductive capacity following cooling is a somewhat separate issue from that of death in the cold.

Even as a separate issue, however, it has important implications for the fuller understanding of the survival of hypothermic bouts, whether forced or voluntary and whether in cold-tolerant or cold-sensitive organisms. In the past most of the attention of those interested in hibernation and hypothermia has been focussed on survival during the period of low body temperature, rather than on the sequellae of recovery. This attitude would be appropriate if maintenance of cellular steady state of energy and mass were the only problem, because in that case survival during the cold and survival during recovery should be functions of the same set of events. But if loss of cellular reproductive capacity occurs as a consequence of other factors and furthermore has a different relationship to temperature, then a new series of issues and questions arises.

One must enquire anew, for example, whether cold-tolerant forms are more resistant in this particular respect than are cold-sensitive ones. It is conceivable that the survival of even hibernators in their habitat during their active periods is adversely affected by their experience of low body temperature, because of debilitation of the immune system, the blood forming organs or the intestinal epithelium. The results cited in Part I on cultured cells militate against this possibility, but it has, so far as I know, hardly been ivestigated in intact organisms. In any case, the nature and severity of the cellular problems faced by organisms enduring shallow torpor, and any adaptations mitigating those problems, will be qualitatively different from those faced by cells of organisms that undergo the more profound cooling of deep hibernation. (Are dividing cells of *C. tereticaudus*, say, more resistant to the effects of cooling on replication than are those of *C. tridecemlineatus*?) If adaptations do exist against these difficulties their mechanisms would have to be quite different from those which protect the cells of the organism from death in the cold. (For example, could the block in G_2 which occurs during hibernation or during cold exposure of peripheral tissues (1) -in addition to the more common metaphase block - be a protective mechanism reducing the number of cells entering mitosis in the cold?)

B. Death in the Cold

With regard to the survival of function of nongrowing tissues, the two candidates most often nominated as critical to the maintenance of the steady state in the cold are energy metabolism (specifically, the production of ATP) and ion regulation (specifically, the maintenance of K, Na and Ca gradients). Each is essential for most of the other activities of cells and, confusingly, each is essential for the other. Even so, the case for connecting either of these with cell death is fragmentary.

1. The ion hypothesis

To begin with the case for ion regulation,

the discussion must first be limited to those ions for which membrane mechanisms are known. Cytoplasmic regulation of H ion and its behavior with change in temperature in mammalian cells are subjects of major significance, only just beginning to receive adequate attention (42). Regulation of calcium ion is only a little better understood. Cytoplasmic Ca of mammalian cells is regulated in part by plasma membrane mechanisms either analogous to or coupled with that of Na/K regulation (8, 75); it increases at temperatures near 0°C, and its rise has various potentially deleterious effects on metabolism and the regulation of other ions (for references, see 68, 75).

The role of Na/K gradient has been somewhat more fully investigated, and the mechanisms of its regulation are better understood. Its effects and cellular roles are also well studied. The best known of the latter is the membrane potential, and it is therefore surprising that in our premier tissue, the cardiac muscle, decline of Na/K gradients has yet to be implicated as the cause of depolarization and conduction block in the cold. In my own laboratory we have been unable to detect significant change in K content of rabbit atria after blocking the beat with cooling. The K and Na concentrations of perfused rat heart after 2-3 hours at 2-6°C are not significantly different from those at 38°C, but they are different from those at 22°C (63). Guinea pig atria held at 5°C for 90 minutes do lose K and gain Na (20), but depolarization would still be faster than these observed changes in concentration. Part of the membrane potential of the heart may arise from the electrogenic activity of the Na/K pump (20, 71) (i.e., pumping more Na out than K in), and some depolarization at low temperature, therefore, could be due to the immediate inhibition of the pump. Such an effect would lead to eventual decline of gradients, but depolarization and blocking would precede that decline.

As documented in Part I, there is no doubt that K concentrations do fall (and Na rises) in many cold sensitive tissues at low temperature. The problem then is to relate this event with other criteria of mortality. Slices of kidney cortex from hamster stored at 5°C for four days retain K better and when rewarmed can reaccumulate K and incorporate amino acid into protein better than those of guinea pig (75). If the cells of hamster kidney, however, are selectively depleted of K in the cold, their accumulation of K and their capacity for protein synthesis after storage is depressed like that in the guinea pig kidney tissue. In cold adapted lines of mouse L cells, Willis and Holeckova (81) found that K retention during exposure to 5°C was better than in the companion unadapted stock line. In several lines of diploid hamster embryonic fibroblasts Willis and Baudysova (76) found that loss of K not only correlated well with survival in the cold, but that loss of K anticipated cell death as measured by dye exclusion.

Thus, so far as it goes, the evidence is consistent with a connection between K retention and survival of cells during and after cold exposure, but it is largely correlative. For the hypothesis to be more fully established requires two further elements. First, it must be shown that otherwise cold-sensitive cells can be made to survive cold better merely by keeping their K concentration from falling (and Na from rising). Most attempts to achieve this objective have so far involved the elevation of extracellular K, a method which has the serious disadvantage of causing swelling and depolarization (36).

The second requirement is crucial and just as difficult of attainment, namely, the demonstration that loss of K (or gain of Na or Ca) is not a consequence of the failure of metabolism to yield ATP.

2. <u>The ATP hypothesis</u>

This is a difficult task partly because ion regulation and metabolism are mutually dependent and partly because assessment of the energy status of the cell is complex. How, for example, should one interpret the lower temperature sensitivity of respiration of slices of cardiac muscle of hamster compared with rat (62)? Mutual dependence prevents a choice between energy releasing pathways and ion regulation (73).

To circumvent this problem one may try to measure mitochondrial respiration in cell-free suspensions with constant ionic milieu provided, and in this particular case the species difference persists -

when pyruvate is provided as a substrate and not succinate (59, 61). In other cell-free preparations results are even more conflicting (for liver compare 37 with 55). Rather than circumvent the mutual dependence one may try instead to measure and account for the temperature sensitivity of metabolism connected with membrane activity, as has been attempted in brain and kidney slices (58, 74). Such attempts have so far been unavailing, partly because the tissues chosen were not as different in temperature sensitivity between species as originally thought (see Parts I and III).

A more direct and possibly more useful approach has been to examine ATP concentrations of cells at different temperatures. Thus, in red blood cells kept at 5°C, K falls even when conditions are provided to prevent a fall in ATP (83). I have recently found that under conditions which lead to the large immediate loss of K at 5°C in cultured kidney cells of guinea pig, ATP fell by no more than 10 per cent by 4 hours in the cold. Such results while tending to minimize the importance of metabolism in cold failure, are also subject to two criticisms. They do not take into account the possible effect of compartmentation (ATP may be high, but trapped, in mitochondria, poor in cytoplasm) nor the presence of a pool of other high energy phosphates nor the total "energy charge" ($ATP/ADP + P_i$) (4). Burlington *et al.* (9) found, for example, that although ATP level falls in ground squirrel hearts at low temperature (6°C) energy charge does not, whereas in rat the converse was the case (i.e., ATP was not lower at the temperature of blocking, about 15°C, but energy charge was). Taken in conjunction with the results on repsiration in heart such findings certainly suggest that at least cardiac muscle of non-hibernators have a problem maintaining metabolism at low temperature, but, even so, one must enquire whether the comparatively modest limitation of energy observed is sufficient seriously to hinder activities essential for cell survival at the low temperature.

III. ANALYSIS OF MEMBRANE FAILURE

Although, as we have seen, documentation of the role of plasma membrane as the critical site of temperature effect is far from complete, there is a straightforward and practical reason for exploring the mechanisms of membrane failure at low temperature. Aside from post-storage functional tests, our ability to judge whether a cell or tissue has survived or is surviving in the cold is based to a very large extent directly upon tests of the integrity of the plasma membrane (excitability, polarization, ion gradients, exclusion of stains or other markers, swelling, hemolysis, loss of enzymes, etc.). Searching for the causes of the loss of this "integrity" is a more defined and manageable - and therefore more hopeful - undertaking than that of trying to understand the larger but vaguer issue of cell death.

The search so far has been in two parts. My own approach has been to try to dissect membrane function, starting at the level of the whole cell and isolating and identifying the roles of the component activities (metabolism, ion pumping, passive leak). In contrast to this reductionist or "hierarchical" approach, others have attempted a more frontal assault of the constituent molecules. Since I summarized the "hierarchical" approach at the last hibernation symposium (78), and since Dr. Charnock will be discussing at least one aspect of the "molecular" approach in this symposium, my purpose here will be confined to bringing the one up to date and the other, down to earth.

A. Hierarchical Analysis Revisited

My version of hierarchical analysis focuses on the Na/K balance of cells, which is central to most other membrane activities, and consists of a series of dichotomous choices. Thus, assuming balance of K ion (the easier ion to analyse in most cases) to be a steady state arising from an equality of influx and efflux, the first question is whether net loss of K with cooling results more from a stringent inhibition of influx (particularly the active, ouabain-sensitive component of influx) or from less inhibition of passive efflux. Then, if stringent inhibition of the pump is a factor, the next question is whether that is a result of direct inhibition of its molecular mechanism or of inhibition of metabolism upon which it depends for ATP. Finally, for the moment, is the question of whether the stringent inhibition of the pump is simply due to slowing (i.e., a decrease in the maximum attainable rate of activity) or due to changes in its response to substrates (K, Na, ATP). In making such comparisons and judgments, hibernator cells, with their superior ability to maintain ion gradients at low temperatures, are the standard against which the specific failure of membranes of cold-sensitive cells is measured. We have sought to resolve this list of questions primarily in two types of cells, erythrocytes and kidney cells, from two species of hibernators (hamsters and ground squirrels) and two species of non-hibernators (rats and guinea pigs).

As presented in 1971 (78), the picture seemed fairly tidy. In both cell types both pump and leak were involved. That is, in hibernator cells, the pump was less inhibited, and the leak (e.g., K efflux) more inhibited with cooling than in cells of non-hibernators. The stringent inhibition of pump by cold in non-hibernators was clearly direct and was not necessarily mediated through inhibition of metabolism. Kinetic analysis of pumping had not been done, but analysis of partial reactions of the Na-K ATPase had begun (78, see also 22).

Since that time the picture has become more complex, in particular the similarity between kidney cells and erythrocytes has declined and the usefulness of the Na-K ATPase has eroded.

Both of these trends hinge on the studies of primary cultures of kidney cells from hamster, ground squirrel and guinea pig (84). Not only did these cells appear to retain the characteristic cold adaptation (or sensitivity) of the donor tissue with respect to K retention at low temperature, but they also permitted meaningful analysis of unidirectional fluxes of K, as a function of temperature, something not feasible in kidney slices. These fluxes revealed that although the difference between species was still apparent with regard to passive loss of K at lower temperature, the active (ouabain-sensitive) component was surprisingly little inhibited in the guinea pig cells at 5°C (7 per cent of the rate at 37°C, compared with 9 per cent in active ground squirrel and hamster). Hence, in this tissue it no longer appeared so certain that there is a large difference in cold sensitivity of the pump as we had previously thought and as there is in erythrocytes (34).

These results resolved a paradox arising from earlier observations with kidney slices of guinea pig in which ouabain-inhibited, Na-stimulated respiration flourished at low temperature even as net K uptake was failing (74). Assuming that renal tubular cells behave like their cultured progeny, we may now infer that active K transport was still occurring, but that it was losing ground to the leak.

Similarly, the new results with unidirectional fluxes reinforced an old finding (82) that cold inhibits actual *in situ* transport much less than it does the activity of the ouabain-sensitive Na-K ATPase in cell-free membrane preparations. This conclusion has now been reconfirmed both in Na-K ATPase of cultured kidney cells of ground squirrel and in erythrocytes of guinea pig, woodchuck and ground squirrel (16,77). The extraordinary inhibition of Na-K ATPase with cooling seems to involve the intactness of the membrane, since the difference disappears in reconstituted erythrocyte ghosts in which ATPase and flux can be measured simultaneously (77).

The earlier conclusion that inhibition of pumping was independent of metabolism rested in small part on the results with guinea pig kidney slices and in very large part on the high cold-sensitivity of Na-K ATPase (7, 12, 78, 82). Neither result seems fully to reflect the true state of affairs at the membrane of the intact cell, and to that extent that conclusion is weakened. It may nevertheless still be true, as discussed in the previous section, but it will take more effort to re-establish it.

In the meantime, until the correspondence between the thermal properties of the Na-K ATPase and those of *in situ* ion transport can be established, either conceptually or experimentally, a safer recourse would seem to be a more detailed kinetic analysis of the Na-K pump in the intact cell. Of the two types of cells with which we have worked, erythrocytes hold the greater promise in this direction because of the greater temperature sensitivity of their pump in non-hibernators and because they are highly suitable to manipulations of ion concentrations on both sides of the membrane. Dr. J. C. Ellory of Cambridge University and I have been collaborating in this work, and so far we have found that decrease in activity at low temperature is not due to a loss in response of the pump either to external K ion nor to internal Na ion (85). On the contrary, affinity for both increases in guinea pig red cells with cooling, and the relative increase is as great as or greater than that in red cells of ground squirrels. (Essentially the same may be said of kidney cells except that in those of ground squirrels an increase in response to altered cytoplasmic K and Na was seen at low temperature which requires further investigation, ref. 84.) Although the analysis is not complete, it seems likely that at least as far as ion stimulation is concerned the loss of activity with cooling and the difference between species will emerge as change in maximal rate ("V_{max}") and not as a change in affinity for the transported species of cations.

B. The Visitation of Molecular Analysis

There is an obvious dichotomy of molecules between which to choose a candidate for differences in the ability of the plasma membrane to function at low temperature, namely, lipids and proteins. The domains of possible interactions and sites of interest were outlined by Gitler (19) at the last hibernation symposium: lipid-lipid, protein-protein, lipid-protein and interactions of both proteins and lipids with water. Since that time interest with regard to the problem of temperature in membranes has focused almost exclusively on the contribution of lipids. The hypothesis that it is differences in fluidity of the membrane as a function of temperature that govern differences in temperature sensitivity of cells and that it is lipid variables (saturation of fatty acids, relative proportion of cholesterol, presence of lysophosphatides) that govern the fluidity, emerges from several roots. These include the effect of membrane fatty acids on activity of membrane enzymes (for review see 18, 26), the effect of temperature on various parameters of viscosity and order in natural and artificial membranes (19), and numerous comparative studies of changes in saturation of fatty acids during cold adaptation (e.g., 17, 32).

Though the notion was growing for a long time, one seed of the recent crystallization was the work of Raison and Lyons (41, 55, 56) showing that the difference between cold-hardy and -sensitive plant and animal cells is often reflected by discontinuities in the Arrhenius plots of membrane-related enzyme activities of the sensitive forms, which could easily be attributed to phase transitions.

With regard to hibernation, changes in saturation of specific fatty acids in heart of ground squirrel (2), brain of hamster (21), and kidney of hamster (72) have been observed with onset of hibernation, but far more striking were the changes in plasmalogens of the brain of hamster and the increase of lysophosphatides in the heart of ground squirrels and in kidney of hamsters. One of the cases cited by Raison and Lyon (55) was a discontinuity in the Arrhenius function of liver succinoxidase of ground squirrels which disappeared with hibernation, but with apparently comparable techniques this was not seen by Liu *et al.* (37).

With regard to specific membrane mechansims under consideration here, several studies indicate that fatty acids in phospholipids affect the temperature sensitivity of Na-K ATPase (33, 66) and correlate the temperature break in activity with a change of order in the membrane (11, 24).

Further review than this hardly seems necessary here, not only because of Dr. Charnock's contribution but also because of the great, possible excessive, popularity of these notions.

It seems inescapable that membrane fluidity, or phase boundaries or fatty acid saturation or *something* has *something* to do with the performance of membranes at low temperature. The question is, what, specifically? To close this discussion I would like to draw attention in this regard to three issues which in the context of the subject of this paper have been neglected or ignored.

First, with respect to hibernators, it must be realized that the only differences that have so far been identified in lipids and membrane fluidity have been between active and hibernating individuals of the same species, whereas, as we have seen, the main differences in resistance are usually between species. Extending the comparison of phospholipids and fatty acids to species, admittedly, will be difficult because any compositional differences observed might be unrelated to differences in temperature sensitivity. Comparisons of changes in microviscosity or order with temperature, would however, still be of use. (In this regard it should be noted that the only differences in physical behavior measured by a membrane probe have been found in mitochondria, presumably inner membranes thereof, which, lacking cholesterol, are a far more likely candidate for a "phase transition" effect than are plasma membranes.)

Secondly, greater attention must be paid to the question of which functional activities of the membrane are to be included in the lipid fluidity hypothesis. For example, both pump and leak are involved in the cold adaptation of hibernator tissues. Are differences in both pump and leak to be laid at the door of lipid fluidity? How do phase transitions or separations affect passive ion leak in intact membranes? In artificial lipid bilayers greater fluidity increases permeation of cations (33). This is in the wrong direction to that required; would pore-mediated diffusion, perhaps through a protein channel, be affected differently? The work with Na-K ATPase is of interest in characterizing the interactions of lipids with an integral membrane enzyme, but if, as it seems, the acute temperature-sensitivity of the Na-K ATPase is an artifact of preparation and isolation, it may be of little relevance to cold adaptation. In any case, the differences observed in pump and Na-K ATPase with hibernation have been of various types, involving both kinetics and catalytic efficiency variably combined (7, 22, 78, 84). The lipid differences as noted are also of various kinds. It remains to be seen whether the two classes of phenomena are actually connected and if so, in what way.

Finally, assuming that fluidity in the region of an enzyme or of a protein associated with passive diffusion is an important determinant of the temperature sensitivity of the process, we must not leap to the conclusion that differences between sensitive and insensitive cells is necessarily due to differences in lipids. It seems likely that site of the most important effect of fluidity will be in the immediate vicinity of proteins and in that region proteins are as important as lipids in establishing order and determining fluidity.

When the current wave of awakened appreciation of what lipids can accomplish has subsided, a return to the original, and still unresolved, question will lead to a more balanced, orderly, detailed and meaningful application of technology and concepts which fostered that awakening.

IV. SUMMARY AND CONCLUSION OF PARTS II AND III

In setting out to discuss "mechanisms of cold tolerance" we immediately find ourselves considering instead the converse and primary problem, the causes of death resulting from cold treatment. Here we discover that death comes in many forms, either immediately to the exposed cell or ultimately to its capacity to reproduce at the optimal temperature. Although these seem to be separable they nevertheless may interact. Metabolic or membrane damage at temperatures near 0°C may be the cause of enhanced loss of reproductive capacity after exposure to such low temperatures. And inability to duplicate and divide successfully may, as in the case of radiation injury, also lead to the death of the cell itself.

For the narrower question of death or loss of function within the cold we are still left with choices between energy metabolism and membrane function and, within each category, amongst several possible alternative activities. The tedious task of sorting and selecting among these categories has been largely superseded by the enthusiasm for membrane fluidity and lipid composition as a pan-hypothesis, which has the great potential of explaining failures in both realms. But, in view of the compositional complexity of eukaryote plasma membranes, the originally simple idea of phase transition has been succeeded, first by that of phase separation, then by that of separations of transitions independently in the two monolayers of the lipid bilayer, and then by a further focussing upon the local or neighboring lipids nearest intrinsic membrane proteins, a zone which is barely accessible at this time to experimental analysis. In short, that collection of possibilities is little more now than a further set of candidates to be interdigitated with the earlier one.

ACKNOWLEDGEMENTS

Original results described herein were obtained in collaboration with J. H. Becker and R. F. Foster, University of Illinois and with J. C. Ellory, Cambridge University, England. This research was supported by a grant, GM11494, from the National Institutes of Health.

This paper is dedicated to Dr. C. P. Lyman whose early work, as described in Part I, is a keystone of the subject.

REFERENCES

1. Adelstein, S. J., C. P. Lyman, R. C. O'Brien, and S. Ito. Cell proliferation kinetics in the tongue and intestinal epithelia of hibernating dormice. Excerpta Med. Monographs 1: 111-119, 1966.
2. Aloia, R. C., E. T. Pengelley, J. L. Bolen and G. Rouser. Changes in phospholipid composition in hibernating ground squirrel, *Citellus lateralis*, and their relationship to membrane function at reduced temperature. Lipid 9: 993-999, 1974.
3. Andrus, W. deW. The effect of temperature on the beating rate of rodent heart cells in culture. Am. Zool. 9: 589, 1969.
4. Atkinson, D.E. Biological control at the molecular level. Science 150: 851-857, 1965.
5. Baudysova, M. Cold resistance of cultured cells from hamster and human embryos. Cryobiology. (In press).
6. Becker, J. H. and J. S. Willis. Changes in properties of K transport in kidney cells during primary growth in culture. The Physiologist 20:7, 1977.
7. Bowler, K., and C. Duncan. The temperature characteristics of brain microsomal ATPases of the hedge-hog: changes associated with hibernation. Physiol. Zool. 42: 211-219, 1969.
8. Brinley, F. J. Calcium and magnesium metabolism in cephalopod axons. Introduction. Fed. Proc. 35: 2572-2573, 1976.
9. Burlington, R. F., G. A. Meininger, and J. T. Thurston. Effect of low temperature on high energy phosphate compounds in isolated hearts from a hibernator and a non-hibernator. Comp. Biochem. Physiol. 55B: 403-407, 1976.
10. Cerny, M., M. Baudysova, and E. Holeckova. Adaptation of mammalian cells to cold. II. Cold-induced endoreduplication and polyploidy. Exptl. Cells Res. 40: 673-677, 1965.
11. Charnock, J. S., and C. L. Bashford. A fluorescent probe study of the lipid mobility of membranes containing sodium- and potassium-dependent

adenosine triphosphatase. Molec. Pharmacol. 11: 766-774, 1975.
12. Charnock, J. S., D. M. Doty, and J. C. Russell. The effect of temperature on the activity of (N_a + K)-ATPase. Arch. Bioch. Biophys. 142: 633-637, 1971.
13. Chatfield, P. O., and C. P. Lyman. Effects of temperature on the ventral caudal nerve of the rat. Am. J. Physiol. 177: 183-186, 1954.
14. Chatfield, P. O., C. P. Lyman, and L. Irving. Physiological adaptation to cold of peripheral nerve in the leg of the herring gull (*Larus argentatus*). Am. J. Physiol. 172: 639-644, 1953.
15. Chevremont, M., R. Bassleer, and E. Baeckeland. Nouvelles recherches sur les acides desocyribonucleique dans des cultures de fibroblasts refroides puis rechauffees. Etudes cytophotometrique et histoautoradiographique. Localization cytoplasmique d'ADN. Arch. Biol. 72: 501-523, 1972.
16. Ellory, J. C., and J. S. Willis. Temperature dependence of membrane function. Disparity between active potassium transport and (Na + K)-ATPase activity. Biochim. Biophys. Acta 443: 301-305, 1976.
17. Fawcett, D. W., and C. P. Lyman. The effect of low environmental temperature on the composition of depot fat in relation to hibernation. J. Physiol. 126: 235-247, 1954.
18. Gennis, R. Protein-lipid interactions. Ann. Rev. Biophys. Bioeng. 6: 195-238, 1977.
19. Gitler, C. Analysis of membrane structure. In: Hibernation and Hypothermia, Perspectives and Challenges. South, F. E., J. P. Hannon, J. S. Willis, E. T. Pengelley, and N. R. Alpert (editors). New York: Elsevier, 1972, p. 239-280.
20. Glitsch, H. E. Ueber des membranpotential des Meerschweinchenvorhofes nach hypothermie. Pflüg. Arch. 307: 29-46, 1969.
21. Goldman, S. S. Cold resistance of the brain during hibernation. III. Evidence of a lipid adaptation. Am. J. Physiol. 228: 834-838, 1975.

22. Goldman, S. S., and R. W. Albers. Cold resistance of the brain during hibernation. Arch. Biochem. Biophys. 169: 540-544, 1975.
23. Goldman, S. S. and J. S. Willis. Cold resistance of the brain during hibernation. I. K transport in cerebral cortex slices. Cryobiol. 10: 212-217, 1973.
24. Grisham, C. M., and R. E. Barnett. The role of lipid-phase transitions in the regulation of the (sodium potassium) adenosine triphosphatase. Biochem. 12: 2634-2637, 1973.
25. Hampel, K. E., and A. Levan. Breakage in human chromosomes induced by low temperature. Hereditas 51: 313-343.
26. Hazel, J. R. The effect of temperature acclimation upon succinic dehydrogenase activity from the epaxial muscle of the common goldfish (*Crassius auratus L.*). II. Lipid reactivation of the soluble enzyme. Comp. Biochem. Physiol. 43B: 836-882, 1972.
27. Holeckova, E. Adaptation in Cell and Tissue Cultures. Publication Czechoslovak Academy of Sciences. Series of Mathematical and Natural Sciences. Vol. 77, No. 8, Prague, 1977.
28. Holeckova, E., M. Baudysova, and O. Cinnervva. Adaptation of mammalian cells to cold. Resistance to cold and multiplication of L, Detroit-6 and HeLa cells adapted to low temperature. Exptl. Cell Res. 40: 396-401, 1965.
29. Hudson, J. W. Thermal sensitivity of isolated-perfused hearts from ground squirrels, *Citellus tereticaudus* and *Citellus tridecemlineatus*. Z. vergl. Physiologie 71: 342-349, 1971.
30. Hudson, J. W. Variations in the patterns of torpidity of small homeotherms. In: Mammalian Hibernation III. Fisher, K. C., A. R. Dawe, C. P. Lyman, E. Shönbaum, and F. E. South (editors). Edinburgh: Oliver and Boyd, 1967, pp.30-46.
31. Jacobs, H. K., and F. E. South. The effects of temperature on cardiac transmembrane potentials. Am. J. Physiol. 230: 403-409, 1976.

32. Johnston, P. V., and B. I. Roots. Brain lipid fatty acids and temperature accliamtion. Comp. Biochem. Physiol. II: 303-309, 1964.
33. Kimelberg, H. K., and D. Papahadjopoulos. Effects of phospholipid acyl chain fluidity, phase transitions and cholesterol on (Na + K)-stimulated adenosine triphosphatase. J. Biol. Chem. 249: 1071-1080, 1974.
34. Kimzey, S. L., and J. S. Willis. Temperature adaptation of active sodium-potassium transport and of passive permeability in erythrocytes of ground squirrels. J. Gen. Physiol. 58: 634-649,1971.
35. Kimzey, S.L., and Willis, J. S. Resistance of erythrocytes of hibernating mammals to loss of potassium during hibernation and during cold storage. J. Gen. Physiol. 58: 620-633, 1971.
36. Leonard, C. D. and B. D. Scribner. Effect of choline and hypothermia upon H^+ production and K^+ loss from isolated kidney. Cryobiol. 8: 290-292, 1971.
37. Liu, C. C., J. L. Frehn, and A. D. LaPorta. Liver and brown fat mitochondrial response to cold in hibernators and non-hibernators. J. Appl. Physiol. 27: 83-89, 1969.
38. Lyman, C. P. The effect of low temperature on the isolated hearts of *Citellus leucurus* and *C. mohavensis*. J. Mammal. 45: 122-126, 1964.
39. Lyman, C. P., and D. C. Blinks. The effect of temperature on the isolated hearts of closely related hibernators and non-hibernators. J. Cell. Comp. Physiol. 54: 53-63, 1959.
40. Lyman, C. P., and E. L. Jarrow. Effect of low temperature on single heart cells from an adult hibernator and non-hibernator. Exptl. Cell Res. 68: 214-217, 1971.
41. Lyons, J. M. Phase transitions and control of cellular metabolism at low temperatures. Cryobiol. 9: 341-350, 1972.
42. Malan, A., T. L. Wilson, and R. B. Reeves. Intracellular pH in cold-blooded vertebrates as a function of body temperature. Respir. Physiol. 28: 29-47, 1976.

43. Marshall, J. M. Effects of low temperatures on transmembrane potentials of single fibers of the rabbit atrium. Circ. Res. 5: 664-669, 1957.
44. Marshall, J. M. and J. S. Willis. The effects of temperature on the membrane potentials of isolated atria of the ground squirrel, *Citellus tridecemlineatus*. J. Physiol. 164: 64-76, 1962.
45. Matsumura, T., T. Takaoka, and H, Katsuta. Survival of cultured cells in cold. A kinetic study with special reference to the effect of serum in culture medium. Exptl. Cell Res. 76: 297-304, 1973.
46. Michl, J., D. Rezacova, and E. Holeckova. Adaptation of mammalian cells to cold. IV. Diploid cells. Exptl. Cell Res. 44: 680-683, 1966.
47. Michl, J., and J. Svobodova. Adaptation of mammalian cells to cold. VII. Molecular basis of cold resistance in LC2 cells. Exptl. Cell Res. 50: 383-387, 1968.
48. Miller, L. K. Caudal nerve function as related to temperature in some Alaskan mammals. Comp. Biochem. Physiol. 21: 679-686, 1967.
49. Miller, L. K. and L. Irving. Alteration of peripheral nerve function in the rat after prolonged outdoor cold exposure. Am. J. Physiol. 204: 359-362, 1963.
50. Murphy, S., and J. R. Lagando. A Comparison of the effects of prior cold incubation on cerebral cortex function in a hibernator (*Cricetus auratus*) and a non-hibernator (*Cavia porcellus*). I. Oxygen consumption in cerebral cortex slices after *in vitro* cold incubation. Comp. Biochem. Physiol. 55B: 567-570, 1976.
51. Nagasawa, H., and W. C. Dewey. Effects of cold treatment on synchronous Chinese hamster cells treated in mitosis. J. Cell Physiol. 80: 89-106.
52. Nardone, R. M. Electrocardiogram of the arctic ground squirrel during hibernation and hypothermia. Am. J. Physiol. 182: 364-368, 1955.

53. Nelson, R. J., and J. Kruuv. Survival of synchronized mammalian cells following exposure to cold. Exptl. Cell Res. 70: 417-422, 1972.
54. Nelson, R. J., J. Kruuv, C. J. Koch, and H. E. Frey. Effect of sub-optimal temperatures on survival of mammalian cells. Exptl. Cell Res. 68: 247-252, 1971.
55. Raison, J. K., and Lyons, J. M. Hibernation: alteration of mitochondrial membrane as a requisite for metabolism at low temperature. Proc. Nat. U.S. Acad. Sci. 68: 2092-2094, 1971.
56. Raison, J. K., J. M. Lyons, and W. W. Thompson. The influence of membranes on the temperature-induced changes in the kinetics of some respiratory enzymes of mitochondria. Arch. Bioch. Biophys. 142: 83-90, 1971.
57. Rao, P. N., and J. Engelberg. Mitotic nondisjunction of sister chromatids and anomalous mitosis induced by low temperature in HeLa cells. Exptl. Cell Res. 43: 332-342, 1966.
58. Roberts, J. C., and R. R. J. Chaffee. Effects of cold acclimation, hibernation and temperature on succinoxidase activity of heart homogenates from hamster, rat and squirrel monkey. Comp. Biochem. Physiol. 44B: 137-144, 1973.
59. Saarikoski, J. Respiratory responses of cerebral cortex slices from a hibernating and a normal homeothermic mammal to electrical stimulation at 10-36°C. Ann. Acad. Sci. Fenn. IVA. 117: 1-22, 1967.
60. Shapiro, I. M., and E. L. Lubennikova. Population kinetics of cells in tissue culture incubated at low temperature. Exptl. Cell Res. 49: 305-316, 1968.
61. South, F. E. Hibernation, temperature and rates of oxidative phosphorylation by heart mitochondria. Am. J. Physiol. 198: 463-466, 1960.
62. South, F. E. Rates of oxygen consumption and glycolysis of ventricle and brain slices, obtained from hibernating and non-hibernating mammals as a function of temperature. Physiol. Zool. 31: 6-15, 1958.

63. Sprach, C. Action du froid sur les teneurs en eau, en sodium et en potassium du coeur du rat perfuse. Arch. Sci. Physiolog. 23: 291-317, 1969.
64. Suadeau, J., and T. Kolobow. Isolated sheep heart hypothermic (5-13°C) perfusion with fresh blood: successful continous ventricular activity. Cryobiology 14: 337-348, 1977.
65. Sushkov, F. V., and V. V. Portugalov. Temperature limits of mitosis in mammalian cell cultures. Bull. Exper. Biol. Med. (U.S.S.R.) 80: 1491-1496, 1975.
66. Tanaka, R., and A. Teruya. Lipid dependence of activity-temperature relationships of (Na, K) activated ATPase. Biochem. Biophys. Acta 323: 584-591, 1973.
67. van Rossum, G.D.V. Net sodium and potassium movements in liver slices prepared from rats of different foetal and post-natal ages. Biochim. Biophys. Acta 74: 1-14, 1963.
68. van Rossum, G.D.V. Relation of intracellular Ca^{2+} retention of K^+ by liver slices. Nature 225: 638-639, 1970.
69. Volfenson, L. G., N. V. Kesemanly, V. A. Filov, E. A. Andreyeva, and F. L. Losina-Losinskaya. The comparative resistance of tissues of some hibernating and non-hibernating rodents to cooling and supercooling. In: The Cell and Environmental Temperature. Troshin, A. S. (editor). Oxford: Pergamon, 1964.
70. Watanabe, I., and S. Okada. Effects of temperature on growth rate of cultured mammalian cells (L5178Y). J. Cell Biol. 32: 309-323, 1967.
71. Weidmann, S. Heart electrophysiology. Ann. Rev. Physiol. 36: 155-169, 1974.
72. Wilkinson, H. L. Lipid involvement in ion transport of the hamster kidney: its role in cold adaptation. Ph. D. Thesis, University of Illinois, Urbana, 1976.
73. Willis, J. S. Cold adaptation of activities of tissues of hibernating mammals. In: Mammalian Hibernation III. Fisher, K. C., A. R. Dawe, C. P. Lyman, E. Schönbaum, and F. E. South (editors). Edinburgh: Oliver and Boyd,

1967, p. 356-381.

74. Willis, J. S. Cold resistance of kidney cells of mammalian hibernators: cation transport vs. respiration. Am. J. Physiol. 214: 923-928, 1968.

75. Willis, J. S. The possible roles of cellular K for survival of cells at low temperature. Cryobiol. 9: 351-366, 1972.

76. Willis, J. S., and M. Baudysova. Retention of K^+ in relation to cold resistance of cultured cells from hamster and human embryos. Cryobiol. (In press).

77. Willis, J. S. J. C. Ellory, and J. H. Becker. Disparity between active Na-K transport of intact cells and Na-K ATPase of isolated membranes. Am. J. Physiol. (In press).

78. Willis, J. S., L. S. T. Fang, and R. F. Foster. The significance and analysis of membrane function in hibernation. In: Hibernation and Hypothermia, Perspectives and Challenges. South, F. E., J. P. Hannon, J. S. Willis, E. T. Pengelley and N. R. Alpert (editors). New York: Elsevier, 1972, p. 123-147.

79. Willis, J. S., R. F. Foster, and C. L. Behrends. Cold-stored kidney tissue of hibernators: effects of brief warming on K retention. Cryobiol. 12: 255-265, 1975.

80. Willis. K. S., S. S. Goldman, and R. F. Foster. Tissue K concentration in relation to the role of the kidney in hibernation and the cause of periodic arousal. Comp. Biochem. Physiol. 39A: 437-445, 1971.

81. Willis, J. S., and E. Holeckova. Retention of K at low temperature in mouse L cells before and after cold adaptation. Cryobiol. 14: 236-239, 1977.

82. Willis, J. S., and N. M. Li. Cold resistance of NaK-ATPase of renal cortex of the hamster, a hibernating mammal. Am. J. Physiol. 217: 321-326, 1969.

83. Wood, L., and E. Beutler. Temperature dependence of sodium-potassium activated erythrocyte adenosine triphosphatase. J. Lab. Clin. Med. 70: 287-294, 1967.

84. Zeidler, R. B., and J. S. Willis. Cultured cells from renal cortex of hibernators and non-hibernators. Regulation of cell K at low temperature. Biochim. Biophys. Acta 436: 628-651. 1976.

REFERENCE ADDED AT PROOF:

85. Ellory, J. C. and J. S. Willis. Temperature dependence of the cation affinities of the sodium pump in red cells from hibernators and non-hibernators. J. Physiol. (London). 274:80P, 1978.

MEMBRANE LIPID PHASE-TRANSITIONS: A POSSIBLE BIOLOGICAL RESPONSE TO HIBERNATION?

John S. Charnock
Department of Pharmacology
University of Alberta
Edmonton, Alberta, Canada T6G 2H7

ABSTRACT

The effect of temperature upon the physical properties of membrane lipids and proteins is briefly reviewed as is the case for interpreting non-linear Arrhenius plots of enzyme activity in terms of lipid phase-transitions in biological membranes. The thermal profiles of cytoplasmic, mitochondrial, plasma and endoplasmic reticular enzymes from poikilotherms, homeotherms and hibernators are compared. It is concluded that whereas lipid phase-transitions play no part in the temperature dependence of the cytoplasmic enzymes lactic dehydrogenase and malic dehydrogenase, the evidence for pyruvate kinase is not yet clear. There can be little doubt of the important role of the lipid phase-transitions in modulating the temperature dependence of mitochondrial succinate oxidation in homeotherms, poikilotherms and during hibernation. Although there is strong evidence that lipid phase-transitions occur in the plasma and endoplasmic reticular membranes from a wide variety of species, and that in some instances these thermionic transitions occur at temperatures which correspond to marked changes in the apparent activation energy of Ca^{++}-ATPase and (Na^{+} + K^{+})-ATPase, there does not appear to be convincing evidence that these parameters undergo significant seasonal variation in hibernating species. Thus no unifying concept of the role of lipid phase-transitions in the biological response to hibernation can be developed at the present time as it appears to vary with the particular enzyme system and its subcellular membrane localization.

ISBN 0-12-734550-7

I. INTRODUCTION

A. Membrane Structure

There has been much interest in the control of metabolism at low temperature, particularly the low body and organ temperatures encountered in some poikilotherms and during the hibernation state in mammals. Quite recently the idea has been growing that enzyme velocity is powerfully modulated by the physical properties of membrane lipids, and that so-called phase-changes in which the fluidity of membranes is greatly altered may provide the means for continuing enzyme function when the organ temperature falls far below the range at which many biochemical processes are expected to cease.

It is therefore my purpose in this review to examine the experimental evidence available in an attempt to decide what role phase-changes in membrane lipids may play in the process of hibernation. However, before looking at some recent experimental evidence from animal studies we should remember that as early as 1962, Luzzati and Husson had noted that the transition temperature (T_C°) of many solid-gel to liquid-crystalline phase-changes in lipoprotein systems seemed to occur very close to the body temperature of the animals from which the lipoproteins were obtained. They speculated that this observation was not coincidence, but that they were observing a very fine biological control mechanism for electrical potential or ion permeability changes in biological membranes (51).

This observation has much greater significance today than it did in 1962 as there has been great progress in recent years in our understanding of membrane structure.

That biological membranes are basically composed of an oriented bilayer of lipids interspersed with proteins has been known for many years (30, 31, 68), yet it has only been since the work of Singer and Nicolson in 1972 that a really satisfactory description of membrane architecture has been available (71). From the information obtained from a very wide variety of techniques which include chemical and immunological analyses as well as such biophysical

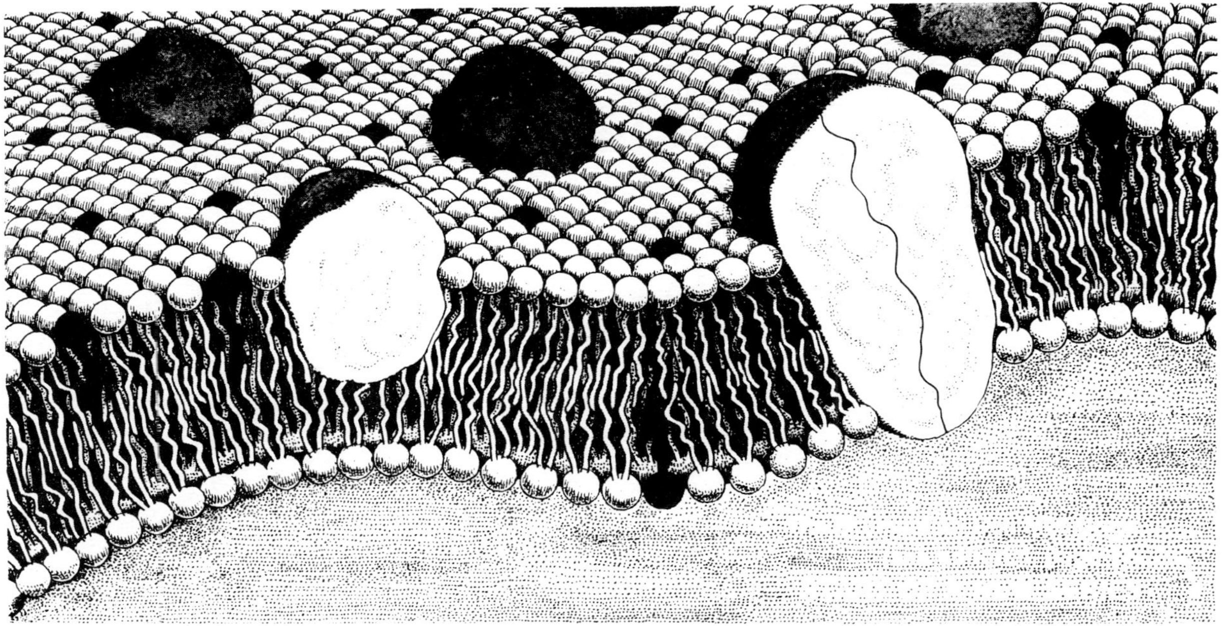

Fig. 1. *The Singer-Nicholson fluid mosaic model of membrane structure. Globular membrane proteins are shown embedded in the lipid bilayer (white) which has occasional cholesterol molecules (black) interspersed between the fatty-acid "tails" of the phospholipids. Some proteins are embedded on only one side of the lipid bilayer with their hydrophobic ends deep within the hydrocarbon core of the membrane, while their hydrophilic ends protrude into the aqueous phase. Other larger transmembrane proteins pass entirely through the lipid bilayer. Illustration adapted from Singer (70).*

procedures of freeze-fracture electron microscopy, X-ray diffraction, circular-dichroism and electron paramagnetic resonance and fluorescent membrane probes, we can be reasonably confident that the now common picture (Fig. 1) of globular protein macromolecules embedded into a bimolecular lipid matrix properly represents the general membrane structure (17, 70).

It is also quite reasonable to suppose that specificity of function takes place not only because of the diversity of chemical composition that can be shown in the lipid and protein components of the system, but also because of the non-homogeneous and asymmetric distribution of these components on either side of the membrane itself (69). In addition, some larger proteins appear to traverse the entire distance of the membrane and possess inherent orientation characteristics of their own (43) which permit vectorial processes to occur.

Furthermore, the work of Edidin and his colleagues (34,36) at Johns Hopkins has dramatically confirmed the dynamic state of the membrane conglomerate in that lateral diffusion of the protein globules in the plane of the membrane is commonplace, while even more recently the work of Chapman has suggested that rotational movement of the protein macromolecules occurs in the microsecond time range (18). In fact, the only molecular movements which appear unlikely are those of the so-called "flip-flop" variety in which either lipids or proteins are translocated from one side of the membrane to the other. These "flip-flop" translocations either do not occur, or the time course of their translocation is so great as to exclude them from consideration in most, but not all, biological processes.

We therefore have a very dynamic view of membrane proteins and hence membrane enzymes functioning in a fairly mobile lipid matrix whose viscosity is subject to many changes in its immediate environment - pressure, temperature, ionic strength and pH, as well as variations which arise as a consequence of changes in the nature (class) and fatty acid composition of the lipid components themselves. Changes in any of these factors are all capable of influencing the physical and chemical properties of the membrane.

In addition, because of the non-homogeneous and asymmetric distribution of membrane components we can now see how one area of a biological membrane may be functioning in a certain mode, whilst another area not too far distant may be exhibiting quite different properties in terms of either permeability, electrical potential and/or enzyme function. It is therefore hardly surprising that vast differences in behaviour are encountered in various membranes from a wide variety of tissues and species as these membranes are required to meet the extraordinary range of situations encountered during the lifetime of the organism, whether it be a relatively simple bacterium or a highly complex multicellular assembly like man himself.

Let us therefore consider only what happens in this complex membrane system when the temperature is varied through the relatively narrow "physiological" range of 0-40°C. One of the most striking features of the behaviour of the lipids in this range is the frequent observation that many of the individual components undergo a change in phase from a highly ordered state in which the hydrocarbon chains of the fatty acids are considered to be relatively rigid with a reduced protein mobility (the so-called solid-gel phase) to a much more fluid liquid-crystalline phase with enhanced protein mobility when the temperature is raised above a certain critical transition temperature (T_c°) (18, 58).

This endothermic transition occurs over a somewhat broader range in biological membranes than in single phospholipid model systems (because of the variety in acyl-chain components as well as polar head-groups of the phospholipids) and is accompanied by a lateral expansion and a decrease in the thickness of the bilayer (3,61).

In this more fluid state the hydrocarbon chains maintain an average orientation perpendicular to the plane of the bilayer but are relatively disordered and undergo rapid rotational isomerization along the chains. This liquid-crystalline state is characterised by the onset of rapid lateral diffusion of both lipid and protein molecules in the plane of the bilayer. One should not forget, however, that both ordered and fluid domains can co-exist at a given temperature and that membrane proteins will tend to preferentially partition into the more fluid parts of the membrane.

B. Arrhenius Analysis - Activation Energy

What of the behaviour of the enzyme proteins embedded in this lipid matrix? How does it alter from the activity of the so-called soluble or cytoplasmic enzymes when the temperature changes from 0-40°C? Before attempting to answer these questions let me briefly review the effect of temperature upon the velocity of any chemical reaction. That the velocity of any chemical reaction is reduced when the temperature falls is taken today as a self-evident axiom, although it was only during the latter half of the nineteenth century when this observation was formally established by van't Hoff who documented that the velocity of a variety of organic reactions increased by a factor of about 2 for every 10°C rise in temperature (74).

However, by 1889 Arrhenius had observed that a number of biological reactions - referred to by him as "life-processes" like the conversion of cane-sugar into glucose and fructose by the enzyme mixture "invertase" -did not properly obey the earlier van't Hoff relationship (5). He therefore studied the invertase reaction in great detail - that is, at a large number of different temperatures - and devised an empirical relationship for the nearly exponential increase in reaction velocity (K) which occurred with the change in absolute temperature (T°) (6).

When the rate constants for the reaction (usually V_{max}) are converted to a log form and the reciprocal of the absolute temperature is employed, a straight line plot (Fig. 2) can be obtained from the Arrhenius equation:

$$Ea = \frac{2.303\ T_1 \cdot T_2\ (\log K_1 - \log K_2)\ R}{T_1 - T_2}$$

where Ea = apparent energy of activation, R is the gas constant (1.987 cals.mole $^{-1}$ deg^{-1}) and K_1 and K_2 are the maximal velocities of the reaction at

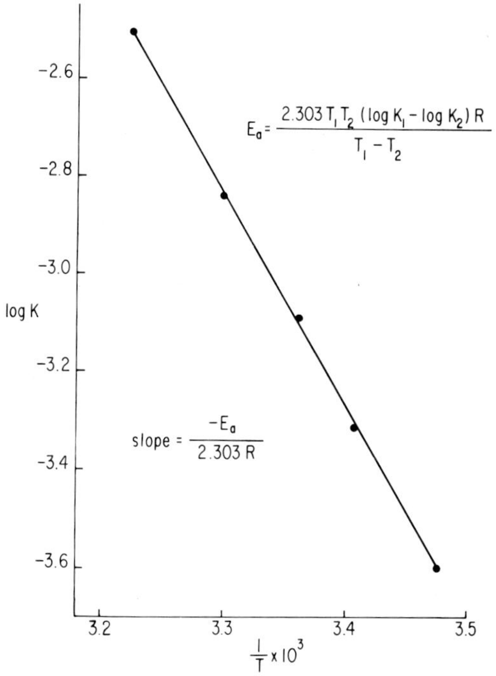

Fig. 2. *Typical linear Arrhenius plot of log initial rate of reaction velocity (K) versus reciprocal of absolute temperature (T°) in degrees Kelvin.*

absolute temperatures T_1 and T_2. Thus the slopes of these plots are equal to -Ea/2.303R and give a measure of the Arrhenius activation energy of the reaction under study. This value is not quite the same as that obtained from direct measurement of the heat content of the reaction, but usually only differs from this value by a few hundred calories/mole (64).

For many years this type of Arrhenius analysis of the effect of temperature upon a biological process was thought to consistently yield a straight line Arrhenius plot and hence a single unique value for the activation energy (Ea), even if it became necessary to disregard some aberrant values or to use what Drost-Hansen (33) has so aptly called "brute force" methods to drive a straight line through fairly obviously non-linear data points. Generally departures from linearity were regarded as deficiencies in experimental data and were not assigned any biological significance (10,46).

However, by the early 1960's sufficient experimental data (Table 1) had been produced for

TABLE 1. PARTIAL LIST OF ENZYMES DEMONSTRATED TO HAVE NON-LINEAR ARRHENIUS PLOTS PRIOR TO 1966

ENZYME	CRITICAL TEMPERATURE ° C	REFERENCE*
Invertase	0	Sizer, et al. (1942)
Trypsin	0	Sizer, et al. (1942)
Pancreatic lipase	0	Sizer, et al. (1942)
D-amino-acid oxidase	8	Massey, et al. (1966)
Phosphorylase b	13	Graves, et al. (1965)
Aldolase	16	Massey, et al. (1966)
Myosin ATPase	16	Levy, et al. (1959)
Fumarase	17	Massey (1953)
β-amylase	20	Piguet, et al. (1952)
Cytochrome C. reductase	21	Vernon, et al (1952)
Salivary amylase	24	Schneyer (1952)
Phosphorylase a	31	Helmreicht, et al. (1964)
Chymotrypsin	34	Havsteen, et al. (1963)
Chymotrypsinogen	44	Brandts (1964)
Ribonuclease	63	Harrington (1956)

* For full details of the references cited see Massey, et al. (56).

some workers to seriously question the absolute truth of the linear "Arrhenius law" and to suggest alternative explanations for the nonlinearity of many Arrhenius plots (56). A typical example of a non-linear Arrhenius plot is shown in figure 3. Naturally, the early heretics who produced such plots were vigorously attacked by the scientific establishment of the day (10,32,46).

Without pursuing this story in detail, it is clear today that in at least some instances, these discontinuities in Arrhenius plots of enzyme activity which lead to more than one value for the activation energy of an enzyme process, are not experimental artifacts but are due to the existence of two different stable conformational forms of the enzyme which can be identified by different sedimentation characteristics, or electrophoretic patterns (56) as well as changes in their thermodynamic properties. In addition, the elegant studies of Raison and Lyons and their colleagues

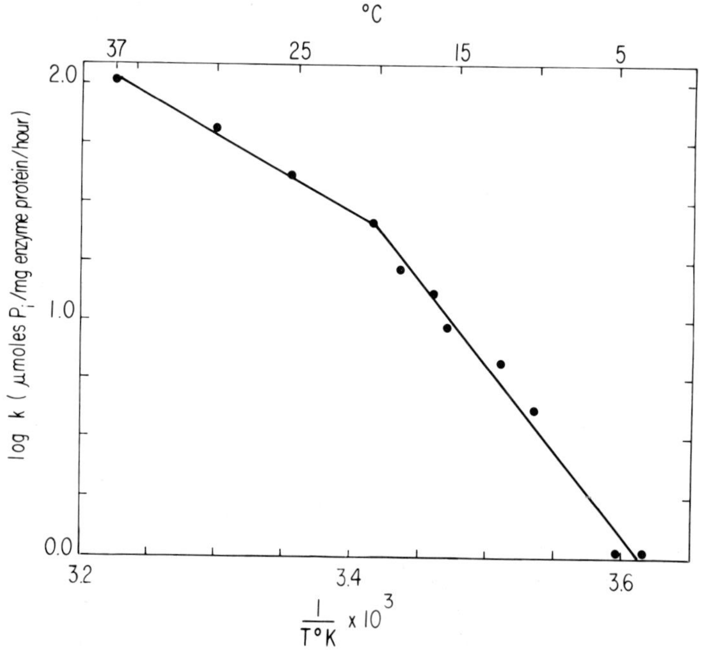

Fig. 3. *Typical non-linear Arrhenius plot of* $(Na^+ + K^+)$*-ATPase activity from 5-37°C. Adapted from Charnock* *et al*. *(22).*

(66) using electron spin probes and those of many others using fluorescent probes of membrane lipid mobilities have shown that quite frequently there is a good correlation between the thermal realtionships of enzyme activity and the physical state of associated membrane lipids (8,20,63).

Although there are still many who do not unequivocally accept this view (32,41,78), there is a great deal of experimental and theoretical evidence to suggest that discontinuities in Arrhenius plots represent thermal transitions between two distinct isothermal phases, which correspond to order-disorder transitions in the mobility of membrane lipids surrounding membrane proteins. In fact, in reviewing the effect of temperature upon cellular metabolism at a symposium of Low Temperature Physiology in 1972, Lyons stated that "one of the most useful and consistent lines of evidence for phase-changes in biological systems has been the presentation of data relating the influence of temperature and reaction velocity (--of enzymes--) as an Arrhenius plot" (52)!

Today there is considerable biochemical evidence that the lipids of biological membranes exert a powerful modulating effect upon membrane enzyme function, and this is particularly noticeable when the temperature of the reaction is changed. By invoking the property of mutual exclusion (49) -that is, that two separate phases exist in the membrane as two separate domains each of constant activity and independent of the amount in each state, we can see how the same enzyme process can function with two markedly different energy requirements in two respective temperature ranges (49,78). The "break" in an Arrhenius plot (T_C°) therefore represents an isokinetic point, that is, the temperature where the value for ΔF^* is the same for the two different transition states (49,64).

With this background let us now look briefly at a number of important enzymes generally involved in energy transduction, and whose temperature profiles have been examined in preparations isolated from homeotherms, poikilotherms and hibernators. Although not an exhaustive list, I will review the information relating to several cytoplasmic enzymes involved in glucose utilization and gluconeogenesis; some membrane-

TABLE 2. ACTIVATION ENERGIES OF CYTOPLASMIC ENZYMES LDH AND MDH DISPLAYING LINEAR ARRHENIUS CHARACTERISTICS

Enzyme	Species	Tissues	Activation Energy kcals/mole	Refs*
	Salmonids	Muscle	10 - 13	42
	Cod	------, liver, heart	15 - 17	60
	Frog	Muscle, liver, heart	11 - 17	60
	#Snake (cold adapted)	Muscle	12	1
LDH	Platypus	Muscle, liver	19	7
	Echidna	Muscle, liver	17	7
	Hedgehog	------, liver, heart, brain	13 - 18	60
	Bat	Muscle, -----, heart	17 - 19	60
	Guinea pig	------, liver, heart, brain	12 - 16	60
	Frog	Muscle, liver, heart	12 - 13	60
	Cod	------, liver, heart	10 - 13	60
	Platypus	------, liver	16.5	7
MDH	Echidna	------, liver	16.5	7
	Avian	------, liver	12 - 15	2
	Hedgehog	------, liver, heart, brain	12 - 14	60
	Bat	Muscle, liver, heart	13	60
	Guinea pig	------, liver heart, brain	12 - 13	60

* Full details of these references are given in the Reference Section to this paper.

The author also describes a non-linear Arrhenius plot having an "inverse break" at 28°C and Ea_I 15.5 kcals/mole and Ea_{II} 6 kcals/mole for a warm-adapted variety of the same reptile.

bound enzymes of mitochondrial oxidative phosphorylation and some plasma membrane enzymes involved in active cation transport.

II. CYTOPLASMIC ENZYMES OF GLUCOSE UTILIZATION

A. Lactic Dehydrogenase (LDH)

In 1968 Hochachka and Somero reported that fish muscle LDH gave linear Arrhenius plots between 5-35°C even when the variety of fish ranged from long-term cold adapted Antarctic fish (-2°C) through lake and brook trout and bluefin tuna (presumably between 10-20°C), to long-term warm adapted South American lungfish whose ambient temperature is normally between 27-30°C (42). In addition, the apparent energies of activation that were obtained for these various muscle LDH preparations were all about 12 kcals/mole and did not correlate with the environmental temperatures of the fish (Table 2).

Similarly, Baldwin and Aleksiuk (7) obtained linear Arrhenius plots for liver and skeletal muscle LDH in the Australian platypus and echidna, both of which are reported to have unusual thermoregulatory properties. The Ea value for the platypus was 19 kcals/mole; Ea value for the echidna was 17 kcals/mole.

However, in an earlier study Aleksiuk (1) had noted an unusual "inverted break" in muscle LDH from a warm-adapted garter-snake collected in Florida, which was not apparent in preparations from similar reptiles collected in Manitoba (see Table 2). Interestingly the Ea value from these latter cold-adapted snakes was again close to 12 kcals/mole.

More recently Borgmann and Moon have examined the kinetics of LDH function from homeotherms (beef heart and muscle) and poikilotherms (flounder muscle) in great detail. Although their purpose was not to obtain data for the usual Arrhenius equation of log V_{max} vs. 1/T°, it is apparent that no major changes were observed in this parameter between 5-30°C, although there were important differences in other thermodynamic properties of these systems between homeotherms and poikilotherms (12,13).

Finally, in another recent study Olsson (60) compared the Arrhenius function of LDH from several organs of a homeotherm (guinea-pig), poikilotherms (frog and cod-fish), and the hibernating hedgehog and bat. Irrespective of the organ (heart, brain, muscle), the species or the season, the LDH Arrhenius plots were all linear between 0-30°C and continued in this fashion for another 20°C in the case of the homeotherm preparations.

The values for Ea obtained by Olsson ranged from 12 kcals/mole for the frog muscle up to 19 kcals/mole for the hibernating bat heart. Again there was no obvious correlation between this thermodynamic parameter and environmental temperature.

B. Malic Dehydrogenase (MDH)

Although less information is available regarding this enzyme than LDH, what is known is very similar to that for LDH (Table 2).

Baldwin and Aleksiuk (7) had found linear Arrhenius plots for liver MDH between 10-40°C for both platypus and echidna, with identical Ea values of 16.5 kcals/mole. This data is very similar to that obtained by Aleksiuk (2) for cytpolasmic MDH from the livers of two species of birds (the common crow and the pintail), where the Arrhenius plots were again linear between 5-35°C, and the values for Ea were 12 kcals/mole and 15 kcals/mole respectively.

In Olsson's major comparative study discussed above for LDH (60) he also found linear Arrhenius plots for MDH between 0-30°C for all the various organs and species examined. Because of heat denaturation of homeothermic enzymes many of his studies were not continued much above 30°C.

C. Pyruvate Kinase (PK)

The data for this enzyme is different for that of LDH and MDH just discussed.

In 1968 Somero and Hochachka (72) showed that while muscle PK from fish (adult rainbow trout) gave linear Arrhenius plots between 5-30°C, with Ea values of 30 kcals/mole, i.e. much higher than the Ea values reported for LDH or MDH, the data for PK from rat skeletal muscle PK between 10-40°C was not linear. There was a sharp "break" in the plot at 25°C with a value for Ea_I of 10 kcals/mole above this temperature, and a value for Ea_{II} of 20 kcals/mole below this temperature! Interestingly these workers also reported an "inverse break" in the data they obtained for the cold adapted (-2°C) Antarctic fish where T_C° was about 10°C, and the values for Ea_I and Ea_{II} were the exact converse of those obtained for the rat (Table 3).

More recently, Borgmann and Moon (11) studied the temperature dependence of liver and flight muscle PK from a species of hibernating bat. These workers also found non-linear Arrhenius plots with PK from both organs whether the bat was hibernating or not. However, there was a marked shift in the transition temperature from 17°C down to 5°C when the animal was in hibernation! It should also be noted that while the values for Ea_I above and Ea_{II} below the transition temperature of 17°C in the non-hibernating bat were 8.6 kcals/mole and 19.5 kcals/mole respectively, the value for Ea above the transition temperature at 5°C for the hibernating bat was 13.7 kcals/mole. That is, a value intermediate between the two values for the non-hibernating animal.

III. MITOCHONDRIAL MEMBRANE ENZYMES OF OXIDATIVE PHOSPHORYLATION

Let us turn now from the enzymes of glucose utilization to those of oxidative phosphorylation located in the mitochondrial membrane.

As early as 1969 Kemp, Groot and Reitsma (47) had demonstrated non-linear Arrhenius plots for succinate oxidation by rat liver mitochondria. The transition temperature they reported was 17°C with values for Ea_I above this of 8.9 kcals/mole and 18.8 kcals/mole for Ea_{II} below the temperature (Table 4).

The following year Lyons and Raison (53) confirmed this non-linearity for succinate oxidation in rat liver mitochondria although their value for T_C° was 23° while that for Ea_I above the transition was 2.5 kcals/mole and that for Ea_{II} below the transition was 23.4 kcals/mole. That is, values very different from those of Kemp, *et al.* (47). Lyons and Raison (53) also studied the temperature dependence of succinate oxidation from mitochondria obtained from the livers of poikilotherms (rainbow trout and channel catfish).

TABLE 3. ARRHENIUS CHARACTERISTICS OF PYRUVATE KINASE FROM VARIOUS SOURCES

Enzyme	Species	Tissue	Arrhenius Plot	T_C°	Activation Energy kcals/mole Ea_I	Ea_{II}	Refs*
	Rainbow Trout	Muscle	Linear	none	30	none	72
	Rat	Muscle	Non-linear	25	10	20	72
PK	Antarctic fish	Muscle	Non-linear#	10	20	10	72
	Bat (euthermic)	Muscle, liver	Non-linear	17	9	20	13
	Bat (hibernating)	Muscle, liver	Non-linear	5	14	44	13

* Full details of these references are given in the Reference Section to this paper.

This non-linearity is described by the authors as an "inverse break" in which the value for $Ea_I > Ea_{II}$.

TABLE 4. ARRHENIUS CHARACTERISTICS OF MITOCHONDRIAL SUCCINATE OXIDATION FROM A VARIETY OF SOURCES

Class	Animals	Tissue	Arrhenius Plot	T_C°	Activation Energy kcals/mole Ea_I	Ea_{II}	Refs*
Poikilotherms	Rainbow trout	Liver	Linear	none	8.3	none	53
	Catfish	Liver	Linear	none	12.2	none	53
	Carp(cold-adapted)	Liver	Non-linear	15	4.8	9.8	76
	Carp(warm-adapted)	Liver	Non-linear	23	5.7	8.9	76
	Cane-toad	Heart	Linear	none	15	none	57
Homeotherms	Rat	Liver	Non-linear	17	8.9	18.8	47
	Rat	Liver,heart	Non-linear	21-24	2.5-2.9	19.5-23.4	53
	Beef	Heart	Non-linear	27	9.1	17.0	50
Hibernators	Ground squirrel (awake)	Liver	Non-linear	22-23	9.1	16.7	65
	Ground squirrel (hibernating)	Liver	Linear	none	13.8	none	65

* Full details of these references are given in the Reference Section of this paper

Both these preparations gave linear Arrhenius plots without any thermal transitions in the 5-35°C temperature range, and with a single Ea value of 8.3 kcals/mole for the trout and 12.2 kcals/mole for the catfish.

All these values were later confirmed in a subsequent paper (67) where it was shown that treatment with detergent converted the discontinuous homeotherm plot of the rat into a linear plot similar to that obtained from poikilotherms.

Very soon after this Raison and Lyons (65) also reported that the non-hibernating ground squirrel gave a non-linear plot for liver succinate oxidation with a T_C^o at about 20°C and values of 9.1 kcals/mole and 16.7 kcals/mole for Ea_I above and Ea_{II} below the transition respectively. However, preparations obtained from a hibernating animal gave a linear plot with no thermal transition between 5-35°C and again an intermediate value of 13.8 kcals/mole for a single Ea. Later, this work was extended by McMurchie and Raison (57) to show similar non-linear data for rabbit heart mitochondria (T_C^o about 21°C; Ea_I 3.4 kcals/mole and Ea_{II} 12.6 kcals/mole) while that from the heart of the poikilotherm cane toad gave a linear response to temperature and hence a single value of 15 kcals/mole for Ea.

Lenaz, *et al.* (50) have also reported non-linear Arrhenius data for beef heart mitochondrial succinate oxidation with a somewhat higher T_C^o at 27°C and values for Ea_I of 9.1 kcals/mole and Ea_{II} of 17.0 kcals/mole above and below this transition temperature.

More recently, Wodtke (76) has reported non-linear Arrhenius plots for both mitochondrial succinate oxidation and cytochrome oxidase activity from the liver and muscles of a species of hatchery-bred carp (*Cyprinus carpio*) which had been maintained at an environmental temperature between 18-20°C. When these fish were exposed to colder water at 10°C for 28 days the value for T_C^o was 15°C, while it rose to 23°C when the fish were warm-adapted to water at 26°C for 21 days. Interestingly the values for Ea_I and Ea_{II} under both cold and warm adapted situations did not change significantly but remained near 5 kcals/mole above and 10 kcals/mole below the T_C^o.

There is some experimental evidence which does not fit this general scheme for mitochondrial enzymes. It is the data of Woodard and Zimny (77) who measured succinic dehydrogenase activity in the hearts and brains of non-hibernating and hibernating ground squirrels. These workers reported only linear Arrhenius plots in both organs and under both physiological states. Unfortunately these investigators studied only four temperatures in their experiments (5°, 13°, 21° and 35°C) and there were huge variations in the standard errors of the mean data points they obtained. Therefore, I do not feel that the data can be regarded as definitive. However, although not the interpretation favoured by the authors, my own assessment of their data from brain preparations is that the non-hibernator probably gave a non-linear Arrhenius plot which became linear during hibernation!

IV. MEMBRANE ENZYMES OF ACTIVE TRANSPORT

Let us now look at enzymes of another form of energy transduction, those involved in the active transport of ions across biological membranes.

A. Ca^{++}-ATPase

In an early study Inesi and Watanabe (45) observed a linear temperature dependence, i.e. a single straight line Arrhenius plot of Ca^{++}- activated ATPase of sarcoplasmic reticulum (SR) preparations from rabbit leg muscle examined between 5-20°C. This plot showed good correlation with $^{45}Ca^{++}$-uptake by the same system. About the same time Yamamoto and Tonomura (79) also reported a linear Arrhenius plot obtained from SR of rabbit dorsal muscle over the wider temperature range of 0-37°C, however these latter workers used a substrate concentration well below the optimum necessary for maximal Ca^{++} uptake (Table 5).

At the 1971 Perth Conference on Muscle Disease, I reported that under experimental conditions of maximal $^{45}Ca^{++}$-uptake by rabbit leg muscle SR preparations (23), we could observe a non-linear Arrhenius plot for Ca^{++}-ATPase activity between 0-37°C with the T_C° at about 10°C. The values for Ea_I

TABLE 5. ARRHENIUS CHARACTERISTICS OF SARCOPLASMIC RETICULUM

Ca^{++}-ATPase FROM A VARIETY OF SOURCES

Class	Animal	Tissue	Arrhenius Plot	T^{o}_{c}	Activation Energy kcals/mole Ea_{I}	Ea_{II}	Refs*
Poikilotherms	Lobster	Skeletal muscle	Non-linear	11.5	10	19.5	54
	Crayfish (cold-adapted)	Abdominal muscle	Linear	None	--	--	28
	Crayfish (warm-adapted)	Abdominal muscle	Linear	None	--	--	28
Homeotherms	Rabbit	Leg muscle	Linear	None	13.8-16.0	none	45
	Rabbit	Dorsal muscle	Linear	None	24	none	79
	Rabbit	Leg muscle	Non-linear	10	16.3	33.4	23
	Rabbit	Skeletal muscle	Non-linear	20	16	29	44
	Rabbit	Skeletal muscle	Non-linear	17	9	22	55
	Rat	Leg muscle	Linear	None	25	none	9
	Rat	Leg muscle	?Non-linear	18.6	19.6	31.6	#
Hibernator	Ground squirrel (awake)	Leg muscle	Linear	None	28	none	9
	Ground squirrel (awake)	Leg muscle	?Non-linear	22.8	19.8	32.1	#

* Full details of these references are given in the Reference Section of this paper.

-- Indicates that values were not given by the original authors.

Recalculated from the original data by the computer assisted procedure described by Charnock, Cook and Casey (22).

and Ea_{II} were 16.3 kcals/mole and 33.4 kcals/mole respectively.

It should be noted that this data was obtained under experimental conditions which yielded a linear Arrhenius plot for Ca^{++}-uptake. That is, the temperature dependence of the enzyme process was different from that of the transport process. Presumably these different thermal profiles reflect two different but very closely associated protein functions in the overall mechanism of the "Calcium Pump".

Subsequently, in a very detailed study of $^{45}Ca^{++}$-uptake and Ca^{++}-ATPase activity of white skeletal muscle of rabbit hind leg, Inesi, Millman and Eletr (44) confirmed that these preparations have non-linear Arrhenius characteristics. They determined a T_c° at 20°C with values for Ea_I and Ea_{II} of 16 and 29 kcals/mole respectively, in their preparations. Similar findings were reported by Madeira and Antunes-Madeira (54) who also observed non-linear Arrhenius characteristics in rabbit skeletal muscle preparations with a T_c° at about 17°C. However, the activation energies reported by these latter workers were somewhat lower in that the values for Ea_I and Ea_{II} were 9 and 22 kcals/mole respectively. A transition temperature at 11.5° has also been reported for sarcoplasmic reticulum preparations of the lobster (55) suggesting that non-linear Arrhenius characteristics for Ca^{++}-ATPase can also be observed in poikilotherms.

More recently, Becker and Willis (9) have reported linear Arrhenius plots for sarcoplasmic reticulum Ca^{++}-ATPase obtained from the leg muscles of the homeothermic rat (*Rattus norvegicus*) and an awake hibernator - the thirteen-lined ground squirrel (*Citellus tridecemlineatus*). Although these observations were made between 5-37°C there were only six temperature points through this range and inspection of the data shows that a non-linear fit could be made to the rat preparation (Fig. 4) with a T_c° about 27°C, while that from the ground squirrel was probably linear near this temperature. In a related study Cossins and Bowler (28) have reported that they did not find any differences in the Arrhenius activation energy for thermal inactivation of Ca^{++}-ATPase activity

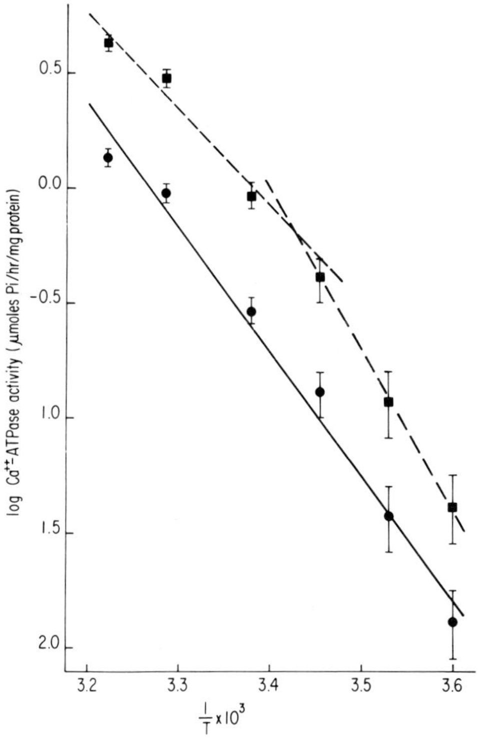

Fig. 4. *Arrhenius plot of rat skeletal muscle sarcoplasmic reticulum Ca++-ATPase. (●); experimental data taken directly from Becker and Willis (9) with a single value for the apparent activation energy of about 25 kcal/mole derived from a straight line fit of the datum points. (■); identical experimental data offset by 0.5 log units and analysed by the computer assisted method described by Charnock, Cook and Casey (22). The calculated transition temperature (T_c^o) is near 18°C. The values for Ea_I above and Ea_{II} below the transition are 19.4 and 32.1 kcals/mole, respectively.*

of sarcoplasmic reticulum from the abdominal muscles of freshwater crayfish which had been temperature adapted to either 4°C or 25°C for several weeks.

Obviously the situation with this particular enzyme is not completely clear and more detailed comparative studies are required before we can finally decide what are the characteristic thermal profiles for Ca^{++}-ATPase in a variety of species, and at variable environmental temperatures.

B. (Na^{+} + K^{+})-ATPase

There is considerably more information available about the other major cation-transporting enzyme system of the plasma and endoplasmic reticular membranes - the (Na^{+} + K^{+})-ATPase of the "Sodium Pump".

Firstly, it is well known that the specific activity of this enzyme is subject to very marked seasonal variations. In poikilotherms (frogs) both Kennedy and Nayler (48) and Csaky and Gallucci (29) have shown dramatic falls in enzyme activity of both the heart and intestine during the winter months. This fall correlates with either the loss of sensitivity to some cardiovascular active agents or to a decrease in sugar and amino-acid transport. Conversely, Fang and Willis (35) reported a significant increase in the specific activity of (Na^{+} + K^{+})-ATPase in the renal cortex of Syrian hamsters during hibernation, and we have recently observed an equally marked increase in brain (Na^{+} + K^{+})-ATPase from ground squirrels during the hibernation state(25).

In addition to these seasonal variations in enzyme activity, there have been many studies of this enzyme's temperature profile from a large variety of animals and tissues. For example, as early as 1966 Gruener and Avi-Dor (40) reported a non-linear Arrhenius plot for (Na^{+} + K^{+})-ATPase obtained from rat brain (Table 6).

This non-linearity in (Na^{+} + K^{+})-ATPase activity of enzymes prepared from the brains of homeotherms has been confirmed repeatedly (15, 59, 73). In my laboratory we have observed the same phenomena in enzymes prepared from guinea-pig, rabbit, sheep and pig kidney (19, 20, 21, 22). By recalculating the data of Willis and Li (75) obtained from an awake hamster, we could also obtain a non-linear Arrhenius plot for the renal cortical (Na^{+} + K^{+})-ATPase of this hibernating species.

TABLE 6. ARRHENIUS CHARACTERISTICS OF MEMBRANE BOUND $(Na^{+} + K^{+})$-ATPase

FROM A VARIETY OF SOURCES

Class	Animal	Tissue	Arrhenius Plot	T°_{C}	Activation Energy kcals/mole Ea_{I}	Ea_{II}	Refs*
Poikilotherms	Toad	Heart	Linear	none	22	none	57
	Frog	Brain	Non-linear	10	9.7	22	14
	Crab	Leg nerve	Non-linear	16	12.5	28.5	24
Homeotherms	Rat	Brain	Non-linear	20	7.8	15.2	40
	Rat	Brain	Non-linear	?	11	80	15
	Beef	Brain	Non-linear	18	18.1	37.0	27
	Rabbit	Kidney	Non-linear	17	15.4	27.5	19
	Rabbit	Heart	Non-linear	22	10	17	57
	Sheep	Kidney	Non-linear	22	15.7	34.4	20
Hibernators	Hedgehog (control)	Brain	Non-linear	?	13.3	37	16
	Hedgehog (hibernating)	Brain	Non-linear	?	13.8	31	16
	Hamster (awake)	Kidney	Non-linear	17	18.7	43.6	75#
	Hamster (control)	Brain	Non-linear	17	19	42	38#
	Hamster (hibernating)	Brain	Non-linear	17	18	34	38#
	Hamster (control)	Brain	Non-linear	16.5	24.5	38.1	39#
	Hamster (hibernating)	Brain	Non-linear	13.6	21.1	40.8	39#

* Full details of these references are given in the Reference Section of this paper.

\# Recalculated from data supplied by authors, using the computer assisted analysis of Charnock, Cook and Casey (22).

McMurchie, Raison and Cairncross (57) have also reported non-linear temperature dependence for $(Na^{+} + K^{+})$-ATPase preparations from rabbit heart at the same time as they detected only linear temperature dependence with an enzyme preparation from the poikilotherm toad. Unfortunately, the enzyme assay system employed by these investigators contained only 40 mM Na^{+} and 40 mM K^{+}, and did not include ouabain or other specific enzyme inhibitors, thus making the assertion that they were measuring $(Na^{+} + K^{+})$-ATPase equivocal. However a similar but not identical change had been reported some years before by Bowler and Duncan (14) who showed that in a brain preparation from a frog kept at 9°C for 1 week prior to their experiment, the microsomal brain preparation of $(Na^{+} + K^{+})$-ATPase had a very much lower value for the energy of activation below T_{C}° than that of a rat. This change in temperature profile they described as representing a characteristic "poikilotherm pattern" was different from that found in enzyme preparations from homeotherms.

Some studies have also been carried out with several hibernating species both awake and during torpor. Bowler and Duncan (16) demonstrated markedly non-linear Arrhenius plots for brain $(Na^{+} + K^{+})$-ATPase from the hedgehog, which by our calculations from their data did not change greatly between non-hibernation and hibernation states.

Similarly when we recalculated the effect of temperature from the brain $(Na^{+} + K^{+})$-ATPase data of Goldman and Willis (38) and Goldman and Albers (39) for hamster preparations, we also obtained markedly non-linear Arrhenius plots whether the animals were awake and active, or whether they had been in hibernation for various periods of time.

We have recently extended these observations in a series of experiments in our own laboratory. Firstly, we compared the temperature dependence of ATP hydrolysis by brain and kidney $(Na^{+} + K^{+})$-ATPase preparations obtained from several common species of homeotherms (rabbit, sheep and beef), an awake hibernating species (the Richardson's ground squirrel, *Spermophilus richardsonii*) common to this area of Alberta, and from the leg nerves of two species of poikilotherms, the Pacific crab (*Cancer magister*) and the Atlantic lobster (*Homarus vulgaris*) (24).

When the Arrhenius plots and the Ea values of all these preparations are examined (Table 7), it is apparent that all the temperature plots are non-linear and there are no striking differences in the temperatures of the thermionic transitions, nor are the values for Ea_{I} and Ea_{II} from a particular tissue very different. The awake hibernator could not be distinguished from a homeotherm on these grounds, although the enzyme preparations from the axons of the poikilotherms gave values for Ea_{I} and Ea_{II} which are lower than any obtained from either homeotherms or an awake hibernator (24).

Apparently the enzyme preparations from poikilotherms are the most thermodynamically efficient when judged on this basis, although they did not produce the straight-line pattern of Arrhenius activity previously reported by McMurchie, *et al.* (57) for their toad heart preparations. Perhaps we should recall that Bowler and Duncan (14) did not observe a linear Arrhenius plot with their frog brain preparations either!

In this particular study we also labelled the membrane enzyme preparations with the fluorescent probe 12-(9-anthroyl) stearic acid (12-AS) which we had used previously to determine the effect of temperature on the molecular mobility of the membrane lipids of sheep and pig kidney (Na^{+} + K^{+})-ATPase preparations (20).

Like our previous findings with the sheep and pig enzymes there was a strong correlation between the non-linear plots of molecular mobility of the lipids of homeotherm brains and non-linear Arrhenius plots of their (Na^{+} + K^{+})-ATPase activity. The T_{c}° values were about 25°C for all preparations (Fig. 5).

The results with poikilotherm membranes were quite different in that the change in fluorescence polarization was only linear between 8-32°C as a thermal transition could be observed above this temperature. That is, the previously observed close correlation between thermal effects on lipid mobility and (Na^{+} + K^{+})-ATPase activity in membranes from kidney and brain of homeotherms could not be so

TABLE 7. APPARENT ACTIVATION ENERGIES OF $(Na^+ + K^+)$-ATPase PREPARATIONS OBTAINED FROM HOMEOTHERMS, AWAKE HIBERNATOR AND POIKILOTHERMS°

Tissue	n	Ea_I*	Ea_{II}*	T_C°	$Ea_I:Ea_{II}$
Beef brain	9	18.1 ± 1.1	37.0 ± 3.4	20.0 ± 1.0	0.49
Ground squirrel brain	6	14.2 ± 1.2	32.4 ± 1.4	19.8 ± 0.4	0.44
Ground squirrel kidney	5	21.4 ± 1.4	45.1 ± 6.2	16.7 ± 2.2	0.47
Rabbit kidney	10	20.4 ± 1.0	44.5 ± 3.4	18.2 ± 1.7	0.46
Sheep kidney	3	16.8 ± 0.2	41.1 ± 7.1	18.2 ± 3.2	0.41
Lobster axon#	5	13.2 ± 0.9	23.7 ± 3.4	22.0 ± 3.3	0.56
Crab axon #	3	12.5 ± 1.5	28.5 ± 1.9	16.2 ± 2.3	0.44

° Data taken From Charnock and Simonson (24).

* Values for Ea_I and Ea_{II} are given as kcals/mole.

0.4 mM ouabain was used to inhibit $(Na^+ + K^+)$-ATPase from homeotherm and awake hibernator preparations, 2.5 mM ouabain was used for the preparations from poikilotherms.

readily observed in membranes from poikilotherms. Whereas there is a marked change in the fluidity of the lipids of homeotherm membranes at about 25°C, no such change could be detected by 12-AS labelling in the fluidity of the membranes from poikilotherms at this temperature, although a change did occur above 32°C.

However, when we compared the slopes of the fluorescence polarization vs. temperature plots from the two species (Fig. 5), it is apparent that the crab membranes are more fluid than those of beef

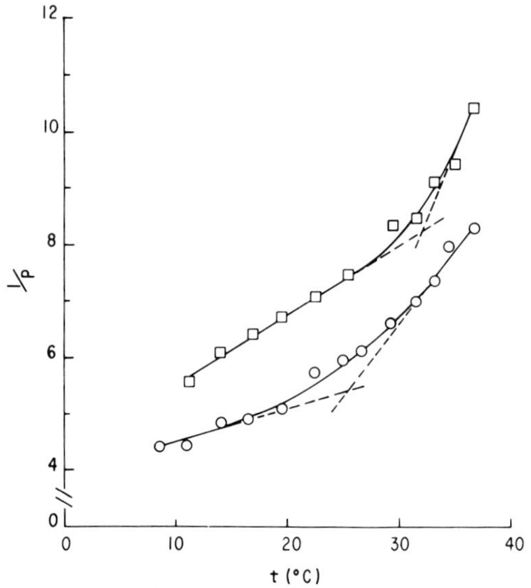

Fig. 5. *Temperature dependence of fluorescence polarization (P) of membrane preparations of (Na^+ + K^+)-ATPase labelled with the lipid probe 12-(9-anthroyl) stearic acid (12-AS). Results are given as 1/P vs. temperature (°C) and are the means of four determinations at each temperature on crab nerve (□) and beef brain (○) enzyme preparations. The transition temperature (T_c^o) for beef membranes is 25°C while that for crab nerve membranes is at 32°C. The temperature dependence of crab nerve preparations is linear below 32°C. Full experimental details are given in Charnock and Simonson (24).*

brain over their normal physiological temperature ranges. This of course is in complete agreement with the lower energies of activation that were found for poikilotherm ($Na^+ + K^+$)-ATPase preparations than for beef brain enzyme preparations at all temperatures studied (24).

In some subsequent experiments we also found that the markedly non-linear Arrhenius pattern of ground squirrel brain ($Na^+ + K^+$)-ATPase did not change even after more than 100 days into their hibernation season (25). Furthermore, the values for T_c° and Ea_I above and Ea_{II} below this critical temperature were not significantly different from those previously reported for enzyme preparations from homeotherms (21) (Table 8).

What we did find which was of great interest to us was that there was a very marked fall in both the rate and amount of inhibitor ([^{3}H]-ouabain) binding to these membrane enzyme preparations after prolonged periods (30-100 days) into their hibernation season. Clearly there were very marked changes in both the rate of [^{3}H]-ouabain binding (affinity) and the total amount of binding, i.e. the number of inhibitor binding sites in this brain tissue during hibernation (27) (Table 9).

Very recently we have completed a similar study of the ($Na^+ + K^+$)-ATPase of the ground squirrel renal cortex (26). Firstly, and unlike a previous study by Fang and Willis (35) using cold adapted and hibernating Syrian Hamsters, we found a highly significant fall in ($Na^+ + K^+$)-ATPase activity in the ground squirrel preparations after both short term (<30 days) and long term (>75 days) hibernation at 5°C (Table 10).

This fall in enzyme activity was matched by a fall in both the rate and amount of ouabain-binding to the membrane enzyme preparations. Thus the fall in kidney enzyme activity during hibernation is unlikely to be due to a change in the functional ability of the enzyme, but is probably due to a loss in enzyme protein under these conditions.

As we have repeatedly seen already in this discussion, another measure of an enzyme's thermodynamic efficiency is to determine the Ea of the system via an Arrhenius analysis. For comparative purposes

TABLE 8.

SEASONAL VARIATION IN THE APPARENT ENERGIES OF ACTIVATION AND TRANSITION TEMPERATURE OF $(Na^+ + K^+)$-ATPase FROM THE BRAINS OF AWAKE AND HIBERNATING GROUND SQUIRRELS°

	Awake animals (June - August) n = 6	After short-term hibernation (Nov - Dec) n = 2	After long-term hibernation (Feb - March) n = 4
Ea_I	14.9 ± 0.91	15.8 ± 1.67	14.3 ± 1.62
Ea_{II}	26.8 ± 0.92	31.3 ± 3.46	30.9 ± 1.68
T°_C	21.2 ± 0.30	18.6 ± 2.22	20.2 ± 1.04

° Adapted from Charnock and Simonson (25).

Values for activation energy (Ea) are given as mean kcals/mole ± S.E., while those for transition temperature (T°_C) are given as the mean °C ± S.E.

TABLE 9.

SEASONAL VARIATION IN THE RATE AND AMOUNT OF $[^3H]$-OUABAIN BINDING TO GROUND SQUIRREL BRAIN $(Na^+ + K^+)$-ATPase COMPARED TO THAT OF OTHER SPECIES

Species	n	Conditions	Specific Activity (mean ± SE)	Rate $[^3H]$-ouabain binding (mean ± SE)	Amount $[^3H]$-ouabain binding (mean ± SE)
#Ground squirrel	5	awake controls	71.3 ± 3.2	5.15 ± 0.30	201 ± 9.7
#Ground squirrel	4	short-term hibernation	62.6 ± 6.3	5.42 ± 0.49	230 ± 19.1
#Ground squirrel	5	long-term hibernation	84.6 ± 3.0	3.97 ± 0.30	166 ± 6.0
*Beef brain	4	---	31.3 ± 1.5	2.42 ± 0.15	133 ± 8.6
*Crab nerve	3	---	12.7 ± 4.2	.041 ± 0.01	6.00 ± 1.15

(#) Data for ground squirrel taken from Charnock and Simonson (25); data for beef brain and crab nerve (*) taken from Charnock and Simonson (24). Specific activity is given as µmoles ATP hydrolysed/mg protein/hour. Rate $[^3H]$-ouabain binding is given as pico moles/mg protein sec^{-1}. Amount $[^3H]$-ouabian bound is given as pico moles/mg protein. All determination were done at 37°C. Short-term hibernation is 30 days or less from the first day of hibernation; long-term hibernation is 75 days or more after onset of the hibernation season.

TABLE 10.

SEASONAL VARIATION IN THE SPECIFIC ACTIVITY AND RATE AND AMOUNT OF $[^3H]$-OUABAIN BINDING TO RENAL CORTEX $(Na^+ + K^+)$-ATPase PREPARATIONS FROM THE GROUND SQUIRREL*

Condition of Animals	n	Specific Activity	Rate $[^3H]$-ouabain Binding	Amount $[^3H]$-ouabain Binding	Ouabain binding / Specific Activity
Awake controls	5	32.2 ± 1.9	2.53 ± 0.24	123.1 ± 6.9	3.82 ± 0.14
<30 days hibernation	5	12.9 ± 1.6	1.25 ± 0.32	46.9 ± 8.4	3.64 ± 0.17
>75 days hibernation	3	13.6 ± 1.6	1.31 ± 0.23	53.6 ± 14.4	3.92 ± 0.26

* Data adapted from Charnock and Simonson (26).

All experiments carried out at 37°C according to the procedures described in detail by Charnock, Simonson and Almeida (27).

Specific activity given as μmoles Pi liberated from ATP/mg protein/hour.

Values for $[^3H]$-ouabain binding are given in pico moles/mg protein/sec for the rate and pico moles/mg protein at equilibrium for the amount.

alone this was most suitable in this case. From the data we obtained in thirteen separate experiments with awake and hibernating ground squirrels, it is also apparent that their renal cortical $(Na^+ + K^+)$-ATPase gave non-linear Arrhenius plots under all experimental conditions (Table 11). In addition, the values for T_C° and Ea_I and Ea_{II} during hibernation were not significantly different from those obtained in the awake summer control group.

Thus it is clear that no changes occurred in any of these thermodynamic parameters during the hibernation season, and that no changes in thermodynamic efficiency of the enzyme could be inferred from this data. Rather, these findings strongly support our earlier proposal that in ground squirrel kidney $(Na^+ + K^+)$-ATPase preparations there is a marked loss of activity due to a significant fall in the amount of "pumping enzyme" available during hibernation. These results are in marked contrast to the concept of "cold resistance" in this issue, which has been proposed by Willis and his colleagues (35,75).

V. DISCUSSION

In discussing these experimental results I think it is clear from Table 2 that for the cytoplasmic enzymes LDH and MDH there is overwhelming evidence that their Arrhenius plots are linear, irrespective of the tissue or species of origin, and apparently the environmental temperature. For pyruvate kinase (PK) the situation may be different in that while it is possible that enzyme preparations from poikilotherms give linear Arrhenius plots, it seems that enzyme preparations from both hibernators and homeotherms could yield non-linear Arrhenius plots (Table 3). In addition, it also seems that biological adaptation to reduced environmental temperature results in a marked reduction in the transition temperature of this enzyme. As PK is part of the complete enzyme mechanism for the breakdown of glycogen which is found in the "soluble" fraction of cells, the non-linearity of PK Arrhenius plots is presumably independent of any membrane effects.

TABLE 11.

SEASONAL VARIATION IN THE APPARENT ENERGIES OF ACTIVATION, TRANSITION TEMPERATURE AND ARRHENIUS ANALYSIS OF OUABAIN-SENSITIVE $(Na^{+} + K^{+})$-ATPase OF THE GROUND SQUIRREL RENAL CORTEX*

Condition of Animals	n	Ea_{I}	Ea_{II}	T_{C}^{o}	Arrhenius Analysis
Awake controls	5	15.7 ± 1.3	31.7 ± 2.4	20.4 ± 0.6	non-linear
<30 days hibernation	3	14.0 ± 1.6	27.4 ± 2.9	20.8 ± 2.5	non-linear
>75 days hibernation	5	16.2 ± 1.1	30.7 ± 1.8	19.8 ± 0.9	non-linear

* Data adapted from Charnock and Simonson (26).

Values for the apparent energies of activation are given as kcals/mole and are the means ± standard errors of the individual experiments.

For the succinate oxidation system of the mitochondrial membranes of either liver or heart, there is some evidence that the enzyme systems from poikilotherms display linear Arrhenius plots, while those mitochondrial enzymes from homeotherms yield non-linear plots. With mitochondrial enzymes from the hibernating ground squirrel there is evidence that the temperature profile seems to change from the non-linearity of a homeotherm in the summer to the linear form of a poikilotherm enzyme during hibernation (Table 4). *In vitro* modification of the lipid content of isloated mitochondrial membranes by detergents strongly suggests that this is a lipid mediated phenomena (64, 67). There is also good evidence that the fatty acid content of mitochondria undergoes major changes during hibernation although the effects seem to be confined to young animals (62).

While there is not yet a great deal of evidence concerning thermal profiles for Ca^{++}-ATPase of either poikilotherms or hibernator preparations, it seems probable that membrane preparations from homeotherms display non-linear Arrhenius plots with a variety of transition temperatures, possibly depending upon the experimental conditions under which the enzyme was assayed (Table 5).

Fortunately, there is a wealth of data concerning the (Na^+ + K^+)-ATPase of coupled sodium and potassium transport across biological membranes. Only the toad heart enzyme has been reported to yield a linear Arrhenius plot (57), and for technical reasons this must be regarded as a somewhat doubtful result. To my knowledge no fish heart (Na^+ + K^+)-ATPase enzymes have been studied in this way. All homeotherm preparations which have been examined yield non-linear Arrhenius plots irrespective of the tissue source of the enzyme (kidney, brain and heart). Even the frog brain preparation of Bowler and Duncan (14) did not yield a linear Arrhenius plot although we should remember that the animal was collected during the spring and had been maintained at 9°C for a week prior to preparation of the enzyme (Table 6).

By recalculating the experimental data of Bowler and Duncan (16), Goldman and Willis (38) and Goldman and Albers (39) obtained from both awake and hibernating hedgehogs and hamsters, we also found non-linear Arrhenius plots for $(Na^+ + K^+)$-ATPase preparations under all these conditions, a result which agrees with our own recent findings for the ground squirrel (25,26). It seems very probable then that these non-linear temperature profiles for $(Na^+ + K^+)$-ATPase are not altered by seasonal changes in the temperature of the environment.

This was certainly true in our own experiments with ground squirrel whether the $(Na^+ + K^+)$-ATPase enzyme was derived from brain or renal cortex (Table 2), as we could not detect any significant change during hibernation. However, we did find major decreases in inhibitor binding to the enzyme preparations (Tables 4 and 10) similar to the changes in ligand and substrate binding to cytoplasmic enzymes during hibernation which had been reported by Borgmann *et al.* (11,12,13).

There is as yet little evidence that the enzymes of different organs in a particular animal display markedly different forms of temperature profiles, although until further work is done it may be that the heart will prove to be different from brain or kidney in this regard.

VI. SUMMARY

Generally linear profiles are obtained with the soluble enzymes LDH and MDH, yet PK is clearly different in that there seems to exist a characteristic pattern for poikilotherms which differs from that of homeotherms. This type of characteristic pattern change is even more apparent with mitochondrial enzymes of oxidative phosphorylation where there is also good evidence for correlation between patterns of activation energy and the molecular mobility of lipids in the membranes.

With the $(Na^+ + K^+)$-ATPase of the plasma and endoplasmic reticular membranes the situation is different again. Experimental evidence is strongly in favour of non-linear Arrhenius plots for enzyme preparations from all tissues and species so far examined. We do not believe that this characteristic temperature profile changes during hibernation.

Thus there is no unifying concept for the role of phase-changes in biological membranes being responsible for the changes in enzyme function during hibernation, for while the organ may not be important in this regard, it seems very likely that the subcellular localization of the enzyme in either the mitochondrial or the plasma membrane makes a most important difference. While there can be little doubt about the profound changes that occur in the lipid composition of both whole brains and hearts of animals during hibernation (4,62), it is therefore not yet clear what effect this has upon the enzymatic function of the organs.

Presumably very detailed reconstitution studies of purified enzymes into liposomes of variable phospholipid composition obtained from both hibernators and euthermic controls will be necessary before this question can be answered, as this would allow us to rigorously control and test the effects of changing membrane lipid composition upon the temperature dependence of selected enzyme systems thought to be important in hibernation.

REFERENCES

1. Aleksiuk, M. An isoenzymic basis for instantaneous cold compensation in reptiles: Lactate dehydrogenase kinetics in *Thamnophis sirtalis*. Comp. Biochem. Physiol. 40B:671-681, 1971.
2. Aleksiuk, M. Temperature-dependent enzyme kinetics during avian ontogeny: Malate dehydrogenase in the common crow and the pintail. Can. J. Zool. 51: 557-565, 1973.
3. Almeida, A. F., and J. S. Charnock. Membrane elasticity as a component of the energy transduction mechanism of ($Na^+ + K^+$)-ATPase. In preparation.
4. Aloia, R. C., E. T. Pengelley, J. T. Bolen, and G. Rouser. Changes in phospholipid composition in hibernating ground squirrel *Citellus lateralis*, and their relationships to membrane function at reduced temperatures. Lipids 9: 993-999, 1974.
5. Arrhenius, S. Zeitschrift für physikalische chemie. 4: 226-239, 1889.
6. Arrhenius, S. In: Theories of Solutions. New Haven: Yale University Press, 1912, pp. 112-130.
7. Baldwin, J., and M. Aleksiuk. Adaptation of enzymes to temperature: Lactate and malate dehydrogenases from platypus and echidna. Comp. Biochem. Physiol. 44B: 363-370, 1973.
8. Bashford, C. L., L. N. Johnson, G. K. Radda, and G. A. Ritchie. Lipid ordering and enzyme activities in chromaffin granule membranes. Eur. J. Biochem. 67: 105-114, 1976.
9. Becker, J. H. and J. S. Willis. Absence of cold resistance in sacroplasmic reticulum Ca-ATPase in a hibernating mammal. Comp. Biochem. Physiol. 52B: 533-535, 1975.
10. Belehradek, J. Physiological aspects of heat and cold. Ann. Rev. Physiol. 9: 59-77, 1957.
11. Borgmann, A. I., and T. W. Moon. Enzymes of the normothermic and hibernating bat *Myotis lucifugus*: Temperature as a modulator of pyruvate kinase. J. Comp. Physiol. 107: 185-199, 1976.

12. Borgmann, U., K. J. Laidler, and T. W. Moon. Kinetics and thermodynamics of lactate dehydrogenase from beef heart, beef muscle and flounder muscle. Can. J. Biochem. 53: 1196-1206, 1975.
13. Borgmann, U., and T. W. Moon. A composition of lactate dehydrogenase from an ectothermic and an endothermic animal. Can. J. Biochem. 53: 998-1004, 1975.
14. Bowler, K., and C. J. Duncan. The temperature characteristics of the ATPases from a frog brain microsomal preparation. Comp. Biochem. Physiol. 24: 223-227, 1968.
15. Bowler, K., and C. J. Duncan. The effect of temperature on the Mg^{2+}-dependent and Na^{+}-K^{+} ATPase of a rat brain microsomal preparation. Comp. Biochem. Physiol. 24: 1043-1054, 1968.
16. Bowler, K., and C. J. Duncan. The temperature characteristics of brain microsomal ATPases of the hedgehog: Changes associated with hibernation. Physiol. Zool. 42: 211-219, 1969.
17. Capaldi, R. A. A dynamic model of cell membranes. Scientific American 230: 27-33, 1974.
18. Chapman, D. In: Biomembranes: Lipids, Proteins and Receptors. Webster Grooves, Mo.: BI-Science Publications, 1975, pp. 65-93.
19. Charnock, J. S., A. F. Almeida, and R. To. Temperature-activity relationships of cation activation and ouabain inhibition of (Na^{+} + K^{+})-ATPase. Arch. Biochem. Biophys. 167: 480-487, 1975.
20. Charnock, J. S., and C. L. Bashford. A fluorescent probe study of the lipid mobility of membranes containing sodium- and potassium-dependent adenosine triphosphatase. Mol. Pharmacol. 11: 766-774, 1975.
21. Charnock, J. S., D. A. Cook, A. F. Almeida, and R. To. Activation energy and phospholipid requirements of membrane bound adenosine triphosphatase. Arch. Biochem. Biophys. 159: 393-399, 1973.

22. Charnock, J. S., D. A. Cook, and R. Casey. The role of cations and other factors on the apparent energy of activation of (Na^+ + K^+)-ATPase. Arch. Biochem. Biophys. 147: 323-329, 1971.
23. Charnock, J. S., and D. Frankel. $^{45}Ca^{++}$ uptake and energy utilization by sacroplasmic reticulum. In: Basic Research in Myology. Amsterdam: Proc. 2nd Internat. Cong. Muscle Diseases. Excerpta Medica, Part I, 1972, pp. 220-225.
24. Charnock, J.S., and L. P. Simonson. Differential lipid control of (Na^+ + K^+)-ATPase in homeotherms and poikilotherms. Comp. Biochem. Physiol. 58B: 381-387, 1977.
25. Charnock, J. S., and L. P. Simonson. Variations in (Na^+ + K^+)-ATPase and Mg^{++}-ATPase activity of the ground squirrel brain during hibernation. Comp. Biochem. Physiol. (In Press), 1978.
26. Charnock, J. S., and L. P. Simonson. Variations in (Na^+ + K^+)-ATPase and Mg^{++}-ATPase of the Richardson ground squirrel renal cortex during hibernation. Comp. Biochem. Physiol. (In Press).
27. Charnock, J. S., L. P. Simonson, and A. F. Almeida. Variation in sensitivity of the cardiac glycoside receptor characteristics of (Na^+ + K^+)-ATPase to lipolysis and temperature. Biochim. Biophys. Acta 465: 77-92, 1977.
28. Cossins, A. R., and K. Bowler. Resistance adaptation of the freshwater crayfish and thermal inactivation of membrane-bound enzymes. J. Comp. Physiol. 111: 15-24,1976.
29. Cśaky, T. Z., and E. Gallucci. Seasonal variation in the active transporting ability and in the membrane ATPase activity of the frog intestinal epithelium. Biochim. Biophys. Acta 466: 521-525, 1977.
30. Danielli, J. F., and H. Davson. A contribution to the theory of permeability of thin films. J. Cell. Comp. Physiol. 5: 495-508, 1935.

31. Davson, H., and J. F. Danielli. In: The Permeability of Natural Membranes (2nd Ed.). London and New York: Cambridge University Press, 1952, pp. 1-365.
32. Dixon, M., and E. C. Webb. In: Enzymes. London: Longman Green and Co., 1958, p. 150.
33. Drost-Hansen, W. Structure and properties of water at biological interfaces. In: Chemistry of the Cell Interface. Brown, H. E. (editor). New York: Academic Press Inc., 1972, Part B, pp. 1-273.
34. Edidin, M. Rotational and translational diffusion in membranes, Ann. Rev. Biophys. Bioenerget. 3: 179-201, 1974.
35. Fang, L. S. T., and J. S. Willis. Increase of Na-K-ATPase activity in renal cortex of hamster (*Mesocricetus auratus*) during prehibernation cold exposure. Comp. Biochem. Physiol. 48A: 687-698, 1974.
36. Frye, C. D., and M. Edidin. The rapid intermixing of cell surface antigens after formation of mouse-human heterokaryons. J. Cell Sci. 7: 319-335, 1970.
37. Goldman, S.S. Cold resistance of the brain during hibernation. III. Evidence of a lipid adaptation. Amer. J. Physiol. 228: 834-838, 1975.
38. Goldman, S.S., and J. S. Willis. Cold resistance of the brain during hibernation. II. Na-K-activated ATPase. Cryobiology 10: 218-224, 1973.
39. Goldman, S.S., and R. W. Albers. Cold resistance of the brain during hibernation: Temperature sensitivity of the partial reactions of the Na^+, K^+-ATPase. Arch. Biochem. Biophys. 169: 540-544, 1975.
40. Gruener, N., and Y. Avi-Dor. Temperature-dependence of activation and inhibition of rat brain adenosine triphosphatase activated by sodium and potassium ions. Biochem. J. 100: 762-767, 1966.
41. Han, M. H. Non-linear Arrhenius plots in temperature-dependent kinetic studies of enzyme reactions. J. Theor. Biol. 35: 543-568, 1972.

42. Hochachka, P. W., and G. N. Somero. The adaptation of enzymes to temperature. Comp. Biochem. Physiol. 27: 659-668, 1968.
43. Hokin, L. E. Purification and molecular properties of the (Sodium + Potassium)-Adenosine triphosphatase and reconstitution of coupled sodium and potassium transport in phospholipid vesicles containing purified enzyme. J. Exptl. Zool. 194: 197-205, 1975.
44. Inesi, G., M. Millman, and S. Eletr. Temperature-induced transitions of function and structure in sarcoplasmic reticulum membranes. J. Mol. Biol. 81: 483-504, 1973.
45. Inesi, G., and S. Watanabe. Temperature dependence of ATP hydrolysis and calcium uptake by fragmented sarcoplasmic membranes. Arch. Biochem. Biophys. 121: 665-671, 1967.
46. Johnson, F. H., H. Eyring, and M. J. Polissar. In: The Kinetic Basis of Molecular Biology. New York: John Wiley & Sons Inc., 1954, p. 187.
47. Kemp, A., G. S. P. Groot, and H. J. Reitsma. Oxidative phosphorylation as a function of temperature. Biochim. Biophys. Acta 180: 28-34, 1969.
48. Kennedy, K. G., and W. G. Nayler. Seasonal variation in the activity of the Na^+ - K^+ activated Mg^{2+} - dependent ATPase enzyme in the membrane-microsomal fraction of the toad cardiac muscle. Comp. Biochem. Physiol. 165: 175-180, 1965.
49. Kumamoto, J., J. K. Raison, and J. M. Lyons. Temperature breaks in Arrhenius plots: A thermodynamic consequence of a phase change. J. Theor. Biol. 31: 47-51, 1971.
50. Lenaz, G., A. M. Sechi, G. Parenti-Castelli, L. Landi, and E. Bertoli. Activation energies of different mitochondrial enzymes: Breaks in Arrhenius plots of membrane-bound enzymes occur at different temperatures. Biochem. Biophys. Res. Commun. 49: 536-542, 1972.
51. Luzzati, V., and F. Husson. The structure of the liquid-crystalline phases of lipid-water systems. J. Cell. Biol. 12: 207-219, 1962.

52. Lyons, J. M. Phase transitions and control of cellular metabolism at low temperatures. Cryobiology 9: 341-350, 1972.
53. Lyons, J. M., and J. K. Raison. A temperature-induced transition in mitochondrial oxidation: Contrast between cold and warm-blooded animals. Comp. Biochem. Physiol. 37: 405-411, 1970.
54. Madeira, V. M. C., and M. C. Antunes-Madeira. Thermotropic transitions in sarcoplasmic reticulum. Biochem. Biophys. Res. Commun. 65: 997-1003, 1975.
55. Madeira, V. M. C., M. C. Antunes-Madeira, and A. P. Carvalho. Activation energies of the ATPase activity of sarcoplasmic reticulum. Biochem. Biophys. Res. Commun. 58: 897-904, 1974.
56. Massey, V., B. Curti, and H. Ganther. A temperature-dependent conformational change in D-amino acid oxidase and its effects on catalysis. J. Biol. Chem. 241: 2347-2357, 1966.
57. McMurchie, E. J., J. K. Raison, and K. D. Cairncross. Temperature-induced phase changes in membranes of heart: A contrast between the thermal response of poikilotherms and homeotherms. Comp. Biochem. Physiol. 44B: 1017-1026, 1973.
58. Melchior, D. L., and J. M. Stein. Thermotropic transitions in biomembranes. Ann. Rev. Biophys. Bionerget. 5: 205-238, 1976.
59. Neufeld, A. J., and H. M. Levy. The steady state level of phosphorylated intermediate in relation to the two sodium-dependent adenosine triphosphatases of calf brain microsomes. J. Biol. Chem. 245: 4962-4967, 1970.
60. Olsson, S. R. Comparative studies on the temperature dependence of lactic dehydrogenase and malic dehydrogenase from a homeotherm, two hibernators and two poikilotherms. Comp. Biochem. Physiol. 51B: 5-18, 1975.

61. Overath, P., T. Lutz, and H. Träuble. Lipid phase transitions and membrane function. Trends Biochem. Sci. 1: 186-189, 1976.
62. Platner, W. S., D. G. Steffen, G. Tempel, and X. J. Musacchia. Mitochondrial membrane fatty acids of liver and heart of the euthermic and hibernating ground squirrel *Citellus tridecemlineatus*. Comp. Biochem. Physiol. 53A: 279-283, 1976.
63. Radda, G. K., and J. Vanderkooi. Can fluorescent probes tell us anything about membranes? Biochim. Biophys. Acta 265: 509-549, 1972.
64. Raison, J. K. The influence of temperature-induced phase changes on the kinetics of respiratory and other membrane-associated enzyme systems. Bioenergetics 4: 559-583, 1973.
65. Raison, J. K., and J. M. Lyons. Hibernation: Alternation of mitochondrial membranes as a requisite for metabolism at low temperature. Proc. Nat. Acad. Sci. U.S.A. 68: 2092-2094, 1971.
66. Raison, J. K., J. M. Lyons, R. J. Mehlhorn, and A. D. Keith. Temperature-induced phase changes in mitochondrial membranes detected by spin labelling. J. Biol. Chem. 246: 4036-4040, 1971.
67. Raison, J. K., J. M. Lyons, and W. W. Thomson. The influence of membranes on the temperature-induced changes in the kinetics of some respiratory enzymes of mitochondria. Arch. Biochem. Biophys. 142: 83-90, 1971.
68. Robertson, J. D. In: Regional Neurochemistry. 4th Intern. Neurochem. Symp. Kety, S. S. and J. Elkes (editors). Oxford: Pergamon Press, 1961, p. 497.
69. Rothman, J. E., and J. Lenard. Membrane asymmetry. Science 195: 743-753, 1977.
70. Singer, S. J. Architecture and topography of biologic membranes. In: Cell Membranes. Weissmann, G. and R. Claiborne (editors). New York: H. P. Publishing Co. Inc., 1975, pp. 35-44.

71. Singer, S. J. and G. L. Nicolson. The fluid mosaic model of the structure of cell membranes. Science 175: 720-731, 1972.

72. Somero, G. N., and P. V. Hochachka. The effect of temperature on catalytic and regulatory functions of pyruvate kinase of the Rainbow trout and the Antarctic fish *Trematomus bernacchii*. Biochem. J. 110: 395-400, 1968.

73. Swanson, P. D. Temperature dependence of sodium ion activation of the cerebral microsomal adenosine triphosphatase. J. Neurochem. 13: 229-236, 1966.

74. van't Hoff, J. H. In: Etudes de dynamique chimique. Amsterdam, 1884.

75. Willis, J. S., and N. M. Li. Cold resistance of Na-K-ATPase of renal cortex of the hamster, a hibernating mammal. Am. J. Physiol. 217: 321-326, 1969.

76. Wodtke, E. Discontinuities in the Arrehnius plots of mitochondrial membrane-bound enzyme systems from a poikilotherm. Acclimation temperatures of carp affects transition temperatures. J. Comp. Physiol. 110: 145-157, 1976.

77. Woodard, C. B., and M. L. Zimny. Energy of activation of succinic dehydrogenase in a hibernator. Am. J. Physiol. 225: 429-431, 1973.

78. Wynn-Williams, A. T. An explanation of apparent sudden changes in the activation energy of membrane enzymes. Biochem. J. 157: 279-281, 1976.

79. Yamamoto, T., and Y. Tonomura. Reaction mechanism of the Ca^{++}dependent ATPase of sarcoplasmic reticulum from skeletal muscle. J. Biochem. 62: 558-575, 1967.

METABOLIC ECONOMY AT THE BIOCHEMICAL LEVEL: THE HIBERNATOR

Hans W. Behrisch
Institute of Arctic Biology
University of Alaska
Fairbanks, Alaska 99701

ABSTRACT

In the course of the year the mammalian hibernator exists in two forms: homeothermic and heterothermic. Because temperature usually exerts large and often disparate effects on homeothermic enzymes, the enzymes of the summer-active (non-hibernating) ground squirrel would seem to be unsuited for function and control in the heterothermic state. Thus before the onset of hibernation the animal appears to anticipate its transition from one metabolic state to another by producing enzyme variants (or isoenzymes) that are designed for function and control under conditions of varying temperatures. In view of available information from enzyme studies on ectothermic organisms, the enzymes that are produced prior to entry into the hibernating state are surprisingly sensitive to temperature. This thermal sensitivity appears to enable the animal to take advantage of the large drop in body temperature in maintaining a very low rate of enzyme function and metabolism. The observed thermal sensitivity of enzyme function seems to be based on a constancy of affinity for physiological ligands (substrates and modulators) and allosteric behavior over a wide temperature range, suggesting that the relative costs of metabolic control are also held constant. The major intracellular cations Na^+, K^+, Mg^{++} and Mn^{++} do not appear to control enzyme activity, but instead function in maintaining an intracellular milieu that permits metabolic regulation to remain independent of temperature over the entire temperature range. This "regulatory homeostasis" is dependent on a circannual cycle of production or interconversion of hibernating or non-hibernating isoenzymes that precedes the animal's transition from one state to the other.

ISBN 0-12-734550-7

I. INTRODUCTION

The ability of certain animals to enter a state of reduced metabolic activity, to withstand long periods of time with little intake of food and to experience substantial and long-term changes in body temperature has long been a subject of interest and an impressive literature on the topic exists. For many animals which hibernate for appreciable periods, the hibernating habit is accompanied (indeed founded upon) torpor, which state is characterized by a pronounced bradycardia, a low rate of respiration and (in the smaller mammals) a quite dramatic fall in body temperatures. In some species, and in the northern or alpine mammals in particular, this state may reach impressive proportions. For example, the Arctic ground squirrel *Citellus undulatus* may hibernate for up to 8 months a year (60). During hibernation this animal may reduce its metabolic rate to about 1/100th of the normothermic rate (72, 73), body temperature may descend to as low as 0°C, and within limits parallels fluctuations in the temperature of the hibernaculum. Within the period of hibernation, this animal arouses spontaneously in cyclic fashion. While the nature of the stimulus for this arousal is at present unknown, it is likely of a metabolic character, and a number of attractive possibilities have been suggested [see the last review on this subject by South *et al.* (118)]. Thus Willis *et al.* (128) have suggested that it may be the slow but steady alteration of the ionic portion of the intracellular environment, while another equally attractive possibility centers around the need of the animal to re-establish levels of blood glucose that have been gradually depleted during the period of deep hypothermia (49), and which reach their lowest values in the hibernating state (72, 131). It is likely that the stimulus for the periodic and spontaneous arousal in the hibernator is a combination of these and other factors. At any rate, central to such a system is the idea of a key process (or processes) that is inhibited or inactivated at low temperature. Such a theory is not without its merits, as there have been identified numerous enzymic systems that become dissociated or inactivated at low temperatures (67, 83, 85). For example, a decrease in temperature has been shown to inactivate the pyruvate kinase from yeast (of the "Budweiser" variety!) which is then followed by a dissociation of the enzyme into its component subunits (85). A somewhat different effect of low temperature is seen in the fructose 1, 6-diphosphatase from liver of the rat; the enzyme loses much of its affinity for its substrate but experiences at the same time a very large increase in affinity for its allosteric inhibitor (119). Further, temperature affects the fluidity and viscosity of the membrane structures of the cell (37, 87, 88, 130, 132), effects which in turn influence the amount of energy the cell must invest in maintaining a proper intracellular environment. Such disparate effects of temperature have usually been observed on homeothermic systems, and it is clear that they do not satisfy the construct of an organism that must adapt seasonally to changing body and environmental temperatures, availability of food and metabolic state. More importantly, the hibernator must (and does) anticipate the cyclic alteration in its condition, as it appears reasonable to postulate that whatever the characters of the adaptive maneuvers it makes, they would not confer any advantage on the organism if it did not possess them before actually going from one state (non-hibernating) to the other (hibernating).

Certain kinds of hibernators can enter hibernation spontaneously and may be collectively called "permissive" hibernators. Conceivably, these animals can avoid making a number of presumably metabolically expensive changes by possessing an elaborate set of metabolic characteristics throughout the year that pre-adapt the animal to a heterothermic existence. In this connection, Orengo and Patenia (98, 99, 100, 101) showed that the fructose, 1, 6-diphosphatase from liver of the hamster has an affinity for substrate and modulator that is lower than that of the fructose diphosphatase from the rat. Based on differences in the circular dichroism of these two enzymes, it appears, not surprisingly, that the two enzymes have somewhat different conformations and that of the two enzymes the hamster enzyme displays a more rigid structure. These authors argue that these differences are evidence of adaptations to the hibernating habit in the hamster. However, it is difficult to see how these data of themselves are evidence of an adaptation to the hibernating habit, without a demonstration of their relation to the animal's ability to regulate its metabolism at low temperature.

In contrast to the "permissive" hibernator, marmots and ground squirrels which may be termed "obligatory" hibernators, undergo an elaborate set of metabolic preparations as they approach entry into the hibernating state (4, 9, 10). From studies in this laboratory on the regulation of carbohydrate metabolism in the Arctic ground squirrel, it appears that prior to entry into (or exit from) the hibernating state, the animal restructures a substantial portion of the enzymes of intermediary metabolism, and generates a complement of enzyme variants or isoenzymes that are peculiarly adapted to heterothermic function and control (9, 10).

Finally, there are the very large hibernators, e.g. bears, that also go into prolonged torpor. Some disagreement about their status within the group exists, as they do not show the marked drop in body temperatures that some of the smaller species do. Further, during hibernation, the bear seems to be fully arousable and capable of defending itself throughout the period of its winter sleep (47, 89, 94, 95). However, although the bear is not a classic hibernator, it is capable of sleeping continously for several months during the winter and is able to do so without the need to eat, drink, defecate or urinate (47, 48, 60, 94). It is evident that the bear achieves these metabolic adjustments by the alteration of several aspects of its intermediary metabolism (48, 86) but whether such changes may take place spontaneously or require an elaborate preparatory period as in the ground squirrels or marmots is unknown.

Within the hibernating mammals as a group there appear to exist at least two classes: (i) those that possess at all times of the year the metabolic machinery necessary for the efficient regulation of metabolism in both the hibernating or torpid states and the non-hibernating (summer-active or homeothermic) states; and (ii) those animals that require seasonally a complex series of biochemical and physiological changes before entry into a successful hibernation (torpid) phase is possible. Because the hibernator alternates between a homeothermic and a hypothermic phase, and because the hypothermic phase is usually quite lengthy, it is tempting to compare hibernating mammal to an ectothermic organism which, largely speaking, cannot avoid the thermal changes in its environment and perforce must adapt. Thus some recent articles on the regulation of enzyme activity in a permissive hibernator, the bat, view the process in part as a period of starvation, because the animal does not feed, and the temperature drops and its associated effects as a "perturbation" of the system, which may be "solved" in a manner similar to the one the ectotherm uses (28, 93). For two reasons this view does not appear to yield much insight into the process. First, the hibernator does not "starve" in the normal sense of the word, having at the onset of the hibernating phase sometimes well over 50% of its body weight as fat, with which it is still well endowed at the time of final arousal. Secondly, the ectothermic system (a non-hibernating one) is active throughout the entire year and its metabolic adaptations are geared for that (23, 24, 61, 69), while those of the hibernator are not.

II. THE INFLUENCE OF TEMPERATURE

What are the effects of a decrease in temperature? The most elementary one (all other factors remaining equal), is a decrease in the rate of chemical reactions. In this connection it is generally known that some temperate ectotherms such as a number of fishes, when acclimated to different temperatures, exhibit a rate compensation of metabolism in such a way as to offset the effect of a decrease in temperature (for review see Hochachka and Somero, 58). However such a route would not seem to lie open to the hibernator, because the hibernating habit in mammals is founded on a massive reduction in metabolic activity. All the same, it is becoming apparent that temperature affects not only the extent of the hibernators metabolism, but also the relative contributions of the various metabolic pathways which contribute to the total metabolism of the intact animal (58). The mechanisms by which this other aspect of adaptation, namely "metabolic reorganization", is effected, is essentially unknown although a number of possibilities have been suggested (53, 58). A further effect of a drop in temperature is an alteration in the permeability of the membrane (2, 87, 88). Associated with this are often (large) changes in the concentrations and activities of the intracellular ions in both the hibernator (126, 127, 128) and in the ectotherm (38, 57, 62, 63). Also in ectotherms a decrease in temperature causes an increase in the pH of the intra- and extracellular fluids, as in these organisms these fluids maintain their relative "alkalinity" (103, 104, 105). This is in marked contrast to the hibernator, which maintains a constant pH of the blood, and presumably also the intracellular fluids, at all temperatures and so becomes markedly acidic at the low temperatures. However, it ought to be mentioned, parenthetically, that in the heterothermic tissues of non-hibernating mammals, it appears that the pH of the blood and intracellular fluids follows the general rule observed by Rahn and his coworkers (103-105) and a number of metabolic adaptations to this change appear to exist (23).

III. METABOLIC REGULATION IN THE HIBERNATOR, AN ANALOGY AND LIKELY SOLUTIONS TO THE PROBLEM

One of the basic problems faced by the hibernator is that most of the aspects of enzyme regulation appear to be directly affected by temperature. Thus in bacterial and mammalian enzyme systems, large and often differential effects of temperature on enzyme-modulator and enzyme-substrate associations can occur (55, 65, 67, 119). Intuitively then, it appears that these types of changes observed for homeothermic enzymes, as well as their magnitude, would be incompatible with survival in a heterothermic systems such as a hibernator. For example, if the extent of the change in enzyme-modulator interaction observed for rat liver fructose diphosphatase (119) should occur in a hibernator like the Arctic ground squirrel, then one would expect that gluconeogenesis would be completely "shut off" at low temperature, whereas in *Citellus spp.* at least, it appears to proceed at low temperatures (33, 120, 131). This is supported by kinetic studies on the fructose diphosphatase (FDPase) from liver of the Arctic ground squirrel that show a population of enzyme variants that are characteristic of either the hibernating or non-hibernating state (9). These hibernating and non-hibernating enzyme variants are further distinguished by their different kinetic behavior over a wide (but physiological) temperature range. The non-hibernating FDPase shows a decrease in affinity for its substrate, as reflected by an increase in the apparent Michaelis constant (Km) for substrate with a decrease in temperature and a large increase in affinity (or decrease in the apparent Km) for its allosteric inhibitor, AMP. These differential effects would make the enzyme unsuited for prolonged function and regulation under heterothermic conditions. In contrast the winter or hibernating FDPase exhibits an increased affinity for both the substrate and the inhibitor at the lower temperature, with the affinity for the inhibitor being affected relatively less by temperature than affinity for substrate. These data would suggest that at least this rate-limiting reaction in hepatic gluconeogenesis remains open in the hibernating state. Thus, gluconeogenesis appears to be an active pathway in the hibernator at all temperatures (33, 79, 131) even though concentrations of glucose in the blood decrease steadily during deep hibernation (48). This is supported by reports of elevated rates of respiration of liver tissue throughout the hibernation phase, while other tissues decrease their rates of consumption of oxygen - and hence do not rapidly deplete stored reserves of energy.

Judging from the differential changes in oxygen consumption of various tissues in the hibernating animal (91) it appears that during hibernation there is some sort of hierarchy among the animal's tissues with respect to function and the consequent need for metabolites and oxygen. During deep hibernation the ground squirrel or marmot is essentially a brain-heart-lung machine with these tissues dividing up the bulk of a vastly reduced circulation. Judging from our own findings and the indirect ones from other workers, the liver also receives a portion of the circulating blood, with very little going to the other tissues.

It is sometimes profitable to draw analogies between groups of organisms which in themselves may be quite different, but are however confronted by the same or similar selective forces. We may thus compare the hibernating ground squirrel or marmot with the breath-holding diver, e.g. the harbour seal (14-17). This breath-hold diver has developed a comprehensive set of biochemical and physiological reflexes that together form an elaborate diving response. The pivotal points of this diving response are a marked bradycardia, and a complete ischemia of all tissues but heart, brain and lung (42, 66, 107). During the dive the seal's perfused tissues are receiving oxygen at progressively lower concentrations, and the ischemic tissues must function anaerobically for virtually the entire dive. Among the ischemic tissues are the kidneys, gut, liver, skeletal muscles and of course the blubber layer around the animal. The diver is noted for its large amount of blood and the very large amounts of oxygen that can be stored in the blood and in the tissues. Further, in the tissues that become ischemic during the dive, appreciable amounts of glycogen are deposited (74). However, these depots of anaerobic energy are finite, and *in vitro* experiments indicate that in an arduous dive they may be seriously depleted (12). It is clear that the diver may prolong its dive according to the length of time the ischemic tissues can survive without oxygen, and the stored oxygen that is available to the perfused tissues. A major way whereby the non-perfused tissues can survive this is the curtailment of all tissue activities, save maintenance metabolism. Thus the survival of these ischemic tissues is directly related to the length of time that anerobic metabolism may be supported. A number of workers in this field (110, 117) have repeatedly put forth the view that a requisite for an extended anaerobic function is a high activity of the enzymes of anerobic metabolism. However, when we examined this question in the harbour seal, a noted breath-holding diver, and compared the data with similar results from the dog (15-17), we obtained results that were distinctly at variance with expectations based on previously published data (117). In virtually every case, the activities of the glycolytic enzymes in the seal tissues that are not perfused during the dive were substantially lower than those in the dog, the non-diver. We reason (17) that the reduction of enzyme activity in the ischemic tissues of the seal is a major means of extending the "anaerobic capacity" of the tissues and the organism. A comparison between the hibernator and the breath-holding diver is not at first glance entirely appropriate, however, as one functions largely anaerobically and remains active, while the other depends almost exclusively on aerobic metabolism and is inactive. However, the selective pressures on both animals are to a certain extent quite similar as it is encumbent on the diver to extend the time over which oxygen and anaerobic energy sources last, while the hibernating ground squirrel or marmot must husband its energy stores in the form of fat so that they extend throughout the hibernating period and still leave a reserve for the animal upon final emergence. In both instances the most efficient method of doing this appears to be a reduction of the metabolic pathways utilizing these stores. The hibernator may then be viewed as having to invoke a very strict metabolic economy. In this context it is significant that in the hibernating ground squirrel there seems to be a specific inhibition of protein synthesis, a factor that would eliminate a very powerful competitor for metabolic energy (78, 124, 125).

IV. PYRUVATE KINASE - A MODEL ENZYME FOR STUDYING ADAPTATIONS TO THE HIBERNATING HABIT

The regulation of hepatic carbohydrate metabolism poses an intriguing and paradoxical problem as on the one hand glycolysis appears to be active at low temperatures in hibernation (29, 33, 34, 120), while on the other hand glycolytic intermediates are largely blocked from entering the tricarboxylic acid cycle (120). Similarly, hepatic gluconeogenesis seems to be favored at low temperatures in the hibernating ground squirrel, although measured concentrations of blood glucose fall steadily during hibernation and are then re-established during arousal (49).

In this laboratory most of our attentions have been devoted to the regulatory enzymes of carbohydrate metabolism, and in particular FDPase and pyruvate kinase (PyK).

Pyruvate kinase occurs in liver of the Arctic ground squirrel in only one form, offering a sharp (and provocative) contrast to the other mammalian systems so far studied, in which both parenchymal and non-parenchymal cells have their own distinct enzyme variants of this enzyme (26, 27). The summer (non-hibernating) PyK has an isoelectric point of 5.3 that is identical with that of liver PyK's from other mammalian sources, while the winter (hibernating) PyK form has an isoelectric point of 5.7 (Fig. 1). The molecular weights of both variants are the same at 243,000 (10, 21, 22). Both appear to bind the two substrates adenosine diphosphate (ADP) and phosphoenolpyruvate (PEP), the inhibitor adenosine triphosphate (ATP), and the activator fructose diphosphate (FDP) in allosteric fashion (21). The major, and almost only, differences between the two variants of the ground squirrel liver PyK lie in the effect of temperature on the binding of these ligands

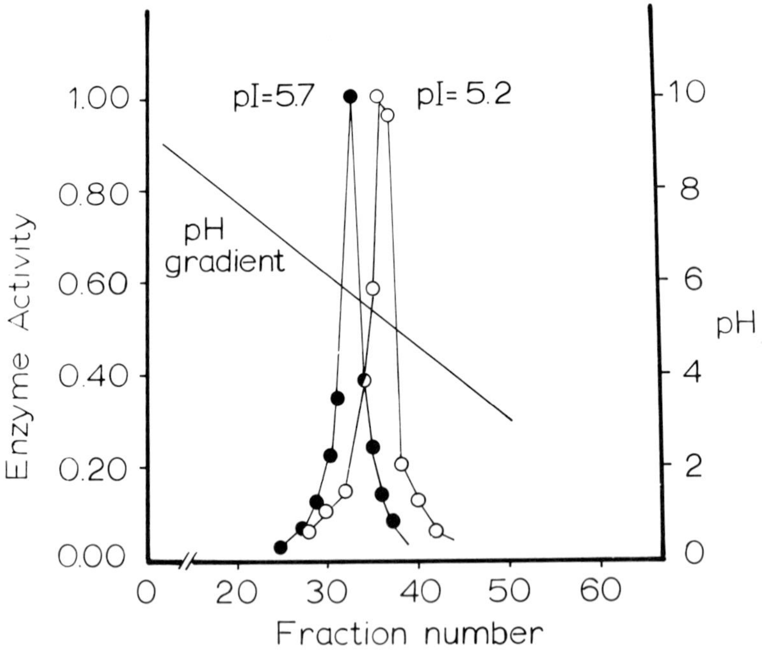

Fig. 1. *Results of isoelectric focussing experiments on liver pyruvate kinase from the Arctic ground squirrel in both the non-hibernating (summer, o) and hibernating (winter, •) phases.*

(Table 1). At low temperature PyK from the non-hibernating animal displays a markedly enhanced (10-fold) affinity (as reflected by apparent Km values) for one substrate (ADP) while the affinity for the other substrate (PEP) is halved. Similarly, a decrease in temperature from 37° to 5°C causes a trebling in values of the apparent Km for the inhibitor ATP, while apparent Km values for the activator FDP remain constant over the entire temperature range. In view of these disparate temperature effects this non-hibernating PyK variant behaves in a manner that is very similar to that of rat liver FDPase (119), and would appear to be unsuited for function and efficient regulation over a wide (but physiological) temperature range. Further, values of the temperature coefficient of the non-hibernating PyK at presumably physiological concentrations of the substrates are in all cases below 2 and generally slightly above unity, indicating a high degree of insensitivity of activity to temperature. In this respect the enzyme is quite similar to the pyruvate kinases from various ectotherms (7, 8, 111-114).

The variant of PyK from the hibernating ground squirrel behaves in a markedly different manner, although the binding of the four ligands PEP, ADP, FDP and ATP is also allosteric. Without exception, values of the apparent Km's for the substrates and modulators are essentially unchanged over the temperature range tested at 37°, 26°, 15° and 5°C; indicating a remarkable degree of stoichiometry of these 4 metabolites to each other at all temperatures (Table 1). In addition to being relatively unaffected by temperature, the apparent affinity for the inhibitor ATP is substantially lower in the hibernating PyK than in the non-hibernating PyK (10). Thus, the hibernating PyK is markedly more sensitive to temperature, on the basis of Q_{10} values that are close to 2.5, over the whole temperature range tested and over the entire concentration range of the metabolites tested. This hibernating PyK is therefore very different from other ecto- or heterothermic PyK's; these tend to show a decrease in the apparent Km's with a fall in temperature and so apparently counter the effect of a decrease in temperature. The resultant increase in affinity, it is argued, reduces the Q_{10} of the reaction and so maintains enzyme and metabolic

Table 1. Comparison of pyruvate kinase isoenzymes from the hibernating and non-hibernating Arctic ground squirrel.

Parameter	Non-hibernating	Hibernating
Molecular weight	243,000	243,000
Isoelectric point	5.2 - 5.3	5.74
Affinity for PEP	Decreases with a fall in temperature	Temperature independent
Affinity for ADP	Increases with a fall in temperature	Temperature-independent
Affinity for FDP	Temperature-independent	Temperature-independent but decreased
Affinity for ATP	Decreases with a fall in temperature	Temperature-independent but decreased
Q_{10}(5-37°C) at 0.5 mM PEP and V_{max} conc. of ADP	1.80	2.51
(5-37°C) at 0.2 mM PEP and V_{max} conc. of ADP	1.27	2.21
(5-37°C) at 0.1 mM PEP and V_{max} conc. of ADP	1.21	2.48
Q_{10}(5-37°C) at 0.08 mM ADP (near K_m) and V_{max} conc. of PEP	1.34	2.55

activity more or less constant over a wide and physiological temperature range (9, 58, 59). The conspicuous thermal stability of the apparent Km's for the substrates and the modulators in the hibernating PyK would of course have to be supported by relatively constant (within limits) concentrations of these intermediates. In this application, incubation of liver slices from the hibernating ground squirrel over the same temperature range showed a surprising constancy in the measured concentrations of these intermediates (Behrisch, unpublished data). The major exceptions to this general rule were the adenine nucleotides, the concentrations of which change rapidly at the higher temperatures, indicating (not surprisingly) a more rapid depletion of the *in vitro* liver cell's energy charge (Behrisch, unpublished). Moreover, the measured concentrations of the metabolites PEP, ADP, ATP and FDP were significantly close to our calculated values of the apparent Km's for these metabolites, suggesting some confidence in our interpretation of the data.

V. THE VALUE OF REGULATORY HOMEOSTASIS

The observed relatively low affinity of the hibernating PyK for the inhibitor ATP also deserves some comment. It is reasoned that this would tend to favor the maintenance of a high energy charge that in turn could support some limited biosynthetic sequences such as gluconeogenesis. Further, the partial desensitization of a key enzyme with a very high $-\Delta G$ to its major inhibitor may be an efficient way of ensuring the continued evolution of small amounts of heat during the deep hibernating state and much greater amounts during the period of arousal when body temperatures must rise rapidly (121). In this context, the liver of the hibernator may well play a significant role in thermogenesis (116). It is of interest that the hibernating PyK also shows a relatively low affinity for its allosteric activator, FDP, that parallels closely its low affinity for ATP, the inhibitor (10). This result is at first glance difficult to interpret as one might have expected a rise in affinity for the activator, and as Flory *et al.* (46) showed the Km concentrations of two inhibitors of PyK, ATP and alanine, together caused virtually complete inhibition of catalytic activity, although we were able to show that at least in the ground squirrel, modulation of this enzyme by alanine does not seem to be a meaningful method of control (21). In this context, Moon and Borgmann (93) put forth the view that a concomitant reduction in affinity for both the activator and the inhibitor would be "counter-productive" and that the organism might be better served by a "strategy" of enhanced affinity for the activator at low temperature. Upon reflection, however, if these metabolites are to exert a regulatory influence on controlling enzymes, it is clear that their concentrations must be able to change according to the demands and supplies of the cell, and do so on a minute-to-minute, or more probably on a second-to-second basis (56) and not seasonally. In addition, it should not be surprising to observe synergism between the effectors of a regulatory enzyme (80, 81). It would seem energetically wasteful to render a regulatory enzyme like pyruvate kinase completely insensitive to the cell's major indicator of its energy charge (3) and inhibitor of energy-yielding reactions. This retention of complete regulatory control of PyK at all temperatures in the hibernator is likely very significant as it appears the glycolytic intermediates are largely blocked from entering the Krebs cycle (120), and the hibernating animal, like the diver, must conserve its high-energy metabolites, of which PEP is the richest. Maintenance of total allosteric control at the level of PyK would seem to be an efficient way of husbanding such energetically valuable intermediates.

The regulation of enzyme activity is necessarily costly, and various estimates have been made of the portion of the cell's budget that is allocated to this integral segment of metabolism (see 96, 97 for review). Whatever the cost in terms of calories or molecules of ATP, it is reasonable to suppose that it is of advantage to the cell or organism to hold constant, relative to the cell's energy budget, the cost of metabolic regulation. This cost is computed in numbers of molecules of the relevant metabolites, and the remarkably constant stoichiometry - regulatory homeostasis - between the substrates and modulators of the PyK reaction in the hibernator would seem to be an efficient mechanism for holding that cost unchanged.

Finally, the striking contrast between the Q_{10} values for the hibernating and non-hibernating PyK's (Table 1) deserves comment. The large effect of temperature on the reaction rate in the winter PyK would seem to be a mechanism that would account for a large portion of the reduction in metabolism at this level as over the range of 0 to 37°C, a Q_{10} effect of 2.5 yields a factor of about 40, vs. a Q_{10} of 1.34 which yields a factor of 3.2.

VI. ROLE OF THE INTRACELLULAR MILIEU

A good deal of evidence indicates that the intracellular concentrations of numerous ions vary during adaptation to temperature in ectotherms (38, 57, 62, 63) and heterothermic mammals (44, 76, 77, 126, 127, 128), and a steadily growing literature on ionic effects on enzyme activity exists (for a comprehensive list see 41, also 5, 8, 35). What is the effect of such alterations in ion concentrations on the regulation of enzyme activity in the hibernating mammal? Do most of these ions affect enzyme activity as modulators, act as substrates, or is their role more that of maintaining an intracellular environment in which the macromolecules of metabolism can function? Most likely the answer is a combination of these, and other alternatives, as a number of metabolites are not accepted as the true metabolites by their enzymes unless they occur as complexes with certain ions (21, 90).

We were able to gain some insight into this question by experiments on the allosteric behavior of the pyruvate kinase of the Arctic ground squirrel (22). The cations K^+, NH_4^+, Mg^{++} and Mn^{++} are recognized as having an influence on activity of PyK's from a variety of sources (40, 98). Of these four, the two bivalent cations Mn^{++} and Mg^{++} appear to play a role directly in the catalytic process as ADP and ATP are bound only as the adenylate-cation complexes and the enzyme requires a bivalent ion for an active conformation (90, 115). Saturation curves for the metabolites PEP, ADP, ATP, and FDP yield allosteric kinetics that are supported by Hill coefficients that show a corresponding degree of homotropic cooperation in the binding of these intermediates (21). As mentioned previously in this communication, the non-hibernating PyK is sensitive to temperature in its affinity for the four intermediates in diverse ways. Thus values of the apparent Km's for some of these metabolites (ADP, PEP) vary considerably over the temperature range tested (Fig. 2). Values of the Hill coefficients are inversely proportional to the apparent Km values, so that when temperature causes a decrease in Km for a particular metabolite, say PEP, the Hill coefficient rises (Fig. 2). This behavior is observed for the binding of all four of PEP, ADP, ATP and FDP. In sharp contrast, saturation curves for the cations K^+, Mg^{++} and Mn^{++} yield Hill values that are directly proportional to calculated values of the apparent Km's (Fig. 3). Thus when the apparent Km for K^+ rises with increasing temperature, the attendant Hill values also rise. When the apparent Km's for Mg^{++} or Mn^{++} rise with increasing temperature they also are paralleled by an increase in the corresponding Hill coefficients, indicating, paradoxically, what appears to be a rise in site-site interaction in the binding of these ions as affinity decreases. Virtually identical behavior is observed in other enzymes from a variety of ectotherms (6, 7, 19, 20). Clearly then,

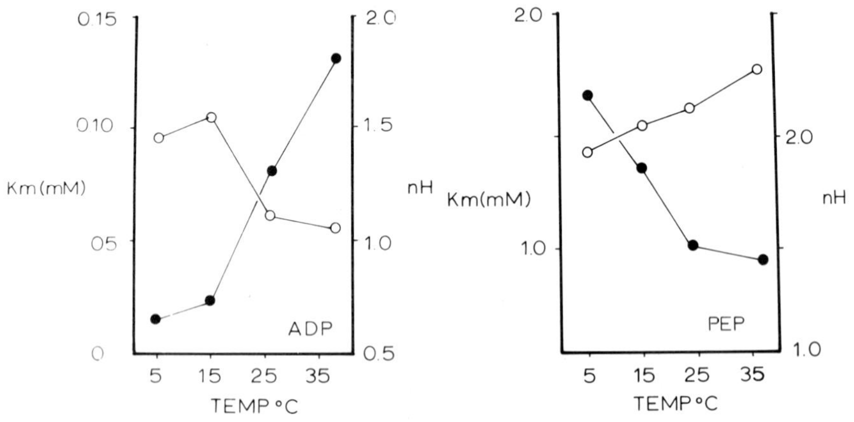

Fig. 2. *Effect of temperature on apparent Michaelis constants (●) for the substrates ADP and PEP and the attendant Hill coefficient (o) in pyruvate kinase from the Arctic ground squirrel.*

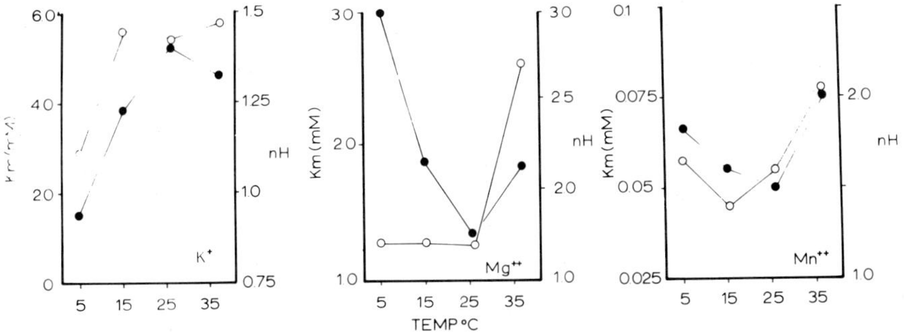

Fig. 3. *Effects of temperature on Km (●) of non-hibernating ground squirrel pyruvate kinase for the cations* K^+ *and* Mg^{++}. *Included are the values of the Hill coefficients (o) for the saturation curves of these ions.*

fundamental differences underlie the responses of this key enzyme to certain metabolites and to the cations, all of which are known to exert an influence on enzyme activity.

The cornerstone of our understanding of enzyme regulation is the principle of allosteric behavior. This in turn rests on the modulation of the degree of interaction or cooperation between the enzyme's component subunits. In their classic discussion of the role of allosterism in enzyme function, Monod, Wyman and Changeux (92) generate a model (to be later borne out in 39) which states that an increase in an enzyme's affinity for a physiological substrate or modulator is accompanied, indeed based on, an enhanced cooperation between the sites or subunits that bind that ligand. Therefore a decline in Km (an increase in affinity) is rooted in an enhanced degree of site-site cooperation in ligand-binding (see for example Fig. 2). The corollary of this principle is that in order to be an effective modulator of enzyme activity the ligand must be able to affect the cooperation between the subunits in such a way that affinity for it (or another ligand) is directly controlled. For example, in the regulation of aspartate transcarbamylase from *Escherichi coli* the inhibitor of the enzyme, cytidine triphosphate, effects a reduction in affinity for substrate (51, 52).

Similarly, ATP brings about a decrease in affinity of the ground squirrel PyK for the substrates (10, 21). It should be noted that in the ground squirrel PyK, only the four metabolites PEP, ADP, ATP and FDP behave in such a manner, while the cations K^+, Mg^{++} and Mn^{++} behave in a non-allosteric fashion (22). Judging from these data it appears that the intermediates play a regulatory role in the integration of metabolism while the ions do not.

In every case of which I am aware, the regulatory mechanisms of intermediary metabolism are of use to the cell or organism only when they are effective under physiological conditions of substrates, modulators and the solutes and solvents of the surrounding medium. In these terms a logical alternate role for major cations such as Na^+, K^+, Mg^{++} and Mn^{++} (which do change in the hibernator) is the maintenance of an intracellular climate that permits continued allosteric function so that control of regulatory enzymes remains equally efficient at all temperatures that the animal experiences. We (18) were able to explore this possibility through the clean work of William Galster at the Institute of Arctic Biology, who determined the time course of changes of various cations in selected tissues of the Arctic ground squirrel throughout the year. In liver of the ground squirrel, concentrations of K^+ decrease and concentrations of Na^+ increase during hibernation; upon arousal they are then restored to levels of the pre-hibernating state. Similarly, Mg^{++} concentrations in the tissue fall while those of Ca^{++} increase during hibernation. It is perhaps significant that these ions are antagonistic or exchange pairs, as Na^+ is the exchange ion for K^+ in many transport situations, and Ca^{++} is often an inhibitor of enzymes that require Mg^{++} either as a cofactor or as part of a substrate-cation complex (6, 19, 20, 75, 90, 129). In this context, Krebs *et al.* (84) found that gluconeogenesis is dependent upon the availability of Ca^{++} in the medium, lending some weight to these data.

Determination of activity of hibernating and non-hibernating Pyk's from the Arctic ground squirrels in the presence of concentrations of Ca^{++}, Mg^{++}, Na^+ and K^+ that are characteristic of the deep hibernating state yields results that are essentially the same as those for the saturation kinetics for the ligands PEP, ADP, ATP and FDP. The only discernible effect of this treatment is an elevation of the temperature coefficient, particularly in the hibernating PyK. Thus under conditions of substrates, modulators and ions that reflect the hibernating state, the Q_{10} of the PyK reaction is about 3.4 for the hibernating PyK and around 2 in the non-hibernating PyK. Thus the hibernating enzyme appears to be altered in such a way that permits the cell to take advantage of the large Q_{10} effect in the reduction of metabolism as the animal enters hibernation (18). It may be noted that this (conversely) is also an apparent adaptation that allows the cell to take advantage of increases in body temperature during the early stages of arousal when metabolism and body temperature must rise rapidly. In the animal entering the hibernating state this involves a decrease in activity to a rate that is less than 1/100th of the rate at 37°C, a factor in striking agreement with the observed reduction in total metabolic rate for this species. In other words the hibernator adapts apparently through the possession of enzyme variants that are able to utilize the change in body temperature.

The allosteric behavior of an enzyme is based on a constellation of weak bonds that render its conformation flexible. Such associations may be between polar and apolar groups. It appears that the proportions of these associations that contribute to the total forces required to hold the enzyme together also determine the thermal relationship of the enzyme. The numerous thermally labile enzymes from various homeothermic sources appear to owe their large loss of catalytic activity and concomitant change in the nature of substrate binding to a low-temperature-induced dissociation of the enzyme molecule (25, 67, 71). It is held that it is the associations between the apolar groups, which are significantly weakened by low temperature, that are responsible for the observed cold inactivation/dissociation (70, 106). The present findings with the hibernating PyK suggest that possibly the number and/or importance of these types of groups are diminished in the regulation of conformation in an enzyme that must function over a wide range of temperatures (10).

These results suggest a formal mechanism for

continued allosteric control at all temperatures. Further, they suggest a plausible method by which the hibernator can take advantage of a large drop in body temperature in maintaining a reduced rate of consumption of energy stores, in this way maintaining a rigid and vital control over the caloric economy of the cell.

VII. A CIRCANNUAL CYCLE IN ISOENZYME PRODUCTION

The mechanisms that have been briefly described here are predicated on a seasonal appearance of enzyme variants or isoenzymes that are designed to allow efficient metabolic control at all temperatures the animal experiences. In this respect at least the obligatory hibernator is like a number of ectotherms that seem to restructure certain cellular components in response to a variety of environmental stimuli (1, 9, 10, 11, 32, 36, 37, 59) while the salient characteristics that are responsible for metabolic control are retained throughout the year.

It seems clear that whatever the adaptive properties of the cellular components that are peculiar to the hibernating or non-hibernating state, they would not confer any advantage on the animal if it did not possess the appropriate alteration before actually entering the hibernating or non-hibernating state. In this connection, hibernating ground squirrels that are preparing to emerge finally from hibernation in about 2 to 3 weeks already possess substantial activities of the (summer active) non-hibernating liver PyK (10), while with one week to go the hibernating PyK has largely disappeared (Fig. 4). A similar but opposite change is observed in animals that are preparing for hibernation. This is felt to be a significant result because it strongly supports the notion that the hibernator "anticipates", apparently without any external stimulus, a cyclic and very dramatic change from one metabolic state to another. Further, it is convincing evidence of a circannual cycle within the animal (see also Ref. 1).

The process of seasonal (pre-) adaptation to, or exit from hibernation appears to follow a definite time-course, and it is likely that it is under tight control. Based on isoelectric focussing of PyK and

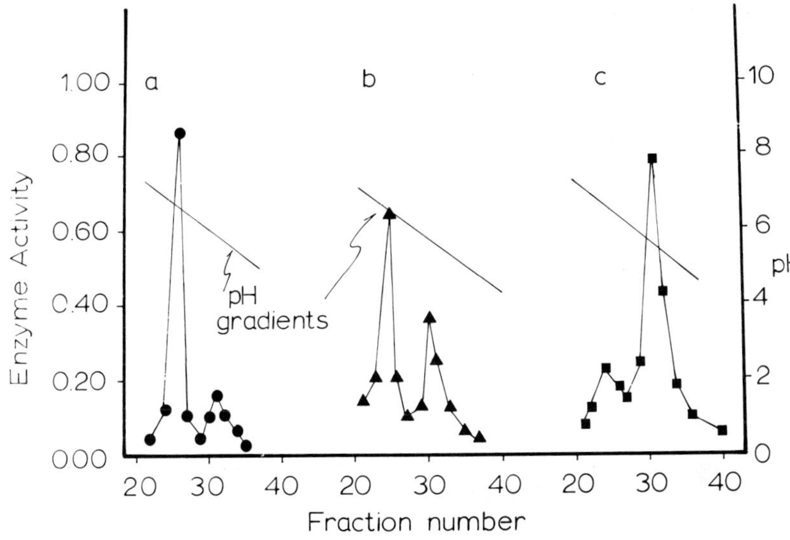

Fig. 4. *Influence of time in the hibernaculum on isoenzyme patterns of pyruvate kinase in the hibernating ground squirrel. The data in (a) are from an animal sacrificed at about 3 weeks before final emergence from hibernation; the results in (b) are from an animal at about 2 weeks prior to emergence; and those from (c) are from a ground squirrel that was judged to be about one week away from final arousal. In each case the total enzyme activity remained unaltered, the hibernating variant of the enzyme disappearing while the non-hibernating variant increased in concentration.*

FDPase from ground squirrels sacrificed at various periods prior to final arousal, it appears that in this animal the molecular adaptations seem to take place within a period of one month, although of course fattening proceeds all through the summer and early fall. At 42 to 35 days before final arousal the non-hibernating variant of PyK or FDPase is completely absent. Then between 35 to 28 days before emergence the non-hibernating PyK variant appears, and continues to increase almost linearly up to the time of arousal. Significantly, there is a parallel decrease in the activity of the hibernating PyK, so that total activity of the enzyme is not altered (Fig. 4) throughout the year, a finding also reported for liver phosphofructokinase from the 13-lined ground squirrel (29). It appears therefore that the breakdown or interconversion of the old (now non-adaptive) isoenzyme is a finely regulated process. A growing body of evidence indicates that the metabolic interconversion through the attachment or detachment of functional groups on the enzyme is a wide-spread method of control of enzyme activity (30, 45, 64, 108). On the other hand the interconversion of isoenzyme forms could entail the alteration of a small portion of a multimeric enzyme, as in FDPase from muscle of the Alaskan king crab (11). In this way a minor alteration of the structure of an enzyme molecule is responsible for a conformational change that renders the entire enzyme molecule more efficient as an allosteric protein in the new thermal regime. In this connection, Pyk also appears to be interconvertible by way of the reversible attachment of an undetermined number of phosphate groups (43, 64). Mechanisms such as these would spare a significant amount of genetic and metabolic energy in rendering a major portion of the cell's constituents functional over a wide range of temperatures.

VIII. METABOLIC STRATEGY VS. PHYSIOLOGICAL POSSIBILITIES

The increasingly frequent use of the term "strategy" in biological processes deserves comment. It does not seem physiologically possible (or genetically fair) to propose that the animal retailor its metabolic make-up to enable it to fit all environmental needs. Clearly it cannot, and relatively few species exist with the ability to adapt to wide ranges in the physical factors of the environment. Also, while on the organismal level the differences between animals appear large and sometimes dramatic, on the molecular level these differences are far more subtle, if at all discernible. For example, it can fairly be said that on the whole the principles of molecular genetics that were first formulated through a series of implosive studies on micro-organisms hold for virtually all living things. Similarly, the factors that regulate the metabolic economy of the cell are part of the metabolic sequences themselves. They evolved as integral parts of the pathways and did not become somehow grafted on to pre-existing sequences at a later stage in evolution. The regulatory mechanisms are therefore conservative and change extremely little, if at all. It seems therefore unsound to propose "strategies" of metabolic regulation that are based on fundamental changes in the scheme and balance of metabolic control. Rather a more plausible mechanism of adaptation is one that preserves the "regulatory homeostasis" of the cell, so that the relative cost of metabolic control is maintained constant and affordable under all conditions. Indeed, it is worth noting that such a mechanism of adaptation is satisfied by the conspicuously invariant indices of enzyme-ligand affinity and allosteric behavior - in other words, control - in the hibernating PyK, a very specific example to be sure. However in spite of this, because the life habits of most organisms are so different from one group to the next or one season to the other, and hence provocative to study, the notion that an observed character is of adaptive significance is very appealing, although clearly little can be said about the "strategic" value of the trait without some considerable knowledge of how it fits into the complex of adaptive characters that is the living organism. Because any adaptation to a change in environmental conditions must fit into all other processes and their regulation, the possibilities that are open to the organism in re-structuring the cell are extremely limited. The term "strategy" is used only teleologically to pose a usually quite vulnerable hypothesis, while the organism is faced with having to satisfy its particular environmental demands with the limited possibilities supplied by its genome.

ACKNOWLEDGEMENTS

This paper is dedicated to Professor Laurence Irving, whose critical encouragement and intellectual honesty have been a constant source of inspiration to the author. The author also wishes to thank Professor Peter Morrison for his encouragement and support during the course of these studies. Funding for the experiments described in this paper came from the National Institutes of Health (NIGMS 10402).

REFERENCES

1. Armitage, K. B. and E. Shulenberger. Evidence for a circannual metabolic cycle in *Citellus tridecemlineatus*, a hibernator. Comp. Biochem. Physiol. 42A: 667-668, 1972.
2. Arthur, H. and K. Watson. Thermal adaptation in yeast: growth temperatures, membrane lipid and cytochrome composition of psychrophilic, mesophilic and thermophilic yeasts. J. Bact. 128: 56-68, 1976.
3. Atkinson, D. E. The energy charge of the adenylate pool as a regulatory parameter. Interaction with feedback modifiers. Biochem. 7: 4029-4034, 1968.
4. Barash, D. P. Pre-hibernation behavior of free-living hoary marmots, *Marmota caligata*. J. Mammal. 57: 182-185, 1976.
5. Bashan, N., S. Moses, Y. Gross and A. Livne. The effect of Na^+ and K^+ on glycolytic enzymes: differential response of pyruvate kinase from dog and human erythrocytes. FEBS Letters 54: 323-326, 1975.
6. Behrisch, H. W. Temperature and the regulation of enzyme activity in poikilotherms: regulatory properties of fish fructose 1, 6-diphosphatase. PhD. Thesis. University of British Columbia, 1969, 152 pp.
7. Behrisch, H. W. Temperature and the regulation of enzyme activity in poikilotherms. Regulatory properties of fructose diphosphatase from muscle of the Alaskan king crab. Biochem. J. 121: 399-409, 1971.
8. Behrisch, H. W. Molecular mechanisms of adaptation to low temperature in marine poikilotherms, some regulatory properties of dehydrogenases from two arctic species. Marine Biology 13: 267-275, 1972.
9. Behrisch, H. W. Molecular mechanisms of temperature adaptation in arctic ectotherms and heterotherms. In: Effects of Temperature on Ectothermic Organisms. Wieser, W. (editor). Berlin: Springer Verlag, 1973, pp. 123-137.

10. Behrisch, H. W. Temperature and the regulation of enzyme activity in the hibernator. Isoenzymes of liver pyruvate kinase from the hibernating and non-hibernating Arctic ground squirrel. Can. J. Biochem. 52: 894-902, 1974.
11. Behrisch, H. W. On the subunit structure of seasonal isoenzymes of fructose 1, 6-diphosphatase from muscle of the Alaskan king crab *Paralithodes camtschatica*. Comp. Biochem. Physiol. 51B: 317-321, 1975.
12. Behrisch, H. W. Molecular adaptation to the diving habit. On the possible role of glycerol in kidney metabolism. Submitted. J. Comp. Physiol.
13. Behrisch, H. W. Molecular evidence for a circannual rhythm in the hibernator. Studies on pyruvate kinase from the Arctic ground squirrel. Submitted. J. Thermal Biol.
14. Behrisch, H. W. and R. Elsner. Fructose diphosphatase from kidney and liver of a diving mammal, the harbour seal. Physiologist 18: 135, 1975.
15. Behrisch, H. W. and R. Elsner. Molecular adaptations to asphyxia. Fructose diphosphatase of a diver, the harbour seal. Submitted. J. Comp. Physiol.
16. Behrisch, H. W. and R. Elsner. Molecular adaptations to the diving habit. Comparison of aspects of enzyme regulation in heart and kidney of the harbour seal and the dog. In preparation.
17. Behrisch, H. W. and R. Elsner. Molecular adaptations to the diving habit. On a possible hierarchy for available substrates among tissues that remain perfused during the dive. In preparation.
18. Behrisch, H. W. and W. A. Galster. Regulation of enzyme activity in the hibernator. Modulation of temperature effects in pyruvate kinase from the Arctic ground squirrel by cations. Submitted. J. Thermal Biol.
19. Behrisch, H. W. and P. W. Hochachka. Temperature and regulation of enzyme activity in poikilotherms. Properties of trout fructose diphosphatase. Biochem. J. 111: 287-295, 1969.

20. Behrisch, H. W. and P. W. Hochachka. Temperature and the regulation of enzyme activity in poikilotherms. Properties of lungfish fructose diphosphatase. Biochem. J. 112: 601-607, 1969.
21. Behrisch, H. W. and C. E. Johnson. Regulatory properties of pyruvate kinase from liver of the summer active Arctic ground squirrel. Can. J. Biochem. 52: 547-559, 1974.
22. Behrisch, H. W. and C. E. Johnson. Regulation of enzyme activity in the hibernator. Role of cations in the control of liver pyruvate kinase in the Arctic ground squirrel. Submitted. J. Thermal Biol.
23. Behrisch, H. W., I. Ortner and W. Wieser. Temperature and the regulation of enzyme activity in heterothermic tissues of an alpine mammal. Lactate dehydrogenase from skeletal muscle of the chamois. J. Thermal Biol. In press, 1977.
24. Behrisch, H. W. and J. A. Percy. Temperature and the regulation of enzyme activity in homeothermic and heterothermic tissues of arctic marine mammals. Some regulatory properties of 6-phosphogluconate dehydrogenase from adipose tissue of the spotted seal *Phoca vitulina*. Comp. Biochem. Physiol. 47B: 437-443, 1974.
25. Bock, P. E., H. R. Gilbert and C. Frieden. Analysis of the cold lability behavior of rabbit muscle phosphofructokinase. Biochem. Biophys. Res. Comm. 66: 564-569, 1975.
26. Bonney, R. J., H. A. Hopkins, P. R. Walker and V. R. Potter. Glycolytic isoenzymes and glycogen metabolism in regenerating liver from rats on controlled feeding schedules. Biochem. J. 136: 115-124, 1973.
27. Bonney, R. J., P. R. Walker and V. R. Potter. Isoenzyme patterns in parenchymal and non-parenchymal cells isolated from regenerating and regenerated rat liver. Biochem. J. 136: 947-954, 1973.

28. Borgmann, A. I. and T. W. Moon. Enzymes of the normothermic and hibernating bat, *Myotis lucifugus*. J. Comp. Physiol. 107: 185-199, 1976.
29. Brabec, M. J. and R. J. McCollough. Phosphofructokinase activity in the brown fat, liver, and heart of hibernating and homeothermic thirteen-lined ground squirrels (*Citellus tridecemlineatus*) Int. J. Biochem. 1: 557-560, 1970.
30. Brand, I. A., M. K. Mueller, C. Unger and H. D. Soeling. *In vivo* and *in vitro* interconversions of active and inactive forms of phosphofructokinase in rat liver. FEBS Letters 68: 271-274, 1976.
31. Bruce, D. S. and J. E. Wiebers. Mineral Dynamics during hibernation and chronic immobility: a review. Aerospace Med. 40: 855-861, 1969.
32. Burlington, R. F., W. D. Bowers, R. C. Daum and P. Ashbaugh. Ultrastructural changes in heart tissue during hibernation. Cryobiology 9: 224-228, 1972.
33. Burlington, R. F. and G. J. Klain. Gluconeogenesis during hibernation and arousal from hibernation. Comp. Biochem. Physiol. 22: 701-708, 1967.
34. Burlington, R. F. and J. H. Sampson. Distribution and activity of lactic acid dehydrogenase isoenzymes in tissues from a hibernator and a non-hibernator. Comp. Biochem. Physiol. 25: 185-192, 1968.
35. Bygrave, F. L. The ionic environment and metabolic control. Nature (London) 214: 667-671, 1967.
36. Caldwell, R. S. Thermal compensation of respiratory enzymes in tissues of the goldfish (*Carassius auratus* L.). Comp. Biochem. Physiol. 31: 79-93, 1969.
37. Caldwell, R. S. and F. J. Vernberg. The influence of acclimation temperature on the lipid composition of fish gill mitochondria. Comp. Biochem. Physiol. 34: 179-191, 1970.
38. Catlett, R. H. and D. R. Millich. Intracellular and extracellular osmoregulation of temperature-acclimated goldfish: *Carassius auratus* L. Comp. Biochem. Physiol. 55A: 261-269, 1976.

39. Changeux, J. P. and M. M. Rubin. Allosteric interactions in aspartate transcarbamylase. III. Interpretations of experimental data in terms of the model of Monod, Wyman and Changeux. Biochem. 7: 553-561, 1968.
40. De Asua, J. E. Rozengurt and H. Carminatti. Some kinetic properties of liver pyruvate kinase (Type L). III. Effect of monovalent cations on its allosteric behavior. J. Biol. Chem. 245: 3901-3905, 1970.
41. Dixon, M. and E. D. Webb. The Enzymes. 2nd Ed. London: Longmans, Green and Co., 1964, 950 pp.
42. Elsner, R., D. Franklin, R. Van Citters and D. Kenney. Cardiovascular defense against asphyxia. Science 153: 941-949, 1966.
43. Ekman, P., U. Dahlqvist, E. Humble and L. Engstrøm. Comparative studies on the L-type pyruvate kinase from rat liver and the enzyme phospholated by cyclic 3'5' AMP stimulated protein kinase. Biochem. Biophys. Acta 429: 374-382, 1976.
44. Ferren, L. G., F. E. South and H. K. Jacobs. Calcium and magnesium levels in tissues and serum of hibernating and cold-acclimated hamsters. Cryobiology 8: 506-508, 1971.
45. Fischer, E., P. Cohen, M. Fosset, L. Muir and J. Saari. Comparative and evolutionary aspects of the control of phosphorylase. In: Metabolic Interconversion of Enzymes. Wieland, O., E. Helmerich and E. Holzer (editors). Berlin: Springer Verlag, 1972, pp 11-29.
46. Flory, W., B. D. Peczon, R. E. Koeppe and H. O. Spivey. Kinetic properties of rat liver pyruvate kinase at cellular concentrations of enzyme, substrates and modifiers. Biochem. J. 141: 127-131, 1974.
47. Folk, G. E. Physiologic observations of subarctic bears under winter den conditions. In: Mammalian Hibernation III. Fisher, K. C., A. R. Dawe and C. P. Lyman (editors). New York: Elsevier Publ. Co., 1967, pp 75-85.
48. Folk, G. E., A. Larson and M. A. Folk. Physiology of hibernating bears. In: Bears: Their

Biology and Management 3rd Int. Conf. on Bear Research and Management. Pelton, M. R., J. W. Lentfer and G. E. Folk (editors). Binghampton and Moscow: 1976, pp. 373-380.

49. Galster, W. A. and P. R. Morrison. Cyclic changes in carbohydrate concentrations during hibernation in the Arctic ground squirrel. Am. J. Physiol. 218: 1228-1232, 1970.

50. Galster, W. A. and P. R. Morrison. Seasonal changes in body composition of the Arctic ground squirrel, *Citellus undulatus*. Can. J. Zool. 54: 74-78, 1976.

51. Gerhart, J. C. Subunits for the control and catalysis in aspartate transcarbamylase. Brookhaven Symp. Biol. 17: 222-231, 1964.

52. Gerhart, J. C. and H. K. Schachman. Allosteric interactions in aspartate transcarbamylase. II. Evidence for different conformational states of the protein in the presence and absence of specific ligands. Biochem. 7: 539-552, 1968.

53. Guillon, G. and J. Massoulie. Multiplicite des forms moleculaires de l'acetylocholinasterase at acclimatation thermique chez le *Carassius auratus*. Biochimie 58: 465-471, 1976.

54. Harris, E. J., C. Berent, and J. Wimhurst. Stimulation of mitochondrial pyruvate carboxylation by Mn^{2+} and the distribution of the products between mitochondrion and medium. FEBS Letters 6(2): 93-98, 1970.

55. Helmreich, E. and D. F. Cori. The effects of pH and temperature on the kinetics of the phosphorylase reaction. Proc. Natl. Acad. Sci. U.S. 52: 647-654, 1964.

56. Hess, B. Biochemical regulations. In: Systems Theory and Biology. Mesarovic, M. D. (editor). New York: Springer Verlag, 1968, 287 pp.

57. Hickman, C. P., R. A. McNabb, R. A. Nelson, E. D. Van Breemen and D. Comfort. Effect of cold acclimation on electrolyte distribution in rainbow trout (*Salmo gairdneri*). Can. J. Zool. 42: 577-597, 1964.

58. Hochachka, P. W. and G. N. Somero. The adaptation of enzymes to temperature. Comp. Biochem. Physiol. 27: 659-668, 1968.

59. Hochachka, P. W. and G. N. Somero. Strategies of Biochemical Adaptation. Philadelphia: W. B. Saunders, 1973, 358 pp.

60. Hock, R. J. Seasonal variations in physiologic functions of Arctic ground squirrels and black bears. In: Mammalian Hibernation. Lyman, C. P. and A. R. Dawe (editors). Bull. Museum Comp. Zool. (Harvard) 124: 155-171, 1960.

61. Hoffman, K. H. Catalytic efficiency and structural properties of invertebrate muscle pyruvate kinases: correlation with body temperature and oxygen consumption rates. J. Comp. Physiol. 110: 185-195, 1976.

62. Houston, A. H. and J. A. Madden. Environmental temperature and plasma electrolyte regulation in the carp, *Cyprinus carpio*. Nature (London) 217: 969-970, 1968.

63. Houston, A. H. and J. A. Madden. Osmoregulatory adaptation of steelhead trout (*Salmo gairdneri*) to sea water. Can. J. Zool. 37: 729-748, 1969.

64. Ibsen, K. H. and P. Trippet. Interconvertible and noninterconvertible forms of fat pyruvate kinase. Biochem. 11: 4442-4450, 1972.

65. Ingraham, J. L. and O. Maaløe. Cold-sensitive mutants and the minimum temperature of growth in bacteria. In: Molecular Mechanisms of Temperature Adaptation. Prosser, C. L. (editor). Washington: AAAS Press, 1967, 297 pp.

66. Irving, L. Changes in the blood flow through the brain and muscle during the arest of breathing. Am. J. Physiol. 122: 207-214, 1938.

67, Iwatsuki, N. and R. Okazaki. Mechanism of regulation of deoxythymidine kinase of *Escherichia coli*. II. Effect of temperature on the enzyme kinetics and activity. J. Mol. Biol. 29: 155-165, 1967.

68. Johnson, G. S., M. S. Kayne and W. C. Deal. Metabolic control and structure of glycolytic enzymes. VII. Reversal of the dissociation of rabbit muscle pyruvate kinase into unfolded subunits. Biochem. 8: 2455-2462, 1969.

69. Joyce, C. M. The effect of environmental temperature on succinic dehydrogenase activity in ear skin of the rabbit. Proc. Royal Irish Acad. 65B: 425-435, 1967.

70. Kauzmann, W. Some factors in the interpretation of protein denaturation. Adv. Protein Chem. 14: 1-63, 1959.

71. Kayne, F. J. and C. H. Suelter. The temperature-dependent conformational transition of pyruvate kinase. Biochem. 7: 1678-1684, 1968.

72. Kayser, C. The Physiology of Natural Hibernation. New York: Pergamon Press, 1961, 236 pp.

73. Kayser, C. and G. Hildwein. La thermoregulation des hibernants a l'etat actif. Arch. Sci. Physiol. 28: 333-360, 1974.

74. Kerem, D., D. D. Hammond and R. Elsner. Tissue glycogen levels in the Wedell seal, *Leptonychotes weddelli*: a possible adaptation to asphyxial hypoxia. Comp. Biochem. Physiol. 45A: 731-736, 1973.

75. Kimmich, G. A. and H. Rasmussen. Regulation of pyruvate carboxylase activity by calcium in intact rat liver mitochondria. J. Biol. Chem. 244: 190-197, 1969.

76. Kimzey, S. L. and J. S. Willis. Resistance of erythrocytes of hibernating mammals to loss of potassium during hibernation and during cold storage. J. Gen. Physiol. 58: 620-633, 1971.

77. Kimzey, S. L. and J. S. Willis. Temperature adaptation of active sodium-potassium transport and of passive permeability in erythrocytes of ground squirrel. J. Gen. Physiol. 58: 634-649, 1971.

78. Klain, G. J. 1976. Hepatic utilization of glutamate during the hibernation cycle. Life Sci. 19: 591-596, 1976.

79. Klain, G. J. and B. K. Whitten. Carbon dioxide fixation during hibernation and arousal from hibernation. Comp. Biochem. Physiol. 25: 363-366, 1968.

80. Klunsøyr, L. and D. E. Atkinson. Regulatory properties of phosphoribosyladenosine triphosphate synthetase. Synergism between

adenosing monophosphate, phosphoribosyl-adenosine triphosphate and histidine. Biochem. 9: 2021-2027, 1970.

81. Klungsøyr, L., J. H. Hageman, L. Fall and D. E. Atkinson. Interaction between energy charge and product feedback in the regulation of biosynthetic enzymes. Biochem. 7: 4035-4040, 1968.

82. Kohl, E. A. and G. L. Cottam. Alteration in liver pyruvate kinase protein and catalytic activity upon starvation and refeeding. Arch. Biochem. Biophys. 176: 671-682, 1973.

83. Kono, N. and K. Uyeda. Chicken liver phosphofructokinase. II. Cold inactivation. J. Biol. Chem. 25: 8003-8009, 1973.

84. Krebs, H. A., D. A. H. Bennett, P. De Gascoyne and T. Yoshida. Renal gluconeogenesis: the effect of diet on the gluconeogenic capacity of rat-kidney-cortex slices. Biochem. J. 86: 22-31, 1963.

85. Kuczenski, R. T. and C. H. Suelter. Effect of temperature and effectors on the conformation of yeast pyruvate kinase. Biochem. 9: 939-945, 1970.

86. Lundberg, D. A., R. A. Nelson, H. W. Wahner and J. D. Jones. Protein metabolism in the black bear before and during hibernation. Mayo Clin. Proc. 51: 716-722, 1976.

87. McMurchie, E. J., J. K. Raison and K. D. Cairncross. Temperature-induced phase changes in membranes of heart: a contrast between the thermal response of poikilotherms and homeotherms. Comp. Biochem. Physiol. 44B: 1017-1026, 1973.

88. Martin, C. E., K. Hiramitsu, Y. Kitamija, Y. Nozawa, L. Skriver and G. Thompson. Molecular control of membrane properties during temperature acclimation. Fatty acid desaturase regulation of membrane fluidity in acclimating *Tetrahymena* cells. Biochem. 15: 5218-5227, 1976.

89. Matson, J. R. Notes on dormancy in the black bear. J. Mammal. 27: 203-212, 1946.

90. Melchior, J. B. The role of metal ions in the pyruvate kinase reaction. Biochem. 4: 1518-1525, 1965.

91. Meyer, M. and P. R. Morrison. Tissue respiration and hibernation in the thirteen-lined ground squirrel, *Spermophilus tridecemlineatus*. Bull. Museum Comp. Zool. (Harvard) 124: 405-420, 1960.
92. Monod, J., J. Wyman and J. P. Changeux. On the nature of allosteric transitions: A plausible model. J. Mol. Biol. 12: 88-118, 1965.
93. Moon, T. W. and A. I. Borgmann. Enzymes of the normothermic and hibernating bat, *Myotis lucifugus*. Metabolites as modulators of pyruvate kinase. J. Comp. Physiol. 107: 201-210, 1976.
94. Nelson, R. A. Winter sleep in the black bear: a physiological and metabolic marvel. Mayo Clinic Proc. 48: 733-737, 1973.
95. Nelson, R. A., H. W. Wahner, and J. D. Jones. Metabolism of bears before, during and after winter sleep. Am. J. Physiol. 224: 491-496, 1973.
96. Newsholme, E. A. and B. Crabtree. Metabolic aspects of enzyme activity regulation. Symosia Soc. Exp. Biol. 27: 429-460, 1973.
97. Newsholme, E. A. and C. Start. Regulation in Metabolism. London: J. Wiley and Sons, 1973, 349 pp.
98. Nowak, T. Conformational changes required for pyruvate kinase as modulated by monovalent cations. J. Biol. Chem. 251: 74-78, 1976.
99. Orengo, A. and D. Patenia. Exploitable molecular mechanisms in hibernation I. Liver diphosphofructose phosphatase of the rat and hamster: a comparison. Comp. Biochem. Physiol. 55B: 283-291, 1976.
100. Orengo, A. and D. Patenia. Exploitable molecular mechanisms in hibernation II. Aromatic circular dichroism of hamster and rat liver diphosphofructose phosphatase. Int. J. Biochem. 7: 414-422, 1976.
101. Orengo, A. and D. Patenia. Exploitable molecular mechanisms in hibernation III. Effects of ligand binding on aromatic circular dichroism of hamster and rat liver diphosphofructose phosphatase. Int. J. Biochem. 7: 423-432, 1976.

102. Pengelley, E. T., S. J. Asmundson, B. Barnes and R. C. Aloia. Relationship of light intensity and photoperiod to circannual rhythmicity in the hibernating ground squirrel *Citellus lateralis*. Comp. Biochem. Physiol. 53A: 273-277, 1976.
103. Rahn, H. PCO_2, pH and body temperature. In: Carbon dioxide and metabolic regulation. Nahas, G. and K. E. Schaefer (editors). New York: Springer Verlag, 1974, pp. 152-162.
104. Rahn, H., R. B. Reeves and B. Howell. Intra- and extracellular pH as a function of body temperature. Proc. 25th Congress of IUPS, New Delhi 10: 56-67, 1974.
105. Reeves, R. B. An imidazaole alphastat hypothesis for vertebrate acid-base regulation: tissue carbon dioxide content and body temperature in bullfrogs. Respir. Physiol. 14: 219-236, 1972.
106. Scheraga, H. A., G. Nemethy and I. Z. Steinberg. The contribution of hydrophobic bonds to the thermal stability of protein conformations. J. Biol. Chem. 237: 2506-2516, 1962.
107. Scholander, P. F. Experimental investigations on the respiratory function in diving mammals and birds. Hvalradts Skrifter Norske Videnskaps Akad. Oslo. No. 22, 131 pp., 1940.
108. Segal, H., Y. Sanada and S. Martin. Properties of purified glycogen synthetase b from liver. In: Metabolic Interconversion of Enzymes. Wieland, O., E. Helmreich and E. Holzer (editors). Berlin: Springer Verlag, 1972, pp. 133-145.
109. Sidell, B., F. R. Wilson, J. Haxel and C. L. Prosser. Time course of thermal acclimation in fish. J. Comp. Physiol. 84: 119-127, 1973.
110. Simon, L., E. D. Robin, R. Elsner, A. L. G. J. Van Kessel and J. Theodore. A biochemical basis for differences in maximal diving time in aquatic mammals. Comp. Biochem. Physiol. 47B: 209-215, 1974.
111. Somero, G. N. Pyruvate kinase variants of the Alaskan king crab. Evidence for a temperature

dependent interconversion between two forms having distinct and adaptive kinetic properties. Biochem. J. 114: 237-241, 1968.

112. Somero, G. N. Molecular mechanisms of temperature compensation in aquatic poikilotherms. In: Hibernation-Hypothermia: Perspectives and Challenges. South, F. E., J. P. Hannon, J. S. Willis, E. T. Pengelley and N. R. Alpert (editors). Amsterdam: Elsevier Publishing Co., 1972, pp. 55-80.

113. Somero, G. N. The role of isoenzymes in adaptation to varying temperatures. In: Isoenzymes II. Markert, C. and G. Whitt (editors). New York: Academic Press, 1975, pp. 221-234.

114. Somero, G. N. and P. W. Hochachka. The effect of temperature on catalytic and regulatory functions of pyruvate kinase of the rainbow trout and the Antarctic fish *Trematomus hernacchii*. Biochem. J. 110: 395-400, 1968.

115. Stammers, D., and H. Muirhead. Three-dimensional structure of cat muscle pyruvate kinase at 6 A° resolution. J. Mol. Biol. 95: 213-225, 1975.

116. Stoner, H. B. The role of the liver in non-shivering thermogenesis in the rat. J. Physiol. 232: 285-296, 1973.

117. Storey, K. B. and P. W. Hochachka. Glycolytic enzymes in muscle of the Pacific dolphin: role of pyruvate kinase in aerobic-anaerobic transition during diving. Comp. Biochem. Physiol. 49B: 119-128, 1974.

118. South, F. E., J. P. Hannon, J. R. Willis, E. T. Pengelley, and N. R. Alpert. Hibernation and Hypothermia: Perspectives and Callenges. Amsterdam: Elsevier Publishing Co., 1972, 743 pp.

119. Taketa, K. and B. Pogell. Allosteric inhibition of rat liver fructose 1, 6-diphosphatase by adenosine 5' monophosphate. J. Biol. Chem. 240: 651-662, 1965.

120. Tashima, L. S., S. J. Adelstein and C. P. Lyman. Radioglucose utilization by active, hibernating and arousing ground squirrels. Am. J. Physiol. 218: 303-309, 1970.

121. Twente, J. W. and J. A. Twente. Concentrations of D-glucose in the blood of *Citellus lateralis* after known intervals of hibernating periods. J. Mammal. 48: 381-386, 1967.

122. Tsukuda, H. Temperature dependency of the relative activities of liver lactate dehydrogenase isoenzymes in goldfish acclimated to different temperatures. Comp. Biochem. Physiol. 52B: 343-345, 1975.

123. Walker, P. R. and V. R. Potter. Allosteric properties of a pyruvate kinase isoenzyme from rat liver cells in culture. J. Biol. Chem. 248: 4610-4616, 1973.

124. Whitten, B. K. and G. J. Klain. Protein metabolism in hepatic tissue of hibernating and arousing ground squirrels. Am. J. Physiol. 214: 1360-1362, 1968.

125. Whitten, B. K., M. A. Posiviata and W. D. Bowers. Seasonal changes in hepatic ribosome aggregation and protein synthesis in the hibernator. Physiologist 13: 339, 1970.

126. Willis, J. S. Cold resistance of kidney cells of mammalian hibernators: cation transport vs. respiration. Am. J. Physiol. 214: 923-928, 1968.

127. Willis, J. S., L. S. T. Fang and R. F. Foster. The significance and analysis of membrane function in hibernation. In: Hibernation-Hypothermia: Perspectives and Challenges. South, F. E., J. P. Hannon, J. S. Willis, E. T. Pengelley and N. R. Alpert (editors). Amsterdam: Elsevier Press, 1972, pp. 123-147.

128. Willis, J. S., S. S. Goldman, and R. F. Foster. Tissue K concentration in relation to the role of the kidney in hibernation and the cause of periodic arousal. Comp. Biochem. Physiol. 39A: 437-445, 1971.

129. Wimhurst, J. M. and K. L. Manchester. Role of bivalent cations in the control of enzymes involved in gluconeogenesis. FEBS Letters 10: 33-37, 1970.

130. Wodtke, E. Discontinuities in the Arrhenius plots of mitochondrial membrane-bound enzyme systems for a poikilotherm: Acclimation temperature

of carp affects transition temperatures. J. Comp. Physiol. 110: 145-157, 1976.

131. Zimny, M. L. and V. Tyrone. Carbohydrate metabolism during fasting and hibernation in the ground squirrel. Am. J. Physiol. 189: 297-300, 1957.

132. Zwingelstein, G., R. Meister and G. Brichon. Temperature d'acclimatation et renouvellement des phospholipides de l'Anguille. J. Physiol. (Paris) 71: 350a, 1975.

ASPECTS OF METABOLIC AND ENDOCRINE CHANGES IN HIBERNATION

R. Agid, L. Ambid, R. Sable, and R. Sicart
Institut de Physiologie
Universite Paul Sabatier
31 Toulouse, France

ABSTRACT

Spontaneous endocrine and metabolic cyclic changes indicate that hibernators are valuable experimental models for physiological studies. In the golden hamster (*Mesocricetus auratus*), thyroid activity governs plasma and liver lipid levels; L-thyroxine, scarcely active on thermogenesis, increases lipid mobilization, and paradoxically decreases glycemia, as does cortisone; thyroidectomy induces atrophy of the digestive tract. In the garden dormouse (*Eliomys quercinus*) during hibernation, atrophy of intestine followed by spontaneous and early restoration correlates with thyroid activity; L-thyroxine increases intestinal RNA and proteins in fasting animals, a potent morphogenetic action. Early exposure to cold from April onwards suppresses autumnal involution of the gonads. A caloric rich protein-free diet infallibly induces torpor in garden dormice after 2 to 7 days, irrespective of season or temperature. Torpor is preceded by inactivity of the adrenal medulla, and by lipid infiltration of liver and brown fat. Anorexia, observed particularly in prehibernation, implies a lack of dietary proteins, leading to fatty liver, inactivity of the adrenal medulla, and finally torpor. In the 13-lined ground squirrel (*Citellus tridecemlineatus*) enlargements of the glomerular basement membrane of the kidney are reversible after hibernation, and possibly correlate with endocrine changes and defective glycoregulation.

ISBN 0-12-734550-7

I. INTRODUCTION

Research into the physiology, biochemistry and endocrinology of hibernating animals is of interest for numerous reasons. In the first place, the changes that are observed, which occur spontaneously, are often dramatic and should lead to a better understanding of the mechanisms involved in hibernation. Furthermore, the physiological and biochemical functions are modified by hypothermia, and to a different extent in different species, so that these animals often provide remarkable models for studying many of the biological processes, such as glucose, lipid and even protein metabolism. In some species, unexpected hormonal changes may occur and these can be instructive. For example, we found that the seasonal cycle of the gonads can be suppressed by cold exposure in the garden dormouse (*Eliomys quercinus*). Whereas the thyroid secretions have little action on calorigenesis, they play an exceptional role in both lipid and protein metabolism in the garden dormouse and the golden hamster (*Mesocricetus auratus*) resulting in protein reorganization and striking morphological changes in the digestive tract. Finally, we draw attention to a peculiar nutritional aspect of the dietary proteins: if they are withdrawn from the food, torpor infallibly and rapidly sets in, in the garden dormouse.

In most of the research described here, the problem arises of whether the metabolic and endocrine changes observed are the causes or effects of hypothermia. In some cases, a solution to this question can be obtained.

II. THE ROLE OF THE THYROID GLAND IN HIBERNATING ANIMALS: METABOLIC EFFECTS AND EVIDENCE FOR PROTEIN REORGANIZATION AND MORPHOLOGICAL CONSEQUENCES

The role of the thyroid in hibernation has always been appreciated and has been reviewed several times (32, 36, 38, 58). This gland undoubtedly exhibits variations of activity that are often very large in hibernating animals (17, 30, 37, 45). In the garden dormouse (30, 46), the activity is very low at the beginning of hibernation, then resumes in January and passes through a maximum during fall hibernation, shortly before the vernal arousal. This observation shows that hibernation is compatible with vigorous activity in this gland even when metabolism is very depressed. Canguilhem (21) shows that at least some L-thyroxine level is essential for the onset of winter hibernation in *Cricetus cricetus*.

We may thus reasonably assume that thyroid secretion has some action on the metabolic events and on the process of hibernation. The research described below shows that the thyroid has a powerful influence on lipid, protein and even carbohydrate metabolism. Its action on the proteins is very important since the thyroid hormones induce striking morphological changes particularly in the digestive tract.

A. Cyclic Variations in Thyroid Activity

Using histologic criteria, it is easy to demonstrate seasonal cycles characteristic of the various species. Figure 4 shows the appearance of these cycles for the garden dormouse from August till April where the differences in activity are particularly striking. Thyroid activity in the golden hamster is plotted in figure 1, the golden hamsters being maintained at a constant temperature (20°C) and exposed to natural daylight.

B. Influence of Thyroid Hormones on Plasma Lipids and Calorigenesis in the Golden Hamster

The evolution of lipemia in parallel with thyroid activity is shown in figure 1. This curve suggests that the plasma lipid levels could be causally related to L-thyroxine secretion. The same is true of the evolution of free fatty acids (FFA). Also, the seasonal variations of plasma lipids are very marked. Two weeks after the ablation of the thyroid however there is a definitive drop in the lipemia and the seasonal cycles disappear. When the thyroid ablation is performed on animals whose lipemia is high, the latter drops to a level of 50% of its normal value.

Measurement of the oxygen consumption ($\dot{V}O_2$; basal metabolism) was performed on all animals. The energy expenditure was decreased by thyroidectomy particularly in spring and autumn, when the normal value of $\dot{V}O_2$ was high. In a group of 9 animals studied

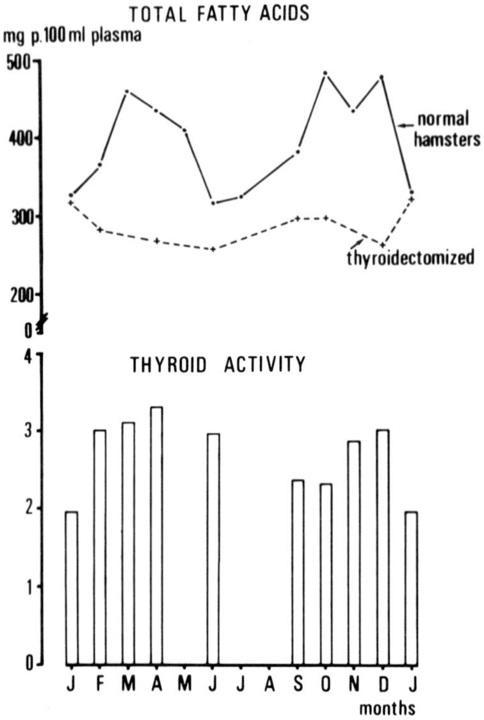

Fig. 1. *Thyroid gland activity and plasma concentration of total fatty acids in the golden hamster (Mesocricetus auratus). The activity of the thyroid gland is established from histological characteristics; the scale is the converse of that of Kampelmann, i.e. from zero to the lowest activity to five at the highest. The whole gland was examined (50 slices per gland). Each data point corresponds to the mean value of at least 7 animals (from 7 to 19). For the lipemia from January to March, $p<0.01$; from April to June, $p<0.01$; from June to October, $p<0.001$; from October to January, $p<0.001$. For the thyroid activity from January to April, $p<0.05$ and from November to January, $p<0.05$.*

in March, $\dot{V}O_2$ dropped two weeks after thyroidectomy from 1160 ± 22 to 930 ± 27 ml/h.kg body weight ($p \leq 0.001$). In another experiment performed in November, the $\dot{V}O_2$ dropped from 1190 ± 24 to 960 ± 14 ($p<0.001$). In January, the $\dot{V}O_2$ of normal hamsters was low (950 ± 36) and thyroidectomy had little effect (935 ± 45).

The role of the thyroid becomes even clearer if we administer L-thyroxine to thyroidectomized animals. Following Tata on his experiments in the rat (73) we have given single injections to thyroidectomized hamsters. The results show that 50 μg/100 g body weight were necessary to obtain, 48 hr after the injection, a 15% increase of $\dot{V}O_2$ in 12 animals. After 3 days the values return to normal. Comparisons were made with rats under the same experimental conditions; the $\dot{V}O_2$ in 6 treated animals was 40% higher ($p<0.01$). Similarly Tata (73) found an increase of 37% after 72 hr with an injection of 40 μg/100 g body weight of L-thyroxine in thyroidectomized rats. The golden hamster, therefore, is much less sensitive to L-thyroxine than the rat. This slight effect of L-thyroxine on calorigenesis in the golden hamster is perhaps species specific since in another hibernator, *C. tereticaudus* (35), physiological doses of L-thyroxine pentahydrate increase the $\dot{V}O_2$ by 100%.

On the other hand, we have observed an important effect of L-thyroxine on the plasma lipids in the golden hamster. A dose of 50 μg/100 g of body weight causes a marked increase of total fatty acid level. The mean value for 55 thyroidectomized hamsters at different seasons was 288 ± 8 mg/100 ml. In another group of thyroidectomized animals treated with a single dose of 50 μg of L-thyroxine the mean value, 48 hr after the injection was 526 ± 38 mg/100 ml, thus significantly higher than the preceding value ($p<0.001$). This increase of the lipemia still persisted 72 hr after injection (391 ± 34 mg/100 ml, $p<0.001$). Figure 2 shows the value of the lipemia at different times after a 50 μg injection in the same animal.

With higher doses of L-thyroxine, for example 500 μg, the plasma appears milky although the animals have been fasting for 15 hr. The lipemia level occasionally reaches 2000-3000 mg of total fatty acids per 100 ml of plasma. We should add that in the rat, a dose of 50 μg/100 g body weight of L-thyroxine is without effect; the mean value of total plasma fatty acids in 10 thyroidectomized rats was 273 ± 13 mg/100 ml and after 48 hr, the level was still unchanged (265 ± 17). In hamsters treated with L-thyroxine we find a modest hepatic steatosis.

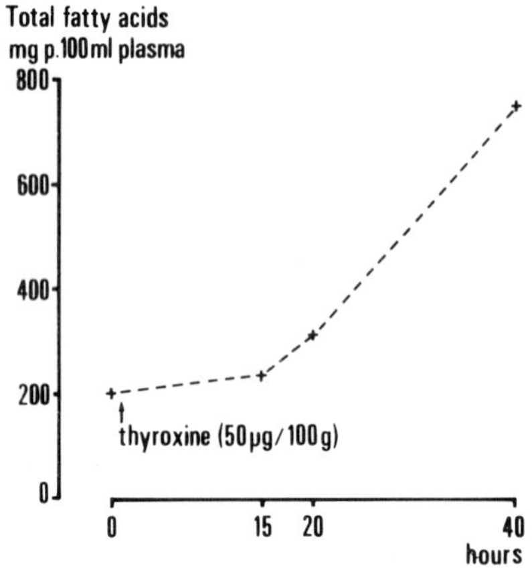

Fig. 2. *Increase of lipemia by administration of L-thyroxine in a golden hamster (M. auratus).*

This increase of lipemia and steatosis in the golden hamster are due, partly at least, to lipid mobilization, as is suggested by the increase of plasma free fatty acids from 17 ± 0.7 to 30 ± 1.4mg/100 ml plasma 48 hr after the injection of 50 µg L-thyroxine ($p<0.001$). Experimentally, the effect of administering L-thyroxine 20 or 40 hr before taking samples is seen as a rise of lipolysis of adipose tissue *in vitro*. These effects of L-thyroxine are suppressed by actinomycin D, a potent inhibitor of enzyme induction.

We should also mention that in ordinary mammals L-thyroxine not only causes peripheral mobilization of glycerides but may also stimulate hepatic lipogenesis (27, 29). The liver of the hamster thus exhibits a marked aptitude for synthesizing fatty acids from acetate (23).

In conclusion, our experiments show that in thyroidectomized golden hamsters the injection of a single but high dose of L-thyroxine (50 µg/100 g body weight) induces a significant lipid mobilization and considerably increases the lipemia. Conversely, the suppression of physiological production of endogenous L-thyroxine by ablation of the thyroid glands causes a permanent 50% decrease of lipemia. This suggests that the thyroid hormones play a major role in the regulation of the tissue lipid levels and also explains the correlation between seasonal cycles of the thyroid and lipemia in the golden hamster. These effects of L-thyroxine and of thyroidectomy are just the opposite of those described in man and other mammals (26, 28, 48). Furthermore, the increase of plasma lipid levels following L-thyroxine administration is by no means general in hibernators since we found it in *Cricetus cricetus* but not in *Eliomys quercinus*.

C. The Thyroid and Glucose Metabolism in the Golden Hamster

1. Paradoxical Hypoglycemic Action of L-thyroxine

L-thyroxine is not usually a hypoglycemic hormone; on the contrary it has hyperglycemic effects (22, 61). It is well known that after repeated injections of this hormone or in cases of hyperthyroidism, diabetes develops (34, 54). In the golden hamster, our results are very different (5, 64). Thyroidectomy of 10 groups of 7 to 13 animals each carried out at different times during the year led to a significant hyperglycemia. The mean value of the glycemia is 109 ± 2 mg% in the control animals and 131 ± 4 mg% in animals thyroidectomized fifteen days previously ($p<0.001$).

The administration of L-thyroxine to thyroidectomized animals not only corrects the thyroidectomy effects, but at doses of 50 μg/100 g body weight results in significant hypoglycemia. Usually, the initial glycemia (131 ± 4) diminishes to 89 ± 6 mg% after 15 hr, then to 86 ± 2 mg% after 48 hr; 3 days later it increases to 98 ± 4 mg%. These three values are significantly different from the initial value ($p<0.001$). It is interesting to note that some plasma glucose levels can be as low as 50 mg%. Hypoglycemia is found even in animals which are not fasting.

The hepatic glycogen is affected differently by L-thyroxine depending on the season: between January and March, it is sharply reduced, but during the rest of the year it is unaffected or increased. In any event, there is no relation between hepatic glycogen and glycemia, since plasma glucose is significantly reduced in all cases.

We can offer no explanation of this unexpected hypoglycemic effect of L-thyroxine which is slow to appear and long-lasting. Since the effect appeared to be peculiar to the golden hamster, we performed experiments to establish whether another notoriously hyperglycemic hormone had a different effect in this species.

2. Paradoxical Hypoglycemia Provoked by Cortisone

In a recent series of experiments done in the month of May, a dose of 0.25 mg of cortisone was administered to several groups of 6-8 golden hamsters weighing between 130-150 g and starved for 15 hr. Glycemia was examined after 15 min, 1 hr and 3 hr, as were the hepatic glycogen and plasma immuno-reactive insulin (IRI). After 3 hr, glucose levels fell from 86 ± 3 to 56 ± 4 mg% on average. This is a very significant change ($p<0.01$). The liver glycogen rose from 470 ± 109 to 1440 ± 212 mg/100 g wet wt. ($p<0.01$).

Cortisone in fasting mammals increases gluconeogenesis, elevating both liver glycogen and plasma glucose, and in the golden hamster the increase of liver glycogen observed is rather significant. In this animal, however, instead of hyperglycemia, we observed a plasma glucose diminution of 40% after cortisone treatment. The plasma IRI on the other hand, changed only very slightly from a value of 14 to 18 μU/ml. This hypoglycemia does not appear to come from an extra secretion of insulin. This effect has not been detected in the garden dormouse and more experiments are needed to clarify this unexpected hypoglycemia induced by cortisone in the golden hamster.

D. The Role of the Thyroid in Protein Metabolism: Morphological Effects

It is well known that thyroxine is capable of stimulating protein synthesis (19, 69, 70) and in particular, growth in young animals. In adult mammals that do not hibernate, however, it is the increase in calorigenesis by this hormone that attracts attention. We have shown that L-thyroxine has a powerful effect on protein distribution in the digestive tract of the golden hamster and garden dormouse.

1. Reduction of Intestinal Proteins after Thyroidectomy in the Golden Hamster

The first experiments were performed on golden hamsters and showed that ablation of the thyroid gland produced atrophy of the whole digestive tract visible to the naked eye. The dry weight fell by 40% on average, relative to the weight of the animals, which changed very little after the operation. This reduction mainly affects the proteins, since its value decreased from 8.1 ± 0.4 to 5.2 ± 0.3 (mg/g b.w.), three weeks after the operation ($p<0.001$). Complementary treatment with L-thyroxine completely prevented this drop in protein. This influence of L-thyroxine on the intestinal protein levels occurs at any age but is particularly striking in young animals.

2. The Effects of Hypophysectomy in the Golden Hamster

Ablation of the pituitary gland causes exactly the same changes to the intestine as thyroidectomy and L-thyroxine treatment is entirely effective in preventing atrophy. Intestinal proteins decreased from 8.09 ± 0.39 (mg/g b.w.) in a group of normal hamsters to 5.64 ± 0.40 in animals hypophysectomized 4 weeks earlier (6 animals); the corresponding figure was 8.12 ± 0.6 for 4 of them which had been treated with L-thyroxine (200 µg per week). Although the number of animals was small, it is likely that, after hypophysectomy, the L-thyroxine deficiency is responsible for intestinal atrophy.

3. The Role of the Thyroid in the Garden Dormouse Hibernating at 15°C

Since thyroid activity is cyclic in the garden dormouse, and atrophy of the digestive tube occurs spontaneously in this species during hibernation, it is of interest to explore the direct relation between thyroid activity and intestinal protein level. Experiments have been performed on numerous groups of animals killed at different times of the year, some of which had undergone L-thyroxine treatment (200 µg i.p. per animal per week for three weeks). The injections awoke the animals for a few hr, but did not interrupt the hibernation season. In order to compare measurements made during different months, the weight of the animal, the dry weight of the intestine and the total protein content of the whole digestive tract are referred to a parameter that is unaltered by hibernation, the length of the leg (femur & tibia). This measurement faithfully reflects the size of the animal ($r = 0.84$, $p<0.01$). There is, moreover, in any given experimental group, an excellent correlation between the weight of the digestive tract and the length of the femur ($r = 0.79$, $p<0.01$). We have also compared the absolute values of tissue protein content in two different groups of animals. In both cases, the conclusions are similar and the differences observed are highly significant. We have also determined the concentration of total ribonucleic acid (RNA) in tissue as mg per 100 g dry tissue.

In normally hibernating, completely starved garden dormice weight is lost regularly from the beginning of November until the vernal awakening at the beginning of April. The intestinal proteins diminish by about 33% until February, after which they increase by 25% in a highly significant manner ($p<0.001$). The digestive tract is therefore reformed even though the animal does not eat and continues to lose weight (Figs. 3, 4).

Measurement of the total RNA of the intestine and liver in other groups of garden dormice reveals low values from September to December, after which values 25% higher are recorded up to the end of hibernation (Fig. 5). The rise in tissue RNA precedes that of the proteins but for these two components, the correlation with thyroid activity is striking. The increase of thyroid secretion which occurs first is followed by a rise in RNA, then by an increase in tissue proteins, finally resulting in hypertrophy of the digestive tract. During hibernation, therefore, the thyroid appears to govern the mass and the morphology of the digestive system.

4. The Influcence of L-Thyroxine on the Hypertrophy of the Digestive Tract in the Garden Dormouse

Further support for this hypothesis is to be found in figures 4 and 5, which represent animals that have received three injections of L-thyroxine. It is clear that thyroid hormone supplements increase significantly both intestinal RNA and protein. In the case of RNA, for example, the rise in RNA even in

December and January reaches and eventually exceeds in three weeks that normally recorded at the end of hibernation.

L-thyroxine thus exerts a powerful effect on the digestive tract, increasing the RNA and the tissue proteins and hence, the mass of the entire digestive tract. The cyclic changes that occur naturally in these animals are apparently directed by the thyroid secretions, the variations of which exactly match the observed metabolic alterations (6).

5. Influence of the Thyroid on Plasma Amino Acids in Garden Dormouse

The cyclic protein changes of the digestive tract caused by L-thyroxine are related to the characteristic varations in plasma amino acids and

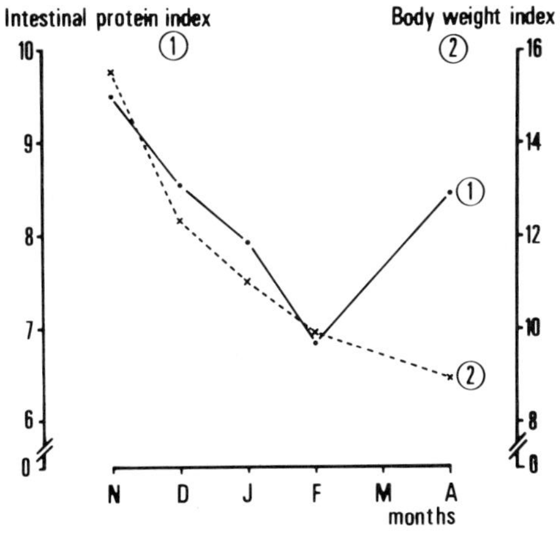

Fig. 3. *Changes of intestinal protein index (1) and body weight index (2) during hibernation in the garden dormouse (Eliomys quercinus L.). Both intestinal protein index and body weight index are expressed in terms of the total length of tibia and femur (see text). Each determination corresponds to the mean value of 7 to 16 animals. Note that proteins are synthetized in the intestine despite the continuous wieght loss due to fasting. Intestinal proteins are thus formed at the expense of other tissues, such as muscles.*

their distribution during both the seasonal cycle and the periodic rewarming. In the garden dormouse (65) hibernating at 15°C, there is a slight increase of total plasma amino acids during each periodic rewarming, from 188 to 225 μmoles/100 ml plasma. Other authors have observed an important rise of the amount of plasma amino acids in the hedgehog hibernating at 4.5°C (43) and in the ground squirrel (41), but in the golden hamster hibernating at 4.5°C the increase of total plasma amino acids was again slight (44).

In the garden dormouse (65) (9 animals hibernating and 11 animals after rewarming), some gluconeogenic amino acids such as serine, glycine, tyrosine and especially alanine increased significantly. Others drop in concentration: valine, leucine and isoleucine. The concentration of lysine is high

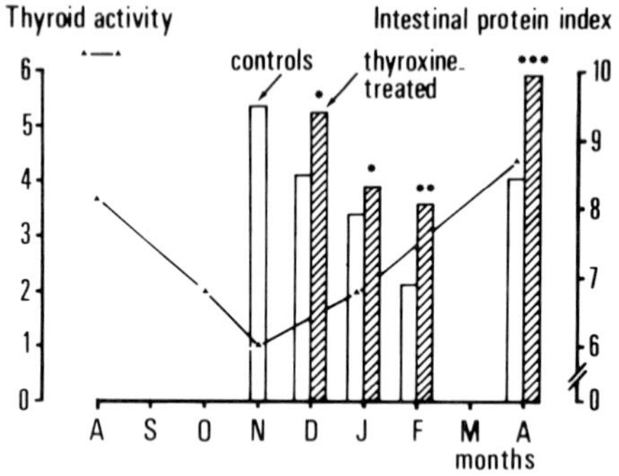

Fig. 4. *Changes of intestinal proteins in the garden dormouse (E. quercinus, L.) and the influence of L-thyroxine. In control dormice, the amount of intestinal proteins decreases regularly during hibernation (November to February) but subsequently increases from February to April, when thyroid activity of control animals (△-----△) resumed in January. Administration of L-thyroxine (200 μg weekly for three weeks) significantly stimulates the deposition of protein in the digestive tract (hatched bars). These changes occur in animals totally deprived of food, indicating that the restoration of intestine takes place at the expense of other tissues proteins.* . $p<0.05$; .. $p<0.01$; ... $p<0.001$.

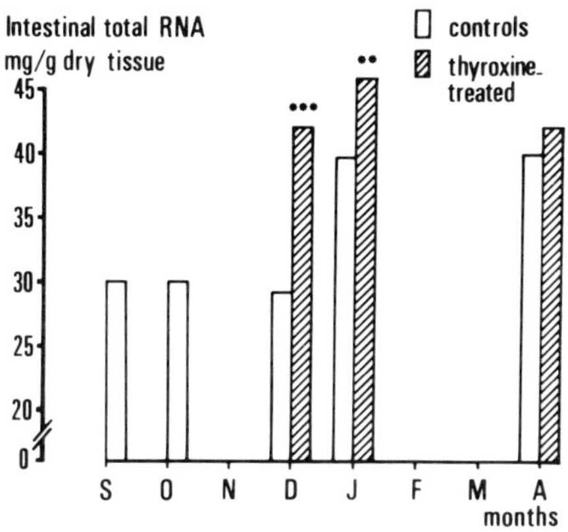

Fig. 5. *Spontaneous seasonal changes of intestinal total RNA in the garden dormouse (E. quercinus, L.) and influence of L-thyroxine. Each bar corresponds to the mean value of at least 7 animals (from 7 to 15). .. p<0.01; ... p<0.001 L-thyroxine was given i.p. at 200 µg/animal per week for 3 weeks.*

but does not change. The rest of the amino acids are present in low concentrations and the differences in their amounts are not significant.

If the animals have been treated with L-thyroxine, the rise in gluconeogenic amino acids during rewarming is less abrupt, whereas the reduction in non-gluconeogenic amino acids such as valine, leucine, isoleucine is accentuated by about 50% ($p<0.01$). After L-thyroxine treatment, lysine also decreased by 50%. During rewarming of L-thyroxine treated animals, therefore, the increase in total plasma amino acids is no longer perceptible since the increase in the amount of some amino acids is balanced by the decrease in amount of others.

6. Discussion and Conclusion

Variations of RNA and tissue protein in the digestive tract, whether spontaneous or provoked by L-thyroxine are quite spectacular in the garden dormouse. In starved or even fed animals, protein synthesis is reduced during the first stage of hibernation and the RNA level is low. These observations agree well with the information to be found in the literature. Thus Whitten *et al.* (80) observed disintegration of the hepatic polyribosomes in the ground squirrel during hibernation and Whitten and Klain (79) found that labelled leucine is incorporated very little into the hepatic proteins. Furthermore, at each arousal the catabolism of proteins is increased, a finding suggested by the rise in circulating creatine (42), and by the increased consumption of amino acids *in vitro* in the hepatic tissue of ground squirrels during arousal (78).

Most authors believe that the reduction in protein synthesis is due to the natural or imposed fast of the animal during hibernation. It is known that in most mammals, lack of proteins in the diet, or total fast, modifies tissue protein synthesis with a reduction in RNA level (71, 81). A drop in circulating amino acids due to the nutritional state would be one of the causes of a partial inhibition of protein synthesis. According to Munro (50), lack of any one of the amino acids (especially tryptophan) would lead to disintegration of polyribosomes into subunits; this would clearly impair the capacity to make protein.

Our results show that when the garden dormice are killed during the second phase of hibernation, protein synthesis is increased at the intestinal level. The interesting feature of our observations is that this protein synthesis is initiated in animals that have been deprived of food for several months. At the intestinal level, this synthesis is revealed by a clearly detectable protein repletion which is reflected by the increase of the total mass of the gut.

These facts seen as a whole recall the development of the gonads in some animal species such as the salmon during migration, at which time the animal is fasting. It is however, rather unusual to observe them in connection with the digestive tract of a mammal, as in our studies. We note too that the variation of plasma amino acids following periodic rewarming is in good agreement with the restoration of the digestive tract, which requires essential amino acids, and also with the reconstitution of hepatic glycogen, which proceeds initially from gluconeogenic amino acids after each periodic arousal.

Above all, the influence of L-thyroxine in the garden dormouse is to strongly stimulate protein synthesis in the digestive tract. The influence of this hormone on protein synthesis has been relatively little studied in hibernating animals although Smith (67) succeeded in showing that the incorporation of labelled amino acids into tissue proteins of young hamsters is accelerated by L-thyroxine, exactly as in the rat; this author also showed that the same doses have similar effects in adult hamsters, in agreement with our findings.

Tata and Widnell (74) showed that L-thyroxine stimulates the synthesis of nuclear RNA, with a concomitant rise in its transfer to the ribosomal fraction. In the rat, Cahilly (20) observed a reduction in the liver RNA of thyroidectomized animals; conversely, L-thyroxine significantly increases the amount of RNA in this tissue. According to these authors, L-thyroxine restores protein synthesis to normal by acting on the RNA metabolism. All these findings are in good agreement with the trophic action of L-thyroxine in the two hibernating species studied.

In summary, our studies show that the main action of L-thyroxine in the golden hamster and the garden dormouse is on protein metabolism, manifested by a shift, or redistribution of amino acids between hibernating and arousal states and the restoration of atrophied digestive tract by increasing its protein and RNA synthesis. The efficient functioning of the thyroid is indispensable for the termination of hibernation. The digestive tract is a good example, since it is needed as soon as feeding recommences. From this point of view, the thyroid is essential for the normal pattern of the cycle of hibernation.

III. SUPPRESSION OF THE SEASONAL CYCLE OF TESTICULAR ACTIVITY BY COLD EXPOSURE IN THE GARDEN DORMOUSE

The cessation of activity and atrophy of the gonads during the period prior to hibernation is a well-known phenomenon in most hibernating species (39, 40, 59, 76). We have given a detailed account of it in the garden dormouse (30), in which we showed that it is preceded by a state of inactivity of the FSH and LH cells of the pituitary gland. A similar seasonal involution has been described in birds, where light is the controlling factor (18,62).

The observation that exposure to cold induces major changes in metabolism and the endocrine system of the garden dormouse at any season led us to enquire whether climatic factors have any effect on the seasonal cycle of testicular activity. In all, the study involved over 200 adult dormice, caught one year earlier. They were placed in cold rooms held at a constant temperature of 6°C and constant humidity of 80%. One room was kept in total darkness, except for 30 min a day of dim light, while the animals were being looked after. In the other room the photoperiod was constant (10L:14D). A commercial animal food was given, with abundant apples (8,11).

A. Results

1. Animals held at 6°C in the Dark From April or June Onwards, Killed 2-8 Months Later

The seasonal difference in weights of the testicles are presented in figure 6. In the control group, under natural conditions of temperature and daylight, the familiar drop in testicular weight in September is clearly seen. Sexual inactivity is typical and the seminiferous tubules, 70-80 μm in diameter, contain only a few rare spermatocytes. In garden dormice kept at 6°C and darkness from April onwards, there is no indication of involution, and the various histological features suggest normal activity. A low temperature from April onwards therefore prevents the usual involution observed in September. This is not so for animals placed at 6°C after June: involution does occur but is less complete than in animals kept in natural conditions. The weight of the testicles is considerably reduced and the diameter of the seminiferous tubules is 130 μm. Seminal development is blocked at the spermatid stage.

2. The Role of Darkness

In the above experiments, we might suspect that the near total darkness played some role. In order to answer this question, garden dormice were kept at 6°C with a constant photoperiod (10L:14D). In this case, the photoperiod had no effect: there was no difference between the groups kept in the dark and those regularly exposed to light; the involution of the gonads was dictated solely by the cold in both cases (Fig. 7). It is important to note that the

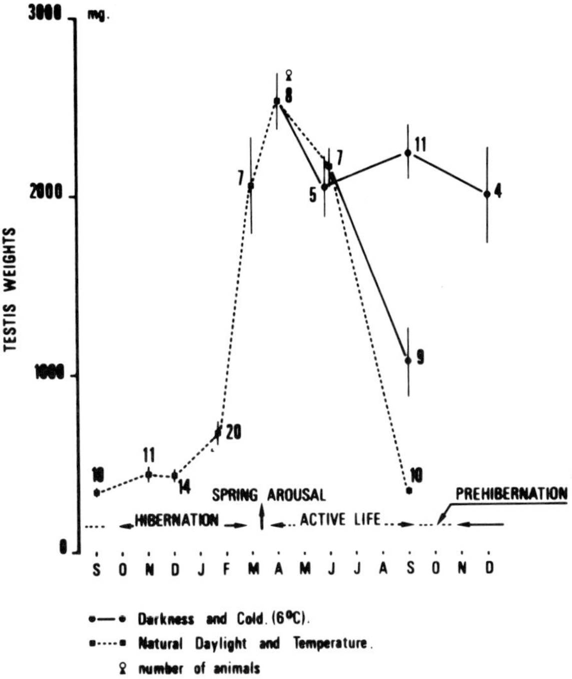

Fig. 6. *Effects of cold exposure on annual changes of testicular weight in the garden dormouse (E. quercinus, L.). The mean S.E.M. and number of animals are shown. The dotted curve shows the normal cycle with the customary near-complete atrophy in autumn. The full curve corresponds to animals exposed to 6°C from April onwards. It is seen that involution does not occur in fall. When the animals are exposed to cold from June onwards, involution is almost as complete as in the control animals.*

animals held at 6°C frequently exhibited torpor. These interludes of hibernation each lasted about 4 days, with body temperature of the animals very near to 6.5°C.

3. Dormice Kept in a Cold Atmosphere from September Onwards

The effect of cold is again very remarkable. In control animals, which hibernate naturally in November, testicular activity recommenced in January and was complete by spring awakening. In the group that was kept cold in September, the rise in testicular weight occurred sooner (Fig. 7) and was more rapid, reaching the same degree of activity by the end of March despite the fact that torpor and reduction in metabolic activity lasted longer in these animals than in the controls.

B. Discussion and Conclusion

Early exposure to cold, from April on, which

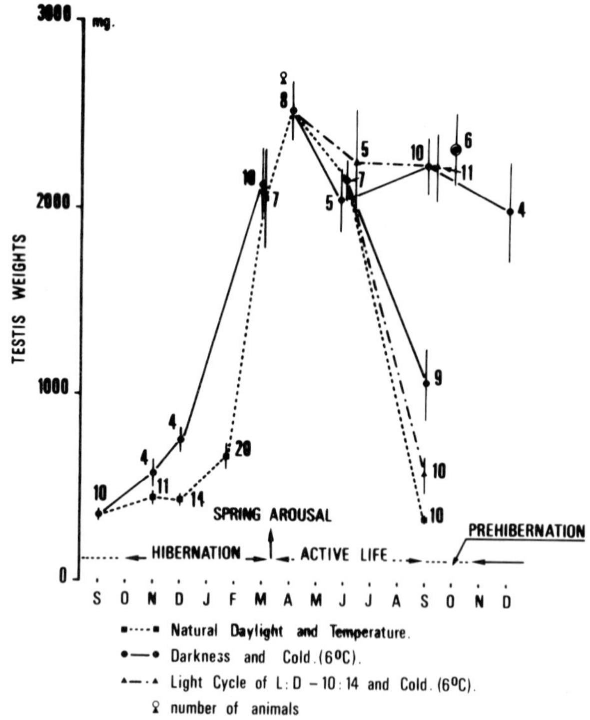

Fig. 7. *Stimulation of reconstitution of the gonads during hibernation by September exposure to cold in the garden dormouse (E. quercinus, L.). The mean S.E.M. and number of animals are shown. The full curve, corresponding to autumn and winter, shows that early exposure to cold in September significantly accelerates gonad reconstitution; this is thus complete at the spring arousal.*

to some extent prolongs hibernation, prevents the involution of the gonads normally observed in September. Light has no influence on involution of the gonads. If the garden dormice are exposed to cold later than April (in June), the gonads will regress normally. Thus involution is absent if the animals are placed in the cold early enough; the animals then exhibited alternating periods of torpor and activity. The garden dormice kept in the cold after June are very active in spring and do not exhibit torpor. Generally speaking, in this group, the durations of periods of continuous activity with elevated body temperature are longer. It seems that an adequate period of activity is necessary in the spring if the normal testicular involution is to occur in September. We can offer no explanation of this. Not all species behave in this way. In *Tamias striatus*, for example, Panuska (55) demonstrated that reproductive activity can be delayed by about two months by keeping the hibernating animal in cold and dark. Wells and Zalesky (77), on the other hand, succeeded in maintaining testicular activity in *Citellus tridecemlineatus* for a certain time, by imposing prolonged cold and dark conditions in the spring after the hibernating season. We have obtained comparable results in *Citellus tridecemlineatus*, namely that maintenance of testicular activity can be achieved by prolonged cold with complete absence of involution in the autumn.

In our experiments with the garden dormouse, light had no influence on the weight of the testicles. In *Mesocricetus auratus*, however, Reiter showed that the photoperiod normally controls the seasonal reproductive rhythm and the effects are mediated by the pineal gland (60). In *Eliomys quercinus* the role of the pineal gland is uncertain since light has no effect.

It is interesting to note that, when the garden dormice are kept early enough in the cold, the thyroid activity is likewise retained in October. Thus cold exposure may prevent both involution of the gonads and of the thyroid. It must be noted that the absence of involution in no way intereferes with the subsequent process of hibernation; the animals exhibit the usual cycles of torpor and activity, despite the presence of active gonads and active thyroid.

Exposure to cold thus exerts an effect at

several levels in the endocrine system. Since the activity of the gonads and of the thyroid are governed by the LH, FSH and TSH cells of the hypophysis, it is clear that the secretions of these cells are normal and persist all year in animals which are exposed to cold in April and which exhibit less activity. If the activity is normal in spring, involution takes place in autumn for both glands, which suggests that the involution of gonads and of the thyroid is an active process whose mechanism is, however, very obscure at the present.

IV. PANCREAS

A. Annual Cyclic Modifications in the Number and Activity of the Pancreatic D Cells of the Garden Dormouse: Possible Metabolic Correlations

We have already described the evolution of the B and A cells of the pancreas during the different seasons and the periodic awakenings of the garden dormouse (4,30). We found a close correlation between the activity of these cells, as measured by histochemical criteria, and the parallel changes of glycemia. In a subsequent paper, dealing for the most part with the same species, Gave and Martoja (31) detected D cells in the pancreatic islets with typical histochemical characteristics. These cells are quite distinct from the B and A cells; they exhibit a characteristic pattern of change during the seasons, both in number and in abundance of argyrophilic secretory product. There is a progressive decrease in secretory activity from January to June, which is followed by a gradual increase, the maximum occurring in prehibernation. These changes of activity of the D cells are not correlated with the changes in glucose metabolism but they are parallel to changes in liver fatty acids, suggesting a possible role in the regulation of lipid metabolism. The increase in activity of D cells, i.e., the increase of secretion, brings about a decrease of liver fatty acids. Thus the diminution of secretion of a hormone from the D cells would explain the fatty liver observed during hibernation in the garden dormouse. However, a close scrutiny of each individual animal does not, in our opinion, confirm that D cells play a role in secreting a hormone that would reduce the amount of liver fat (such as "lipocaic hormone").

A possibility exists that the hormone somatostatin is also secreted by the D cells of the pancreas and it is now well established that somatostatin inhibits the A and B cells, thus influencing glucose metabolism (25). This hypothesis needs more support than the simple suggestion that the activities of the D cells, A and B cells and glycemia are related. Moreover, the cause of these changes of activity of the D cells, whatever the secretion, is unknown.

B. Possible Role of the Pancreas and of Defective Glycoregulation in the Enlargement of the Glomerular Basement Membrane of the Kidney

Some hibernating species are known to exhibit anomalies of renal function during hibernation (glycosuria, albuminuria, desquamation cells); these are accompanied by characteristic morphological changes the most striking of which is the enlargement of the basement membrane (52, 53, 56, 82). A comparative study of the problem, in the garden dormouse and the ground squirrel, has been made in our laboratory in collaboration with K. Ondrick and J. A. Panuska (in Washington D.C.), with a view to relating the variable importance of the modifications to the magnitude of the metabolic and hormonal changes, themselves variable, of these two hibernating species. It was hoped that some light would be shed on the very obscure origins of these disorders. Enlargement of the glomerular basement membrane is common in aging, and diabetes (66), but the mechanisms are unknown. These modifications of the basement membrane are unexpected in hibernators, and one may speculate about their meaning and enquire whether they play some role during the process of hibernation. An interesting interpretation has been proposed by Zimny and Rigamer (82), who assert that the enlargement of the basement membrane alters its permeability. The distribution of ions between the surfaces of the cell membranes would thus be maintained particularly when active transport is decreased by hypothermia. It is also possible that the modifications of the basement membrane are correlated with the parallel changes in the endocrine glands and in various metabolic processes.

The research has been difficult since the ground squirrel (*Citellus tridecemlineatus*) had far fewer renal lesions than those studied in Washington, even though they came from the same stock. Nevertheless, one point has been established, namely, that the lesions can be reversed in a few weeks after the hibernation seasons (53). Furthermore, these lesions are less frequent in the garden dormouse than the ground squirrel. We therefore embarked on a comparative study of glycemia, free fatty acids, plasma glycerol, and plasma insulin in two species, *Eliomys quercinus* and *Citellus tridecemlineatus* (12).

The changes in the plasma FFA and glycerol are very similar in the two species at the periodic arousals and the same is true of the seasonal variations, except that in the dormouse the FFA levels remain high during prehibernation (Fig. 8). Conversely, the large variations in plasma glucose level associated with hibernation are not found in the ground squirrel (Fig. 9). We therefore measured the plasma I.R.I. in both species in various physiological conditions (Fig. 10).

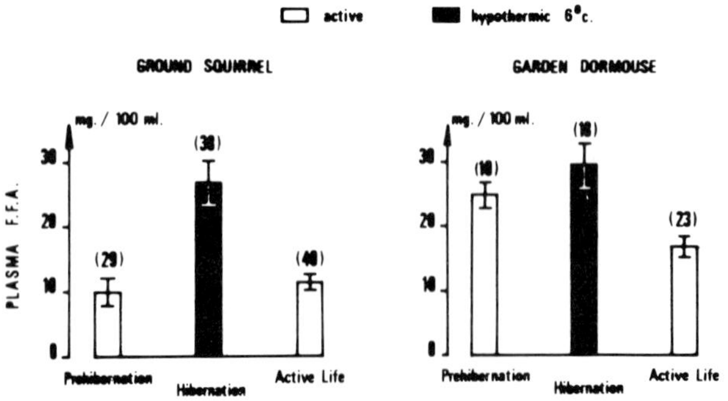

Fig. 8. *Plasma free fatty acids in the garden dormouse (E. quercinus, L.) and in the ground squirrel (Citellus tridecemlineatus) during different seasons. The mean, S.E.M. and number of animals (in parenthesis) are shown. During hibernation, the highly significant increase in FFA found in the ground squirrel is not observed in the garden dormouse.*

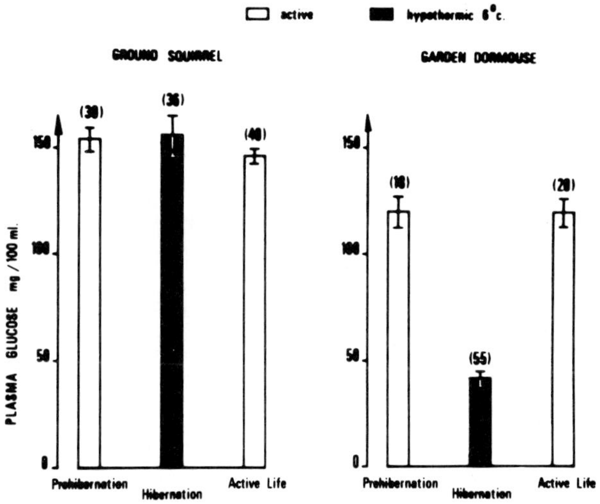

Fig. 9. *Plasma glucose in the garden dormouse (E. quercinus, L.) and in the 13-lined ground squirrel (C. tridecemlineatus) during different seasons. The mean, S.E.M. and number of animals are shown. Glycemia in the hibernating ground squirrel remains high whereas in the garden dormouse, it is very low.*

We see that plasma I.R.I. does not fluctuate in the ground squirrel between hibernation and active states unlike the case in the garden dormouse. In all cases, however, the variations are insufficient to suggest that the pancreatic hormones have any major influence in controlling glycemia. It is probably the variations in glycemia and in body temperature that influence the hormonal activity of the pancreas (7, 13, 33, 47).

V. DIET, NUTRITION AND THERMOREGULATION IN HIBERNATORS

A. Torpor Rapidly Induced in the Garden Dormouse by Deprivation of Dietary Proteins, Irrespective of Season or Ambient Temperature

We demonstrated earlier (2,10) that either cold or two to seven days of fasting induces torpor in the garden dormouse at any season, irrespective of the

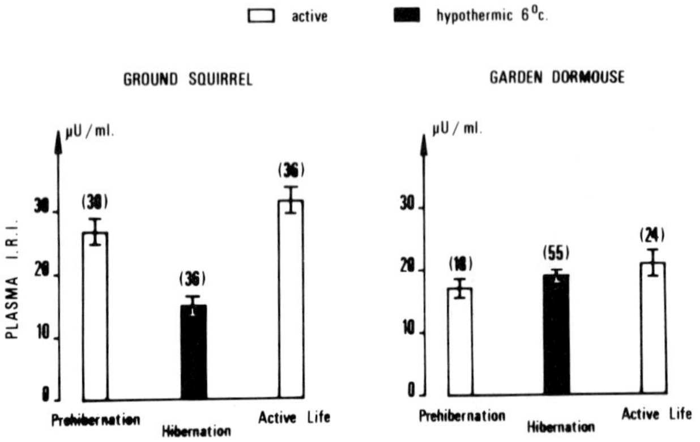

Fig. 10. *Plasma insulin in the garden dormouse (E. quercinus, L.) and in the 13-lined ground squirrel (C. tridecemlineatus) during different seasons. The mean, S.E.M. and number of animals are shown. The concentration of plasma I.R.I. in the ground squirrel drops in hibernation. This fall is not found in the garden dormouse.*

ambient temperature. This is accompanied by a characteristic increase of lipids in the heart, brown adipose tissue and particularly in the liver.

The fatty acids are mobilized from the adipose tissue and the amount of FFA in the plasma increases enormously. A moderate degree of fatty liver after fasting is not uncommon in most mammals but in the case of both *Eliomys quercinus* and *Glis glis*, the amount of lipids in the liver may attain values as high as 30% after only two to seven days fasting. As for the plasma lipids, the total fatty acids are on the contrary significantly lower than in fed animals and a diminution of β and pre- β-lipoproteins can be detected.

These observations explain the fatty liver which has been noted at the period of hibernation in many species (51, 57, 75) particularly when they are deprived of food, and also explain the disappearance of the fatty liver after spring arousal, when the animals resume feeding.

Several observations, and above all the

characteristic decrease of lipoproteins in the plasma, suggest that the causal factor of the fatty liver and perhaps of the incidence of hibernation could be a perturbation of the protein metabolism.

Diets poor or completely deficient in protein lead to practically the same results as fasting. Experiments are at present in progress to shed further light on this. Several groups of garden dormice have been studied: normally fed animals; animals fed on a diet of apples, which contain little protein; animals given no protein at all; and animals given a normal diet to which a protein synthesis inhibitor, ethionine, has been added. The results are as follows:

1. The Normal Diet

This consists of a balanced commercial foodstuff intended for rats. We add unlimited apples, which the animals consume in large quantities. Some animals exhibit torpor in June and July, as we have described elsewhere (3). Both in July and in autumn initiation of torpor is always preceded by loss of appetite and cessation of feeding for a few days; thus there appears to be voluntary fasting in the animals.

2. The Low-Protein Diet, Consisting Wholly of Apples

The composition of the apples used is as follows (per 100 g of apple):calorific value, 52 kcal; water, 84 g; proteins 0.30 g; lipids 0.35 g; carbohydrates, 12 g; ascorbic acid, 4.8 mg; small quantities of mineral salts and other vitamins.

The animals eat on average, 40 g of apples each per day. The total caloric intake is therefore about 50% higher than that of normally fed animals, but despite this, all the garden dormice enter a state of torpor in less than seven days, usually within two or three days. This torpor is accelerated if the temperature is lowered but nevertheless still occurs at 25 - 27°C; it is unaffected by the season. In figure 11, we show the results of an experiment conducted in August, 1975. In the control animals on normal diet, there were no cases of torpor (the ambient temperature varying between 19 and 23°C) whereas all the animals fed on a diet of apples alone fell into a state of

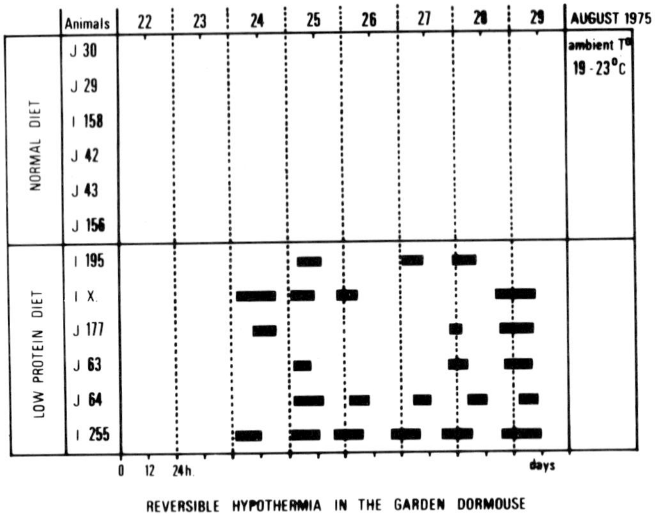

Fig. 11. *Torpor caused by low protein diet during August at a temperature of 19°-23°C. The periods of torpor are shown as black rectangles. It is seen that normally fed control animals (above) never became torpid. Dormice receiving low protein diet (below) exhibit torpor with body temperature about 1° higher than the ambient temperature. The effect sets in after 76 hr.*

torpor.

The length of torpor in the garden dormouse is a function of the ambient termperature both in winter and in spring; as the temperature rises, the length shortens, from 10-12 days at 6°C down to a few hr at 25°C (1). Further series of experiments with this diet in May and June yielded the same results - torpor in every case.

In these animals fed exclusively on apples, only a slight rise in plasma FFA is observed, but a fatty liver of variable intensity is developed. If the animals are killed before the anticipated entry into torpor while the body temperature is still normal, a characteristic accumulation of pheochrome granules is observed in the adrenal medulla. This may be interpreted as a state of inactivity, exactly as in animals deprived of all food, or at the onset of natural hibernation (16).

3. Protein-free Diet

Like the garden dormice fed on a low-protein diet, the animals on a protein-free diet also become torpid and in fact resemble starved animals. Mobilization of the lipids is clear and a hepatic steatosis results. Finally, the adrenal medulla exhibits an accumulation of pheochrome granules in the hr prior to the onset of torpor. Glycemia and hepatic glycogen remain at high levels.

4. Normal Diet Combined With a Protein Synthesis Inhibitor Ethionine

The above results underline the importance of dietary proteins and suggest that hibernation as well as lipid irregularities are a consequence of some disturbance of the protein metabolism. We have therefore attempted to block protein synthesis by means of ethionine, a methionine analogue in which the methyl radical is replaced by an ethyl. This toxic compound was administered in September to six garden dormice at a dose of 75 mg/100 g; the dose was injected in three sessions, once every twelve hr. With ethionine, the lipid mobilization is intense in the garden dormouse, as is hepatic steatosis. Furthermore, hypoglycemia is unambiguous. These various metabolic effects are well known in other mammals (72).

The interesting point is that animals treated with ethionine for these short periods also exhibit a fall in central temperature and even pass through phases of torpor with body temperature very near the ambient temperature (up to 27°C). Furthermore, the decline in activity of the adrenal medulla as measured by the accumulation of pheochrome granules, is very striking.

It seems, therefore, that blocking protein synthesis leads to torpor. Nevertheless, further experiments are needed to confirm this, since ethionine has toxic effects and in particular, it has a tendency to reduce the appetite, thus reducing food intake. It is thus possible that the compound acts by this mechanism, resulting in starvation-induced torpor. Other drugs (cyclophosphamide and cycloheximide) are being tested, but these substances are even more toxic in physiological doses; for the moment, the results are inconclusive and the experiments will need to be performed for a longer period. Finally, more work is needed to measure accurately the degree of incorporation of amino acids into the tissue proteins in the various groups.

What emerges, independent of any working hypothesis, is the incontestable fact that, in the garden dormouse, protein deficiency is routinely sufficient to induce torpor, whatever the ambient conditions and the physiological state of the animal. It should be stressed also that these last results are only applicable to garden dormice (*Eliomys quercinus*). However, in another species of garden dormouse (*Glis glis*), starvation also induces torpor.

B. Lipid Disturbances Induced by Starvation or by a Low-Protein Diet

The changes in lipid level that we have been describing merit further discussion. Generally speaking, the liver, the brown adipose tissue and the heart acquire an excess of lipids; at the same time there is a marked drop in β and pre-β-lipoproteins in the plasma despite the obvious stimulation of lipid mobilization which is seen as a rise in the plasma FFA (15).

The drop in lipoproteins is particularly striking during fasting or when the diet is low in protein. The characteristic feature is the promptness with which the changes occur. Figure 12 shows the variation of the plasma lipoprotein level in a single animal, initially fed normally (solid line) and then 96 hr after being fed on a non protein diet (dotted line). The drop, essentially in the low and very low density lipoproteins, suggests that their hepatic synthesis is decreased.

The increase in glycerides in the heart and, above all in the brown adipose tissue, due to these diets, is also characteristic. The rise is of the order of 30% after a few days of starvation, and precedes torpor. Each periodic rewarming in the garden dormouse is associated with a fall of 25-30% of the glycerides in the brown adipose tissue, with reconstitution of the glycerides during the hr following awakening and prior to the next torpor (9, 14). These cyclic variations in

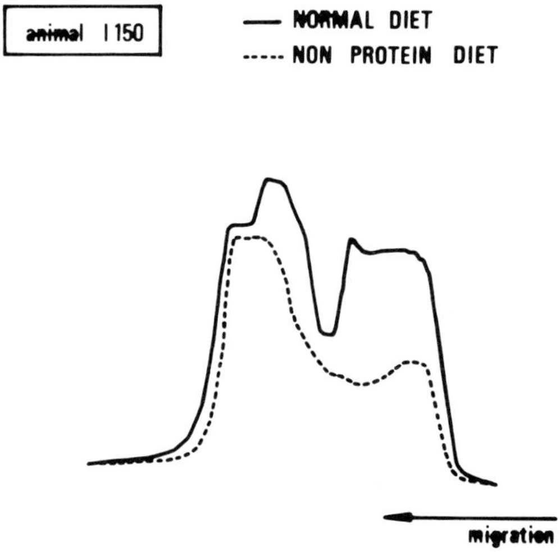

Fig. 12. *Representative separation of serum lipoproteins by electrophoresis on cellulose acetate. The full curve corresponds to the normally fed animal, the broken curve to the same animal after 96 hr of non-protein diet. Note the abrupt decrease in pre-beta and beta-lipoproteins, on the right part of the graph.*

the glycerides of the brown adipose tissue have already been described in detail in other species (68). Here, however, we see that deprivation of food or proteins produces very similar changes in the lipid pattern.

Infiltration of the liver by glycerides can be extremely intense. Figure 13 gives an example, from animals starved for four days. On average, the total fatty acids are more than doubled, but it is worth noting that some very high individual values are found. There too, a fatty liver preceded hibernation. It seems likely that lipid infiltration of the liver results from the absence of lipoprotein synthesis by the liver; the plasma lipoprotein level drops and the liver, incapable of eliminating the excess, becomes overloaded.

Analogous results have been obtained during seasonal hibernation; the fatty liver, which is always

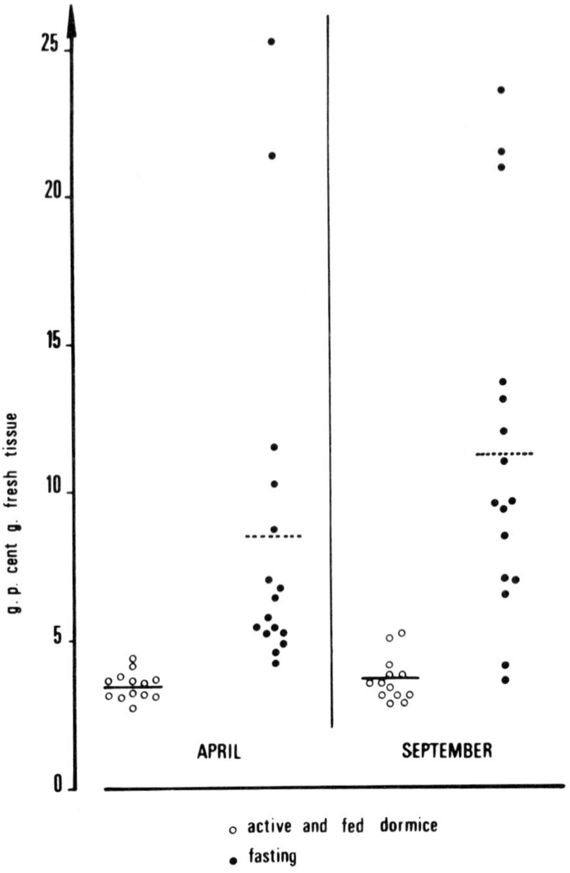

Fig. 13. *Formation of fatty liver in the garden dormouse (E. quercinus, L.) after 96 hr of total starvation. Each circle corresponds to a single animal. Similar results are obtained in April and September. Notice the extremely high liver fatty acid levels in some individuals after fasting. The same results are obtained with a low protein and with a protein-free diet.*

found in the garden dormouse and the ground squirrel, is more noticeable in animals deprived of food than in those which have food at their disposal. The existence of this fatty liver is thus explained by the self-imposed starvation to which hibernating animals are naturally subjected. When garden dormice resume feeding after the vernal awakening, the fatty liver immediately reverts to normal. The experiments described here thus provide the explanation of an intriguing observation, the presence of a fatty liver during seasonal hibernation and its disappearance when the animals resume feeding. It must be added that the fatty liver is not entirely due to triglycerides, since the rise in cholesterol esters in the liver is very significant. The hepatic cholesterol level alters on average from 340 mg to 430 mg/100 g fresh weight, some values attaining 600 mg/100 g or more. This is an unusual feature of these hibernating animals for, in ordinary hepatic steatosis, the increase in lipids consists essentially of triglycerides. In a strain of golden hamsters isolated in our laboratory the amount of hepatic cholesterol, when the animals were exposed to cold and fasting, reached very high levels of 12 g/100g fresh liver as compared to the normal, although very high value, of 2 g/100g. We cannot at present say whether such values are actually related to hibernation (63).

C. Inactivation of the Adrenal Medulla, Induced by Low-protein Diets or by Ethionine

Like starvation, diets low in protein or the administration of ethionine rapidly inactivate the adrenal medulla, even in animals with normal body temperature which have hence not yet begun to exhibit torpor (16). This inactivation is visible as an accumulation of pheochrome granules in the medulla.

Very similar results are obtained with protein-free diets, diets consisting largely of apples, or with ethionine treatment. The accumulation of pheochrome granules consisting of catecholamines is interpretable as a drop in the circulating levels of these hormones. However, we have not been able to determine the level of circulating catecholamines in these experiments. This interpretation should be capable of corroboration by other signs of hypo-activity visible at the ultra-structural level. If our interpretation is correct, it is reasonable to suppose that the reduction in catecholamine secretion has the effect of reducing most of the tissue oxidization, resulting in a drop in metabolic activity, particularly as these animals are exceptionally sensitive to adrenalin. The mechanism whereby starvation or protein deficiency leads to inactivation of the adrenal medulla remains unexplained.

D. Discussion

It appears beyond dispute that in the garden dormouse, deprivation of dietary protein is always sufficient to induce torpidity. Starvation has the same effect but it is not the absence of caloric intake that is important but the protein deficiency, as is shown by the experiments with an apple diet and with protein-free rations, in which the caloric intake was higher than normal. It remains to be established whether there is a reduction in protein synthesis as seems probable or whether the overall deficiency in protein is decisive, or whether specific amino acids govern the mechanism.

It seems reasonable to suppose, tentatively at least, that some hibernating species conserve their tissue proteins as much as possible; new protein synthesis may be difficult or impossible without dietary intake.

The different changes induced by lack of dietary protein or by total starvation duplicate very closely those associated with hibernation and in particular, those preceding torpor in garden dormice: accumulation of lipids in the liver, heart and brown adipose tissue and accumulation of catecholamines in the adrenal medulla, all phenomena that are useful to the animal, for at each periodic awakening, the animal releases adrenalin from the medulla and fatty acids from the brown adipose tissue. The role of proteins in the case of normally fed animals might seem more doubtful but, in reality, observations show that the onset of hibernation is always preceded by reluctance or refusal to eat for a few days. This can be observed at all periods of the year and the cause is undoubtedly the relative loss of appetite which develops abruptly, for reasons that are unknown. We believe that a fundamental cause of hibernation is the loss of appetite, because it results in a very reduced ingestion of proteins and torpor then infallibly ensues.

Under starvation or protein deficiency, lipid disturbances, especially hepatic steatosis, are easily understood in terms of the reduction of lipoprotein synthesis. Such changes in lipid metabolism, however, do not play a vital role in the induction of torpidity. We cannot propose any clearcut mechanism, but we may speculate that the disturbances observed in the liver as a result of dietary protein deficiency could also affect the central nervous system, rendering modifications of the enzymes involved in biosynthesis of biogenic amines. This hypothesis is suggested by the inactivation of the adrenal medulla, indicating a reduced sympathetic activity in the whole organism prior to onset of torpor. Draskoczy and Lyman (24) have already established that a comparable pheonomenon occurs in the brain; a general decrease in the sympathetic activity precedes hibernation.

The research described here is still at an early stage but our results suggest that in some species, protein deficiency alone might cause modifications of the nervous system with subsequent torpor. Further research concerning the role of protein metabolism on the induction of torpor in other species should thus be fruitful. The influence of fasting on torpor is also found in another garden dormouse (*Glis glis*) that we have studied. The effect of fasting on the ease with which hibernation sets in in another species (*Citellus lateralis*) has already been described by Mrosovsky and Barnes (49) and we too have observed this in the 13-lined ground squirrels.

VI. GENERAL CONCLUSIONS

The exact role of the endocrine glands in hibernation is sometimes clear, but occasionally quite obscure. There is no doubt that, during arousal and rewarming, the secretions of the adrenals and of the A and B cells of the pancreas explain the chronologic changes observed in the glucose metabolism in the garden dormouse at least. Other hibernators such as the ground squirrel and the hamster do not exhibit such changes in glycemia, however. So far as the pancreas is concerned, the seasonal modifications differ according to species, and glucose metabolism thus varies greatly from one animal to another. It is interesting to notice that the activity of the D cells of the pancreas in the garden dormouse is cyclic. If these D cells do secrete the hormone somatostatin (which needs further verification), and since somatostatin is known to inhibit secretion from A and B cells, to illuminate the possible role of D cells in the regulation of glucose metabolism in hibernators would be an interesting aspect. The differences between species are so great that we cannot believe that the role of the endocrine glands is dominant.

Considering now seasonal variations, we find that the states of activity of the glands or the levels of hormones in the blood do not necessarily indicate the role such hormones play in the process of hibernation. For example, the autumnal involution of the gonads and thyroid inactivity in garden dormice are prevented by early exposure to cold in spring; nevertheless, these animals hibernate normally, exactly like those with involuted gonads and thyroid inactivity. Thyroid activity may be very high, particularly during the last phase of hibernation, without interrupting hibernation. In fact, the influence of L-thyroxine on energy expenditure was found to be slight.

Some effects of the thyroid gland in the golden hamster are unexpected, since thyroidectomy decreases plasma lipids and increases glycemia, L-thyroxine having the reverse effect. Comparable results were not obtained in garden dormice however and they are probably peculiar to the golden hamster.

Some of the above results may seem curious or unexpected and their importance or their relation to the process of hibernation is sometimes questionable. On the other hand, two other results deserve more consideration. The thyroid gland, which exhibits a spontaneous and cyclic activity, governs to a great extent the distribution of proteins in the digestive tract and possibly in other tissues as well, even in fasting animals. L-thyroxine stimulates the synthesis of RNA and of proteins in these tissues and thus facilitates the restoration of organs that are necessary to the animal at the moment of spring arousal.

The second result demonstrates that deprivation of dietary proteins rapidly induces torpor in garden dormice. During seasonal hibernation, the deprivation of food and hence of proteins is a consequence of the self-imposed fast which always precedes torpor at any season. Whatever the mechanism, it is probable that characteristic modifications of protein metabolism play a major role in the process of hibernation.

REFERENCES

1. Agid, R., and L. Ambid. Effects of corporeal temperature on glucose metabolism in a homeotherm, the rat, and a hibernator, the garden dormouse. In: Depressed Metabolism. Musacchia, X.J., and J.F. Saunders (editors). New York: Elsevier, 1969, 119-157.
2. Agid, R., and L. Ambid. Hypothermia and metabolic perturbations induced by fasting in the garden dormouse (*Eliomys quercinus* L.). Cryobiology 10: abst. 36, 1973.
3. Agid, R., L. Ambid, P. Laporte, et R. Sicart. Estivation du Lérot (*Eliomys quercinus* L.). C.R. Acad. Sci. (Paris) 260:677-679, 1965.
4. Agid, R., M. Gabe, et M. Martoja. Modifications histologiques des glandes endocrines au cours des réveils périodiques du Lérot hibernant. C.R. Soc. Biol. 161:459-463, 1967.
5. Agid, R., et R. Sicart. Rôle de la thyroïde dans la régulation glycémique chez le Hamster doré. C.R. Acad. Sci. (Paris) 269:1551-1553, 1969.
6. Agid, R., et R. Sicart. Rôle de la thyroïde dans l'involution et la restauration du tube digestif d'animaux hibernants. C.R. Acad. Sci. (Paris) 269:1866-1869, 1969.
7. Ambid, L. Aspects endocrines de l'hibernation. Bull. Soc. Ecophysiol. (Paris) 1:30-39, 1976.
8. Ambid, L. Suppression de l'évolution saisonniére de l'activité testiculaire par exposition au froid (6°C), chez un hibernant le Lérot (*Eliomys quercinus* L.). Ann. Endocrin. (Paris) 37:493-494, 1976.
9. Ambid, L., et R. Agid. Modifications périodiques des lipides de la graisse brune et du plasma chez un hibernant, le Lérot. J. Physiol. (Paris) 61:198-199, 1969.
10. Ambid, L., et R. Agid. Hypothermie réversible et modifications du métabolisme lipidique induites par privation de nourriture chez

le Lérot, (*Eliomys quercinus* L.). J. Physiol. (Paris) 65:337A, 1972.

11. Ambid, L., and R. Agid. Suppression of the summer involution of gonads by April exposure to cold in the garden dormouse (*Eliomys quercinus* L.). Cryobiology 11:581, abst. 139, 1974.
12. Ambid, L., and R. Agid. Seasonal changes in plasma immunoreactive insulin in the garden dormouse (*Eliomys quercinus* L.) and in the ground squirrel (*Citellus tridecemlineatus*). J. Thermal Biol. In press, 1978.
13. Ambid, L., D. Abadie, et P. Bouissou. Acides gras libres et insuline plasmatiques chez le Spermophile (*Citellus tridecemlineatus*) au cours du cycle annuel. J. Physiol. (Paris) 72:61A, 1976.
14. Ambid, L., M. Berlan, et R. Agid. Contribution de la graisse brune à la mobilisation des lipides au cours des réveils périodiques chez un hibernant, le Lérot (*Eliomys quercinus* L.). J. Physiol. (Paris) 63:505-522, 1971.
15. Ambid, L., et R. Montoya. Hypothermie et stéatose hépatique chez le Lérot (*Eliomys quercinus* L.). J. Physiol. (Paris) 71:258A, 1975.
16. Ambid, L., et R. Montoya. Mise au repos de la médullo-surrénale précédant l'hypothermie induite par suppression de l'alimentation chez un hibernant, le Lérot (*Eliomys quercinus* L.). J. Physiol. (Paris) 12:62A, 1976.
17. Bauman, T.R., and R.R. Anderson. Thyroid activity of the ground squirrel (*Citellus tridecemlineatus*) using a cannula technique. Gen. Comp. Endocrinol. U.S.A. 15:369-373, 1970.
18. Benoit, J. Stimulation du dévelopment testiculaire par l'éclairement artificiel. C. R. Soc. Biol. 118:664-672, 1935.
19. Buchanan, J., and D.F. Tapley. Stimulation by thyroxine of amino acid incorporation into mitochondria. Endocrinol. U.S.A. 79:81-89, 1966.

20. Cahilly, G.M. The effect of thyroxine on protein biosynthesis and ribonucleic acid metabolism in the rat. Dissert. Abst. U.S.A. 26:3610, 1966.
21. Canguilhem, B. Effets de la radiothyroïdectomie et des injections d'hormone thyröidienne sur l'entrée en hibernation du Hamster d'Europe, *Cricetus cricetus*. C. R. Soc. Biol. 164:1366-1369, 1970.
22. Christophe, J., and J. Mayer. Effects of hormones on glucose utilization in fasted rats. Endocrinol. U.S.A. 65:475-486, 1959.
23. Denyes, A., and J.D. Carter. Utilization of acetate $1^{14}C$ by hepatic tissue from cold exposed and hibernating hamsters. Am. J. Physiol. 200:1043-1046, 1961.
24. Draskoczy, P.R., and C.P. Lyman. Turnover of catecholamines in active and hibernating ground squirrels. J. Pharmacol. Exp. Ther. 155:101-111, 1967.
25. Dubois, M.P. Immuno reactive somatostatin is present in discrete cells of the endocrine pancreas. Proc. Nat. Acad. Sci. 72: 1340-1343, 1975.
26. Ellefson, R.D., and H.L. Mason. Effects of thyroid hormone on lipid metabolism in the rat. Endocrinol. U.S.A. 71:425-430, 1962.
27. Fain, J.N., and A.E. Wilhelm. Effects of adrenalectomy, hypophysectomy, growth hormone and thyroxine on fatty acid synthesis *In vivo*. Endocrinol. U.S.A. 71:541-548, 1962.
28. Fleischmann, W., H.B. Shumacker Jr., and L. Wilkins. Effect of thyroidectomy on serum cholesterol and basal metabolic rate in the rabbit. Amer. J. Physiol. 131:317-324, 1940.
29. Fletcher, K., and N.B. Myant. Effect of thyroxine on the synthesis of lipid in rat liver. Endocrinol. U.S.A. 71:870-877, 1962.
30. Gabe, M., R. Agid, M. Martoja, M.C. Saint Girons, et H. Saint Girons. Données histophysiologiques et biochimiques sur l'hibernation

et le cycle annuel chez *Eliomys quercinus* L. Arch. Biol. Liège 75: 5-92, 1964.
31. Gabe, M., et M. Martoja. Contribution à l'histologie du pancréas endocrine d'*Eliomys quercinus* L. Arch. Histol. Jap. 30:123-147, 1969.
32. Hoffman, R.A. Terrestrial animals in cold: hibernators. In: Handbook of Physiology, Adaptation to Environment. Washington D.C.: Am. Physiol. Soc. 1964, sect. 4, chap. 24, 319-403.
33. Hoo Paris, R., Ch. Castex, et B. Sutter. Variations de l'insulinémie du Hérisson en hiver. J. Physiol. (Paris) 72:43A, 1976.
34. Houssay, B.A. Thyroid and metathyroid diabetes. Endocrinol. U.S.A. 35;158-172, 1944.
35. Hudson, J.W. Ineffectiveness of exogenous L-thyroxine in preventing torpor in *Citellus tereticaudus*. Am. Zool. 8:(4), 159, 1968.
36. Hudson, J.W. Torpidity in mammals. In: Comparative Physiology of Thermoregulation. Whittow, G.C. (editor). New York: Academic Press, 1973, Vol. III, 97-165.
37. Hulbert, A.J., and J.W. Hudson. Thyroid function in a hibernator, *Spermophilus tridecemlineatus*. Amer. J. Physiol. 230:1211-1216, 1976.
38. Kayser, C. The physiology of natural hibernation. New York: Pergamon Press, 1961.
39. Kayser, Ch., et M. Aron. Cycle d'activité saisonniére des glandes endocrines chez un hibernant, le Hamster (*Cricetus frumentarius*). C.R. Soc. Biol. 129:225-228, 1938.
40. Keever Mc, S. In: Comparative Biology of Reproduction in Mammals. Symp. Zool. Soc. London. New York: Academic Press, 1966, 365.
41. Klain, G.L., and B.K. Whitten. Plasma free amino acids in hibernation and arousal. Comp. Biochem. Physiol. 27:617-619, 1968.
42. Kristoffersson, R. Hibernation of the hedgehog (*Erinaceus europaeus* L.). Blood creatinine and creatine levels in relation to seasonal and hibernation cycles. Ann. Acad. Sc. Fenn. A, IV, 95:1-10, 1965.

43. Kristofferson, R., and S. Broberg. Free amino acids in blood serum of hedgehogs in deep hypothermia and after spontaneous arousals. Experientia 24:148-150, 1968.
44. Kristoffersson, R., and S. Broberg. Free amino acids in blood of hibernating and non-hibernating hedgehogs and golden hamsters, and in common frogs acclimated to different ambient temperatures. Ann. Acad. Sc. Fenn. A, IV, 130:1-22, 1968.
45. Lachiver, R. Cycle Saisonnier de l'iodémie d'un hibermant, le Lérot (*Eliomys quercinus* L.). C.R. Soc. Biol. 146:245-248, 1952.
46. Lachiver, F., M. Olivereau, et Ch. Kayser. L'activité de la thyroïde d'un hibernant le Lérot (*Eliomys quercinus* L.) en hiver et au printemps. C. R. Soc. Biol. 151:653-656, 1957.
47. Laurila, M., and P. Suomalainen. The changes in the insulin level induced by seasons and hibernation cycle. Ann. Acad. Sci. Fenn. A, IV, 201:1-40, 1974.
48. Man, E.B., E.F. Gildea, and J.P. Peters. Serum lipids and proteins in hyperthyroidism. J. Clin. Invest. 19:43-49, 1940.
49. Mrosovsky, N., and D.S. Barnes. Anorexia, food Deprivation and hibernation. Physiol. Behav. 12:265-270, 1974.
50. Munro, H.N. Role of amino acid supply in regulating ribosome function. Fed. Proc. 27:1231-1237, 1968.
51. Musacchia, X.J., and C.G. Wilber. Studies on the biochemistry of the arctic ground squirrel. J. Mammal. 33:356-362, 1952.
52. Ondrick, K. Aspects of renal structure and function in hibernating ground squirrels (*Citellus tridecemlineatus*). Ph.D. Georgetown University, Washington, 1973.
53. Ondrick, K.A., T.F. Albert, and J.A. Panuska. Preliminary findings on glomerular ultrastructure and function in hibernating 13-Lined ground squirrels (*Citellus tridecemlineatus*). Cryobiology 11:580, abst. 137, 1974.

54. Orsetti, A., A. Basseres, et. J. Macabies. Etudes des variations insulino glucidiques chez le chien thyroidectomisé traité par des extraits thyroidiens. J. physiol. Paris 61:362-363, 1969.
55. Panuska, J.A. Delay of spring spermatogenesis in cold exposed *Tamias striatus*. Amer. Zool. 5:739, 1965.
56. Petry, G., H. Amon, V. Herzog, et W. Kuhnel. Vergleichende untersuchungen an der niere des Siebenschläfers (*Glis glis* L.). Z. Zellforsch 64:827-846, 1964.
57. Platner, W.S., B.C. Patnayak, and X.J. Musacchia. Seasonal changes in the fatty acid spectrum in the hibernating and non-hibernating ground squirrel, *Citellus tridecemlineatus*. Comp. Biochem. Physiol. 42A: 927-938, 1972.
58. Popovic, V.P. Endocrines in hibernation. Bull. Mus. Comp. Zool. Harward Coll. 124:105-130, 1960.
59. Ramussen, A.T. Seasonal changes in the interstitial cells of the testis in the woodchuck (*Marmota monax*). Amer. J. Anat. 22:475-515, 1917.
60. Reiter, R.J. Pineal control of a seasonal reproductive rhythm in male golden hamsters exposed to natural daylight and temperature. Endocrinology 92:423-430, 1973.
61. Rinaudo, M.T., C. Giunta, R. Bruno, A. Guardabassi, M. Olivero, and P. Clerici. Carbohydrate metabolism in normal and thyroidectomized *Bufo bufo* adult males. Comp. Biochem. Physiol., G.B. 29:1079-1086, 1969.
62. Rowan, W. Relation of light to bird migration and developmental changes. Nature 115:494, 1925.
63. Sable - Amplis, R., R. Sicart, and R. Agid. Spontaneous high storage of liver cholesterol in a strain of golden hamsters (*Mesocricetus auratus*). J. Thermal Biol. (In Press), 1978.
64. Sicart, R., et R. Agid. Action hypoglycémiante de la thyroxine chez le Hamster doré. J. Physiol. (Paris) 63:95-96, 1971.

65. Sicart, R., et R. Agid. Influence de la thyroïde sur la répartition des acides aminés plasmatiques et les remaniements protidiques au cours de l'hibernation du Lérot. J. physiol. (Paris) 65:304-305A, 1972.

66. Siperstein, M.D., R. H. Unger, and L.L. Madison. Studies of muscle capillary basement membranes in normal subjects, diabetic and prediabetic patients. J. Clin. Invest. 47:1973-1999, 1968.

67. Smith, E.R. Effects of the thyroid gland on protein metabolism in the golden hamster (*Mesocricetus auratus*) as measured by the in vivo incorporation of S 35 L methionine. Dissert. Abst. U.S.A. 26: 1133, 1965.

68. Smith, R.E., and B.A. Horwitz. Brown fat and thermogenesis. Physiol. Rev. 49:330-423, 1969.

69. Sokoloff, L., and S. Kaufman. Thyroxine stimulation of amino acid incorporation into protein. J. Biol. Chem. 236:795-803, 1961.

70. Sokoloff, L., S. Kaufman, P.L. Campbell, C.W. Francis, and H.U. Gelboin. Thyroxine stimulation of amino acid incorporation into protein. Localisation of stimulated step. J. Biol. Chem. 238:1432-1437, 1963.

71. Summers, J.D., and H. Fisher. Nutritional requirements of the protein-depleted chicken. II. Effect of different protein depletion regimens on nucleic acid and enzyme activity in the liver. J. Nutr. 76:187-198, 1961.

72. Tani, H., K. Ogata, and T. Itatsu. Effect of ethionine on carbohydrate and lipid metabolism. J. Lipid Res. 14:32-40, 1973.

73. Tata, J.R. Biological action of thyroid hormones at the cellular and molecular levels. In: Actions of Hormones on Molecular Processes. Litwack, G. and D. Kritchevsky (editors). New York: J. Wiley and Sons, Inc., 1964, Chap. 2, 58-131.

74. Tata, J.R., and C.C. Widnell. Ribonucleic acid synthesis during the early action of thyroid hormones. Biochem. J. 98:604-620, 1966.

75. Weill, J.D., et Ch. Kayser. Les réserves glycogéniques et lipidiques du foie de la Marmotte en état d'activité au cours du sommeil hivernal. C. R. Soc. Biol. (Paris) 151:374-377, 1957.

76. Wells, L.J. Seasonal sexual rhythms and its experimental modification in the male of the 13-lined ground squirrel (*Citellus tridecemlineatus*). Anat. Rec. 62:409-447, 1935.

77. Wells, L.J., and M. Zalesky. Effects of low environmental temperatures on the reproductive organs of male mammals with annual aspermia. Amer. J. Anat. 66:429-447, 1940.

78. Whitten, B.K., and R.F. Burlington. Amino acid catabolism during arousal from hibernation. Fed. Proc. 30:483-495, 1971.

79. Whitten, B.K., and G.J. Klain. Protein metabolism in hepatic tissue of hibernating and arousing ground squirrels. Amer. J. Physiol. 214:1360-1362, 1968.

80. Whitten, L.E., L.E. Schrader, R.L. Huston, and G.R. Honold. Hepatic polyribosomes and protein synthesis: seasonal changes in a hibernator. Int. J. Biochem. 1:406-408, 1970.

81. Williams, J.N. Response of the liver to prolonged protein depletion. I. Liver weight, nitrogen and deoxyribonucleic acid. J. Nutr. 73:199-209, 1961.

82. Zimny, M.L., and E. Rigamer. Glomerular ultra-structure in the kidney of a hibernating animal. Anat. Rec. 154:87-94, 1966.

HIBERNATION TRIGGER RESEARCH UPDATED

Albert R. Dawe, Ph.D.
Adjunct Professor of Physiology
Department of Physiology
Loyola University Medical Center
Stritch School of Medicine
2160 South First Avenue
Maywood, Illinois 60153

ABSTRACT

A film, "Trigger for Hibernation" (1975) by A. R. Dawe and W. A. Spurrier is shown. Purposes and results of two conferences (March, 1974, and February, 1977) on trigger research are also described. A new theoretical model is put forth based on interactions of three presumed substances: "Trigger", "Antitrigger", and "Hibernation Substrate".

ISBN 0-12-734550-7

I. THE FILM: "TRIGGER FOR HIBERNATION" BY DAWE-SPURRIER, 1975

A rationale for producing an audiovisual record of the methodology which Miss Spurrier and I had found to be successful for induction of hibernation using "trigger" substance became apparent to us as early as 1968. This was shortly after we had discovered the existence of a trigger substance in blood of hibernating animals (10). Rather quickly after we announced our initial transfusional induction successes, other investigators indicated that they wished for us to provide them with trigger substance. Invariably, their plan was to use it for administration to animal recipients of physiological status unknown to us. We provided such material in several instances but, essentially, no positive feedback was received. Actually, almost immediately after our first successes, we became crucially aware of the fact that the administration of trigger by itself, entirely without consideration for the fulfillment of certain conditions, was nonuseful. We were dealing with a donor-recipient phenomenon wherein the physiological status of the *recipient* was of equal or greater importance to a production of hibernation induction than was the use only of the biologically effective substance obtained from a donor (11, 12, 13, 14, 43). Thus, for success to be achieved, we began to emphasize repeatedly, both orally and in publications, first, the nature of the extracted material from donor hibernation blood had to be obtained by a method proven correct; second, the physiological status of the recipient was definitive to success; and third, environmental conditions for each step in the procedure were critical. Even beyond these essentials, we found that induction of hibernation unquestionably depended on factors such as species, time of year, sex, cold adaptation.

In sum, there is need for a very precise methodology. Thus, in order to give to our colleagues a way whereby they, too, could induce out-of-season hibernation at their own laboratories, we settled upon filming what we found to be a particular set of procedures which guaranteed success as we had seen it to be. At that time, our pre-film experience was five years of research on this matter. To have the film made, required administrative and fiscal approval of Dr. Walter C. Randall who then chaired the Physiology Department at the Stritch School of Medicine in Maywood, Illinois, and who was also Principal Investigator on National Institutes of Health Research Grant NO. HE-08682 which grant provided funds to us for hibernation research, including this film-making. The Office of Naval Research provided about twenty percent of my work time to direct that Stritch hibernation research project.

The triggering methodology shown in the film is in no way claimed to be an exclusive method for inducing hibernation out-of-season, nor is it claimed that all of the procedures must be used to successfully attain hibernation induction. The film does claim, however, that if the method as shown is duplicated by a colleague, otherwise unattainable out-of-season hibernation will be produced in the thirteen-lined ground squirrel (*Citellus tridecemlineatus*) or in the woodchuck (*Marmota monax*). The film exists only because we wished to provide an exact audiovisual instruction for our colleagues. Emphasis, as you will see, is approximately equal between footage on the extraction procedure from a proper donor and footage on the proper transfusional (injection) procedure to a proper recipient. The restricted methodology as here illustrated guarantees success. We certainly, however, do not fully understand why seemingly small deviations from this method negate such successes. To elucidate this more narrowly, we cannot give adequate explanation, for example, as to why intravenous transfusion into a recipient seems to be an essential to achieve hibernation induction, whereas transfusion (injection) via other portals gives questionable or negative results. As another example, we also cannot explain to our own satisfaction why recipients that have been cold-adapted prior to transfusion seem to resist the trigger effect, whereas warm-adapted recipient animals apparently are more amenable to it.

Results translated into data soon led us to a conclusion that a typical successful procedure should be documented in a sound motion picture. The major effort which went into making the film itself was contributed largely by my colleague, Wilma Spurrier. She prepared film sequences, wrote the script, and supervised putting the film sequences into a unitary whole. Mrs. Margaret

Conneely, head of the Photolab at the Stritch School of Medicine, did all filming and provided photo technology, as needed.

Finally, it should be mentioned that the film also includes a sequence on what I believe is the first time infant hibernation has been demonstrated, and also includes a sequence illustrating our first models pointing toward a theory of tissue hibernation.

II. EVENTS CONCURRENT OR SUBSEQUENT TO THE FILM-MAKING, RELEVANT TO TRIGGER RESEARCH

In recent years, several of our colleagues expressed interest in attempts at confirmation and/or extension of our research. We anticipated rightly many difficulties they would encounter in these attempts. Hence, we convened at the Stritch School of Medicine two rather small ONR-sponsored roundtables or workshops. At the first workshop we endeavored to explain to scientists who had shown an interest, our technology as well as the scope of our trigger research. The second workshop was intended to summarize all scientific work underway anywhere wherein trigger substance was used or studied. The first workshop convened on March 23, 1974; and the second workshop was held on February 12, 1977. At the first workshop, six biochemists and four physiologists learned about and discussed with one another the status of research on trigger. That workshop's summary result and primary advice to us was to push ahead toward better biochemical identification of the trigger substance itself. Later, in 1976, largely consequent to that recommendation, Dr. Peter Oeltgen of the Biochemistry Department at Stritch received an ONR contract wherein he precisely directed his efforts toward that end. Indeed, at this present conference he reports his latest findings which follow onto our original work. His research moves swiftly ahead, and interested scientists will find what he has to say to be of great interest in this context.

That second workshop in February, 1977, endeavored to bring together into one room all persons who, to date, had experimented with trigger substance *per se* as well as several persons of whom we were aware who had primary interests in extending this research. We pooled our understandings and compared notes on alternative methodologies. Seventeen scientists were present, eight of whom had had actual experience with trigger substance. Three of these eight neither positively confirmed nor denied trigger induction effectiveness when used in their experiments. Significantly and almost inevitably we noted their use of crucially different procedures than we were using. Two laboratories (Dr. David Bruce's at Wheaton College in Illinois and Dr. G. Edgar Folk's, with Martin Steiner, at the University of Iowa Medical School) enhanced the discussion by happily citing indications of positive results confirming the trigger effect. Dr. Peter Oeltgen of Stritch and Dr. David Polis of the Naval Air Development Command Research center of Warminster, Pennsylvania, described experiments they have considered or have undertaken looking toward better biochemical identification of hibernation trigger. Though that route is a hard one to pursue, Dr. Oeltgen is fully embarked in that direction (37). As adjunctive research, Dr. Henry Swan of Colorado State University discussed his experiments subsequently published in Science (45) wherein he made an extract from brains of hibernating ground squirrels which upon being introduced into white rats would successfully lower their body temperatures. The Swan substance, which he names "anabalone", has a curious physiological effect which appears to be related to that of trigger substance insofar as both result in hypothermia in recipients. But it is not probable that Swan's effective substance is the same material as trigger since its effect does not result in a bout of hibernation. More correctly, I believe, Swan's anabalone may be biochemically and physiologically closer in its action to a substance I will theorize about in a moment and which I call "hibernation substrate". Other persons present at that meeting were three investigators who were then considering embarking upon trigger research.

Today I think it is correct to say that although very little research on trigger is underway anywhere, interest in the subject is growing among members of the scientific community. The exacting restrictions and the long time-courses we have discovered to be needed for experiments (usually a year or more in length) always create prospective burdens which most investigators who write to us or visit us are unable to or do not care to shoulder. Consequently, duplicative or advanced research along this line has seldom got off the ground. For example, and in contrast to others observations (46), one of the crucial requirements we have found to be absolutely essential for success is intravenous transfusion of cold material into recipients. Oral, intracardiac, intraperitoneal, subcutaneous, or intramuscular combined or not with introduction of material warmed above 20°C invariably we find to give negative results. The only exception was use of the intraperitoneal portal in infant animals wherein we did indeed obtain positive results. The extreme difficulty in quickly transfusing 5°C dialysate solution into the hair-thin saphenous vein of a ground squirrel requires painstaking experience. To my knowledge only two persons have ever left our laboratory after witnessing and trying this method and then gone home having mastered it. In these two cases unquestioned success was later achieved by one of them in being able to induce hibernation, and cautiously optimistic results were obtained by the other.

The major result of that second workshop coincided with my personal interest in seeing this research move ahead. I have spent many, many hours theorizing with respect to what it is we really have observed physiologically. The answer which comes to me again and again is that if hibernation can be induced out-of-season by this endogenously-produced triggering substance given intravenously into an active warm room-adapted recipient, then body tissues of the recipient must be amenable or ready to take on the hibernation state. It is true that if trigger substance is properly introduced, it will set in motion the profound physiological change which is called "hibernation". The tissues themselves must make this response (53). It seems only reasonable that tissues of the animal's body "read some kind of signal" which trigger sends to them. If true, why not experimentally examine that particular hibernating tissue for which data pertaining to such a model already exists, i.e., the heart (2, 6, 17, 20, 32, 48)? A heart puts out electrical discharges which experts can interpret and understand. In addition, the heart is made up largely of muscle. Muscle moves. Thus, electrical and/or mechanical outputs from the heart of an animal which hibernates can be observed as a response to perfusion (44). I would especially hope to examine possible effects of trigger or antitrigger (15). Clues are obtained as to metabolic changes purportedly initiated when these substances naturally exhibit capabilities for beginning or ending the hibernation of the whole animal. The intent of my personal research now is to show, using isolated hearts, that coronary-perfused trigger or antitrigger substances do or do not have an effect on the a) myocardium or b) conductile system of the heart of an animal which hibernates. Furthermore, I hope to learn whether tissue which hibernates *in vivo* can be made to hibernate *in vitro*.

III. THEORETICAL CONSIDERATIONS

Over the past nine years, my experiences at workshops, discussions, and my own routine research efforts involving trigger have all contributed to my reconsideration of those hypotheses which may be useful in explaining what we have seen experimentally. I personally enjoy such theorizing. Everyone of us can set up his own premises, of course. Each of us can easily "take shots" at any new theory model-making ventures. Let me set up a "paper tiger" (as Dr. Wang would say) for your critical consideration. My model addresses itself only to what the 13-lined ground squirrel [*Citellus (or Spermophilus) tridecemlineatus*] and the woodchuck (*Marmota monax*) do in the course of their annual hibernation cycles. I will introduce references by others as illustrative of research which substantiates some of the theoretical components of this model which are in addition to our substantiations. Let me boldly strike forth then and in a theoretical sense assume the following to be true. I approach this theoretical model by positing three substances, each I regard as unique and each I regard as significant to the occurrence of hibernation. By this I do not mean that these three are unknown substances. Rather, they may well be substances that are known but are as yet unidentified biochemically in the context of our experiments. The three substances whose actions I would briefly describe for you are as follows:

1. A small (less than 5,000 Daltons) molecular substance, possibly a small peptide (16, 28) which we call "trigger". It is possible that this substance is derived from the pineal gland (4, 39, 41). The substance is unstable insofar as it is denatured at temperatures above about 20°C. It is stable at cold temperatures: 10°C and below. Trigger has an avidity for combining with another substance we call "antitrigger" (described below) in blood. I herewith also presume that trigger has a potential for combining with "hibernation substrate" (also described below) within metabolizing cells of a hibernating species. When trigger is complexed in these two ways, these complexes maintain stable configurations at all physiological temperatures encountered, i.e., about 0°C to 40°C.

2. A larger molecule than trigger (more than 5,000 Daltons) called "antitrigger" which is probably secreted at the adrenal cortex or at the gonads. Its secretion may be noted as a response to the influence of circannual light cycling and/or sexual cycle changes (8, 27). This substance, in contrast to trigger, is heat-stable and cold-labile. Antitrigger "lives its lifetime" in the bloodstream. It does not react with trigger within metabolizing cells as is the case for hibernation substrate. The major function of antitrigger is at the blood where it will combine with any free trigger that may be there. In this way antitrigger draws trigger substance out of action and away from cells, hence preventing the activation of hibernation substrate, which, in turn, prevents cells from hibernating. In view of the fact that we have shown that rather soon trigger can be denaturable by heat, yet is only effective when it is cold and entering a warm animal body, there is here an apparent physiological contradiction. This contradiction can only be resolved by presuming that cold trigger substance must almost instantaneously combine

with any of this antitrigger substance which may be available in blood. It can also be hypothesized that cold trigger would within a very few moments enter cells and there combine almost equally rapidly with hibernation substrate. This latter reaction could only occur on those occasions in which there is no antitrigger available to present a blood barrier to trigger.

3. A fat or fat derivative I will call "hibernation substrate" (30). Its primary characteristic would be to have a capability to be metabolized at an extremely slow rate, but such catabolism to occur only at a time when it would be complexed with trigger. This material (hibernation substrate) would seem likely to be derived from brown fat, a fat which is in larger than usual supply in hibernating species. The concept here would be that in addition to the well known usage by an animal of brown fat as its high energy source for arousal from hibernation (31, 40), brown adipose would also be a source for a *low energy* fat derivative as well. In 1941, Hook and Guzmann-Barron demonstrated the extraordinary metabolic (heat) source derivable from brown fat (21). Since that time, the certain utilization by hibernators of brown fat as the "flame" for arousal heat has been very well documented. It seemed not unlikely to me that brown fat may also provide a second energy source to hibernators, but this one would be a material which only "smolders" (does not "flame"). Thus, the so-called smoldering substance would be the low energy source needed to sustain hibernation in the cold. I believe furthermore and I theorize herewith that this is an inactive material except when it is bound to trigger. Thus, it is at this point in the metabolic scheme that trigger would be effective--that is, trigger has the capability to activate this fat derivative, this "hibernation substrate". In contrast to blood-borne anti-

trigger (or antitrigger complexed to trigger) I conceive of the location for hibernation substrate to be inside of metabolizing cells. To see it in blood would be incidental only to noting its travel from a bodily fat depot to cells (29, 34). When trigger enters a cell it would then combine with any of this hibernation substrate which may be present in the cell. The complex which thereby forms--i.e., trigger linked to the hibernation substrate--would be a very labile working metabolic material whereby, by its slow metabolism, the cell is permitted to continue functioning but with its utilization of oxygen at a very, very low level indeed--almost anaerobic. Evolution of heat is minimal. The metabolism of this complex would represent the metabolism of hibernation. When this animal is in a cool environment the result would be that there would not be enough heat generated by breakdown of the trigger-hibernation substrate complex to keep body temperature steady. The animal body would cool, i.e., the animal now induces into hibernation. Since the metabolized substance is a fat derivative, RQ is at about 0.7 which agrees with experimental observations. Very small amounts of carbon dioxide and water would be produced. In this process, in a cold hibernating animal body, bound trigger would again be loosened from the complex and freed into the cold cell. When trigger--so freed--attains high enough concentration in that cell, it would then diffuse into the cold hibernation blood hence increase the trigger concentration in such blood in direct proportion to the length of time of that particular bout of hibernation. Our data indicates this to be the case.

If we permit ourselves to suppose that the three substances premised above truly exist and that they are secreted or formed at rates that I will describe hereafter, remarkably enough one is able to account for many aspects of hibernation as interdigitated into an annual cycle (3). The first induction in the fall and the bouts of hibernation in the winter can be accounted for. Hibernation bouts which at first lengthen until midwinter followed by bouts which shorten until a final arousing in the spring can be explained. One can deduce no hibernation during the breeding season in the spring and one can account for lack of hibernation in the summer months. Causality for these events can be described in terms of the activities of trigger, antitrigger, and hibernation substrate as well as activities of two complexes, i.e., trigger complexed with antitrigger and trigger complexed with hibernation substrate. Seasonal hibernation models such as this have been devised by others primarily invoking hormonal activities as Kayser has done (26).

To make the story complete we must hypothesize rates of manufacture or secretion rates for the three above-outlined substances. The following rates for them to be secreted gives rationale for the events we observe during annual hibernation cycles:

1. Antitrigger can be thought of as a substance secreted into blood at zero rate in midwinter (19), but it begins then to be secreted at an accelerating rate until early spring when it reaches a maximum rate. This maximum rate is achieved as the hibernation season ends and the breeding period begins (8). A gradual fall-off in rate of secretion then occurs, first rapidly and then more slowly as the year progresses into late spring, summer, fall, and early winter.
2. Trigger can be thought of as a substance which is being constantly slowly secreted into blood throughout the year, but only with an exception of a somewhat lowered rate of secretion during each hibernation bout, when anabolic processes (including the manufacture of trigger) would be at profoundly slower rates. Trigger--a heat labile substance--would be rapidly metabolized if it were secreted when the animal body is euthermic unless sufficient free antitrigger were present in blood for it to be bound to, and/or unless sufficient hibernation substrate were immediately available at

cells to be bound and to be activated by trigger.

3. The hibernation substrate would be thought of as an energy producer, a fat derivative prepared at fat depots in the body and then moved into the cells within this animal which has a capability for hibernation. Enzymological response may possibly speed up this process when ambient temperature lowers (50). This manufacture and movement of hibernation substrate would occur at all times except during hibernation bouts. At the time that a hibernation bout begins, hibernation substrate in cells would have been linked or complexed with trigger. The trigger-hibernation substrate complex would thus become amenable to its own slow metabolic destruction. At some critical level (possibly signaled cybernetically by an excess of trigger in blood, having bound all available antitrigger and hibernation substrate) this breakdown of the trigger-hibernation substrate complex would begin to provide low levels of energy necessary to sustain the life of the animal in hibernation.

I give a brief tabulation of annual or circannual hibernatory events in these two species which would then be dependent upon these three substances. Investigators have speculated in similar ways as to the events of hibernation related to other substances and other species (1, 5, 7, 18). This theory employs the following time course:

1. SPRING: Antitrigger secretion increases tremendously in step with the breeding season. The small amounts of trigger secreted into the blood complexes with antitrigger to form a relatively stable complex. Thus, no free trigger is available to enter cells and tissues. Hibernation substrate enters cells, but no trigger is there available for complexing. Hence, since it would be presumed that hibernation substrate only can be activated when complexed with trigger it remains physiologically inactive at this time. Antitrigger

production begins to taper off at the ending of spring. Since whatever trigger may be present during this period is bound to antitrigger in blood, trigger there is not free to enter the tissue cells and to complex with hibernation substrate. Thus, there is no hibernation of cells (or thereby of the animal) in spring.

2. SUMMER: The rate of antitrigger secretion goes down. Hence its level also goes down in blood, but there is always more than enough antitrigger to bind whatever trigger will be secreted into the blood. Hence there is no summer hibernation.

3. FALL: Rate of antitrigger secretion falls (22) to such a low level that not all of the trigger secreted can be bound to it. Some excess trigger "spills over" into the cells. Such extra trigger quickly passes the cellular membranes and enters cells where unbound hibernation substrate is present. The fundamental molecular hibernatory complex of trigger with hibernation substrate is thereby formed. The cell preferentially permits this complex to be metabolized. One reason may be that it requires very little oxygen. Therefore, at some cybernetically critical level, it begins to split apart in the process yielding very small amounts of energy. The tissues give this splitting reaction preference and turn down more usual metabolic processes. Thus, greatly lowered heat production occurs which, in the face of any environmental cooling, will then result in a facultative hypothermia, i.e., the first induction of the animal into hibernation in the fall. Gradual breakdown of this trigger-hibernation substrate complex occurs as long as any of it is available at the cells. Actually, the quantity of this complex in the cells determines how long each bout of hibernation will last. Considering other substances, the same rationale has been used in explaining bout arousals. Thus, loss of K^+ ion from

excitable tissue (51) or glucose utilization by adipose tissue (9) have been posited as arousal *modi vivendi*.

4. EARLY WINTER: During hibernation of the now-cold animal body, some trigger in cells is continuously being freed from its hibernation substrate complex. Freed trigger moves from that splitting complex in cells back to the cold blood. Since hibernation blood carries no free antitrigger with which trigger could complex, the trigger concentration of hibernation blood climbs. As we have experimentally determined, the concentration of trigger in hibernation blood increases the longer the bout of hibernation. When trigger-hibernation substrate is used up at hibernating metabolizing cells, the hibernation bout ends. The cells revert to normal (25) metabolic processes: much oxygen is called for, much heat is generated, and an arousing from hibernation begins. At each arousing, any free trigger in the blood is denatured by heat or (like urea) could be excreted (38). Thus, the only trigger left in the animal's body after an arousing process is to be found at either (i) the trigger-secreting organ (whatever that may be), or (ii) complexed or bound to antitrigger in the blood. There is, of course, at early winter little or no antitrigger present. During the winter hibernation season, at those periods of time when the animal is in the aroused state (between hibernation bouts), newly-formed trigger moves instantly into tissue because little or no free antitrigger is available in blood to link to it and hence to impede its progress toward cells. When trigger enters cells, it binds to hibernation substrate. The result is, that it can be said that during each arousal period all of the necessary conditions are quickly set up for a new induction into hibernation. During early winter, rate of secretion of antitrigger continues to go down, hence less and less antitrigger is available in blood to form

trigger-antitrigger complex. This allows more trigger to enter cells and to complex with more hibernation substrate. Since length of a bout of hibernation depends on the amount of trigger-hibernation substrate complex available in the cells, there is obviously concomitantly a lengthening of hibernation bouts during this early winter period.

5. MIDWINTER: At midwinter, the longest bout of hibernation occurs. At this time there is no antitrigger in blood. Following this longest bout antitrigger again would be secreted at increasingly faster rates into the blood at arousals (42). Along this line such a restoration of endocrine activities in the middle of a hibernation season also has been demonstrated for the thyroid (23). All such new antitrigger produced will bind to some of the trigger being formed. The midwinter step-up in secretion of antitrigger progressively determines more and more trigger will be bound to it. This, in turn, makes less trigger available to be complexed with hibernation substrate in cells. The final result is that each hibernation bout progressively becomes shorter. This is true since length of a bout of hibernation is determined by the amount of trigger-hibernation substrate complex that is available.

6. LATE WINTER (FINAL AROUSING): The final arousing occurs when the balance of antitrigger vs. trigger tips in favor of more antitrigger in the blood than there is trigger available to complex with it. Free trigger then no longer is available for entering the tissues, and the hibernation season ends.

IV. SUMMARY

A. Elaboration of the Theory

1. The theory of hibernation presented herein was developed out of experiments at the Dawe-Spurrier laboratory whereat the existence was pointed to of a "trigger" substance obtainable from bloods of animals

in hibernation, and the existence of an "antitrigger" substance was pointed to from bloods of hibernators when not in a state of hibernation. A third substance was introduced only theoretically and is named "hibernation substrate". That substance is presumed to exist based on theory alone.

2. Both trigger and hibernation substrate are presumed to be secreted throughout the year at relatively constant rates, whereas antitrigger is thought of as having a sharply circannual secretion pattern in which rate begins from zero in the middle of the winter hibernation season, increases to its fastest rates by the time of the spring breeding season and then tapers off through late spring and summer and back to low then zero levels by late fall and early winter (13). If, contrary to the above, hibernation substrate were to be moved from fat depots to cells at rates inversely proportional to the degree of ambient temperature (and in accord with species specificity) one could elaborate on this theory in such a way as to give a more mechanistic rationale for observed differences between "permissive" and "obligate" hibernators. Secondly, if trigger (presumed to be secreted at the brain, at other parts of the nervous system, or at the pineal) also had a modified secretion rate based upon light cycling changes, one could develop more cogent explanations for observed fall inductions and spring arousals in the obligate hibernators (49).

3. It is inaccurate to say that trigger is a substance which induces hibernation. It is far more correct to say that a system exists for hibernation to occur. This system has in part been unmasked, wherein it has been discovered that a number of critically important substances are involved in a "dance of the molecules" which ends in hibernation induction. In the laboratory methodology we have employed, an artificial situation was created in the form of the donor animal. In truth the bodily system which must be understood is at the recipient animal. It is the recipient animal which does the hibernating, and it is that animal's bodily system which must be understood in order to know about the total biological system ending in hibernation wherein trigger activity is only one facet thereof.

B. Disclaimer

I find that just to get started in research it is almost always more satisfying than not, indeed it is almost always necessary (even momentarily) to accept some explanatory hypothesis. Such can serve as a backdrop to one's actions upon a new stage of personal research activity. In the theory I have presented here, I have opened up to you my innermost present predilections for my own plan for future research. I hope that some young investigators (who have enough years ahead of them) may be inspired to try some of the experiments which suggest themselves.

The "initiated" will recognize immediately that I have omitted whole foundational chunks in delineatory comprehensions essential to any up-to-date understandings of hibernation. As examples, I have left out the nervous system entirely. I am sure you agree that any fool knows that hibernation must and does deeply involve nervous mechanisms. Most prominent of all probably are mechanisms attributable to functions of the hypothalamus. Second, well understood hormonal mechanisms have only been alluded to, never named directly, as part of this theoretic structure. This occurs largely, I plead, because we really do not know exactly what substances are involved or from what bodily organs they derive. Conspicuously absent are the details of thyroidal (24, 36, 47), digestive (33, 35) and excretory (52) activities in hibernation which must apply.

There are also no ethological assertions whatever in the theory, yet most of us have now been exposed to ecological and behavioral animal science and recognize its needful portion in the "grand scheme" for understanding hibernation.

Last, as one will note, no visualization nor any account whatever is proposed by me to explain mechanisms set in motion as a result of forced arousal out of hibernation, either by application of heat or otherwise.

DEDICATION

I dedicate this paper to Professor Edward F. Adolph, most inspiring physiologist.

REFERENCES

1. Akhmetov, I. Z. and K. S. Khairutdinov. Seasonal characteristics of functional thyroid activity in Rodentia. (Trans. from Russ.) Ekologiya 4(2): 108-110, 1973.
2. Aloia, R. C., E. T. Pengelley, J. L. Bolen and G. Rouser. Changes in phospholipid composition in hibernating ground squirrel, *Citellus lateralis*, and their relationships to membrane function at reduced temperatures. Lipids 9 (12): 993-999, 1974.
3. Armitage, K. B. Evidence for a circannual metabolic cycle in *Citellus tridecemlineatus*, a hibernator. Comp. Biochem. Physiol. A. Comp. Physiol. 42(3): 667-688, 1972.
4. Binkley, S., E. Kluth and M. Menaker. Pineal function in sparrows: Circadian rhythms and body temperature. Science 174(4006): 311-314, 1971.
5. Boulouard, R. Study of the adrenal cortex function during the annual cycle in a hibernating animal, the garden dormouse. I. Activity during the summer: reproduction period. (Trans. from Fr.) Gen. Comp. Endocrinol. 16(3): 465-477, 1971.
6. Burlington, R. F., W. D. Bowers, Jr., R. C. Daum and P. Ashbaugh. Ultrastructural changes in heart tissue during hibernation. Cryobiology 9(3): 224-228, 1972.
7. Castex, C. Seasonal variations of the fatty acids and the lipogenetic activity of some organs in a hibernant, the common dormouse, active and normothermic. (Trans. from Fr.) C. R. Seances Soc. Biol. Fil. 166 (2/3): 359-361, 1972.
8. Christian, J., J. E. Steinberger and Ted D. McKinney. Annual cycle of spermatogenesis and testis morphology in woodchucks. J. Mammal. 53(4): 708-716, 1972.
9. Daudova, G. M. and I. B. Soliternova. Effect of insulin on the glucose uptake by the adipose tissue of the suslik *Citellus suslicus* during hibernation and waking. Zh. Evol. Biokhim. Fiziol. 8(4): 449-452, 1972.

10. Dawe, A. R. and W. A. Spurrier. Hibernation induced in ground squirrels by blood transfusion. Science 163: 298-299, 1969.
11. Dawe, A. R., W. A. Spurrier and J. A. Armour. Summer hibernation induced by cryogenically preserved blood trigger. Science 168: 497-498, 1970.
12. Dawe, A. R. and W. A. Spurrier. The blood-borne trigger for natural mammalian hibernation in the 13-lined ground squirrel and woodchuck. Cryobiology 9: 163-172, 1972.
13. Dawe, A. R. and W. A. Spurrier. Evidences for blood-borne substances which trigger or impede natural mammalian hibernation. Proc. Symp. on Biological Clocks. Pengelley, E. T. (editor). San Francisco: Academic Press, 1974.
14. Dawe, A. R. and W. A. Spurrier. Summer hibernation of infant (6-week old) 13-lined ground squirrels, *Citellus tridecemlineatus*. Cryobiology 11: 33-43, 1974.
15. Dawe, A. R. and W. A. Spurrier. Effects on cardiac tissue of serum derivatives from hibernators. Proc. Symp. Temp. Reg. and Drug Action. Lomax, P., E. Schöenbaum and J. Jacob (editors). Paris, France: Karger, Basel, 1975.
16. Emirbekov, E. Z. and M. I. Mukailov. Dynamics of the concentration of glutamic, aspartic and gamma-aminobutyric acids in the brain of susliks in the early period of hibernation. (Trans. from Russ.) Biolnauk 14(2): 38-39, 1971.
17. Ferren, L. G., F. E. South and H. K. Jacobs. Calcium and magnesium levels in tissues and serum of hibernating and cold-acclimated hamsters. Cryobiology 8(5): 506-508, 1971.
18. Garcin, H. Isotopic study [^{45}Ca] of calcium metabolism of the garden dormouse during the seasonal cycle. C. R. Seances Soc. Biol. Filiales (Trans. from Fr.) 164(10): 2004-2007, 1971.
19. Garcin, H. and R. Hoo-Paris. The study of some blood and urine constituents of the garden dormouse in different physiological conditions. (Trans. from Fr.) C. R. Seances Soc. Biol. Filiales 164(10): 2007-2010, 1971.

20. Hagopian, M., M. D. Gershon and E. A. Nunez. Accumulation of lipid droplets in ventricular cardiac muscle of the bat following administration of parachlorophenylalanine: the influence of hibernation. Anat. Rec. 175(4): 747-756, 1973.

21. Hook, W. E. and E. S. Guzman-Barron. The respiration of brown adipose tissue and kidney of the hibernating and non-hibernating ground squirrel. Am. J. Physiol. 133(1): 56-63, 1941.

22. Hoo-Paris, R. Hibernation and ACTH in the hedgehog. (Trans. from Fr.) Ann. Endocrinol. 32(6): 743-752, 1971.

23. Hudson, J. W. and D. R. Deavers. Thyroid function and basal metabolism in ground squirrels *Ammospermophilus leucurus* and *Spermophilus spp*. Physiol. Zool. 49(41): 425-444, 1976.

24. Hulbert, A. J. and J. W. Hudson. Thyroid function in a hibernator. Am. J. Physiol. 230(5): 1211-1216, 1976.

25. Il'in, V. S., G. M. Daudova and M. A. Shuets. Non-esterified fatty acids in the blood of the ground squirrel *Citellus suslicus* under different physiological conditions. (Trans. from Russ.) Zh. Evol. Biokhim. Fiziol. 9(3): 235-239 1973.

26. Kayser, C. The seasonal cycle of thermoregulation in the garden dormouse (*Eliomys quercinus*). Arch. Anat. Histol. Embryol. 58: 41-59, 1975.

27. Kolaeva, S. G., M. G. Kolpakov and M. V. Robinson. Dynamics of the periodic changes of the size of nucleoli in adrenal cortex cells of sousliks during entry into hibernation. (Trans. from Russ.) Doki. Akad. Nauk. SSR Ser. Biol. 216(1): 199-201, 1974.

28. Kristoffersson, R. and S. Broberg. The distribution of free amino acids in the brain tissue of non-hibernating and hibernating hedgehogs (*Erinaceus europaeus* L.). Ann, Acad. Sci. Fenn, Ser. A, Iv. Biol. 174: 1-11, 1970.

29. Krulin, G. S. and J. S. Sealander. Annual lipid cycles of the gray bat *Myotis grisescens*. Comp. Biochem. Physiol. A. Comp. Physiol. 42(2): 537-549, 1972.

30. Laukola, S. and P. Suomalainen. Studies on the physiology of the hibernating hedgehog: 13 Circum-annual changes of triglyceride fatty acids in white and brown adipose tissue of the hedgehog. Ann. Acad. Sci. Fenn. Ser. A. IV. Biol. 180: 1-11, 1971.

31. Linck, G., G. Teller and A. Petrovic. Identification by gas chromatography and mass spectrometry of 1-monoglycerides in European hamster brown adipose tissue and plasma. Eur. J. Biochem. 58(2): 511-516, 1975.

32. Lyman, C. P. and E. L. Jarrow. Effect of low temperature on single heart cells from an adult hibernator and non-hibernator. Exp. Cell. Res. 68(1): 214-219, 1971.

33. Montoya, A. R., L. Ambid and R. Agid. Circannual cycle of oxygen consumption (QO_2) in a hibernator, the garden dormouse: decrease of QO_2 and hypothermia induced by starvation. C. R. Seances Soc. Biol. Fil. 170(3): 680-686, 1976.

34. Morton, M. L. and H. L. Tung. The relationship of total body lipid to fat depot weight and body weight in the Belding ground squirrel. J. Mammal. 52(4): 839-842, 1971.

35. Mrosovsky, N. and D. C. Barnes. Anorexia, food deprivation and hibernation. Physiol. Behav. 12(2): 265-270, 1974.

36. Nunez, E. A., J. Wallis and M. D. Gershon. Secretory processes in follicular cells of the bat thyroid: III The occurrence of extracellular vesicles and colloid droplets during arousal from hibernation. Am. J. Anat. 141(2): 179-201, 1974.

37. Oeltgen, P. R. and W. A. Spurrier. Isolation of a hibernation inducing trigger(s) in the plasma of hibernating woodchucks and characterization of hemoglobins in the 13-lined ground squirrel. Fed. Proc. 36(3): 424, Abstract #822, 1977.

38. Passmore, J. C., E. W. Pfeiffer, and J. R. Templeton. Urea excretion in the hibernating Columbian ground squirrel (*Spermophilus columbianus*). J. Exp. Zool. 192(1): 83-86, 1975.

39. Popova, N. K., S. G. Kolaeva and I. L. Dianova. State of the pineal gland during hibernation. (Trans. from Russ.) Byull. Eksp. Biol. Med. 79(4): 116-177, 1975.
40. Rauch, J. C. and D. D. Beatty. Comparison of regional blood distribution in *Eptesicus fuscus* (big brown bat) during torpor (summer) hibernation (winter) and arousal. Can. J. Zool. 53(2): 207-214, 1975.
41. Reiter, R. J. Pineal control of a seasonal reproductive rhythm in male golden hamsters exposed to natural daylight and temperature. Endocrinology 92(2): 423-430, 1973.
42. Saboureau, M. Lag of the endocrine and spermatogenetic activities of the testicle during spring awakening in the male hedgehog. (Trans. from Fr.) C. R. Seances Biol. Filiales 164(11): 2364-2367, 1970.
43. Spurrier, W. A., G. E. Folk, Jr. and A. R. Dawe. Induction of summer hibernation in the 13-lined ground squirrel shown by comparative serum transfusions from Arctic mammals. Cryobiology 13: 368-374, 1976.
44. Spurrier, W. A. and A. R. Dawe. Electrical and mechanical sustained automaticity of the isolated heart of the 13-lined ground squirrel: both circannually and arousing from hibernation. Comp. Biochem. Physiol. 57A: 457-469, 1977.
45. Swan, H. and C. Schatte. Antimetabolic extract from the brain of the hibernating ground squirrel, *Citellus tridecemlineatus*. Science 195(4273): 84-85, 1976.
46. Twente, Janet and John W. Twente. Distribution of [^{14}C] epinephrine and [^{131}I] albumin in regions of the brain and other tissues of hibernating *Citellus lateralis* after intraperitoneal injection. Comp. Gen. Pharmacol. 2(6): 138-144, 1971.
47. Velicka, V. and J. Velicky. A study on the thyroid gland of bat in winter and early spring. I. Observations in the light microscope. Folia Morphol. (Prague) 20(4): 337-347, 1972.

48. Wells, L. A. Circulatory pattern of hibernators. Am. J. Physiol. 221(5): 1517-1520, 1971.
49. Wells, L. A. Permeability of the blood-brain barrier system to rubidium in euthermia, hibernation, and hypothermia. Comp. Biochem. Physiol. A. Comp. Physiol. 42(2): 551-557, 1972.
50. Wenberg, G. M. and J. C. Holland. The circannual variations in the total serum lipids and cholesterol with respect to body weight in the woodchuck (*Marmota monax*). Comp. Biochem. Physiol. A. Comp. Physiol. 44(2): 577-583, 1973.
51. Willis, J. S., S. S. Goldman and R. F. Foster. Tissue K concentration in relation to the role of the kidney in hibernation and the cause of periodic arousal. Comp. Biochem. Physiol. 39(3A): 437-445, 1971.
52. Zatzman, M. and F. E. South. Concentration of urine by the hibernating marmot. Am. J. Physiol. 228(5): 1336-1340, 1975.
53. Zimny, M. L. and E. D. Levy, Jr. Ultrastructure of mesangial and juxtaglomerular cells in the kidney of a hibernator. Z. Zellforsch. Mikeoskop. Anat. 118(3): 326-332, 1971.

SESSION IV

Thermogenesis and Man in Cold

THE BIOCHEMICAL MECHANISM OF THERMOGENESIS IN BROWN ADIPOSE TISSUE

Barbara Cannon, Jan Nedergaard,
Lennart Romert, Ulf Sundin
and Jan Svartengren
Wenner-Gren Institute
University of Stockholm
Norrtullsgatan 16
S-113 45 Stockholm, Sweden

ABSTRACT

Non-shivering thermogenesis in brown adipose tissue is initiated by norepinephrine released from sympathetic nerve terminals within the tissue. The heat production is entirely due to oxygen consuming reactions. There are 150,000 receptors for norepinephrine per cell. As a consequence of norepinephrine interaction with the plasma membrane, lipolysis is activated. The majority of fatty acids formed are exported. Only a fraction is required as substrate for mitochondrial respiration. During thermogenesis there is a requirement for both lipolysis and glycolysis. This can be explained as a need for anaplerotic reactions for the citric acid cycle (pyruvate carboxylation to oxaloacetate). It is suggested that the heat production is localized to the mitochondria and is a consequence of a loosening of coupling between ATP synthesis and respiration. The quantitative significance of brown fat thermogenesis requires further investigations.

ISBN 0-12-734550-7

I. INTRODUCTION

For more than fifteen years now (47), brown fat has been recognized as one of the sites of non-shivering thermogenesis. During this time the relative contribution of brown fat to total thermogenesis has risen from a mere 8% (26) to at least 65% (17). Time will tell which estimate is the more correct. During most of this time our laboratory has been investigating the biochemical processes underlying thermogenesis in brown fat. When our results are combined with those from other laboratories, quite a full picture of biochemical thermogenesis can be given. This review will concern itself with these biochemical processes of thermogenesis in brown fat.

In *in vivo* and *in vitro* determinations of heat production, the parameter most commonly measured is oxygen consumption. This is then equated with thermogenesis, and a number of assumptions are made to obtain an actual value for the amount of heat generated. Calorimetric measurements afford a direct means of determining heat evolution and thus a control as to whether processes other than oxygen consuming reactions contribute quantitatively to thermogenesis. We have, therefore, employed a microcalorimeter to directly measure heat production in cell suspensions of brown adipocytes under the influence of the calorigenic hormone norepinephrine (32). In figure 1 oxygen consumption measurements are compared with calorimetric determinations. The qualitative course of both is identical, and good quantitative agreement is also obtained. The direct caloric determinations give a value of 0.1 W/gram tissue, whereas calculations from oxygen uptake give 0.15 W/gram tissue. It is therefore possible to exclude significant anaerobic heat production in brown fat. If we use these values, we can calculate that in an insulated environment brown fat could increase its own temperature by about 100°C per hr, an impressive figure. However, assuming that a 100 gram hamster possesses 2 gram brown fat, this would only be equivalent to an increase of 2°C per hr for the whole hamster. (For these calculations we assume a specific heat of 1 cal/°C·g and neglect heat loss to the surroundings.) Thus, whereas the intrinsic capacity for heat production

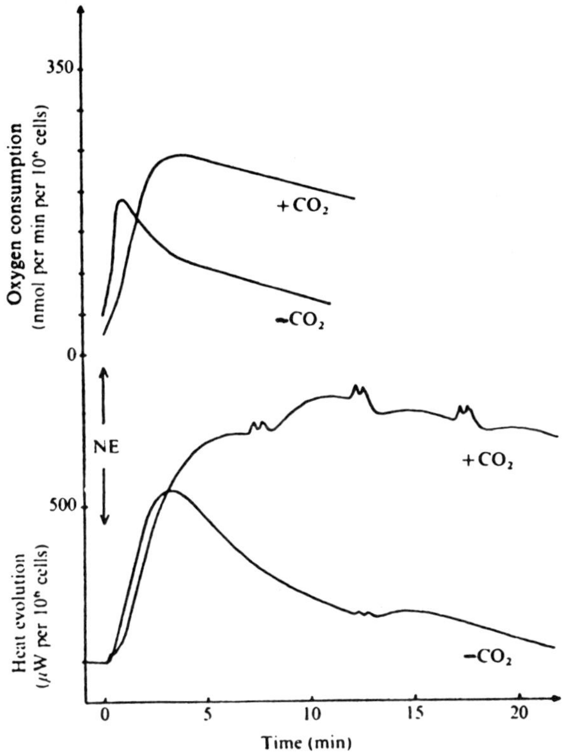

Fig. 1. *Comparison between oxygen consumption rates and heat production in isolated brown adipocytes. 140,000 cells were incubated in 1.4 ml modified Krebs Ringer phosphate buffer at 37°C and the rate of utilization of oxygen measured (37). 1μM norepinephrine (NE) was added. Where stated the buffer was bubbled with 5% CO_2 in air. For microcalorimetric method see (32). Reproduced from (32) by permission.*

is extremely high, the relative importance of this heat for a whole animal is open to discussion. Additionally, a hibernating hamster arouses (from 5° to 37°C) in about two hrs in a cold environment, but the brown fat heating capacity (2°C/hr) is measured at 37°C and must be much lower at the body temperature of hibernating hamsters. It appears therefore that brown fat could play a main role during some crucial phase of arousing, rather than heating the whole hamster.

II. OXYGEN CONSUMPTION *IN VIVO* AND *IN VITRO*

Since we conduct our biochemical studies in *in vitro* systems, the validity of our results should be questioned as to whether thermogenesis (= oxygen consumption) is impaired *in vitro* relative to *in vivo*. The highest estimates of *in vivo* oxygen consumption were reported recently by Foster (17). In Table 1 we have compared these results with our *in vitro* data, and the agreement between the different systems is good. Naturally the approximations necessary for this comparison may be questioned, but they should be correct within a factor of 2. This essentially good correlation has convinced us that we are studying the main events in non-shivering thermogenesis.

Most of our investigations have been conducted on isolated brown adipocytes from hamsters or on brown fat mitochondria from hamsters and rats. In the following we will discuss the reactions outlined in figure 2: first at the cellular membrane, then the cytoplasmic events, the release of fatty acids, their oxidation in the mitochondrial matrix and finally the production of heat at the mitochondrial membrane.

Table 1. Comparison of Maximum Oxygen Consumption in brown adipose tissue *in vivo* and *in vitro*

	$\frac{\text{nmol Oxygen}}{\text{min} \cdot \text{mg mito} \cdot \text{prot.}}$
Cold-acclimated rat (NE-infusion) (17)	800[1]
Hamster brown adipocytes (NE-stimulation, 37°C)	600-800[2]
Hamster brown fat mitochondria (uncoupled palmitoyl carnitine + malate respiration, 37°C)	1000

1. assuming 15% of the wet weight to be mitochondrial protein
2. assuming that 10^6 cells contain ca. 1.5 mg mitochondrial protein

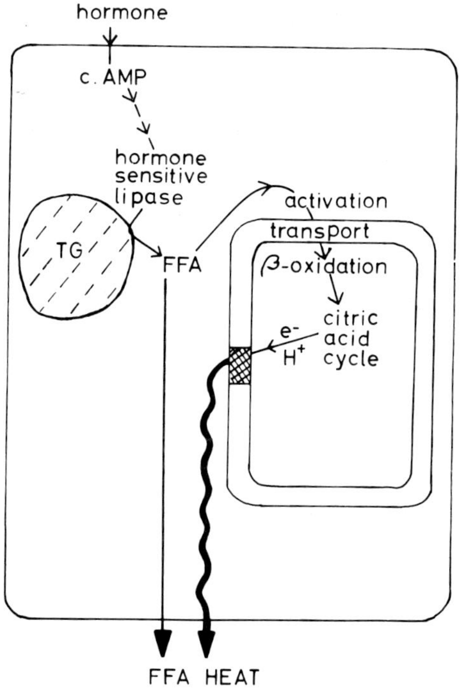

Fig. 2. *A schematic representation of the biochemical events of thermogenesis in brown adipose tissue.*

III. BIOCHEMICAL EVENTS OF THERMOGENESIS IN BROWN ADIPOSE TISSUE

A. At the Cellular Membrane

Norepinephrine is the effector hormone of thermogenesis. Physiologically, it is released from the sympathetic nerve terminals within the brown fat (48). The thermogenic effect of norepinephrine infusion is caused by its increase in concentration in the synaptic areas, thus mimicking sympathetic stimulation.

The isolated cells respond to catecholamine stimulation as if the receptor for thermogenesis were a β-receptor, i.e. isoproterenol (10 nM) is as effective as norepinephrine (100 nM) and epinephrine (1000 nM) (37). The α-agonist phenylephrine cannot elicit a full response, and the doses needed to give any response are in the millimolar range (24). Furthermore, propranolol (a β-antagonist) totally blocks the norepinephrine stimulated thermogenesis, whereas phentolamine (an α-antagonist) only has a very weak effect, that can be classified as unspecific (37, our unpubl. obs.). We therefore conclude that only β-receptors are responsible for the physiological increase in oxygen consumption.

We have quantitated the number of β-receptors on the cell membrane by measuring the specific binding of a labelled β-antagonist (dihydroalprenolol). We found as many as 150,000 receptors per adipocyte (Fig. 3). This number per cell is somewhat higher than the highest value so far reported on any cell [80,000 per myoblast (2)], and the binding to isolated plasma membranes is much higher than the corresponding estimate from a white adipocyte (49). The figure is an indication

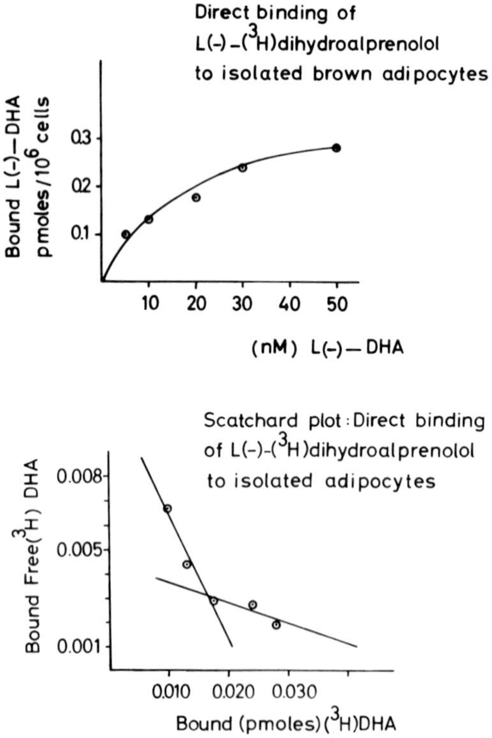

Fig. 3. a. *Direct binding of L(-)-(^{3}H)-dihydroalprenolol to isolated hamster brown adipocytes. Determined in principle as in (49). The medium was modified Krebs Ringer phosphate buffer.* b. *Scatchard plot of binding determined in a.*

of the high sensitivity of brown fat to norepinephrine.

The stimulatory effect of norepinephrine can under certain circumstances be inhibited by insulin (Fig. 4) (16). Even unphysiologically high concentrations of insulin do not affect the respiration seen in hamster brown adipocytes after maximal stimulation with norepinephrine. Using suboptimal norepinephrine doses, however, insulin can reduce or abolish the norepinephrine response, even at concentrations as low at 50μIU/ml (our unpubl. obs.). This pattern of insulin action on brown adipocyte respiration is very similar to its effect on white fat lipolysis (29), and insulin presumably works via the same mechanism in both cases.

We have also examined effects of these two "antagonistic" hormones on the rate of protein synthesis in brown adipocytes, measured as incorporation of labelled amino acids. The common view has been that norepinephrine has, in itself, a trophic effect on brown fat and that persistent sympathetic stimulation of brown fat during cold exposure leads successively to an increase both in number of cells and in their mitochondrial content (21). Accordingly an increased

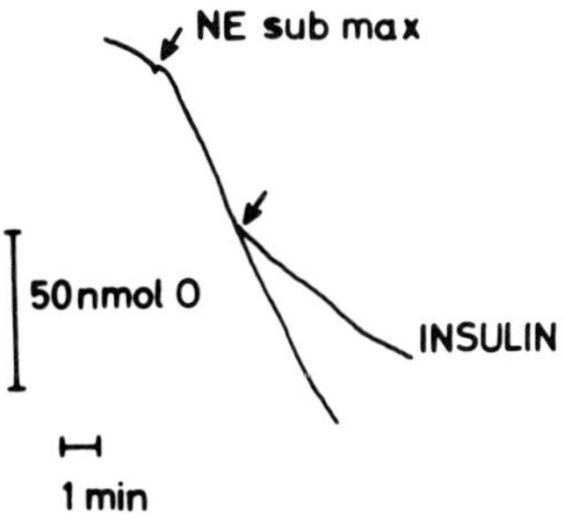

Fig. 4. *Inhibition by insulin of respiration induced by a submaximal dose of norepinephrine in hamster brown adipocytes. The medium was Krebs Ringer bicarbonate buffer. Norepinephrine concentration was 50nM; insulin concentration 500μIU/ml.*

amino acid incorporation would be expected also during norepinephrine stimulation of isolated adipocytes. However, the "catabolic" norepinephrine does not stimulate the rate of protein synthesis, whereas the "anabolic" insulin increases the rate (our unpubl. obs.). These effects are difficult to reconcile with a simple trophic effect of norepinephrine, and the effect seen *in vivo* may either be indirect and/or norepinephrine may stimulate the differentiation of some adipocyte precursor cells in the tissue.

Apart from the hormones norepinephrine and insulin, some substrates were shown to stimulate respiration in brown adipocyte preparations: free fatty acids, succinate and glycerol-3-phosphate (40,50, 51). Pettersson, Lundberg and Bernson in our laboratory have, however, demonstrated that succinate respiration is of extracellular origin and is probably caused by some broken membrane contamination (38). They infer that the same is true for glycerol-3-phosphate respiration. However, the free fatty acids in micromolar concentrations can, in themselves, stimulate respiration of the adipocytes. This is explained in terms that free fatty acids mimic the effect of norepinephrine.

In summary then: only catecholamines and insulin - both of which affect lipolysis - and free fatty acids themselves can influence respiration (thermogenesis) in brown fat cells.

B. In the Cytoplasm

The first process to occur upon norepinephrine binding seems to be an increased ion conductance across the cell membrane (23), and a depolarization of the membrane potential at higher hormone concentrations (18). Simultaneously or consecutively with these processes there is a rapid and transient rise in cyclic AMP concentration (37). Cyclic AMP activates a specific protein kinase (30) which in its turn is responsible for activation, via phosphorylation, of the hormone sensitive lipase. Both formation of cyclic AMP and phosphorylation by protein kinase demand ATP and probably also a high energy potential. Concurrently with these biochemical events, the larger fat droplets in the cells become smaller, thus increasing their surface area to permit maximal susceptibility to the lipases (1).

The free fatty acids formed as the product of the hormone sensitive lipase can proceed via three different pathways: they can be reesterified, they can leave the cell as free fatty acids, and they can enter the mitochondria for combustion.

1. Reesterification

Reesterification was once suggested to be the futile cycle responsible for brown fat thermogenesis (3, 48), but it cannot play any quantitative role, as glycerol kinase is absent in hamster cells (31), and as it is possible to account for 3 fatty acids (either for export or for combustion) for every glycerol released (our unpubl. obs.)

2. Fatty Acid Export

Fatty acid export has generally been thought to be a minor function of brown fat, and it has actually been proposed that a competition exists between fatty acid release and fatty acid combustion (=thermogenesis) (7). We have however found a very high capacity for fatty acid export; it amounts to 100 nmoles per million cells per minute in a Krebs Ringer bicarbonate buffer where 4% fatty acid free albumin is present (Fig. 5). The same buffer system gives the highest respiratory rates and the two functions do not seem to be competitive. This high rate of fatty acid export is not merely a reflection of a spill-over (i.e. that only those fatty acids which the mitochondria could not combust are released): figure 5 shows that even at low norepinephrine concentrations the amount of fatty acids released is at least as great as the amount combusted. The actual figure (100 nmol/min) is as high as that which is currently found in white adipocytes (45), but it must be remembered that brown fat cells are much smaller than white fat cells: 10 - 20 μm in diameter compared to 40 - 50 μm. This means that there are at least ten times more cells per gram wet weight in brown fat that in white fat, and thus its fatty acid releasing capacity would be at least ten times higher on the same basis.

In other words, 2 gram of brown fat (as found in a hamster) are at least as effective in

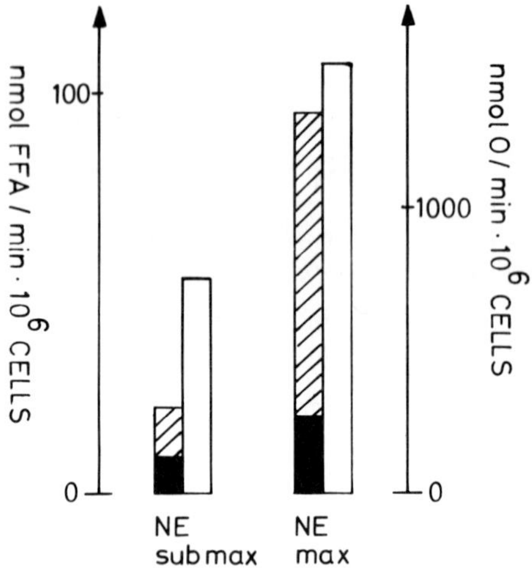

Fig. 5. *The lipolytic capacity of hamster brown adipocytes in relation to their oxidative capacity. The hatched bars represent lipolysis rates in 50 nM (submax) and* 1 *μM (max) norepinephrine. The open bars represent respiration rates. The black areas show the theoretical amount of fatty acid required to sustain these respiration rates. The medium was Krebs Ringer bicarbonate buffer and contained 4% fatty acid free bovine serum albumin.*

fatty acid release as 20 gram of white fat in a situation where there is a need for a very high rate of supply of energy - as occurs during a cold stress. This lends support to the hypothesis that brown fat produces some of "its" heat by delivering substrate for combustion in other tissues (27, 32).

3. Combustion

Prior to combustion in the mitochondria, the fatty acids must first be activated, i.e. they must combine with CoA and form acyl-CoA. This is an energy-requiring process, where two phosphate bonds are split (ATP→AMP), and because of this, the cell cannot oxidise fatty acids if it does not contain sufficient ATP. The acyl-CoAs enter the mitochondria through a shuttle system, where they are first transferred to carnitine, then the acyl-carnitine passes the mitochondrial inner membrane and there acyl-CoA is resynthesized. Acyl-CoA is the substrate for β-oxidation. We have indications from studies on isolated mitochondria that acyl-CoA in the intermembrane space may be the intracellular trigger which initiates thermogenesis (*vide infra*).

C. In the Mitochondrial Matrix

Through the process of β-oxidation acyl-CoA is initially degraded to acetyl-CoA. By this process one acyl-CoA, e.g. palmitoyl-CoA, will form 8 acetyl-CoA, 7 $FADH_2$ and 7 NADH + 7 H^+. This represents a partial oxidation of the fatty acid and some energy may be liberated through oxidation of the cofactors, but for complete combustion of fatty acids to CO_2 and H_2O, the acetyl-CoA formed must enter the citric acid cycle (Fig. 6). A sufficient amount of condensing partner (i.e. oxaloacetate) is thus needed. When respiration is stimulated 20 times - as happens when the adipocytes are transferred into a thermogenic state - the need for an increased amount of oxaloacetate is evident. Pettersson in our laboratory found (36) that bubbling of a Krebs Ringer phosphate buffer with 5% CO_2 in air resulted in norepinephrine-stimulated respiration which was both higher than in the absence of CO_2 and was maintained at this high level [without CO_2 the respiration rate drops

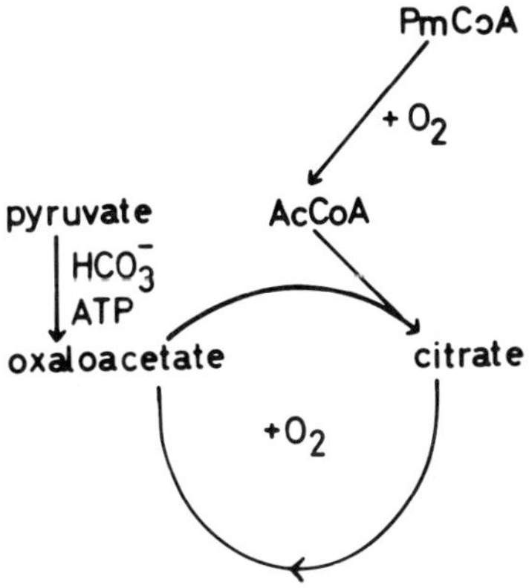

Fig. 6. *The sites of oxygen consumption in fatty acid oxidation in relation to the function of pyruvate carboxylase.*

within a few minutes (Fig. 1)]. We have now evidence (9) that this beneficial action of CO_2 is primarily caused by its participation in an anaplerotic process: the carboxylation of pyruvate (by pyruvate carboxylase) to yield oxaloacetate (Fig. 6). The decrease with time in the rate of respiration (Fig. 1) seems to be an artefact caused by loss of intermediates from the citric acid cycle. They possibly act as substrates for amino acid synthesis, since the decrease can be overcome if amino acids are added to the medium. They cannot be replaced by adding citric acid cycle intermediates, as the permeases for citrate, succinate and malate are very poorly developed (our unpubl. obs.) Thus, in order to increase the level of citric acid cycle intermediates, pyruvate carboxylase must be active, and activation is achieved by acetyl-CoA (Fig. 7). The cells will therefore form oxaloacetate when needed.

Pyruvate carboxylase is present in the mito-

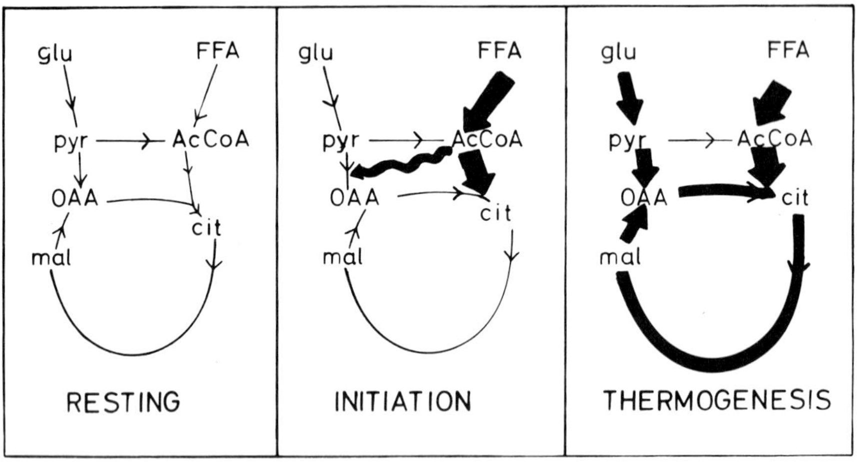

Fig. 7. *The requirement of both glycolysis and lipolysis for thermogenesis. Note that high acetyl-CoA concentrations activate pyruvate carboxylase.*

chondrial matrix, and can be depleted of substrate if pyruvate entry into the mitochondria is inhibited. Pyruvate entry can be specifically inhibited by UK 5099 (Pfizer, Sandwich, Kent) (20). Addition of UK 5099 does not influence the norepinephrine-induced respiration in the absence of CO_2. In a CO_2-bubbled buffer - when pyruvate carboxylase should be active - the same concentration of UK 5099 diminished the respiration down to the level it had when CO_2 was absent - as expected if pyruvate carboxylase was responsible for the beneficial effect of CO_2 (Fig. 8). This inhibition can be overcome by addition of a high concentration of pyruvate. These results indicate that pyruvate carboxylation is required to supply oxaloacetate in order to support the high respiratory rate in norepinephrine-stimulated cells. Physiologically, the pyruvate needed may come from lactate in the circulation or from glycolysis. It has earlier been shown that both lipolysis and glycolysis are required in parallel to maintain thermogenesis (25, 39).

During citric acid cycle oxidation, substrate

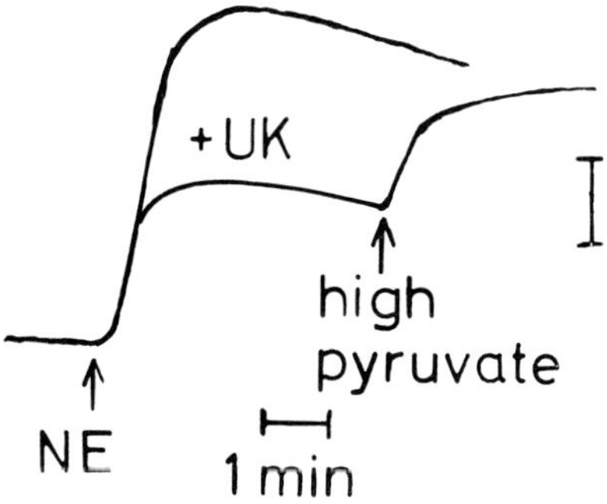

Fig. 8. *Inhibition by UK 5099 of norepinephrine-stimulated respiration in hamster brown adipocytes and its release by high concentrations of pyruvate. The medium was Krebs Ringer phosphate buffer bubbled with 5% CO_2 in air. Norepinephrine 1 μM and pyruvate 10 mM. The vertical bar represents 100 nmol O/min · 10^6 cells.*

level phosphorylation occurs in the step between 2-oxoglutarate and succinate and GTP (ATP) is formed. The importance of this non-oxidative process has been much discussed (43), as it provides a means of forming ATP even if oxidation is totally uncoupled from phosphorylation. The ATP formed here can be considered more than sufficient for fatty acid activation. If some ATP is also formed by oxidative phosphorylation during maximal respiration [as is e.g. confirmed by oligomycin (41) and FCCP (our unpubl. obs.) inhibition of cellular respiration], a role for extra ATPases to take care of all this ATP may be postulated. This could be the importance of e.g. a stimulated Na^+/K^+-ATPase.

That brown adipocytes are well equipped to control their ATP level is confirmed by our recent (unpubl.) studies of creatine metabolism. Brown adipose tissue contains high concentrations of creatine and creatine phosphate and creatine kinase activity is high (6). The amount of total creatine is in the same order of magnitude as the total adenine nucleotides. The ratio (creatine/ creatine phosphate) presumably reflects the cytoplasmic energy status. During norepinephrine-induced respiration in brown adipocytes, the ATP/ADP ratio decreases only slightly (36) and it may be buffered by the creatine phosphate system. Upon cold adaptation in hamster (Fig. 9) the total amount of creatine increases, and the creatine phosphate potential decreases, reflecting higher energy turnover and lower energy potential in the cytoplasm. It appears therefore that creatine metabolism is related to the physiological state of the animal and that it buffers the adenine nucleotides in the cells.

D. At the Mitochondrial Membrane

When the fatty acids have passed through β-oxidation and through the citric acid cycle, their redox energy is found as $NADH + H^+$ and as $FADH_2$. The transfer of this energy to heat takes place at the mitochondrial membrane.

According to the chemiosmotic hypothesis for oxidative phosphorylation, the redox energy found in $NADH + H^+$ and $FADH_2$ is transformed into potential energy (Fig. 10). When the electrons pass through the

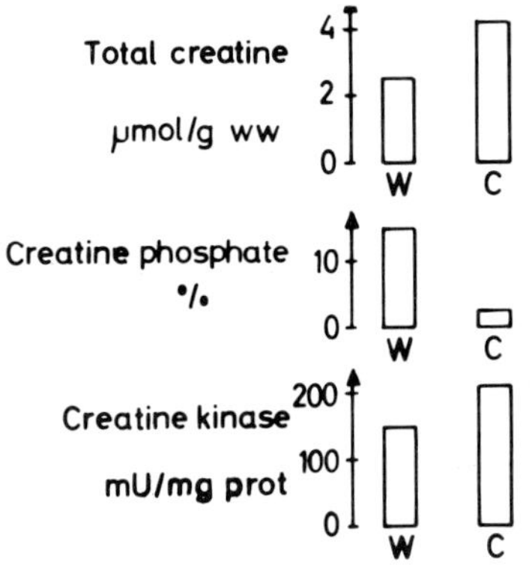

Fig. 9. *Alterations in total creatine, creatine-phosphate: creatine ratio and creatine kinase activity in hamster brown adipose tissue upon cold adaptation. W: animals kept at room temperature. C: animals kept at +5° for at least three months. Methods as in (5).*

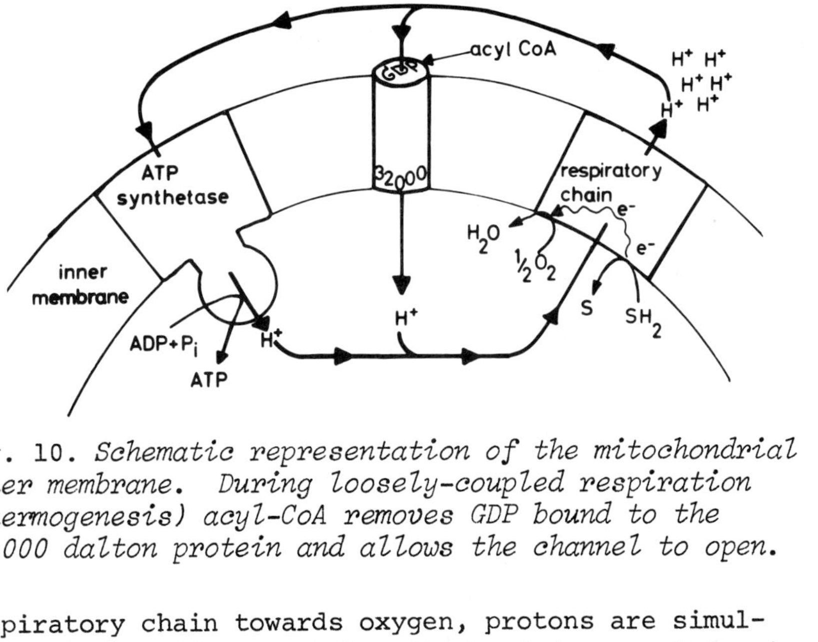

Fig. 10. *Schematic representation of the mitochondrial inner membrane. During loosely-coupled respiration (thermogenesis) acyl-CoA removes GDP bound to the 32,000 dalton protein and allows the channel to open.*

respiratory chain towards oxygen, protons are simultaneously pumped out of the mitochondria, resulting in the appearance of a "proton motive force". The potential energy is retransformed to chemical energy when the protons pass down their concentration gradient through the ATP synthetase and ATP is formed. In most tissues mitochondrial oxygen consumption is stoichiometrically coupled to ATP synthesis; indeed oxygen is only consumed as a consequence of a need for ATP, and respiration only occurs as long as ADP is supplied to the ATP synthetase. ATP synthesis is close to 100% efficient (46), and thus very little heat is generated by this process until ATP is hydrolysed directly - or - indirectly until the breakdown of any product into which its energy has been incorporated. When thermogenesis occurs in brown fat, the respiration increases 20-fold, and if oxygen consumption should still be coupled to ATP formation, some process must be activated which would rapidly hydrolyse all ATP to ADP, resulting in a respiration which releases heat. Indeed such "futile cycles" have been found in some tissues. In "obligatory" or thyroid thermogenesis, the Na^+/K^+-pump of the cellular membrane is activated and consumes extra ATP (15), and in insects during preparation for flight phosphofructokinase and fructose-1,6,-diphosphatase are activated in parallel and use ATP (33). In brown fat no such cycle has been demonstrated to date which could have quantitative significance in thermogenesis.

Further, in order to account for the maximum oxygen consumption rates observed *in vivo*, the maximum capacity of the mitochondrial respiratory chain demonstrated *in vitro* is required. Thus regardless of the capacity of any postulated ATPase or its location, in tightly coupled mitochondria the mitochondrial ATP synthetase must have an activity which parallels the respiratory activity under maximal conditions. However, the respiratory rate in tightly coupled mitochondria is far lower than in mitochondria which are uncoupled or loosely coupled (Fig. 11). We have recently shown (10) that this result is not a consequence of an inhibited ATP synthetase, but that the enzyme (which is structurally normal) is present in very much lower concentration than in other mitochondria. There appears to be only about 40% as many ATP synthetase units as in rat liver mitochondria (or 20% as many as in beef heart mitochondria). There are however five times more respiratory chain units in hamster brown fat mitochondria than in rat liver (all expressed per mg mitochondrial protein). It seems therefore impossible for the mitochondria to remain tightly coupled during thermogenesis when respiration proceeds at its maximum rate. This does not

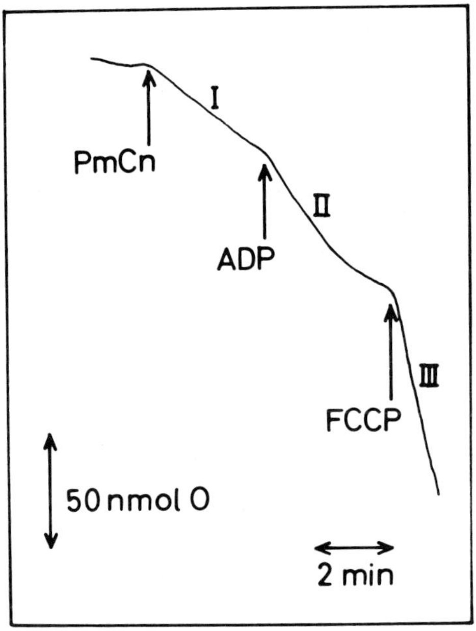

Fig. 11. *Comparison between ADP-stimulated and uncoupled respiration rates in hamster brown fat mitochondria. The medium was 100 mM KCl, 20 mM K-TES, 2 mM $MgCl_2$, 4 mM KH_2PO_4 and 1 mM EDTA. 1% bovine serum albumin and 1 mM GDP were present to couple the mitochondria. 0.1 mM ADP and 4µM FCCP were added. Reproduced from (8) with permission.*

exclude the possibility that there is some ATPase activity which could hydrolyse ATP synthesized during loosely coupled respiration (*vide infra*) or ATP from substrate level phosphorylation.

An alternative and perhaps more expedient means of heat production in brown fat could be achieved if the coupling between ATP synthesis and mitochondrial respiration was loosened, such that oxygen consumption could continue at maximum rate without concomitant stoichiometric ATP production. This does not mean that no ATP will be produced, merely that the stoichiometry is less than that found in tightly coupled mitochondria.

The mitochondria will become loosely coupled if a means of dissipating the proton gradient is provided in addition to the ATP synthetase, e.g. if the mitochondrial inner membrane is permitted, in a regulated way, to become permeable to protons (or equivalently to hydroxyl ions) (Fig. 10). These will then move so as to cancel the gradient.

There is now good evidence that a form of physiological loose coupling occurs in brown fat mitochondria and that the extent of loose coupling can be correlated to the thermogenic requirements of the tissue. [For review of following see Nicholls (34) and Cannon and Lindberg (8)]. In a hibernator the potential for loose coupling is constitutive, whereas in a non-hibernator it is present at birth, declines with age, and returns upon cold acclimation. This pattern is in agreement with the capacity of the animal to exhibit norepinephrine-induced non-shivering thermogenesis. In isolated mitochondria, it is possible to control the degree of coupling by the extent of binding of a purine nucleotide di- or triphosphate (e.g. GDP, GTP, ADP, ATP) to the mitochondria. When no GDP (e.g.) is bound to the mitochondria, they show high anion permeability and no coupling. GDP binding to the outside of the inner membrane inhibits anion permeability and the mitochondria become able to conserve energy. The purine nucleotides bind to a polypeptide of molecular weight 32,000 dalton (35) (distinct from the adenine nucleotide translocase) and in rat brown fat mitochondria a polypeptide of the same molecular weight increases dramatically upon cold adaptation (44).

This can also be followed by measuring GDP binding to mitochondria. Figure 12 shows that GDP binding increases in situations where thermogenesis is needed: after birth or during cold acclimation. The same pattern has been found in guinea pig (42). Another marker protein for the mitochondrial inner membrane, the DCCD-binding protein of the ATP synthetase, does not change in parallel, indicating a specific synthesis of the GDP-binding protein (our unpubl. obs.).

It may thus be visualised that in thermogenic brown fat mitochondria, this 32,000 dalton protein is involved in a "pore" or "channel" for hydroxyl ions. This channel becomes closed when GDP (or other purine nucleotide) binds to the protein.

Because of the high concentrations of cyto-

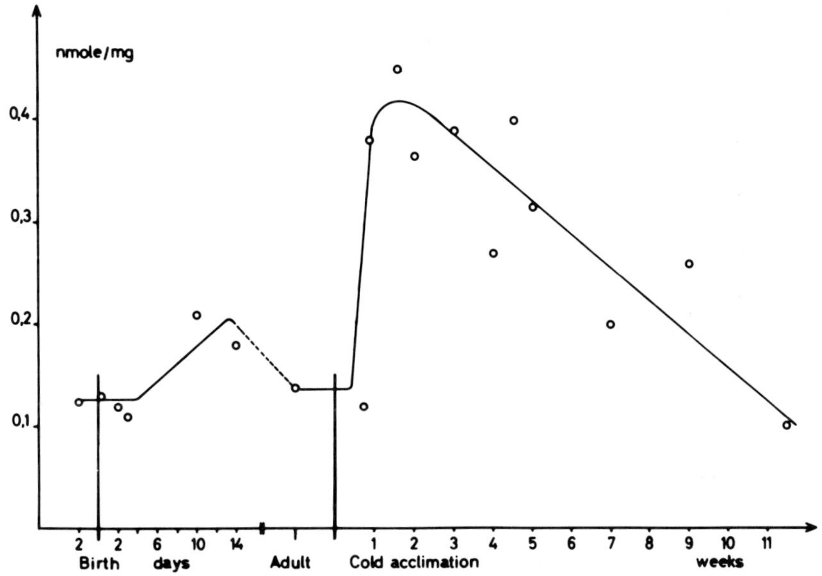

Fig. 12. *GDP binding to rat brown fat mitochondria. Binding was measured as previously reported (12). The points represent the average of 4 determinations. The successive decrease in GDP binding after the second week in the cold probably proceeds in parallel with a decrease in binding in the adult warm animal, and is related to age and size and not to acclimation state (our unpubl. obs. and J. Himms Hagen, pers. comm.).*

plasmic purine nucleotides, it was necessary to postulate a physiological antagonist to nucleotide-binding in brown fat mitochondria in order to regulate the loose-coupling. We have recently shown (12) that palmitoyl-CoA at very low concentrations can reintroduce anion permeability in GDP-treated mitochondria. Addition of excess GDP can then competitively reinhibit permeability. A similar phenomenon is demonstrated for GDP-binding in figure 13.

We therefore conclude that acyl-CoA is a suitable candidate for the role as physiological antagonist to purine nucleotides and, as such, can function not only as a substrate but also as the trigger which initiates mitochondrial respiration.

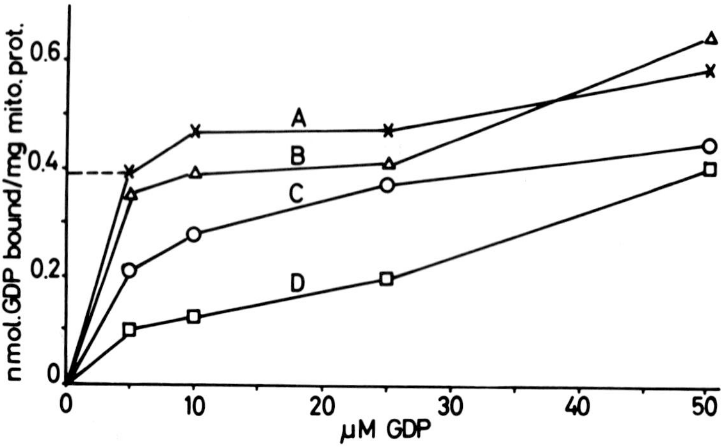

Fig. 13. *GDP binding to brown fat mitochondria, and the influence of palmitoyl-CoA. A: without palmitoyl-CoA. B: + 2μm palmitoyl-CoA. C: + 5μM palmitoyl-CoA. D: + 10 μM palmitoyl-CoA. Reproduced from (12) by permission.*

IV. A WORKING HYPOTHESIS FOR THERMOGENESIS

Thermogenic mitochondria possess a high amount of the 32,000 dalton protein and through this a high potential capacity for loose coupling and thermogenesis. Nonetheless, they remain coupled in the resting state because purine nucleotides are bound to this protein. When thermogenesis is initiated by norepinephrine, free fatty acids are formed which become activated to acyl-CoA esters. These can then remove the bound nucleotides, so opening the ion channel. Acyl-CoA serves as the substrate for the mitochondrial respiration. When thermogenesis ceases, lipolysis decreases, the acyl-CoA concentration is reduced, purine nucleotides rebind and block the ion channel, the mitochondria become again tightly coupled and respiration returns to resting levels.

V. A CONTINUUM OF FATS

If brown fat is responsible for non-shivering thermogenesis, it is of importance to understand what is meant by the term "brown adipose tissue". In Table 2 we have listed some possible criteria of brown fat. Unfortunately, for most of these, exceptions can readily be found. Even the colour is dependent upon the nutritional status!

Daniel and Derry (13) suggested "double innervation" as a criterion of brown fat i.e. that single adipocytes are innervated as well as blood vessels. However, some tissues which are normally considered white do have innervated adipocytes e.g. mesenteric adipose tissue in the dog (4).

The absence of elementary particles in negatively stained mitochondria can be suggested as a criterion, but perirenal fat in lamb possesses these (11) and would then not be classified as brown fat, although it fulfils most other criteria. Similar considerations can be made about all the cited properties.

We think that a classification of all fatty tissues into only two clearly separate groups is an oversimplification. It seems more realistic to envisage a spectrum of adipose tissues, with perhaps epididymal white adipose tissue in rat and brown adipose tissue in the newborn rabbit forming the extremities, whereas perirenal adipose tissue in lamb and ox, and mesenteric and subcutaneous adipose tissue in dog and human appear in intermediate positions. Further the possibility of interconvertibility of these fatty tissues may be present.

Table 2. Criteria of Brown Adipose Tissue

brown
double innervation
no "elementary particles"
densely packed cristae
purine nucleotide binding
carnitine acetyl transferase (19)
carnitine palmitoyl transferase (19)
heat production

VI. NON-SHIVERING THERMOGENESIS, BROWN FAT AND ADULT MAN

Heldmaier (22) concluded that no or only little non-shivering thermogenesis should be present in mammals weighing more than 10 kg. He found that the increase in respiration evoked by NE decreased with the logarithm of the weight of the animal and that the regression line intercepted the abscissa below 10 kg.

Adult man can nonetheless adapt to cold. In a cold chamber man will increase his metabolism ca 25%. Initially this occurs by shivering, but after a few weeks shivering ceases although the increase in oxygen consumption persists (14). This is mainly a facultative thermogenesis, as the increase in basal metabolic rate is very small. This adaptation is quite parallel to that seen in a rat or any other small rodent. Furthermore, under these circumstances man develops an increased sensitivity to norepinephrine. When infused, norepinephrine is able to increase oxygen consumption about 20% in cold-adapted man (28), approximately the same as the physiological increase seen during cold exposure.

The relatively low increase in oxygen consumption in man, only ca 20% compared to several hundred percent in a rat, is presumably a consequence of his more favourable surface/volume ratio. Man seems therefore to have the same physiological capacity as other mammals to resist cold environments. As this non-shivering thermogenesis is very similar to that defined as "facultative thermogenesis" in laboratory animals, and as facultative thermogenesis in these animals does involve (or is equivalent to) brown fat thermogenesis, it is tempting to suggest that even man may show hypertrophy of brown fat during cold adaptation, and that he uses the biochemical mechanisms of thermogenesis outlined in this review.

ACKNOWLEDGEMENTS

The authors are grateful to Professor Olov Lindberg for stimulating discussions. The work reported from our laboratory was supported by a grant from the Swedish Natural Science Research Council. Skillful assistance was provided by Agneta Bergström, Marianne Karlsson and Barbro Kyte.

REFERENCES

1. Ahlabo, I. and T. Barnard. A quantitative analysis of triglyceride droplet structural changes in hamster brown adipose tissue during cold exposure and starvation. J. Ultrastruct. Res. 48: 361-376, 1974.
2. Atlas, D., E. Hanski and A. Levitzki. Eighty thousand β-receptors in a single cell. Nature 268: 144-146, 1977.
3. Ball, E. G. and R. L. Jungas. On the action of hormones which accelerate the rate of oxygen consumption and fatty acid release in rat adipose tissue *in vitro*. Proc. Natl. Acad. Sci. (U.S.) 47: 932-941, 1961.
4. Ballard, K., T. Malmfors and S. Rosell. Adrenergic innervation and vascular patterns in canine adipose tissue. Microvascular Res. 8: 164-171, 1974.
5. Bergmeyer, H.U. (editor). Methods of Enzymatic Analysis. New York and London: Verlag Chemie Weinheim, Academic Press, Inc., 1974, Vol. IV: 1772-1781, Vol. II: 784-788.
6. Berlet, H. H., I. Bonsmann and H. Birringer. Occurrence of free creatine, phosphocreatine and creatine phosphokinase in adipose tissue. Biochim. Biophys. Acta 437: 166-174, 1976.
7. Bieber, L. L., B. Pettersson and O. Lindberg. Studies on norepinephrine-induced efflux of free fatty acid from hamster brown adipose tissue cells. Eur. J. Biochem. 58: 375-381, 1975.
8. Cannon, B. and O. Lindberg. Mitochondria from brown adipose tissue: isolation and properties Methods in Enzymology, Biomembranes. Part D.I., (In press).
9. Cannon, B. and J. Nedergaard. Energy dissipation in brown fat. In: Effectors of Thermogenesis. Girardier, L. and J. Seydoux (editors). Basel: Experientia, in press.
10. Cannon, B. and G. Vogel. The mitochondrial ATPase of brown adipose tissue. Purification and comparison with the mitochondrial ATPase from beef heart. FEBS Lett. 76: 284-289, 1977.

11. Cannon, B., L. Romert, U. Sundin and T. Barnard. Morphology and biochemical properties of perirenal adipose tissue from lamb (*Ovis aries*). A comparison with brown adipose tissue. Comp. Biochem Physiol. 56B: 87-99, 1977.
12. Cannon, B., U. Sundin and L. Romert. Palmitoyl coenzyme A: a possible physiological regulator of nucleotide binding to brown adipose tissue mitochondria. FEBS Lett. 74:43-46, 1977.
13. Daniel, H. and D.M. Derry. Criteria for differentiation of brown and white fat in the rat. Can. J. Physiol. Pharmacol. 47:941-945, 1969.
14. Davis, T. R. A. Chamber cold acclimatization in man. J. Appl. Physiol. 16: 1011-1015, 1961.
15. Edelman, I. S. and F. Ismail-Beigi. Thyroid thermogenesis and active sodium transport. Recent Prog. Hormone Res. 30: 235-257, 1974.
16. Fain. J. N. and N. Reed. A mechanism for hormonal activation of lipolysis and respiration in free brown fat cells. Lipids, 5: 210-219, 1970.
17. Foster, D. O. Brown adipose tissue: the dominant site of non-shivering thermogenesis in the rat. In: Effectors of Thermogenesis. Girardier, L. and J. Seydoux (editors). Basel: Experientia, in press.
18. Girardier, L., J. Seydoux and T. Clausen. Membrane potential of brown adipose tissue. A suggested mechanism for the regulation of thermogenesis. J. Gen. Physiol. 52: 925-940, 1968.
19. Hahn, P., C. H. Beatty and R. M. Bocek. Carnitine transferases in brown fat of newborn and developing primates. Biol. Neonate 30: 30-34, 1976.
20. Halestrap, A. P. The mitochondrial pyruvate carrier. Kinetics and specificity for substrates and inhibitors. Biochem J. 148: 85-96, 1975.
21. Heick, H. M. C., C. Vachon, M. A. Kallai, N. Bégin-Heick, and J. LeBlanc. The effects of thyroxine and isopropylnoradrenaline on cytochrome oxidase activity in brown adipose

tissue. Can. J. Physiol. Pharmacol. 51:751-758, 1973.
22. Heldmaier, G. Zitterfreie Wärmebildung und Körpergrösse bei Säugetieren. Z. vergl. Physiologie, 73: 222-248, 1971.
23. Horowitz, J.M., B.A. Horwitz and R.E. Smith Effects *in vivo* of norepinephrine on the membrane resistance of brown fat cells. Experientia, 27: 1419-1421, 1971.
24. Horwitz, B. A. Plasma membrane involvement in brown fat thermogenesis. In: Effectors of Thermogenesis. Girardier, L. and J. Seydoux (editors). Basel: Experientia, in press.
25. Hull, D. and M. J. Hardman. Brown adipose tissue in newborn mammals. In: Brown Adipose Tissue. Lindberg, O. (editor). New York: American Elsevier, 1970, pp. 97-115.
26. Imai, Y., B. A. Horwitz and R. E. Smith. Calorigenesis of brown adipose tissue in cold-exposed rats. Proc. Soc. Exp. Biol. Med. 127:717-719, 1968.
27. Joel, C. D. The physiological role of brown adipose tissue. In: Handbook of Physiology. Renold, A. E. and G. F. Cahill Jr. (editors). Washington, D. C.: American Physiological Society, Vol. 5, 1965, pp 59-85.
28. Joy, R. T. J. Responses of cold-acclimatized men to infused norepinephrine. J. Appl. Physiol 18: 1209-1212, 1963.
29. Jungas, R. L. Metabolic effects on adipose tissue *in vitro*. In: Experimental Handbook of Pharmacology, 32-2. Born, G. W. R. *et al.* (editors). Berlin: Springer Verlag, 1975, pp. 371-412.
30. Knight, B. L. and J. P. Skala. Protein kinases in brown adipose tissue of developing rats. J. Biol. Chem. 252: 5356-5362, 1977.
31. Lindberg, O., L. L. Bieber and J. Houštěk. Brown adipose tissue metabolism: an attempt to apply results from *in vitro* experiments on tissue *in vivo*. In: Regulation of Depressed Metabolism and Thermogenesis. Janský, L. and X. J. Musacchia (editors). Springfield, Ill.: Charles C. Thomas Publ. Co., 1976, pp. 117-136.

32. Nedergaard, J., B. Cannon and O. Lindberg. Microcalorimetry of isolated mammalian cells. Nature 267: 518-520, 1977.
33. Newsholme, E. A., B. Crabtree, S. J. Higgins, S. D. Thornton and C. Start. The activities of fructose diphosphatase in flight muscles from the bumble-bee and the role of this enzyme in heat generation. Biochem. J. 128: 89-97, 1972.
34. Nicholls, D. G. The bioenergetics of brown adipose tissue mitochondria. FEBS Lett. 61: 103-110, 1976.
35. Nicholls, D. G. and G. M. Heaton. The identification of the component in the inner membrane of brown adipose tissue mitochondria responsible for regulating energy dissipation. In: Effectors of Thermogenesis. Girardier, L. and J. Seydoux (editors). Basel: Experientia, in press.
36. Pettersson, B. CO_2-mediated control of fatty acid metabolism in isolated hamster brown fat cells during norepinephrine stimulation. Eur. J. Biochem. 72: 235-240, 1977.
37. Pettersson, B. and I. Vallin. Norepinephrine-induced shift in levels of adenosine - 3':5' monophosphate and ATP parallel to increased respiratory rate and lipolysis in isolated hamster brown fat cells. Eur. J. Biochem. 62: 383-390, 1976.
38. Pettersson, B., P. Lundberg and V. S. M. Bernson. Succinate oxidation in isolated hamster brown adipocytes. In: Effectors of Thermogenesis. Girardier, L. and J. Seydoux (editors). Basel: Experientia, in press.
39. Portet, R., M. C. Laury, R. Bertin, C. Senault, M. T. Hluszko and L. Chevillard. Hormonal stimulation of substrate utilization in brown adipose tissue of cold acclimated rats. Proc. Soc. Exp. Biol. Med. 147: 807-812, 1974.
40. Prusiner, S. B., B. Cannon and O. Lindberg. Oxidative metabolism in cells isolated from brown adipose tissue. 1. Catecholamine and fatty acid stimulation of respiration. Eur. J. Biochem. 6: 15-22, 1968.

41. Prusiner, S. B., B. Cannon, T. M. Ching and O. Lindberg. Oxidative metabolism in cells isolated from brown adipose tissue. 2. Catecholamine regulated respiratory control. Eur. J. Biochem. 7: 51-57, 1968.
42. Rafael, J. and H. W. Heldt. Binding of guanine nucleotides to the outer surface of the inner membrane of guinea pig brown fat mitochondria in correlation with the thermogenic activity of the tissue. FEBS Lett. 63: 304-308, 1976.
43. Rafael, J. and E. Wrabetz. Brown adipose tissue mitochondria: recoupling caused by substrate level phosphorylation and extramitochondrial adenosine phosphates. Eur. J. Biochem. 61: 551-561, 1976.
44. Ricquier, D. and J.- C. Kader. Mitochondrial protein alteration in active brown fat: a sodium dodecyl sulfate-polyacrylamide gel electrophoretic study. Biochem. Biophys. Res. Comm. 73: 577-583, 1976.
45. Rosak, C. and K. J. Hittelman. Characterization of lipolytic responses of isolated white adipocytes from hamsters. Biochim. Biophys. Acta 496: 458-474, 1977.
46. Slater, E. C., J. Rosing and A. Mol. The phosphorylation potential generated by respiring mitochondria. Biochim. Biophys. Acta 292: 534-553, 1973.
47. Smith, R. E. The thermogenic activity of the hibernating gland in the cold acclimated rat. The Physiologist 4: 113, 1961.
48. Smith, R. E. and B. A. Horwitz. Brown fat and thermogenesis. Physiol Rev. 49: 330-425, 1969.
49. Williams. L. T., L. Jarett and R. J. Lefkowitz. Adipocyte β-adrenergic receptors. Identification and subcellular localization by (-)-^{3}H dihydroalprenolol. J. Biol. Chem. 251: 3096-3104, 1976.
50. Williamson, J. R. Control of energy metabolism in hamster brown adipose tissue. J. Biol. Chem. 245: 2043-2050, 1970.

51. Williamson, J. R., S. Prusiner, M. S. Olson and M. Fukami. Control of metabolism in brown adipose tissue. Lipids 5: 1-14, 1970.

BIOCHEMICAL ASPECTS OF NONSHIVERING THERMOGENESIS

Jean Himms-Hagen
Department of Biochemistry
University of Ottawa
275 Nicholas Street
Ottawa, Canada K1N 9A9

ABSTRACT

Nonshivering thermogenesis (NST), a facultative thermogenesis that is rapidly switched on and off by the sympathetic nervous system in response to a change in environmental temperature, occurs to a major extent in hibernators and in certain nonhibernators after acclimation to cold. The major site is brown adipose tissue (BAT) with other tissues, particularly skeletal muscle, also playing a role. Because the very large calorigenic response to noradrenaline, known to be the basis of NST in cold-acclimated animals, is not readily demonstrable in tissues *in vitro*, the process of NST is not easily accessible for study. In this review the approach taken is to compare the composition and metabolic characteristics of those components thought to participate in NST in warm-acclimated animals with those in cold-acclimated animals and in hibernators, in order to identify the mechanism involved. There are qualitative and quantitative changes in BAT mitochondria of cold-acclimated nonhibernators which result in a greater capacity of the tissue for heat production and a greater capacity of the mitochondria for respiring in the loosely-coupled state. Mitochondria of the BAT of hibernators appear always to have a high capacity for respiring in the loosely-coupled state and the principal change during acclimation to cold is an increase in the total mitochondrial mass. Muscle mitochondria of cold-acclimated rats undergo changes which result in greater capacity for coupled respiration. Changes in the plasma membrane are either minor (sarcolemma in cold-acclimated rats) or have not been studied. The changes described are consistent with the loose-coupling hypothesis for thermogenesis in BAT but do not exclude the participation of an ATP-ase type of thermogenic mechanism.

ISBN 0-12-734550-7

I. OCCURRENCE OF NONSHIVERING THERMOGENESIS

The term nonshivering thermogenesis is defined in this review as that increase in heat production over and above the basal metabolic rate which is not due to mechanical activity of muscle and which can be rapidly switched on and off in response to a change in environmental temperature. Nonshivering thermogenesis thus defined does not include the basal metabolic rate, which can be described as *obligatory thermogenesis* (44) even when the basal metabolic rate is elevated as in hyperthyroidism. It corresponds in classification, although not in mechanism, to shivering thermogenesis, and may be described as regulatory thermogenesis (44), or as *facultative* thermogenesis, i.e. thermogenesis which may or may not be activated according to the circumstances. The control mechanisms for facultative thermogenesis would be expected to differ from the control mechanisms for obligatory thermogenesis, although both would of course ultimately depend upon the same metabolic machinery.

The basic mechanism for thermogenesis by cells is the combustion of fuel. In considering biochemical aspects of nonshivering thermogenesis, it is necessary to consider in which organ or organs the combustion is occurring, how the effect of a decrease in ambient temperature is transmitted as a stimulus to these organ(s) and how the regulation of metabolic processes within the cells, already adjusted in accordance with their contribution to obligatory thermogenesis, is modified in such a way that thermogenesis increases in response to the stimulus. In addition to these considerations it is necessary to discuss the species, age and previous environmental history of the animal in question, because the mechanism for nonshivering thermogenesis is not only facultative but also adaptive in certain species and constitutive in others and a review of its biochemical mechanism must also take these factors into account.

Nonshivering thermogenesis can occur in hibernators such as hamsters, hedgehogs and ground squirrels during exposure to cold and during arousal from hibernation (44); the capacity to use nonshivering thermogenesis appears to be constitutive in these animals when they are adult. In certain non-hibernators, such as the rat, guinea pig, mouse and rabbit, nonshivering thermogenesis appears to be an adaptive process in that in the adult it is only present to any appreciable extent after adaptation to living in the cold. However, in these and some other species (e.g., man, sheep) the newborn animals are born in the "adapted" state and lose this state during development. Once fully grown some non-hibernators, such as man and sheep, appear to be unable to undergo to any appreciable extent the necessary adaptation for nonshivering thermogenesis (13).

II. THE SYMPATHETIC NERVOUS SYSTEM AND NONSHIVERING THERMOGENESIS

There is considerable evidence that a decrease in ambient temperature initiates a stimulus which is transmitted via the sympathetic nervous system to the organs involved in nonshivering thermogenesis (34) and that the capacity to use nonshivering thermogenesis, whether adaptive or constitutive, is associated with a capacity to respond to catecholamines by a large increase in metabolic rate (calorigenic response). Thus, the calorigenic response to noradrenaline is large in the warm-acclimated hamster, a hibernator, and only approximately doubles after acclimation to cold (38, 44, 71). Other hibernators also have a large response to noradrenaline (44). In contrast, in the warm-acclimated rat the calorigenic response to noradrenaline is quite small but it increases as much as 4-5 times after acclimation to cold (34).

III. SITES OF NONSHIVERING THERMOGENESIS

The major sites of nonshivering themogenesis appear to be brown adipose tissue and skeletal muscle (34). There has been argument for many years about the quantitative contribution of each site, most authors estimating 10-20% for the contribution of brown adipose tissue in cold-acclimated nonhibernators and up to 80% in newborn nonhibernators (see 34, 44). However, recent work by Foster (25, 26) has made necessary a considerable upward revision of estimates of the contribution by brown adipose tissue in the cold-acclimated rat to approximately

60%.

Comparison of selected nonhibernators and hibernators shows that in the former the amount of brown adipose tissue (per unit of body weight) is largest at or immediately after birth and decreases progressively as maturation occurs whereas in the latter the proportion of brown adipose tissue is small at birth and increases progressively as the animal matures (4). In those adult nonhibernators which are able to adapt to cold by increasing their capacity to use nonshivering thermogenesis the small amount of brown adipose tissue remaining grows in response to cold, as also occurs in hibernators. Some adult nonhibernators appear to lose the brown adipose tissue they had at birth (man, sheep), do not develop this tissue when exposed to cold, and do not undergo any adaptation for nonshivering thermogenesis.

The existence of a large calorigenic response to noradrenaline is always associated with the presence of abundant deposits of brown adipose tissue and this tissue usually undergoes changes in size, composition and metabolic characteristics when the capacity of the animal to use nonshivering thermogenesis changes (13, 67). Thus, the brown adipose tissue of the hamster not only doubles in size during acclimation to cold (from 25° to 6°) but its content of mitochondria (as judged by its cytochrome oxidase content) increases twelve-fold (57). The brown adipose tissue of the rat also doubles in size during cold-acclimation (from 30° to 6°) but its content of mitochondria increases only seven-fold (57). However, there appears to be no direct correlation between amount of brown adipose tissue or mitochondrial mass of brown adipose tissue and the capacity of the animal to respond to noradrenaline. For example, the rat acclimated to 24°C has the same total amount of mitochondrial material per unit body weight in its brown adipose tissue as the hamster acclimated to the same temperature (57) and yet the latter has a much larger response to noradrenaline. This discrepancy is most probably due to a difference in the properties of the mitochondria themselves, those of the hamster acclimated to this temperature having a greater capacity (constitutive) to produce heat in response to the appropriate signal than those of the rat, in which the mitochondria acquire this capacity only after adaptation to cold. This will be discussed further below.

Not only do the mitochondria of brown adipose tissue change in nature as well as number during acclimation of the rat to cold but so also do the mitochondria of skeletal muscle (5). These become smaller and more numerous and exhibit altered respiratory properties. In considering the biochemical mechanism of nonshivering thermogenesis it is thus useful to compare the biochemical characteristics of the mitochondria and other cell constituents of the organs involved in nonshivering thermogenesis in animals in the cold-acclimated state with those of animals in the warm-acclimated state since any changes could be expected to be related to the altered capacity for nonshivering thermogenesis. Such a comparison is made in Section V of this review. A further comparison of these biochemical characteristics of cold-acclimated nonhibernators with those of hibernators is also useful in considering the mechanism of nonshivering thermogenesis. Such a comparison is made in Section VI.

Before reviewing the biochemical characteristics of the tissues involved in nonshivering thermogenesis it is appropriate to review mechanisms of cellular thermogenesis in general to provide a background for the subsequent discussion: this is done in Section IV.

IV. MECHANISMS OF THERMOGENESIS

The basic mechanism for thermogenesis is the combustion of fuel, i.e., the complete oxidation of the foodstuffs available to animal cells, mainly glucose and fatty acids. The rate of such combustions, i.e., the rate of respiration, is generally controlled at the mitochondrial level either by the phosphorylation state ratio ($[ATP]/[ADP]\cdot[P_i]$), the phosphorylation of ADP resulting in an increased rate of respiration, or by the availability of ions which are transported into and out of the mitochondrion, e.g. calcium, the uptake of such ions resulting in an increased rate of respiration. In order for increased heat production to occur in such mitochondria in which the rate of respiration is coupled to the rate of ADP phosphorylation or to the rate of ion translocation there must be increased production of ADP and phosphate from ATP (referred to below as an ATPase type of thermogenic mechanism) or an increased availability of ions for translocation, either by release from intracellular stores or by entry from outside the cell (referred to below as an ion-cycling type of thermogenic mechanism). In coupled mitochondria it is the production and maintenance of a high proton electrochemical gradient by the functioning of the electron transport chain in pumping protons out of the mitochondria, together with the low mitochondrial permeability to protons, which provides not only the driving force for ATP synthesis (from the proton electrochemical gradient) and for calcium translocation [from the membrane potential, itself determined by the proton electrochemical gradient (9)] but also the control mechanism for respiration, the rate of which is inversely related to the magnitude of the gradient (see 50). It is also possible to envisage a mechanism by which proton re-entry could be regulated independently of ATP synthesis and ion translocation. If, for example, proton permeability were to be specifically increased there would result a decrease in the proton electrochemical gradient and a consequent increase in the rate of respiration independent of changes in the availability of ADP or of ions. Such a state may be described as loosely-coupled (partial discharge of the gradient; some ATP synthesis still possible) or uncoupled (complete discharge of the gradient; no ATP synthesis possible) (see 50). A regulated proton leakage pathway has been proposed by Nicholls (50) to provide a control mechanism for heat production in brown adipose tissue mitochondria (referred to below as loose-coupling type of mechanism; this term is used in preference to a more exact label, such as proton-leakage type of mechanism, because it can be applied to results of numerous reported studies which do not include measurements of proton conductance).

Thus at least three classes of thermogenic mechanism involving altered control of mitochondrial function can be envisaged: 1) ATPase type; 2) ion-cycling type; 3) loose-coupling type. The current hypotheses for the mechanism of nonshivering thermogenesis in brown adipose tissue and skeletal muscle can be classified in the same way. Thus, the concept that the activity of the plasma membrane $Na^{+}K^{+}$-ATPase is increased by noradrenaline, either via a direct action or via altered ion permeability of the plasma membrane, which has been proposed by Horwitz and co-workers (41) for nonshivering thermogenesis in brown adipose tissue, would be classified as an ATPase type of mechanism. This concept is derived from the hypothesis of Edelman (see 20, 21) for the mechanism of thyroid thermogenesis, also based upon altered activity or amount of the sodium pump, and is now being extended by others (32, 43, 69, 70) to nonshivering thermogenesis in other tissues. Cycling of calcium, an ion-cycling type of thermogenic mechanism, has been proposed as a mechanism for nonshivering thermogenesis in brown adipose tissue mitochondria (15, 39) and in skeletal muscle mitochondria (30). A loose-coupling type of mechanism is proposed by a number of different groups for nonshivering thermogenesis in brown adipose tissue mitochondria (see 12, 23, 50, 58).

In comparing the biochemical characteristics of organs involved in nonshivering thermogenesis in the cold-acclimated state and in hibernators with these characteristics in warm-acclimated nonhibernators (Sections V, VI and VII) it is important to bear in mind the class of mechanism in which a particular component might participate.

V. BIOCHEMICAL CHANGES IN BROWN ADIPOSE TISSUE AND SKELETAL MUSCLE IN ANIMALS ADAPTED TO USE NONSHIVERING THERMOGENESIS (NONHIBERNATORS)

The two principal species used by most investigators for such studies have been the rat and the guinea pig and this discussion will concentrate on these animals. In both species the brown adipose tissue grows during acclimation to cold and develops a higher concentration of mitochondria.

The mitochondria of the brown adipose tissue of cold-acclimated rats have an increased capacity for respiration as indicated by their increased rate of oxygen uptake in the presence of uncoupling agents (51). Their rate of state 4 respiration (in the absence of ADP) not only increases markedly (51, 68) but also becomes equal to the rate of uncoupled respiration (compared to 67% of that rate in the warm-acclimated rat) (51). They are nevertheless capable of phosphorylating ADP when this compound is supplied. These characteristics suggest the existence of a more loosely-coupled state (51). A marked effect of purine nucleotides to tighten coupling in such mitochondria can be demonstrated (40, 51, 65) for all temperatures of acclimation below 30° (65). The capacity to bind purine nucleotides is greatly increased (Desautels and Himms-Hagen, unpublished results). The rate of synthesis of protein by isolated mitochondria is increased after two weeks of adaptation to cold (8). However, once the rats are fully acclimated the rate of protein synthesis returns to the warm-acclimated level (8).

In parallel with these alterations in function some changes have been observed in the composition of the mitochondria in the cold-acclimated state. The calcium content decreases (30). The increase in chain length and in unsaturation of the fatty acids of the mitochondrial phospholipids seen in one study (60) has not been observed in another study (10). The concentration of the major respiratory proteins may increase (45, 63) and there is a large increase in the amount of a polypeptide of molecular weight 32,000 daltons (37, 59) as well as lesser changes in amounts of certain other polypeptides (37). The recent identification by Nicholls and Heaton (52) of a 32,000 dalton polypeptide as the binding site for purine nucleotides in hamster and guinea pig brown adipose tissue mitochondria suggests that the increase in amount of such a polypeptide in the mitochondria of the cold-acclimated rat and the increase in the amount of binding of purine nucleotides are both associated with the increased extent of loose-coupling in these mitochondria and its susceptibility to control by purine nucleotides.

Thus, there are both qualitative and quantitative changes in brown adipose tissue mitochondria of cold-acclimated rats. The qualitative changes are associated with a greater capacity to respire in the loosely-coupled state and with a greater capacity for coupling to be controlled by purine nucleotides.

The changes in brown adipose tissue mitochondria of the cold-acclimated guinea pig are similar to those in the cold-acclimated rat [see review by Flatmark and Pedersen (23)]. Thus there is a reduction in ADP/O ratio and in respiratory control, both sensitive to purine nucleotides (23). The rate of ADP translocation is unchanged (17). The rate of state 4 respiration (51) and the "leakage" of the energy potential (22) are increased as also is the rate of uncoupled respiration (51). At the same time the effect of uncouplers is lost (51). These characteristics indicate a loosely-coupled type of respiration.

Purine nucleotide binding by mitochondria (58) and the amount of purine nucleotide binding polypeptide (52) are both high in brown adipose tissue mitochondria of the newborn guinea pig and diminish as the animal matures. In the young animal the purine nucleotide binding polypeptide reappears after only four days of acclimation to cold (52), a time known to be sufficient for the development of loose-coupling (3, 53) although not for the proliferation of the mitochondria in the tissue, which takes 12 - 14 days (23, 24). The large increase in tissue content of a specific iron-sulfur flavoprotein, identified as a component of the acyl dehydrogenase complex and present in high concentration in brown adipose tissue mitochondria, appears to be largely due to proliferation of the mitochondria and not to any selective change in its concentration (24). Its lack of a specific role in the development of the loosely-coupled state is indicated by the longer time required for its increase than that needed for the appearance of loose-coupling (24). This flavoprotein, of molecular weight 70,000 daltons (62), differs from the one polypeptide which is known to change (52). Other possible alterations in polypeptide and fatty acid composition have not been studied but two other changes have been reported in mitochondria from cold-acclimated guinea pigs: the buoyant density is unchanged (66) and the calcium content is decreased (15).

Thus, in brown adipose tissue mitochondria of the cold-acclimated guinea pig there are both qualitative and quantitative changes. The former, as in the cold-acclimated rat, involve the development of a greater capacity for respiration in the loosely-coupled state.

In another species capable of adapting to use non-shivering thermogenesis, the artic lemming, respiratory control of brown adipose tissue mitochondria is low in the absence of purine nucleotides except in animals kept at 30° (16). From the limited information available it would appear that changes occur in this species similar to those described for the rat and the guinea pig. The main difference between these three species is the temperature at which the adaptation is initiated.

Little information is available about possible adaptive changes in the plasma membrane of brown adipose tissue, the other site thought to be important in the development of the capacity for nonshivering thermogenesis. While no change has been observed in total fluoride-stimulated adenylate cyclase activity a reduction has been demonstrated in catecholamine-stimulated adenylate cyclase activity in brown adipose tissue of cold-acclimated rats (46, 64). Although catecholamines have been shown to stimulate the $Na^{+}K^{+}$-ATPase of this tissue (42) so far the effect of cold-acclimation on this stimulation has not been studied. The resting membrane potential of the intact tissue is lower in the cold-acclimated state (28, 29), suggesting an alteration in the properties or the functioning of the plasma membrane.

Not only brown adipose tissue mitochondria but also skeletal muscle mitochondria undergo changes during acclimation of rats to cold. In cold-acclimated rats the skeletal muscle mitochondria are smaller and more numerous (5). The rates of ADP-stimulated (36) and calcium-stimulated (30) respiration are increased, changes apparently due to an increased adenine nucleotide translocase activity (Desautels and Himms-Hagen, unpublished results) and an increased rate of calcium transport (30). There are no changes in ADP/O or respiratory control ratios and no change in concentration of the major respiratory enzymes (5, 36). There are no great changes in polypeptide composition of these altered mitochondria and there is no major polypeptide component of 32,000 daltons as there is in brown adipose tissue mitochondria (37). The capacity to bind purine nucleotides is extremely low and is unchanged by acclimation to cold (Desautels and Himms-Hagen, unpublished results). The fatty acid composition of the phospholipids is changed in that there is an increase in 18:2 and a decrease in 20:4 fatty acids consistent with reduced $\Delta 6$ desaturase activity (37). The calcium content is unchanged (30). Thus the changes in muscle mitochondria are qualitative in that the same mass of mitochondrial material is distributed in a different way and has a greater capacity for coupled respiration.

Increases in both $Na^{+}K^{+}$-ATPase and Mg^{2+}-ATPase of muscle plasma membrane occur in the cold-acclimated rat but it has not been possible to demonstrate any effect of catecholamines on these enzymes (Cerf and Himms-Hagen, unpublished results). This does not of course exclude an increased rate of action of the muscle $Na^{+}K^{+}$-ATPase in the intact animal in response to catecholamines. Catecholamines are known to increase the action of this enzyme in intact muscle (61). No change in catecholamine-stimulated adenylate cyclase occurs in skeletal muscle of cold-acclimated rats (47).

In summary, the major changes in tissues of cold-acclimated nonhibernators are an increase in the capacity of the brown adipose tissue mitochondria to assume a loosely-coupled state with a concomitant increase in the capacity of the mitochondria to be regulated by purine nucleotides, and an increase in the capacity of muscle mitochondria for coupled respiration. The mitochondrial mass increases in brown adipose tissue and is distributed differently in muscle. The relation of these changes to the mechanism of nonshivering thermogenesis will be discussed further in section VIII.

VI. BIOCHEMICAL PROPERTIES OF BROWN ADIPOSE TISSUE IN HIBERNATORS

The principal animal studied has been the hamster and the discussion concerns the brown adipose tissue of this species unless otherwise mentioned.

The rate of state 4 respiration of hamster brown adipose tissue mitochondria is always high and is unaffected by uncouplers regardless of the state of acclimation of the animal (51). Thus these mitochondria would appear always to have a high capacity to be loosely coupled. The coupled state can be restored by purine nucleotides regardless of the state of acclimation and the 32,000 dalton polypeptide, identified as the purine nucleotide binding site, is always present (52). Binding of purine nucleotides is high and these compounds reduce the very high proton conductance of the mitochondria

(48, 49, 50).

Mitochondria of the brown adipose tissue of the hedgehog also appear to be always loosely coupled and to be susceptible to the coupling effect of purine nucleotides (65).

Thus, in hibernators although there are quantitative changes in the amount of brown adipose tissue mitochondria during acclimation to cold (4) there appears to be no change in the biochemical properties of the mitochondria themselves. The control of gene expression in this organ thus resembles that in nonhibernators in that growth of the tissue and proliferation of its mitochondria can occur but differs in that there do not appear to be any selective changes in synthesis of mitochondrial proteins, this synthesis being always programmed to produce mitochondria capable of a high rate of thermogenesis.

VII. BIOCHEMICAL PROPERTIES OF BROWN ADIPOSE TISSUE IN NONHIBERNATORS THAT DO NOT ADAPT IN ADULT LIFE TO USE NONSHIVERING THERMOGENESIS

Relatively little is known about the biochemical characteristics of brown adipose tissue mitochondria in those species which possess this tissue and the capacity to use nonshivering thermogenesis at birth but are unable in later life to redevelop it or to develop the capacity for nonshivering thermogenesis to any appreciable extent.

One report (11) indicates that the brown adipose tissue of the newborn lamb has mitochondria which are morphologically similar to those of other species except that elementary particles are present on the inner membranes. However, these mitochondria have a greater degree of coupling, a lower proton conductance and a much greater capacity to synthesize ATP than would be expected from studies of brown adipose tissue of hibernators and cold-acclimated nonhibernators. Some control of coupling by purine nucleotides has however been demonstrated (11).

Thus, in the lamb, the basic design of the mitochondria appears to differ from that in hibernators and in the nonhibernating species which are capable of adapting to use nonshivering thermogenesis. However, they display some features believed to be associated with nonshivering thermogenesis, such as purine nucleotide control of the state of coupling. As already pointed out, the newborn lamb is known to be capable of nonshivering thermogenesis (1, 2).

VIII. MECHANISMS OF NONSHIVERING THERMOGENESIS

The main conclusion which can be drawn from the survey of biochemical characteristics outlined above is that for brown adipose tissue the correlation between the properties and amount of brown adipose tissue mitochondria and the existence of a large calorigenic response to noradrenaline is in keeping with a loose-coupling type of mechanism. The control of thermogenesis in the mitochondrion would appear to be exerted by the opposing effects of purine nucleotides, which reduce the innately high proton leakage by binding to a specific site on the outer surface of the inner membrane (48, 49, 50, 52) and fatty acids (33) or acyl CoA (12) which increase proton conductance. A competition between acyl CoA and GDP for the binding site has been demonstrated (12). The mechanism of switching on of nonshivering thermogenesis is proposed to involve noradrenaline-induced activation of adenylate cyclase, cyclic AMP-induced activation of triglyceride lipase, displacement by acyl CoA of purine nucleotides bound to the mitochondria and thus an increase in proton leakage into the mitochondria, a consequent decrease in the proton electrochemical gradient and a compensatory increase in the rate of respiration and hence thermogenesis (11, 49, 50).

The information available about biochemical characteristics of the plasma membrane of cold-acclimated animals does not shed much light on the postulated ATPase-type mechanism for nonshivering thermogenesis in brown adipose tissue. There is no doubt that the ion movements induced by noradrenaline must result in the utilization of ATP and thus contribute to the noradrenaline-induced increase in thermogenesis [see (35) for review; the hypothesis is outlined in more detail by Horwitz elsewhere in this Symposium] but the proportion of the total nonshivering thermogenesis which can be ascribed to this process is uncertain [see

(14)]. It is quite likely that the proportion may vary, depending on the species and on the state of acclimation of the animal.

As far as muscle is concerned, the comparison made of biochemical properties of mitochondria and plasma membrane of cold-acclimated animals with those of warm-acclimated animals would suggest an ATPase-type mechanism or a calcium-cycling mechanism but only if the necessary concomitant increases in an ATPase or in calcium availability could be demonstrated.

IX. FUTURE DEVELOPMENTS

It is clear that a complete description of the biochemical processes involved in nonshivering thermogenesis requires the development of an *in vitro* system in which the process operates as it does *in vivo*. The only one currently available is the isolated cell preparation of hamster brown adipose tissue and, even with this preparation, it has only recently been possible to demonstrate a maintained rather than a transient effect of noradrenaline on respiration (54, 55).

Although the response *in vivo* of interscapular brown adipose tissue of the cold-acclimated rat to noradrenaline differs from that of the warm-acclimated rat [greater blood flow per gram (25, 26, 56); retention of fatty acids produced in lipolysis within the tissue rather than their release into the blood (56)] it has not been possible to demonstrate any increase in lipolytic or calorigenic response of slices (18, 19), of isolated cells (6, 7), or of isolated tissue fragments (27). Also, although the calorigenic response of leg muscle of the cold-acclimated rat to infused noradrenaline is greater than that in the warm-acclimated rat (31) and the increase in blood flow is also greater (25, 26) it has not been possible to demonstrate any increase in the response of slices of this tissue to noradrenaline. There would appear to be some damage to the cells and/or some deficiency in the incubation conditions which prevents a response *in vitro* to noradrenaline by these tissues comparable in magnitude to their response *in vivo*.

The emerging picture of the brown adipose tissue as an organ in which not only the concentration of mitochondria but also the composition of the mitochondria is regulated in accordance with the need for its thermogenic function shows that this organ will probably prove to be a useful model system for the study of the regulation of mitochondrial biogenesis as well as for the study of the regulation of nonshivering thermogenesis.

REFERENCES

1. Alexander, G. and A. W. Bell. Quantity and calculated oxygen consumption during summit metabolism of brown adipose tissue in newborn lambs. Biol. Neonate 26: 214-220, 1975.
2. Alexander, G., A. W. Bell and J. R. S. Hales. Effects of cold exposure on tissue blood flow in the newborn lamb. J. Physiol. 234: 65-77, 1973.
3. Anderson, H. T., E. N. Christiansen, H. J. Grav and J. I. Pedersen. Thermogenesis of brown adipose tissue reflected in mitochondrial respiratory control. Acta Physiol. Scand. 80: 1-10, 1970.
4. Barnard, T. and J. Skala. The development of brown adipose tissue. In: Brown Adipose Tissue. Lindberg, O. (editor). New York: American Elsevier, 1970, pp 33-72.
5. Behrens, W. A. and J. Himms-Hagen. Alteration in skeletal muscle mitochondria of cold-acclimated rats: association with an enhanced metabolic response to noradrenaline. J. Bioenergetics Biomembranes 9: 41-64, 1977.
6. Bertin, R. Glycerokinase activity and lipolysis regulation in brown adipose tissue of cold-acclimated rats. Biochimie (Paris) 58: 431-434, 1976.
7. Bertin, R. and R. Portet. Effects of lipolytic and antilipolytic drugs on metabolism of adenosine 3':5'-monophosphate in brown adipose tissue of cold-acclimated rats. Eur. J. Biochem. 69: 177-183, 1976.
8. Bukowiecki, L. and J. Himms-Hagen. Alterations of mitochondrial protein metabolism in liver, brown adipose tissue and skeletal muscle during cold-acclimation. Biochim. Biophys. Acta 428: 591-599, 1976.
9. Bygrave, F. L. Mitochondrial calcium transport. In: Current Topics in Bioenergetics. Sanadi, D. R. (editor). New York: Academic Press, 1977, 259-318.

10. Cannon, B., C. F. Polnaszek, K. W. Butler, L. E. G. Eriksson and I.C.P. Smith. The fluidity and organization of mitochondrial membrane lipids of the brown adipose tissue of cold-adapted rats and hamsters as determined by nitroxide spin probes. Arch. Biochem. Biophys. 167: 505-518, 1975.
11. Cannon, B., L. Romert, U. Sundin and T. Barnard. Morphology and biochemical properties of perirenal adipose tissue from lamb (*Ovis aries*). A comparison with brown adipose tissue. Comp. Biochem. Physiol. 56B: 87-99, 1977.
12. Cannon, B., U. Sundin and L. Romert. Palmitoyl CoA: a possible physiological regulator of nucleotide binding to brown adipose tissue mitochondria. FEBS Letters 74: 43-46, 1977.
13. Chaffee, R. R. J. and J. C. Roberts. Temperature acclimation in birds and mammals. Ann. Rev. Physiol. 33: 155-202, 1971.
14. Chinet, A., T. Clausen and L. Girardier. Microcalorimetric determination of energy expenditure due to active sodium-potassium transport in the soleus muscle and brown adipose tissue of the rat. J. Physiol. 265: 43-61, 1977.
15. Christiansen, E. N. Calcium uptake and its effects on respiration and phosphorylation in mitochondria from brown adipose tissue. Eur. J. Biochem. 19: 276-282, 1971.
16. Christiansen, E. N. Respiratory properties of brown adipose tissue mitochondria from the Norwegian lemming. Comp. Biochem. Physiol. 56B: 19-24, 1977.
17. Christiansen, E. N., Z. Drahota, J. Duszynski and L. Wojtczak. Transport of adenine nucleotides in mitochondria from the brown adipose tissue. Eur. J. Biochem. 34: 506-512, 1973.
18. Dorigo, P., R. M. Gaion and G. Fassina. Lack of correlation between cyclic AMP synthesis and free fatty acid release in brown fat of cold-adapted rats. Biochem. Pharmacol. 23: 2877-2885, 1974.

19. Dorigo, P., I. Maragno, A. Bressa and G. Fassina. Reduced lipolytic response *in vitro* to catecholamines, ACTH and cyclic adenosine monophosphate in brown fat of cold-acclimated rats. Biochem. Pharmacol. 20: 1201-1211, 1971.
20. Edelman, I. S. Transition from the poikilotherm to the homeotherm: possible role of sodium transport and thyroid hormone. Fed. Proc. 35: 2180-2184, 1976.
21. Edelman, I. S. Thyroid hormone: thermogenesis and the biosynthesis of Na^+ pumps. In: Biogenesis and Turnover of Membrane Macromolecules. Cook, J. S. (editor). New York: Raven Press, 1976, pp. 169-177.
22. Flatmark, T. and J. I. Pedersen. Studies on the energy state of isolated brown adipose tissue mitochondria. Effect of adenine nucleotides and oligomycin on the generation and dissipation of the "energy potential". Biochim. Biophys. Acta 292: 64-72, 1973.
23. Flatmark, T. and J. I. Pedersen. Brown adipose tissue mitochondria. Biochim. Biophys. Acta 416: 53-103, 1975.
24. Flatmark, T., F. J. Ruzicka and H. Beinert. The pattern of iron-sulfur centers in brown adipose tissue mitochondria: preponderance of ETF dehydrogenase and invariance with the thermogenic state. FEBS Letters 63: 51-55, 1976.
25. Foster, D. O. Calorigenic potential of rat brown adipose tissue and muscle re-evaluated from blood flow measurements with tracer microspheres. Physiologist 19: 194, 1976.
26. Foster, D. O. and M. L. Frydman. Brown adipose tissue: The dominant site of nonshivering thermogenesis in the rat. In: The Effectors of Thermogenesis. Girardier, L. and J. Seydoux (editors). Experientia, (In press).
27. Friedli, C., A. Chinet and L. Girardier. Comparative measurements of the calorigenic response of brown adipose tissue from control and cold-adapted rats. In: The Effectors of Thermogenesis. Girardier, L. and J. Seydoux (editors). Experientia, (In press).

28. Girardier, L. and J. Seydoux. Le contrôle de la thermogenêse du tissue adipeux brun. J. Physiol. (Paris) 63: 147-186, 1971.
29. Girardier, L., J. Seydoux and T. Clausen. Membrane potential of brown adipose tissue. A suggested mechanism for the regulation of thermogenesis. J. Gen. Physiol. 52: 925-940, 1968.
30. Greenway, D. C. and J. Himms-Hagen. Increased calcium uptake by muscle mitochondria of cold-acclimated rats. Am. J. Physiol.: Cell Physiol., (In press).
31. Grubb, B. and G. E. Folk Jr. Effect of cold acclimation on norepinephrine stimulated oxygen consumption in muscle. J. Comp. Physiol. B. 110: 217-226, 1976.
32. Guernsey, D. L. and E. D. Stevens. The cell membrane sodium pump as a mechanism for increasing thermogenesis during cold acclimation in rats. Science 196: 908-910, 1977.
33. Heaton, G. M. and D. G. Nicholls. Hamster brown-adipose-tissue mitochondria. The role of fatty acids in the control of the proton conductance of the inner membrane. Eur. J. Biochem. 67: 511-517, 1976.
34. Himms-Hagen, J. Role of the adrenal medulla in adaptation to cold. In: Adrenal Gland. Blaschko, H., G. Sayers and A. D. Smith (editors). Washington: American Physiological Society, 1975, pp. 637-665.
35. Himms-Hagen, J. Cellular thermogenesis. Ann. Rev. Physiol. 38: 315-351, 1976.
36. Himms-Hagen, J., W. Behrens, A. Hbous and D. Greenway. Altered mitochondria in skeletal muscle of cold-acclimated rats and the adaptation for nonshivering thermogenesis. In: Regulation of Depressed Metabolism and Thermogenesis. Janský, L. and X. J. Musacchia (editors). Springfield: C. C. Thomas, 1976, 243-260.
37. Himms-Hagen, J., J. Cerf, M. Desautels and G. Zaror-Behrens. Thermogenic mechanisms and their control. In: Effectors of Thermogenesis. Girardier, L. and J. Seydoux (editors). Experientia, (In press).

38. Hissa, R. and P. Hirsimäki. Calorigenic effects of noradrenaline and Inderal in the temperature-acclimated golden hamster. Comp. Gen. Pharmacol. 2: 217-224, 1971.
39. Hittelman, K. J., A. S. Fairhurst and R. E. Smith. Calcium accumulation as a parameter of energy metabolism in mitochondria of brown adipose tissue. Proc. Nat. Acad. Sci. 58: 697-702, 1967.
40. Hohorst, H. J. and J. Rafael. Oxydative Phosphorylierung durch Mitochondrien aus braunem Fettgewebe. Z. Physiol. Chem. 349: 268-270, 1968.
41. Horwitz, B. A. Role of active sodium transport in brown fat thermogenesis. Israel J. Med. Sci. 12: 1086-1089, 1976.
42. Horwitz, B. A. and M. Eaton. The effect of adrenergic agonists and cyclic AMP on the Na^+/K^+ ATPase activity of brown adipose tissue. Eur. J. Pharmacol. 34: 241-245, 1975.
43. Horwitz, B. A. and M. Eaton. Ouabain-sensitive liver and diaphragm respiration in cold-acclimated hamster. J. Appl. Physiol. 42: 150-153, 1977.
44. Janský, L. Nonshivering thermogenesis and its thermoregulatory significance. Biol. Rev. 48: 85-132, 1973.
45. Lindgren, G. and T. Barnard. Changes in interscapular brown adipose tissue of rat during perinatal and early postnatal development and after cold-acclimation. IV. Morphometric investigation of mitochondrial membrane alterations. Exp. Cell Res. 70: 81-90, 1972.
46. Muirhead, M. and J. Himms-Hagen. Changes in the amount and properties of adenyl cyclase in brown adipose tissue during acclimation of rats to cold. Can. J. Biochem. 49: 802-810, 1971.
47. Muirhead, M. and J. Himms-Hagen. Changes in the adenyl cyclase system of skeletal muscle of cold-acclimated rats. Can. J. Biochem. 52: 176-180, 1974.

48. Nicholls, D. G. Hamster brown-adipose-tissue mitochondria. The control of respiration and the proton electrochemical potential gradient by possible physiological effectors of the proton conductance of the inner membrane. Eur. J. Biochem. 49: 573-583, 1974.
49. Nicholls, D. G. Hamster brown-adipose-tissue mitochondria. Purine nucleotide control of the ion conductance of the inner membrane, the nature of the nucleotide binding site. Eur. J. Biochem. 62: 223-228, 1976.
50. Nicholls, D. G. The bioenergetics of brown adipose tissue mitochondria. FEBS Letters 61: 103-110, 1976.
51. Nicholls, D. G., B. Cannon, H. J. Grav and O. Lindberg. Energy dissipation in nonshivering thermogenesis. In: Dynamics of Energy-transducing Membranes. Ernster, L., R. W. Estabrook and E. C. Slater (editors). Amsterdam: Elsevier, 1974, pp. 529-537.
52. Nicholls, D. G. and G. M. Heaton. The identification of the component in the inner membrane of brown-adipose tissue mitochondria responsible for regulating energy dissipation. In: Effectors of Thermogenesis. Girardier, L. and J. Seydoux (editors). Experientia, (In press).
53. Pedersen, J. I. and H. J. Grav. Physiologically-induced loose coupling of brown-adipose-tissue mitochondria correlated to endogenous fatty acids and adenosine phosphates. Eur. J. Biochem. 25: 75-83, 1972.
54. Pettersson, B. CO_2-mediated control of fatty acid metabolism in isolated hamster brown-fat cells during norepinephrine stimulation. Eur. J. Biochem. 72: 235-240, 1977.
55. Pettersson, B. and I. Vallin. Norepinephrine-induced shift in levels of adenosine 3':5'-monophosphate and ATP parallel to increased respiratory rate and lipolysis in isolated hamster brown-fat cells. Eur. J. Biochem. 62: 383-390, 1976.

56. Portet, R., M. C. Laury, R. Bertin, C. Senault, M. T. Hluszko and L. Chevillard. Hormonal stimulation of substrate utilization in brown adipose tissue of cold-acclimated rats. Proc. Soc. Exp. Biol. Med. 147: 807-812, 1974.

57. Pospisilova, D. and Janský, L. Effect of various adaptational temperatures on oxidative capacity of the brown adipose tissue. Physiol. Bohemoslov. 25: 519-527, 1976.

58. Rafael, J. and H. W. Heldt. Binding of guanine nucleotides to the outer surface of the inner membrane of guinea pig brown fat mitochondria in correlation with the thermogenic activity of the tissue. FEBS. Letters 63: 304-308, 1976.

59. Ricquier, D. and J. C. Kader. Mitochondrial protein alteration in active brown fat: a sodium dodecyl sulfate-polyacrylamide gel electrophoretic study. Biochem. Biophys. Res. Comm. 73: 577-583, 1976.

60. Ricquier, D., G. Mory, and P. Hemon. Alterations of mitochondrial phospholipids in the rat brown adipose tissue after chronic treatment with cold or thyroxine. FEBS Letters 53: 342-346, 1975.

61. Rogus, E. M., L. C. Cheng and K. Zierler. β-Adrenergic effect on Na^+ + K^+ transport in rat skeletal muscle. Biochim. Biophys. Acta 464: 347-355, 1977.

62. Ruzicka, F. J. and H. Beinert. A new membrane iron-sulfur flavoprotein of the mitochondrial electron transfer system. Biochem. Biophys. Res. Comm. 66: 622-631, 1975.

63. Skala, J., T. Barnard and O. Lindberg. Changes in interscapular brown adipose tissue of the rat during perinatal and early postnatal development and after cold-acclimation - II. Mitochondrial changes. Comp. Biochem. Physiol. 33: 509-528, 1970.

64. Skala, J., E. Novak, P. Hahn and G. I. Drummond. Changes in interscapular brown adipose tissue of the rat during perinatal and early postnatal development and after cold-acclimation -

V. Adenyl cyclase, cyclic AMP, protein kinase, phosphorylase, phosphorylase kinase, and glycogen. Int. J. Biochem. 3: 229-242, 1972.

65. Skaane, O., E. N. Christiansen, J. I. Pedersen and H. J. Grav. Oxidative properties of brown adipose tissue mitochondria from rats, guinea pigs and hedgehogs. Comp. Biochem. Physiol. 42B: 91-107, 1972.

66. Slinde, E., J. I. Pedersen and T. Flatmark. Sedimentation coefficient and buoyant density of brown adipose tissue mitochondria from guinea pigs. Analyt. Biochem. 65: 581-585, 1975.

67. Smith, R. E. and B. A. Horwitz. Brown fat and thermogenesis. Physiol. Rev. 49: 330-425, 1969.

68. Smith, R. E., J. C. Roberts and K. J. Hittelman. Nonphosphorylating respiration of mitochondria from brown adipose tissue of rats. Science 154: 653-654, 1966.

69. Stevens, E. D. The evolution of endothermy. J. theoret. Biol. 38: 597-611, 1973.

70. Stevens, E. D. and M. Kido. Active sodium transport: A source of metabolic heat during cold adaptation in mammals. Comp. Biochem. Physiol. 47A: 395-397, 1974.

71. Vybiral, S. and L. Janský. Nonshivering thermogenesis in the golden hamster. Physiol. Bohemoslov. 23: 235-243, 1974.

NEUROHUMORAL REGULATION OF NONSHIVERING THERMOGENESIS IN MAMMALS

B. A. Horwitz
Department of Animal Physiology
University of California
Davis, California 95616 U.S.A.

ABSTRACT

This paper focuses on current postulated mechanisms for nonshivering thermogenesis and the neurohumoral factors regulating these mechanisms in mammals exposed to short-term as well as to prolonged periods of cold. During short-term exposures, the major neurohumoral regulatory agent appears to be norepinephrine which interacts at the target cell membrane such that an intracellular signal is generated appropriate to the stimulation of cellular pathways leading to increased substrate oxidation (and therefore increased heat production). In brown fat and muscle, two major nonshivering thermogenic tissues, available data suggest several cellular processes (e.g. altered mitochondrial coupling, enhanced ion transport across the plasma membrane, and enhanced ion transport across the mitochondrial membrane) as potential contributors to the norepinephrine-induced thermogenesis. Evidence relevant to these processes is considered along with experimental data dealing with the roles played by catecholamines and thyroid hormones in the development and maintenance of the trophic changes induced by exposure to extended periods of cold.

ISBN 0-12-734550-7

I. INTRODUCTION

Exposure of mammals to ambient temperatures below the thermoneutral zone evokes compensatory mechanisms that include an enhanced rate of conversion of chemical energy to heat. Although some of this heat generation is attributable to shivering, in many mammals a considerable portion derives from metabolic pathways independent of shivering. The relative involvement of these two modes of heat production depends on the species, age of the animal, and on the environmental conditions to which it has been previously exposed. By the same token, the quantitative contributions of various organ systems to cold-induced (i.e., regulatory)

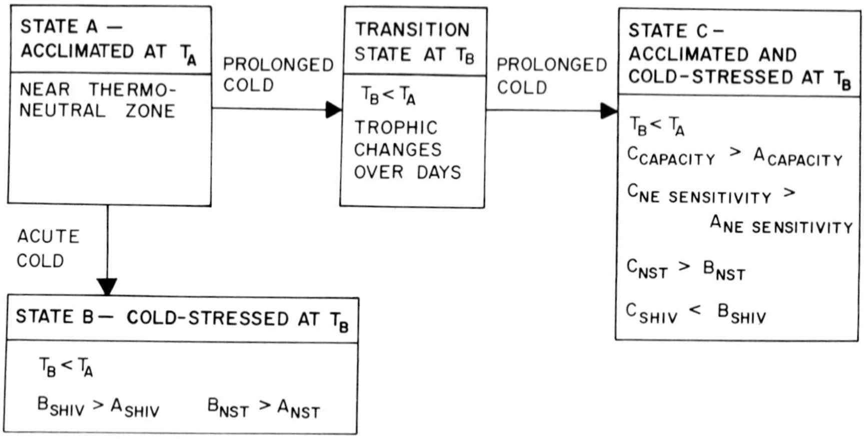

Fig. 1. *Effect of cold exposure on the thermogenic responses of mammals. Under conditions of acute or short-term (minutes to hours) cold exposure (State B), the animal may respond via shivering (B_{SHIV}) as well as nonshivering (B_{NST}) modes of heat production. As cold exposure at the ambient temperature T_B is prolonged (days), trophic changes are induced, and a new steady state is reached (State C). This state of acclimation is characterized by increased NE sensitivity and increased capacity for NST, with shivering thermogenesis being replaced by nonshivering heat production.*

nonshivering thermogenesis (NST) also vary with the types and history of the animal, although it is currently believed that most of this regulatory NST reflects the heat generated by only 4 organ systems -- namely, skeletal muscle, brown fat, liver and gut (46).

In considering the factors that regulate this cold-induced NST, it is useful to distinguish between those that are associated with the rapidly-occurring responses initiated when the animal is acutely exposed to cold (State B, Fig. 1) from those that are associated with the changes induced by prolonged cold exposure (State C, Fig. 1). The NST response elicited initially upon cold exposure (e.g., State B) appears to be mediated by the sympathetic nervous system which rapidly activates appropriate metabolic pathways in effector organs. When exposure to cold is prolonged over hours and days (Transition State and State C, Fig. 1), the target tissues undergo trophic changes that result in an increased capacity of the animal for NST and an increased sensitivity to the sympathetic neurotransmitter. Thus, the cold-acclimated animal (State C) has achieved a new steady state in which the amount of heat being generated by NST at the temperature of acclimation has increased (relative to that observed upon initial cold exposure) as has the magnitude of its calorigenic response to low as well as to high concentrations of administered catecholamines. In the ensuing discussion, the neurohumoral agents implicated in the changes evoked by acute as well as prolonged cold exposure will be considered. For the most part, this discussion will focus on the responses of laboratory-maintained rodents (rats, mice, hamsters) from which has derived most of the data related to the regulation of NST.

II. REGULATION OF THE NST CHANGES INDUCED BY PROLONGED COLD EXPOSURE

The neurohumoral factors responsible for the adaptive changes in NST [e.g., the increased NST capacity, norepinephrine (NE) sensitivity and NST at the temperature of acclimation] are not completely defined although there currently is evidence implicating the involvement of both the catecholamines and the thyroid hormones (see below).

Thyroid hormones: The notion that thyroxine (T_4) and/or triiodothyronine (T_3) play an important role in the metabolic changes induced by cold acclimation has waxed and waned in popularity over the last two decades, a major concern being the question of the level of thyroid activity in the cold-exposed animal (cf. 8 for review). Although recent use of radioimmunoassay techniques has revealed elevated plasma levels of T_3 in the cold-acclimated (CA) rat (62) these rats do not appear to be hyperthyroid (30) -- that is, their resting metabolic rate in the thermoneutral zone (RMR) is not greatly elevated (cf. 30) and thyroidectomized rats treated with small amounts of T_4 survive indefinitely in the cold (65). On the other hand, the fact that untreated thyroidectomized rats die after only a few days in the cold demonstrates that some level of thyroid hormone is essential for the acclimation process to occur. Moreover, thyroid hormones also appear to be necessary for the maintenance of the acclimation state as indicated by the observation that rats thyroidectomized after cold acclimation: (i) will not survive indefinitely in the cold [dying within 3-5 weeks (65)]; and (ii) exhibit a diminished NST response to cold, the magnitude of the decrease being correlated with time after thyroidectomy (44).

Elucidation of the precise nature of the role of thyroid hormones in cold acclimation has been difficult partly because the trophic responses induced by cold seem to reflect the interaction of several neurohumoral agents (cf. 68) and thus treatment of animals with any one factor (e.g., the thyroid hormones) does not result in changes identical to those seen in the CA animal. However, even if thyroid hormones play only a permissive role in the acclimation process (27, 45), it is necessary to consider the mechanisms underlying their action on cellular thermogenesis since these mechanisms will determine the baseline upon which will be built the responses induced by cold.

Current views on the events responsible for the calorigenic effects of the thyroid hormones are based on the concept that electron-transport-linked oxygen consumption is regulated by the energy status[1] of the cell -- i.e., when the energy requirements of the cell are low, high energy elements (e.g., ATP) able to support biological work will be available, and substrate oxidation will proceed slowly. In contrast, respiration rates will be enhanced under conditions of increased energy demand and accompanying decreased availability of such high energy elements. Thus, substances such as the thyroid hormones or catecholamines may stimulate rates of oxygen consumption by: (i) increasing the energy demands of the target cell; or (ii) releasing cellular respiration from its dependency on energy status via an uncoupling of mitochondrial oxygen consumption from the synthesis of high energy elements such as ATP. In situation (i), respiration remains dependent on the phosphorylation state ratio, while in situation (ii), oxygen consumption will be controlled by substrate availability.

The uncoupling of oxygen consumption from oxidative phosphorylation was one of the early biochemical theories proposed to explain the calorigenic effects of thyroid hormones. It evolved in the 1950's from observations such as: (i) liver mitochondria isolated from rats treated with thyroxine synthesized less ATP per mole of oxygen consumed (i.e., had low P/O ratios) than did mitochondria from untreated rats; (ii) isolated mitochondria treated with thyroid hormones *in vitro* could be uncoupled; and (iii) thyroid hormones, like the uncoupling agent dinitrophenol, stimulated mitochondrial ATPase activity (see 72 for historical discussion). Among the criticisms of these studies was the fact that they involved the administration of large doses of hormone. Subsequent experiments in which rats were treated with low (physiological) levels of T_4 resulted in a calorigenic effect in the intact animal, but the isolated mitochondria, when well washed, showed normal respiratory control and P/O ratios (13). These latter observations, along with the demonstration of thyroid hormone-induced protein synthesis that corresponded temporally with the increased oxygen consumption led to the conclusion that the calorigenic effect of the hormones was not due primarily to thyroid-induced uncoupling, but, rather may have involved protein synthesis (cf. 72). This effect on protein synthesis has since been well documented and may account for the thyroid-induced stimulation of the activities of several mitochondrial enzymes associated with substrate turnover and respiration [e.g., α-glycerol phosphate dehydrogenase, succinoxidase, cytochrome c, cytochrome oxidase (72)]. Thus, one important aspect of thyroid hormone function appears to be the modulation of the oxidative (thermogenic) capacity of the mitochondria via regulation of the activities and/or levels of respiratory components.

Thyroid hormones also affect a major energy-requiring process within the cell -- namely, the plasma membrane mechanism for the active transport of Na^+ and K^+. That the activity of this system varies with the thyroid status of the rat has been demonstrated by Edelman and his co-workers in an extensive series of experiments that began with their observations that: (i) the increased respiration of tissues (liver, kidney, diaphragm) from rats treated with T_3 was sensitive to inhibition of the Na^+/K^+ membrane pump, the degree of inhibition varying with the tissue; (ii) the Na^+/K^+ ATPase activity of membrane preparations from these tissues was also enhanced in T_3-treated rats; (iii) the temporal changes in the ATPase activity were similar to changes in oxygen consumption; and (iv) the intracellular levels of Na^+ as well as the ratio of $[Na^+]_i/[K^+]_i$ fell in tissues (liver, diaphragm, cardiac muscle) from T_3-treated rats (cf. 11). These findings led to the proposal that thyroid hormones exert at least a portion of their calorigenic effect by enhancing the energy utilized for the active transport of Na^+ and K^+ (11).

This conclusion has been confirmed in recent *in vitro* experiments using monolayer cultures of nondividing hepatocytes. Within 24 hr after addition of physiological levels of T_3 to these cultures, parallel increases in the rates of oxygen consumption and in the Na^+/K^+ ATPase activity were detectable (12).

Experiments designed to examine the basis for the elevated pump activity have demonstrated that in kidney cortex and intestine from T_3-treated rats, there is a rise in the amount of ^3H-ouabain bound to plasma membrane preparations, an increased amount of phosphorylated intermdeiate associated with the Na^+/K^+ membrane pump and an increase in the V_{max} for ATP (53, 55, 56). Such results suggest that thyroid hormones induce the synthesis of more pump units (Fig. 2). Moreover, since there

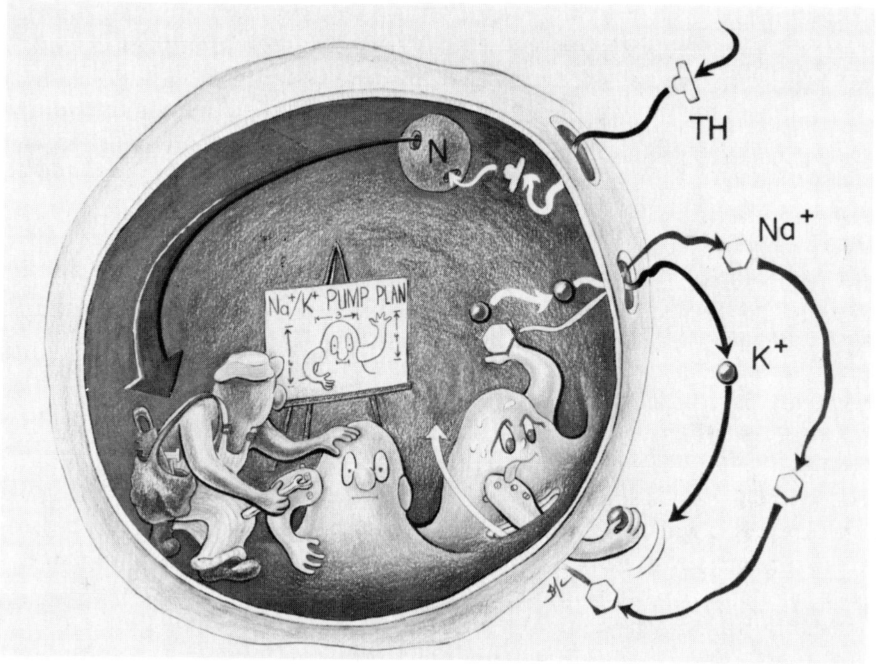

Fig. 2. *Proposed mechanism wherein thyroid hormones (TH) affect the Na^+/K^+ transport system of the plasma membrane by inducing the synthesis of more pump units (N = nucleus). This energy-requiring transport system has been suggested to serve as pacemaker for the resting metabolism of various target cells (11, 12, 30).*

occurred no changes in apparent Km or $K_{1/2}$ for the renal cortical Na^+/K^+ ATPase, the enzyme molecules induced by T_3 appear to be equivalent to those present before T_3 treatment (56).

Thus, current evidence indicates that a significant portion of the thyroid hormone-dependent resting oxygen consumption of several tissues in the rat can be attributed to the increased ATP turnover attending the transport of Na^+ and K^+ across the cell membrane. This increased energy turnover reflects the operation of newly synthesized pumps and, to some extent, an increase in membrane permeability to Na^+ and K^+ (1).

The extent to which these mechanisms contribute to the enhanced NST capacity of the CA animal is not yet clear. Several studies have reported that tissues removed from CA rats (24, 74), hamsters (40), and mice

(70) exhibit elevated rates of oxygen consumption, with a significant portion of this increase being sensitive to pump blockade. For example, the respiratory rates of CA hamster liver and diaphragm provided with exogenous glucose averaged about 24% greater than did tissue from nonacclimated hamster, and 60-70% of this increase was ouabain sensitive (40). In addition, the activity of the Na^+/K^+ ATPase of rat muscle appears to be enhanced after cold acclimation (31). In view of the results of Edelman and his colleagues, these responses of the CA tissues can be interpreted in support of there being some increase in thyroid activity during cold exposure. Such an interpretation agrees with the finding of increased plasma T_3 levels in the CA rat (62), and implies that there should also be some increase in the RMR of these animals. Whether such an elevation in resting metabolism does in fact occur after cold acclimation is still an open question. Although some investigators report increased rates (e.g., 49, 71), others have found no changes (e.g., 30, 51). However, in light of the demonstration that the manner in which the metabolic rate is expressed (i.e., as some function of body weight, surface area, or metabolically active tissue) can drastically alter the data interpretation (43), resolution of the effects of cold acclimation on resting metabolism awaits a technique that allows the comparison of oranges to apples. In lieu of this, it is probably safe to conclude that any increase in the RMR of the CA animal is relatively small. Thus, at present, the data are not inconsistent with there being some increase in thyroid activity during cold exposure with a concomitant increase in the resting metabolic rate. However, it is unlikely that thyroid hormones can, by themselves, account for the large increase in NST capacity of the CA animal.

By the same token, although the possibility that thyroid hormones contribute to the increased NE sensitivity of the CA animal is supported by studies such as those demonstrating enhanced calorigenic responsiveness to NE of T_4-treated rats (51), the magnitude of the thyroid hormone contribution or the mechanisms involved are still uncertain.

Catecholamines: Also thought to be involved in the trophic changes occuring during prolonged cold are the catecholamines. Excretion of these substances is enchanced by cold exposure (cf. 68), and daily treatment of rats with either NE or isoproterenol (a β-adrenergic agonist) results in an elevated sensitivity to NE (49, 50) similar to that seen in CA animals (i.e., a given concentration of NE elicits a larger increase in oxygen consumption from the catecholamine-treated rats than from the nontreated animals). Another characteristic of cold acclimation is enhanced cold resistance [reflected by the survival time and/or the rate of fall of the core temperature when animals are exposed to severe (e.g., -25°C) cold (51)]. This parameter is also altered by pretreatment of the rats with daily injections of catechols. However, even after 35 days of such treatment, the cold resistance of the injected rats is still below that of the CA rats although their NE sensitivity is the same (51). Interestingly, rats treated with catecholamines and T_4 exhibit greater cold resistance and NST sensitivities than do rats treated with either of the agents alone (51).

In addition to their effect on NE sensitivity and cold resistance, catecholamine-(as well as T_4-) treated rats have enlarged interscapular brown fat pads as, in fact, do the CA rats. Again, the response to the combined administration of T_4 and NE is greater than that observed with NE or T_4 alone (51). Whether these increases in brown fat mass reflect changes in composition comparable to those occurring in cold acclimation is unknown [e.g., the hypertrophy of brown fat induced by thyroid hormones is not similar to that evoked by cold in that the latter involves increased tissue water, protein and lipid, while thyroid hormone treatment results primarily in lipid accumulation (69)]. Also of interest would be a comparison of the mitochondria from brown fat and skeletal muscle of CA rats with those from rats treated with catecholamines with and without T_4 particularly with respect to those parameters (e.g., size, number, respiratory enzymes, membrane proteins) that have been reported to be altered in the CA rat (30).

Notwithstanding this need for a more extensive comparison of the effects of cold acclimation and pretreatment with catecholamines, the available data are consistent with the suggestion that the catecholamines play a central, although not solo, role in the development of the trophic changes accompanying extended periods of cold (50). However, the mechanisms underlying the induction of such changes are currently unknown.

Brown adipose tissue: That the development of the cold-induced NE sensitivity of the CA animal is related to the presence of brown adipose tissue has been suggested on the basis of: (i) the striking correlation between the amount of brown fat (relative to body weight) and the magnitude of the NE-evoked calorigenesis (8); and (ii) experiments that have examined the effects of surgical removal of a portion of the tissue (cf. 17, 30 for detailed review). The general finding from most of the extirpation studies is that removal of the interscapular brown fat from CA rats and mice is followed by a significant reduction (30-60%) of the calorigenic response to injected NE as measured during the first postoperative week. Since this reduction was thought to be too large to be accounted for by the small amount of tissue removed (usually less than 1% of the body weight), it was suggested that the brown fat, in some way, modulated the NE-sensitivity of the other thermogenic tissues in the animal. Two mechanisms were proposed to explain this modulating action: (i) that brown fat served an endocrine function, secreting into the circulation some factor that affected the calorigenic responses of other tissues (29); and (ii) that heat generated from the interscapular and cervical brown fat was applied to thermosensitive regions of the spinal cord, thereby providing a signal to the brain such that shivering thermogenesis was progressively inhibited and NST was concomitantly enhanced (38).

Although experimental data relating to both of these hypotheses are available (e.g., 29, 38), the issue has been clouded by the sometimes contradictory findings of various workers (for review cf. 17, 30). Moreover, it has recently been suggested that brown fat may contribute much more heat to the intact CA rat than had previously been estimated [as much as 60% of the total according to Foster (18)], and if this turns out to be correct, the decreased oxygen consumption following removal of the brown fat may actually reflect the metabolic rate of the extirpated tissue.

Other factors: The possible involvement of agents other than the thyroid hormones and catecholamines (e.g., ATCH, TSH, glucocorticoids, insulin, growth hormone, etc.) in the cold-induced changes in NST has been considered to varying degrees. However, despite some evidence that such factors may be important in the cold-exposed animal, there is as yet no definitive role assigned for any of them in terms of the specific changes observed in the NST responses of the CA animal (46, 68).

III. ACTIVATION OF NONSHIVERING THERMOGENESIS

Central neural control: The central mechanisms associated with initiation of NST upon cold exposure appear, in at least two mammalian species, to be distinct from those signalling the onset of shivering. That is, experiments with both guinea pigs (3) and rats (19) support the view that sensory inputs from cervical and cutaneous thermoreceptors converge on the neural thermocontroller to regulate the magnitude of shivering activity. The neural controller also receives encoded information from thermoreceptors in the hypothalamus and skin, and this information, after processing, appears to be channeled to effectors of NST. In other words, inputs received by the central neurocontroller from thermoreceptors located in various areas are processed to provide for the specific regulation of shivering and nonshivering thermogenesis. Such specificity allows the cold-exposed animal to first invoke NST to meet its thermal demands. Shivering, the less economical mode of heat production[2], then comes into play when NST is inadequate (3).

Target tissues: Sympathetic activation of NST may involve direct and/or indirect effects on the effector tissue (57). For those tissues where catecholamines are thought to act directly on the target cell to stimulate heat production (e.g. brown fat, skeletal muscle), an initial step in the process is interaction of the catecholamine at specific membrane receptors. This is followed by the generation of an intracellular signal appropriate for stimulation of mitochondrial oxygen consumption (and thus cellular heat production).

These events have received considerable attention in brown adipose tissue and to a much lesser degree in muscle, and will be discussed below. Mention will also be made of the possibility that adrenergic stimulation of tissue oxygen consumption may, in some cases, occur without a direct action of the catecholamine on the heat producing cell, such an indirect effect having been suggested for liver (57).

Regulation of brown fat thermogenesis -- nature of the extracellular signal and membrane receptor: The establishment of brown adipose tissue as an important site of heat production in newborn mammals, arousing hibernators and some warm- as well as cold-acclimated nonhibernators is based upon work from several laboratories, the majority of which followed the proposal by Smith (67) that thermogenesis was a major function of this tissue. Much of this documentation as well as a description of the tissue's anatomy, histology, embryonic development, and phylogenetic distribution has been extensively reviewed (e.g., 54, 69); and the ensuing discussion will focus on recent investigations relating to the cellular events underlying catecholamine regulation of brown fat NST.

As indicated above, cold-induced brown fat thermogenesis is initiated following interaction of the catecholamine (principally NE) at the plasma membrane. The adrenergic agent diffuses to the brown adipocyte after its release from sympathetic nerve terminals near the cell, and *in situ* experiments indicate that the tissue responds thermogenically in a graded manner to neural stimulation (15). Moreover, as is the case for several other adrenergically-activated systems, it is the temporal and spatial pattern of NE release that represents the extracellular signal regulating the cellular response.

That recognition of this signal at the brown adipocyte membrane occurs via β-adrenergic receptors has been suggested from experiments dealing with the nature of the membrane-neurotransmitter interaction. Actually, the involvement of a β-receptor (i.e., a membrane binding site associated with the adenyl cyclase-cAMP system) in the response of brown fat is a well-established concept built upon a variety of observations. For example, among the *in vivo* results in support of a β-pathway are those showing that: (i) injection of NE elicits an increase in brown fat oxygen consumption that is inhibited by propranolol, a β-blocker (26); (ii) the magnitude of the elevation of brown fat temperature evoked by nerve stimulation or NE injection is depressed by propranolol (42); and (iii) the enhancement of brown fat temperature induced by isoproterenol, a β-agonist, is inhibited by propranolol but not by the α-antagonist, phentolamine (15). Moreover, *in vitro* experiments have demonstrated that the oxygen consumption and lipolysis of isolated brown fat cells is stimulated by isoproterenol (25, 36), theophylline (35, 61) and cAMP (25, 35). Such observations as these provide the basis for the long-standing proposal that the β-adrenergic pathway plays a central role in brown fat thermogenesis.

More recent is the view that the brown adipocyte also possesses an α-adrenergic component that contributes to the cell's thermogenic response. This idea evolved from the finding that α- as well as β-blockers inhibited one of the initial events thought to be associated with physiological activation of the cells -- namely, the redistribution of ions across the plasma membrane as manifested by a membrane depolarization following electrical stimulation of the nerves to the tissue (42). Subsequent experiments *in vivo* (15) as well as *in vitro* (14) have shown that either the α-agonist, phenylephrine, or the β-agonist, isoproterenol is able to evoke membrane depolarization. Moreover, the depolarization elicited by phenylephrine is not significantly affected by propranolol but is inhibited by phentolamine (14) whereas that induced by isoproterenol is sensitive to propranolol but not to phentolamine (14). Thus, there appear to be distinct α- and β-receptors on the membrane of the brown adipocyte.

That these α-receptors are associated with stimulation of brown fat heat production has been suggested in view of the observation that phenylephrine enhances the oxygen consumption of isolated hamster brown adipocytes (36). The fact that this enhancement is completely inhibited by concentrations of phentolamine that have no depressant action on the respiratory response to isoproterenol (37) indicates that the α-stimulation is not occurring via interaction with membrane sites (i.e., receptors) to which β-agonists bind; i.e., the two types of receptors appear to be distinct. However the finding that the respiratory responses to submaximum concentrations of isoproterenol and phenylephrine are not additive (37) suggests that although the agonists may bind to separate membrane receptors, the intracellular thermogenic pathways activated via each receptor type have some events in common. Nonetheless, that the pathways are not identical is indicated by the observation that the β-agonist evokes significant glycerol release from isolated brown adipocytes while the α-agonist does not (25) a difference that may, in part, explain the greater thermogenic response of the cells to the β-agonist.

Regulation of brown fat thermogenesis -- nature of the intracellular events: Interaction of norepinephrine with the receptors at the adipocyte plasma membrane is followed by a transformation of the extracellular information to intracellular signals responsible for promoting the availability of substrate [i.e., free fatty acids (FFA)] and the rate at which this substrate is oxidized by the mitochondria.

With respect to substrate mobilization, the currently accepted view is that the catecholamines enhance the rate of lipolysis via activation of the adenyl cyclase-cAMP system. The synthesized cAMP modulates the activity of a protein kinase which in turn activates an enzyme that catalyzes the hydrolysis of the stored triglyceride to glycerol and free fatty acids. Experiments with α- and β-agonists (25) as well as antagonists (32) support the concept that the lipolytic response to NE is mediated primarily via the β-adrenergic pathway. [Although cGMP is also capable of stimulating glycerol release from brown adipocytes (25), the physiological significance of such an effect has not yet been demonstrated.]

In contrast to the general agreement on the pathway for substrate mobilization, some controversy still exists regarding the events regulating the rate of mitochondrial oxidation. One school of thought ascribes the increased oxygen consumption to an altered coupling between fatty acid oxidation and ATP synthesis via oxidative phosphorylation (e.g., 4-6, 54, 58-60), while a second point of view (e.g., 28, 35-37,

41) considers the occurrence of an increased rate of energy turnover to be a regulatory component of the mitochondrial respiratory rate (see footnote 1). As indicated below, experimental evidence exists in support of both viewpoints, although it is not yet clear to what degree the results from the *in vitro* systems studied can be extrapolated to the *in vivo* situation. Nevertheless, it should be noted that the studies relevant to each of the proposed mechanisms have served to clarify important cellular properties of the brown adipocyte and are thus of interest beyond their significance to *in vivo* NST. [For example, brown fat mitochondria appear to have unique permeability characteristics (see below) making them an interesting model system for studying proton conductance and conditions modifying this conductance. By the same token, the sensitivity of the electrical and transport properties of the plasma membrane to adrenergic agents provides a system for studying mechanisms relating to recognition of extracellular information and its recoding to appropriate intracellular signals.]

Promulgation of uncoupling as the primary mechanism underlying brown fat thermogenesis originated with the finding that mitochondria isolated from this tissue exhibited rapid rates of respiration, low P/O ratios and lack of respiratory control (see 16, 69 for review). Although these early experiments suggested that the mitochondria were not coupled and were, perhaps, even incapable of oxidative phosphorylation, subsequent studies demonstrated that in the presence of albumin and purine nucleotides (ATP, ADP, GDP or GTP) respiratory control could be re-established (16, 69). The mitochondria could also be recoupled by addition of carnitine and ATP, thus suggesting that the uncoupling of freshly isolated mitochondria resulted from FFA bound to the mitochondria and could be reversed upon mitochondrial oxidation of these FFA. These observations led to the proposal that NE exerted its calorigenic effect by promoting lipolysis and increasing the intracellular concentrations of FFA which then served as an uncoupling agent as well as substrate (16). However the continuing failure to detect significant changes in intracellular FFA levels during NE-stimulated brown fat respiration (2, 76) is onc factor that has made questionable the likelihood that FFA regulate coupling characteristics under physiological conditions.

Recently, the uncoupling hypothesis has been reformulated by Nicholls (58-60) in terms of Mitchell's chemiosmotic theory of oxidative phosphorylation. Mitchell has proposed that during mitochondrial substrate oxidation (represented in figure 3 by the reaction NADH + $H^+ \rightarrow H_2O$), protons flow out of the matrix across the inner mitochondrial membrane (pathway a, Fig. 3) and in coupled mitochondria, re-enter the matrix via a route (path c, Fig. 3) linked to the generation of ATP. If the rate of ATP turnover is low, the rate of proton flow along path c will also be low (resulting in decreased ATP synthesis) and this in turn will inhibit flow along pathway a. [In other words, the rate of oxygen consumption will be dependent on the rate of ATP synthesis.]

Nicholls has postulated the existence of a second pathway for proton return (or OH^- exit, the electrical equivalent of H^+ entry) to the matrix. This additional route (path b, Fig. 3), visualized as a channel that would allow re-entry of protons independent of ATP synthesis and would thereby provide a means for increased substrate oxidation without an equivalent enhancement of ATP synthesis, derives from observations made on isolated brown fat mitochondria subjected to various incubation conditions. Among such observations is the finding that uncoupled brown fat mitochondria exhibit high permeabilities to a number of ions (including protons) and that restoration of respiratory control (recoupling) by addition of purine nucleotides and/or albumin is accompanied by a decrease in the magnitude of the ion conductance (a measure of ion permeability). Examination of the characteristics of this proton conductance pathway in isolated brown fat mitochondria has indicated that: (i) although both fatty acid removal and nucleotide addition are needed for "optimal coupling", the nucleotide effect differs from that of albumin and appears to involve interaction at different membrane sites (58); (ii) the affinity of binding of the purine nucleotides to the mitochondria is pH dependent as is their effect on the proton conductance (less binding and greater conductance being seen at alkaline pH) (58); (iii) the site to which the

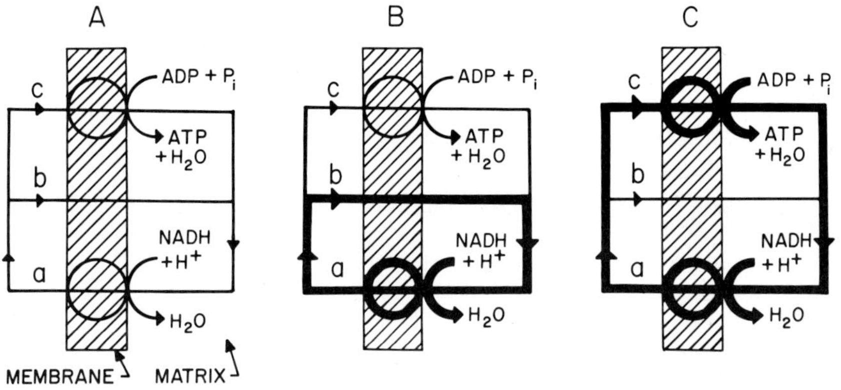

Fig. 3. *Diagrammatic representation of mitochondrial respiration (NADH + H^+ → H_2O) and oxidative phosphorylation (ADP + P_i → ATP + H_2O) in terms of the proton circuits proposed to exist in brown fat. In all 3 modes shown (A-C), circuit "a" represents the path for the flow of protons out of the matrix during substrate oxidation; "b", the leakage pathway allowing protons to re-enter the matrix without driving ATP synthesis; and "c", the path for return via the proton-translocating-ATP synthetase system. Mode A represents the resting brown adipocyte where the rates of oxygen consumption and ATP synthesis are slow and membrane conductance is high (leak is minimal). Mode B depicts the condition of loose coupling (see text) where the respiration rate is enhanced and most (but not all) proton return is independent of ATP synthesis. In mode C, the rate of oxygen consumption is elevated, accompanied by an increased rate of oxidative phosphorylation* [*modified after Nicholls (58-60)*].

nucleotide binds is not sensitive to atractyloside (an inhibitor of the adenine nucleotide translocase responsible for the exchange of ADP and ATP across the mitochondrial membrane), thus suggesting that the binding site is distinct from the normal pathway for adenine nucleotide translocation across the inner membrane (58, 59); and (iv) this atractyloside-insensitive binding site is on the outer surface of the inner membrane and has been reported to be a protein with an apparent molecular weight of 32,000 daltons (60). Observations as these have led to the proposal that brown fat mitochondria possess a proton re-entry (or hydroxyl exit) pathway through which flow can be regulated by low concentrations of purine nucleotides and which represents the mechanism whereby the calorigenic effects of NE are mediated (58-60, Fig. 3B).

However, additional studies are required to explain how such a mechanism might operate in the intact cell. For example, the plentiful supply of endogenous intracellular purine nucleotides and the failure to detect any changes in their levels correlated with the thermogenic activity of the brown adipocyte places in question the regulatory role of such agents. In recognition of this and the finding that the nucleotide binding is pH sensitive and is competitively inhibited by long chain fatty acyl CoA molecules, factors such as pH (10) and/or fatty acyl CoA molecules (5) have been nominated as candidates for the role of modulator of purine nucleotide binding in the intact cell. Such possibilities, however, remain to be experimentally examined for *in vivo* conditions. A further difficulty with the hypothesis that a loosening of mitochondrial coupling represents the primary thermogenic event in the activated brown adipocyte is reconciliation of such an event with the observed anticalorigenic effects resulting from blockade of the Na^+/K^+ membrane pump (see below). Thus, at this point in time, the extent to which the findings from isolated mitochondrial systems can be applied to the intact brown adipocyte and the physiological activation of brown fat thermogenesis are uncertain.

An alternative explanation for the thermogenic action of NE is the postulation that activation of the brown adipocyte is accompanied by an increased intracellular energy demand (and therefore energy turnover) which in turn, provides the driving force for the increased rate of substrate oxidation (represented by the proton flows diagrammed in Fig. 3C). This idea has prompted the examination of several energy-requiring processes over the last 10 years. Some of these reactions have been evaluated and deemed quantitatively insignificant in terms of their contribution to the overall thermogenic response [e.g., the futile cycles involving fatty acid oxidation/resynthesis and triglyceride hydrolysis/re-esterification (69)]. For others, such as the mitochondrial uptake of Ca^{++} (33) or K^+ (61), the data are insufficient to allow assessment of their involvement in the response of the intact cell. However, available evidence does point to at least one energy-requiring process -- namely, the Na^+/K^+ transport system across the plasma membrane -- as playing a significant role in brown fat thermogenesis.

The involvement of this membrane pump in the thermogenic response was suggested initially by Girardier *et al.* (21) following their demonstration that NE elicited membrane depolarization in cells of the interscapular fat pad. Their *in vitro* observations have since been confirmed in a number of laboratories (14, 48, 75) and similar changes have also been seen *in vivo* subsequent to electrical stimulation of the nerves innervating the interscapular fat pad (42) or to intravenous injection of NE (15, 42), isoproterenol (15) or phenylephrine (15).

This redistribution of ions across the adipocyte membrane occurs prior to any increase in brown fat temperature (42) or oxygen consumption (66) and is accompanied by an altered membrane resistance (34), decreased intracellular K^+ levels (20) and increased intracellular Na^+ concentrations (20), changes consistent with the occurrence of enhanced membrane permeability to ions such as Na^+ and K^+. An increased permeability, in turn, implies a compensatory enhancement of the activity of the Na^+/K^+ membrane pump to provide for the return of the ion distribution to prestimulus conditions, a restoration that is indeed observed *in vivo* (15, 42) as well as *in vitro* (20, 66). Thus, associated with the physiological activation of brown fat thermogenesis is an enhancement of the activity of the Na^+/K^+ membrane pump. [This increased pump activity may be mediated in part by the altered ion fluxes resulting from the greater membrane permeability as well as from an action mediated by cAMP, the latter agent being able to stimulate the Na^+/K^+ ATPase activity of brown fat membrane preparations (39)].

That the enhanced activity of the Na^+/K^+ pump does contribute significantly to the thermogenic response of the brown adipocyte has been inferred from experiments evaluating the effects of pump inhibition. Experiments have shown that addition of ouabain to adipocytes isolated from hamsters or rats decreases the NE-induced respiratory response, the magnitude of this depression averaging about 60% for the hamster cells (35). Comparable inhibition is observed when pump blockade is achieved by replacing extracellular NaCl with choline chloride (resulting in removal of one of the substrates for the transport system) rather than by ouabain addition (35). Similarly the oxygen consumption elicited by either isoproterenol or by phenylephrine is diminished approximately 70% in the presence of ouabain (37). In view of the finding that the cAMP-induced oxygen consumption is also ouabain sensitive (35), the observed reduction of the NE response does not appear to be attributable to inhibition of cAMP synthesis. Furthermore, since ouabain also decreased brown adipocyte respiration in the presence of palmitic acid, the inhibitory effect of the cardiac glycoside is not easily explained by an action on lipolysis (35). Rather, it would appear that the antithermogenic effects of ouabain result primarily from its influence on the Na^+/K^+ transport system.

The possibility that the respiratory inhibition evoked by ouabain [which in isolated rat cells has been shown to be maximal after 4 min of preincubation (28)] is ancillary to changes in intracellular Na^+ and/or decreased intracellular K^+ has been judged unlikely in view of several observations regarding the system under study (41). For example, the anticalorigenic effect of pump inhibition occurs regardless of whether intracellular Na^+ levels increase or decrease. That is, pump blockade via ouabain addition (where intracellular Na^+ levels increase) and that achieved by replacing extracellular NaCl with choline chloride (thereby resulting in decreased intracellular Na^+ concentrations) are equally effective in diminishing the adrenergic-induced respiration (see above). On the other hand, both methods of pump inhibition lead to lower values of intracellular K^+, prompting the suggestion that the respiratory inhibition reflects this K^+ loss (30).

However, the fact that intracellular K^+ is also decreased during NE stimulation of brown fat oxygen consumption (20) indicates that respiratory inhibition does not necessarily follow a drop in intracellular K^+.

These considerations have led to the view that the respiratory inhibition observed upon pump blockade does not result from changing Na^+ or decreasing K^+ levels, but rather reflects the fact that the reduced ATP turnover represents a decrease in the generation of an intracellular signal able to stimulate mitochondrial substrate oxidation. [Consistent with this latter interpretation is the recent finding that the isoproterenol-induced thermogenesis of the adipocytes is atractyloside sensitive, although it is not yet clear to what degree this sensitivity reflects the inhibition of the adenine nucleotide translocase.]

Despite the evidence for the importance of the Na^+/K^+ transport system in the thermogenic response of the brown adipocyte (e.g., the 60-70% ouabain sensitivity of the adrenergic-induced respiration), evaluation of the basis for the pump involvement requires further study. That is, it has been argued that brown fat mitochondria do not have a sufficiently active ATP synthetase to generate the amount of ATP needed to support the magnitude of the observed NE-evoked respiration (6); and although this conclusion may be challenged[3], quantitation of the work done by the pump during adrenergic activation is needed to determine if the role of the pump is indeed related to its ATP utilization or to its generation of an intracellular signal unrelated to the adenylate charge (see below).

Thus, at present, there is a lack of agreement as to the physiological mechanisms responsible for the heat generation in the activated brown adipocyte. Studies relevant to two possible explanations [increased flow over pathway b or c (Fig. 4)] have implicated events at the plasma membrane as well as at the mitochondrial membrane. In the past, these studies have been interpreted in terms of either loose coupling or increased energy turnover, but future work may bring together these observations with a unifying mechanism. Such a mechanism, however, must take into account the dependence of the thermogenic response on the operation of the Na^+/K^+ membrane pump. One could visualize the

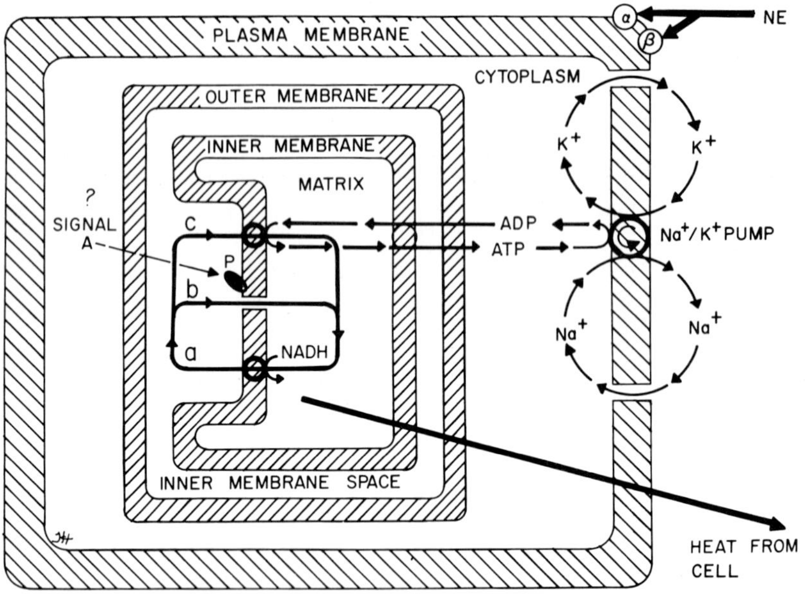

Fig. 4. *Relationship between the putative mitochondrial proton circuits and other cellular components involved in brown fat thermogenesis. Within the adipocyte mitochondrion, pathway "a" represents the substrate-oxidation-driven proton flow from the matrix, with the concurrent establishment of a proton gradient across the inner membrane. The flow of protons back down their electrochemical gradient over pathway "b" is postulated (58-60) to be independent of ATP synthesis and to be regulated by signal A acting on protein P located on the outer surface of the inner membrane[in the intact cell, the nature of signal A is unknown (see text)]. In contrast, during the movement of protons down their electrochemical gradient over pathway "c", energy is conserved in the synthesis of ATP, and flow over this path is regulated by energy-requiring cellular processes such as the Na^+/K^+ pump at the plasma membrane. These increased proton flows are accompanied by enhanced conversion of chemical energy to heat as depicted by the arrow for heat flow from the cell. The extracellular signal that results in enhancement of the proton flows is norepinephrine (NE) acting on α- and β-plasma membrane receptors.*

coexistence of increased Na^+/K^+ transport activity and altered coupling if: (i) the degree of loose coupling were small enough to allow some respiratory control (37, 41); (ii) the enhanced pump activity were supported by ATP from substrate (rather than oxidative) phosphorylation and were necessary to prevent TCA cycle enzymes (and therefore substrate oxidation) from being inhibited by increased adenylate charge (30); or (iii) the pump generated an intracellular signal that resulted in the loosening of mitochondrial coupling [e.g., cytosol alkalinization (10)]. Although these possibilities are speculative they emphasize the fact that elucidation of the intracellular mechanisms underlying the norepinephrine regulation of brown fat thermogenesis requires additional experimentation.

Regulation of muscle thermogenesis: Unlike brown fat, skeletal muscle does not possess detectable sympathetic innervation. Thus, the source of catecholamine available to activate the tissue during cold exposure is in question. Among the possibilities is the circulating catecholamine secreted by the adrenal medulla and/or the "spill-over" of neurotransmitter released from the innervation to nearby smooth muscle cells of the vasculature.

However, regardless of the source of the catecholamines, if skeletal muscle is a significant contributor to the NST response of the cold-exposed animal, its rate of heat production should be sensitive to adrenergic agents. Such a sensitivity, although not apparent in isolated muscle preparations *in vitro* has been demonstrated in various *in situ* hind limb preparations. For example, *in vivo* administration of NE results in increased oxygen consumption of the isolated rat gracilis anticus (57) as well as the partly isolated leg muscles of the cold-acclimated rat (22, 47, 57), the noncold-acclimated rat (23), and the dog (64). Although vascular changes sometimes accompany these adrenergic-induced increases in oxygen uptake, the results from several studies indicate that respiratory stimulation can occur independently of vascular effects (22, 23, 47, 57, 64). Such findings are consistent with the view that catecholamines stimulate muscle NST via a direct action on the target cells (57).

Neither the cellular pathway nor the nature of the receptor whereby NE exerts its calorigenic effects on skeletal muscle are known. In recent experiments, phenylephrine was found to be as effective as NE and epinephrine in stimulating the oxygen consumption of isolated perfused rat muscle preparations (23) while isoproterenol elicited no such calorigenic response. On the other hand, both the α-blocker phentolamine, and the β-blocker, propranolol, (at high concentrations) were able to completely antagonize the NE-induced response (23). These results were interpreted as indicating that in contrast to brown fat, the predominant component of the increased oxygen consumption of skeletal muscle may be mediated by an α-adrenergic pathway. In view of the evidence that the cold-induced thermogenic response of the intact animal is primarily a β-adrenergic phenomenon and can be evoked by administration of isoproterenol (46), the significance of the α-induced response in skeletal muscle is not yet clear. It may indicate that the quantitative contribution of skeletal muscle thermogenesis to that of the intact rat is relatively small as has been recently suggested by Foster (18).

Catecholamines also appear able to alter skeletal muscle membrane properties, and at least in some preparations, they elicit depolarization (30, 73). For example, in the anesthetized hamster, intravenous injection of NE is followed by a small (7-9%) but significant decrease in potential across the sarcolemma of the sartorious and gracilis anticus muscles (73). This depolarization is generally accompanied by a decreased membrane resistance (Teskey, unpublished) implying a concomitant increase in membrane permeability. Thus, in these muscles, as in brown fat, NE does evoke changes in ion distribution. However, the magnitude of the membrane changes may be smaller in muscle than in brown fat where, depending on the concentration of NE, the potential may fall 40% or more (15, 20, 42, 75).

In addition to these changes in ion distribution, catecholamines also enhance the Na^+/K^+ ATPase activity of a sarcolemma fraction from rat skeletal muscle (9). This enhancement is not mediated by cAMP or phenylephrine, but rather, may reflect a structural change in the membrane resulting from the binding of catechols and their orthoquinones (9). In contrast, the reported catecholamine-induced enhancement of net Na^+ efflux and K^+ influx in rat skeletal muscle does appear to be a specific β-adrenergic response mediated through cAMP (63).

The relationship of these demonstrated membrane changes to muscle NST has not yet been established, and experiments are required to determine if these alterations serve as a trigger for initiating an appropriate intracellular signal leading to the increased oxygen consumption and/or as a significant site of heat production (e.g., via enhanced operation of a membrane pump). Also requiring further evaluation is the relationship between the known metabolic effects of the catecholamines (particularly with respect to the mobilization of substrate) and the calorigenic response of the tissue.

Thus, at present, experimental studies are beginning to focus on a description of the cellular events associated with the transduction of the extracellular information carried presumably by the catecholamines to the intracellular signal effecting NST in the skeletal muscle.

Regulation of NST in other tissues: Although other tissues (liver, brain, gut) have been implicated in the NST response of cold-exposed rodents, their contribution to the overall effect has recently been questioned (18). Moreover, there is little evidence suggesting any direct action of catecholamines on the oxygen consumption of these tissues. In fact, perfusion of the rat liver with NE failed to elicit an enhanced rate of oxygen uptake although glycogenolysis was stimulated (57). On the other hand, liver oxygen consumption does appear, under certain circumstances, to depend on the concentration of substrates in the blood and on the oxygen supply; and catecholamine-induced alterations of such parameters during cold exposure might stimulate tissue heat production (57).

IV. CONCLUSION

This review has been concerned with regulatory nonshivering thermogenic mechanisms in mammals and the nature of the neural and humoral agents activating these mechanisms. Available data for the two major thermogenic tissues, brown fat and muscle, point to several cellular pathways (in particular, altered mitochondrial coupling, increased ion transport across the plasma membrane, and increased ion transport across the mitochondrial membrane) as potential contributors to the cold-induced thermogenesis, with the extent that these events underlie the physiological response being in various stages of evaluation. By the same token, although the catecholamines clearly play a central role in the rapid thermal regulation evoked upon cold exposure, their participation and the mechanisms mediating their action during the acclimation process are still being examined, as is the function of the thyroid hormones in the development and maintenance of the trophic changes induced by extended periods of cold. Thus, among the areas meriting further attention are included: (i) analysis of the interaction of catecholamines with other agents (e.g., thyroid hormones, ACTH, glucocorticoids, TSH, etc.) in terms of possible mediation of the responses characteristic of cold acclimation; (ii) evaluation of the quantitative contributions of various cellular events to thermogenesis under *in vivo* conditions; and (iii) elucidation of the intracellular pathways involved in signalling the observed responses to cold exposure.

ACKNOWLEDGEMENTS

The author is grateful to Dr. John M. Horowitz for his skillful art work and his perceptive comments during the preparation of this manuscript. Portions of the research reported herein were supported by grants from the National Space and Aeronautics Administration (NGR-05-004-099), the National Science Foundation (GB-30594), the National Institutes of Health (AM-16716), and the University of California at Davis (D816).

FOOTNOTES

1. One way to express the energy status of the cell is by the phosphorylation potential, defined as the ratio of: $[ATP]/[ADP][P_i]$. Alternatively, the value of the adenylate (or energy) charge of the system can be used as an indication of its energy status (52), this charge being equal to the value of: $(0.5)([ADP] + 2[ATP])/([AMP] + [ADP] + [ATP])$. Both ratios reflect the concept that the concentration of ATP relative to that of ADP is a measure of the potential of the cell for supporting ATP-requiring processes, and in tightly coupled mitochondria regulates the rate of substrate oxidation (i.e., higher ratios promote lower rates of substrate oxidation and lower ratios promote higher oxidation rates).

2. Shivering is less economical (only 10-15% efficient) than NST in terms of the heat transferred to the animal per calorie expended (7). This probably reflects the fact that the major effectors of NST are further from the external environment than are the skeletal muscles associated with shivering and thus less heat is lost from the surface of the organism during NST.

3. Estimates for the amount of ATP synthetase activity of brown fat mitchondria vary depending on the nature of the preparation (and the degree of biochemical manipulation). For example, recoupled isolated brown fat mitochondria exhibit a DNP-stimulated ATPase that is about 40% lower than that of liver mitochondria (4); however, whether this difference in activity is sufficient to limit the *in vivo* rate of oxidative phosphorylation has not been demonstrated. Recently, crude sonicates of frozen brown fat mitochondria were reported to contain only 18% (total activity) of that seen in similar preparations from beef heart and in more purified preparations, only 10% of the activity was recovered (6).

REFERENCES

1. Asano, Y. Increased cell membrane permeability to Na^+ and K^+ induced by thyroid hormone in rat skeletal muscle. In: Effectors of Thermogenesis. Girardier, L. and J. Seydoux (editors). Basel: Experientia (In press).
2. Bieber, L. L., B. Pettersson, and O. Lindberg. Studies on norepinephrine-induced efflux of free fatty acid from hamster brown-adipose-tissue cells. Eur. J. Biochem. 58: 375-381, 1975.
3. Brück, K. and W. Wünnenberg. "Meshed" control of two efferent systems: nonshivering and shivering thermogenesis. In: Physiological and Behavioral Temperature Regulation. Hardy, J.D., A. P. Gagge, and J. A. J. Stolwijk (editors). Springfield, Ill: Thomas, 1970, pp. 562-580.
4. Bulychev, A., R. Kramar, Z. Drahota, and O. Lindberg. Role of a specific endogenous fatty acid fraction in the coupling-uncoupling mechanism of oxidative phosphorylation of brown adipose tissue. Expt. Cell Res. 72: 169-187, 1972.
5. Cannon, B., U. Sundin, and L. Romert. Palmitoyl coenzyme A: a possible physiological regulator of nucleotide binding to brown adipose tissue mitochondria. FEBS Letters 74: 43-46, 1977.
6. Cannon, B. and G. Vogel. The mitochondrial ATPase of brown adipose tissue. Purification and comparison with the mitochondrial ATPase from beef heart. FEBS Letters 76: 284-289, 1977.
7. Carlson, L. D. and A. C. L. Hsieh. Control of Energy Exchange. New York: Macmillan, 1970, pp. 81-82.
8. Chaffee, R. R. J. and J. C. Roberts. Temperature acclimation in birds and mammals. Ann. Rev. Physiol. 33: 155-202, 1971.
9. Cheng, L. C., E. M. Rogus, and K. Zierler. Catechol, a structural requirement for $(Na^+ + K^+)$-ATPase stimulation in rat skeletal muscle membrane. Biochim. Biophys. Acta 464: 338-346, 1977.
10. Chinet, A., C. Friedli, J. Seydoux, and L. Girardier. Does cytoplasmic alkalinization trigger

mitochondrial energy dissipation in the brown adipocyte? In: Effectors of Thermogenesis. Girardier, L. and J. Seydoux (editors). Basel: Experientia (In press).

11. Edelman, I. S. and F. Ismail-Beigi. Thyroid thermogenesis and active sodium transport. Recent Prog. Hormone Res. 30: 235-257, 1974.

12. Edelman, I. S. Control mechanisms of energy dissipation in the cytomembrane and cytosol. In: Effectors of Thermogenesis. Girardier, L. and J. Seydoux (editors). Basel: Experientia (In press).

13. Fairhurst, A. S., J. M. Maher, and R. E. Smith. Effect of physiological levels of thyroxine on phosphorylation in rat-liver mitochondria. Biochim. Biophys. Acta 31: 296-297, 1959.

14. Fink, S. A. and J. A. Williams. Adrenergic receptors mediating depolarization in brown adipose tissue. Am. J. Physiol. 231: 700-706, 1976.

15. Flaim, K. E., B. A. Horwitz, and J. M. Horowitz. Coupling of signals to brown fat: α- and β-adrenergic responses in intact rats. Am. J. Physiol. 232: R101-R109, 1977.

16. Flatmark, T. and J. I. Pedersen. Brown adipose tissue mitochondria. Biochim. Biophys. Acta 416: 53-103, 1975.

17. Foster, D. O. Evidence against a mediatory role of brown adipose tissue in the calorigenic response of cold-acclimated rats to noradrenaline. Can. J. Physiol. Pharmacol. 52: 1051-1062, 1974.

18. Foster, D. O. and M. L. Frydman. Brown adipose tissue: the dominant site of nonshivering thermogenesis in the rat. In: Effectors of Thermogenesis. Girardier, L. and J. Seydoux (editors). Basel: Experientia (In press).

19. Fuller, C. A. J. M. Horowitz, and B. A. Horwitz. Spinal cord thermosensitivity and sorting of neural signals in cold-exposed rats. J. Appl. Physiol: Respirat. Environ. Exercise Physiol. 42: 154-158, 1977.

20. Girardier, L. and J. Seydoux. Cytomembrane phenomena during stimulation of brown fat thermogenesis by norepinephrine. In: Nonshivering Thermo-

genesis. Janský, L. (editor). Prague: Academia, 1971, pp. 255-269.

21. Girardier, L., J. Seydoux, and T. Clausen. Membrane potential of brown adipose tissue. A suggested mechanism for the regulation of thermogenesis. J. Gen. Physiol. 52: 925-940, 1968.

22. Grubb, B. and G. E. Folk, Jr. Effect of cold acclimation on norepinephrine-stimulated oxygen consumption in muscle. J. Comp. Physiol. 110: 217-226, 1976.

23. Grubb, B. and G. E. Folk, Jr. The role of adrenoceptors in norepinephrine-stimulated $\dot{V}O_2$ in muscle. Eur. J. Pharmacol. 43: 217-223, 1977.

24. Guernsey, D. L. and E. D. Stevens. The cell membrane sodium pump as a mechanism for increasing thermogenesis during cold acclimation in rats. Science 196: 908-910, 1977.

25. Hamilton, J. and B. A. Horwitz. Brown fat lipolysis: effects of adrenergic agonists and cyclic nucleotides. Fed. Proc. 36: 579 (Abstract), 1977.

26. Heim, T. and D. Hull. The effect of propranolol on the calorigenic response in brown adipose tissue of new-born rabbits to catecholamines, glucagon, corticotrophin and cold exposure. J. Physiol. 197: 271-283, 1966.

27. Hemon, P., D. Ricquier, and G. Mory. A role for thyroid hormones in the response of brown adipose tissue to chronic cold. In: Regulation of Depressed Metabolism and Thermogenesis. Janský, L. and X. J. Musacchia (editors). Springfield, Ill.; Thomas, 1976, pp. 174-195.

28. Herd, P. A., R. P. Hammond, and M. W. Hamolsky. Sodium pump activity during norepinephrine-stimulated respiration in brown adipocytes. Am. J. Physiol. 224: 1300-1304, 1973.

29. Himms-Hagen, J. The role of brown adipose tissue in the calorigenic effect of adrenaline and noradrenaline in cold-acclimated rats. J. Physiol. 205: 393-403, 1969.

30. Himms-Hagen, J. Cellular thermogenesis. Ann. Rev. Physiol. 38: 315-351, 1976.

31. Himms-Hagen, J., J. Cerf, M. Desautels, and G. Zaror-Behrens. Thermogenic mechanisms and their control. In: Effectors of Thermogenesis. Girardier, L. and J. Seydoux (editors). Basel: Experientia (In press).
32. Hittelman, K. J., R. Bertin, and R. W. Butcher. Cyclic AMP metabolism in brown adipocytes of hamsters exposed to different temperatures. Biochim. Biophys. Acta 338: 398-407, 1974.
33. Hittelman, K. J., A. S. Fairhurst, and R. E. Smith. Calcium accumulation as a parameter of energy metabolism in mitochondria of brown adipose tissue. Proc. Nat. Acad. Sci. (U.S.) 58: 697-702, 1967.
34. Horowitz, J. M., B. A. Horwitz, and R. Em. Smith. Effect *in vivo* of norepinephrine on the membrane resistance of brown fat cells. Experientia 27: 1419-1421, 1971.
35. Horwitz, B. A. Ouabain-sensitive component of brown fat thermogenesis. Am. J. Physiol. 224: 352-355, 1973.
36. Horwitz, B. A. Adrenergic receptor involvement in brown adipose tissue activation. In: Drugs, Biogenic Amines and Body Temperature. Cooper, K.E., P. Lomax, E. Schönbaum (editors). New York: Karger, 1977, pp. 160-166.
37. Horwitz, B. Plasma membrane involvement in brown fat thermogenesis. In: Effectors of Thermogenesis. Girardier, L. and J. Seydoux (editors). Basel: Experientia (In press).
38. Horwitz, B. A., J. F. Detrick, and R. Em. Smith. Norepinephrine-induced thermogenesis: effect of interscapular brown fat. Experientia 28: 284-285, 1972.
39. Horwitz, B. A. and M. Eaton. The effect of adrenergic agonists and cyclic AMP on the Na^+/K^+ ATPase activity of brown adipose tissue. Eur. J. Pharmacol. 34: 241-245, 1975.
40. Horwitz, B. A. and M. Eaton. Ouabain-sensitive liver and diaphragm respiration in cold-acclimated hamster. J. Appl. Physiol.: Respirat. Environ. Exercise Physiol. 42: 150-153, 1977.

41. Horwitz, B. A. and J. M. Horowitz. Cation transport and brown fat thermogenesis. Fed. Proc. (In press).
42. Horwitz, B. A., J. M. Horowitz, and R. Em. Smith. Norepinephrine-induced depolarization of brown fat cells. Proc. Nat. Acad. Sci. (US) 64: 113-120, 1969.
43. Hsieh, A. C. L. The basal metabolic rate of cold-adapted rats. J. Physiol. 169: 851-861, 1963.
44. Hsieh, A. C. L. and L. D. Carlson. Role of the thyroid in metabolic response to low temperature. Am. J. Physiol. 188: 40-44, 1957.
45. Hsieh, A. C. L., C. W. Pun, K. M. Li, and K. W. Ti. Circulatory and metabolic effects of noradrenaline in cold-adapted rats. Fed. Proc. 25: 1205-1209, 1966.
46. Janský, L. Nonshivering thermogenesis and its thermoregulatory significance. Biol. Rev 48: 85-132, 1973.
47. Janský, L. and J. S. Hart. Participation of skeletal muscle and kidney during nonshivering thermogenesis in cold-acclimated rats. Canad. J. Biochem. Physiol. 41: 953-954, 1963.
48. Krishna, G., J. Moskowitz, P. Dempsey, and B. B. Brodie. The effect of norepinephrine and insulin on brown fat cell membrane potentials. Life Sci. (Part 1) 9: 1353-1361, 1970.
49. LeBlanc, J. and J. Cote. Increased vagal activity in cold-adapted animals. Can. J. Physiol. Pharmacol. 45: 745-755, 1967.
50. LeBlanc, J. and M. Pouliot. Importance of noradrenaline in cold adaptation. Am. J. Physiol. 207: 853-856, 1964.
51. LeBlanc, J. and A. Villemaire. Thyroxine and noradrenaline on noradrenaline sensitivity. Am. J. Physiol. 218: 1742-1745, 1970.
52. Lehninger, A. L. Biochemistry. New York: Worth, 1975, 1104 pp.
53. Liberman, U. A., Y. Asano, C-S. Lo, and I. S. Edelman. The mechanism of the respiratory and transport effect of thyroid hormone on small intestinal mucosa. In: Effectors of

Thermogenesis. Girardier, L. and J. Seydoux (editors). Basel: Experientia (In press).
54. Lindberg, O. (Editor) Brown Adipose Tissue. New York: Elsevier, 1970, 337 pp.
55. Lo, C-S, T. R. August, U. A. Liberman, and I. S. Edelman. Dependence of renal $(Na^{+} + K^{+})$-adenosine triphosphatase activity on thyroid status. J. Biol. Chem. 251: 7826-7833, 1976.
56. Lo, C-S. and I.S. Edelman. Effect of triiodothyronine on the synthesis and degradation of renal cortical $(Na^{+} + K^{+})$-adenosine triphosphatase. J. Biol. Chem. 251: 7834-7840, 1976.
57. Mejsnar, J. and L. Janský. Mode of catecholamine action during organ regulation of nonshivering thermogenesis. In: Regulation of Depressed Metabolism and Thermogenesis. Janský, L. and X. J. Musacchia (editors). Springfield, Ill.: Thomas, 1976, pp. 225-242.
58. Nicholls, D. G. The bioenergetics of brown adipose tissue mitochondria. FEBS Letters 61: 103-110, 1976.
59. Nicholls, D. G. Hamster brown-adipose-tissue mitochondria. Purine nucleotide control of the ion conductance of the inner membrane, the nature of the nucleotide binding site. Eur. J. Biochem. 62: 223-228, 1976.
60. Nicholls, D. G. and G. M. Heaton. The identification of the component in the inner membrane of brown-adipose tissue mitochondria responsible for regulating energy dissipation. In: Effectors of Thermogenesis. Girardier, L. and J. Seydoux (editors). Basel: Experientia (In press).
61. Reed, N. and J. N. Fain. Potassium-dependent stimulation of respiration in brown fat cells by fatty acids and lipolytic agents. J. Biol. Chem. 243: 6077-6083, 1968.
62. Reichlin, S., J. Bollinger, I. Nejad, and P. Sullivan. Tissue thyroid hormone concentration of rat and man determined by radioimmunoassay: biologic significance. Mt. Sinai J. Med. N. Y. 40: 502-510, 1973.

63. Rogus, E. M., L. C. Cheng, and K. Zierler. β-adrenergic effect on Na^+-K^+ transport in rat skeletal muscle. Biochim. Biophys. Acta 464: 347-355, 1977.
64. Schmitt, M., P. Meunier, A. Rochas, and J. Chatonnet. Catecholamines and oxygen uptake in dog skeletal muscle in situ. Pflügers Arch. 345: 145-158, 1973.
65. Sellers, E. A. and S. S. You. Role of the thyroid in metabolic responses to a cold environment. Am. J. Physiol. 163: 81-91, 1950.
66. Seydoux, J. and L. Girardier. Control of brown fat thermogenesis by the sympathetic nervous system. In: Effectors of Thermogenesis. Girardier, L. and J. Seydoux (editors). Basel: Experientia (In press).
67. Smith, R. E. Thermogenic activity of the hibernating gland in the cold-acclimated rat. The Physiologist 4: 113 (Abstract), 1961.
68. Smith, R. E. and D. J. Hoijer. Metabolism and cellular function in cold acclimation. Physiol. Rev. 42: 60-142, 1962.
69. Smith, R. E. and B. A. Horwitz. Brown fat and thermogenesis. Physiol. Rev. 49: 330-425, 1969.
70. Stevens, E. D. and M. Kido. Active sodium transport: a source of metabolic heat during cold adaptation in mammals. Comp. Biochem. Physiol. 47A: 395-397, 1974.
71. Strømme, S. B. and H. T. Hammel. Effects of physical training on tolerance to cold in rats. J. Appl. Physiol. 23: 815-824, 1967.
72. Tata, J. R., L. Ernster, O. Lindberg, E. Arrhenius, S. Pedersen, and R. Hedman. The action of thyroid hormones at the cell level. Biochem. J. 86: 408-428, 1963.
73. Teskey, N., B. Horwitz, and J. Horowitz. Norepinephrine-induced depolarization of skeletal muscle cells. Eur. J. Pharmacol. 30: 352-355, 1975.
74. Videlia, L., K. V. Flattery, E. A. Sellers, and Y. Israel. Ethanol metabolism and liver oxidative capacity in cold acclimation. J. Pharmacol. Exptl. Therap. 192: 575-582, 1975.

75. Williams, J. A. and E. K. Matthews. Effects of ions and metabolic inhibitors on membrane potential of brown adipose tissue. Am. J. Physiol. 227: 981-986, 1974.

76. Williamson, J. R. Control of energy metabolism in hamster brown adipose tissue. J. Biol. Chem. 245: 2043-2050, 1970.

SIGNIFICANCE AND POSSIBLE CENTRAL MECHANISMS OF THERMOREGULATORY THRESHOLD DEVIATIONS IN THERMAL ADAPTATION

Kurt Brück and Eugen Zeisberger
Institute of Physiology
Justus-Liebig University
D-6300 Giessen
Federal Republic of Germany

ABSTRACT

Threshold deviations for the actuation of the control processes occur in the course of thermal adaptation. Depending on the relative directions of the threshold deviations for heat production and heat dissipation, a widening or narrowing of the interthreshold zone results; this influences precision and economy of temperature regulation. A downward shift of the threshold for thermogenesis is the basis of "hypothermic" or "tolerance" adaptation. Threshold deviations for thermogenesis can also be caused by microinjection of noradrenaline (NA) and an adrenergic alpha-receptor blocking agent into a "NA-sensitive" area of the anterior hypothalamus, situated caudally to the "thermoreceptive" preoptic area. It is assumed that these drugs act on some interneurons interconnected in the thermoregulatory integrative system. The activity of these interneurons appears to be modulated by a catecholaminergic pathway ascending from the midbrain reticular formation. Moreover, a sustained decrease in shivering threshold can be evoked in guinea pigs when the peripheral NA level is increased by means of NA infusion. This and other results suggest that a negative feedback mechanism exists, on the basis of which the cold-adaptive threshold deviation may be interpreted.

ISBN 0-12-734550-7

I. INTRODUCTION

The stability of deep body temperature in homeothermic animals and in man is based on the action of a regulatory (negative feedback) system. A central controller, thought to be localized in the hypothalamus, receives information on the temperature from at least two sites of the body: from the body surface and from the body core. Body surface temperature is measured by the cutaneous thermal receptors, core temperature by receptors within the preoptic area, the spinal cord and possibly some other places (22, 42); this temperature information is processed in the central controller. Effector signals, which are the outcome of this processing, then control the various effector systems: thermogenesis, vasomotor and evaporative heat dissipation; behavioral actions are elicited in a similar way.

For the discussion of the present problem - adaptive threshold deviation - some further cybernetic considerations may be useful: In grossly generalized models of biological thermoregulatory systems the input to the controller, i.e. the "actuating signal", "deviation" or "error signal", Δx, is the difference between the actual temperature of the controlled system (or a special site of it), x, and "reference" or "set" temperature, x_o, which represents the "desired value of the controlled variable." The output signals y_i (i = 1, 2 ....) controlling the various effector systems may be, in the simplest case, a linear function of Δx. The output signals, y_i, may not become greater than zero, however, before a certain threshold value of the derivative, Δx_{th} or Δx_d has been exceeded, as shown in figure 1. Moreover, the effectiveness of the system is determined by the "gain" (degree of amplification), K. The output signals, y_i, may not only be a function of Δx but also of its first derivative: $y = f(\Delta x, dx/dt)$, i.e. the activation of the effector system depends on both the magnitude of temperature deviation and the rate of change of temperature. In other words, the control system possesses the properties of a proportional (response to magnitude change) and differential (response to rate change) control system (PD-Control System).

In a multiple input system, each temperature signal, x_i, might be subtracted from a corresponding reference

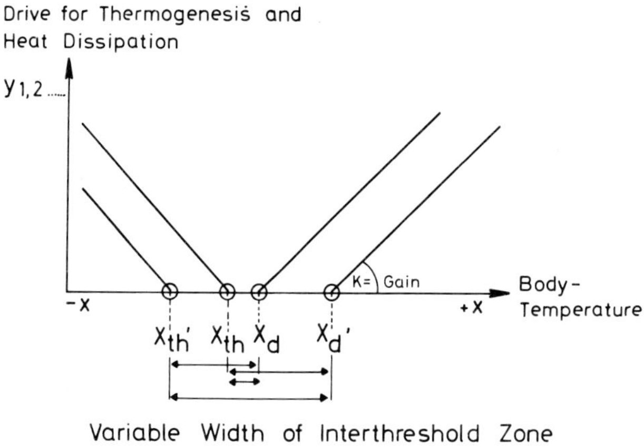

Fig. 1. *Schematic representation of the input - output relation in the thermoregulatory control system with variable threshold temperatures for thermogenesis (X_{th}, X'_{th}) and heat dissipation (X_d, X'_d).*

signal, x_{oi}; the differences, Δx_i, would then be computed to produce the error signal, Δx. Or the computed temperature signals, x_i, might be subtracted from a common reference signal, x_o. It has also been suggested that an error signal is constituted by the counteraction of warm and cold receptors. With a balanced activity indicating the biological null-point, the error signal becomes zero and increases when the activity of either the warm or cold receptors dominates. No definite experimental evidence for the one or the other proposition is available (for review see 4). Furthermore, for the computation of each output signal, y_i, different actuating signals, Δx_i, as computed from special combinations of the temperature input signals, x_i, might be used. Some experimental evidence for this was obtained from studies in the guinea pig (11): nonshivering thermogenesis (NST) is preferentially determined by skin temperature ($\overline{T}_{sk}$) and hypothalamic temperature (T_{hy}), shivering thermogenesis by $\overline{T}_{sk}$ and spinal cord temperature (T_{sc}) (cf. Fig. 16).

Since the actuation of an effector system, e.g. NST, is determined by at least two temperatures, body surface and body core temperature, the threshold conditions for the elicitation of this effector system are described by a curve in a two dimensional coordinate system (Fig. 2). Shivering threshold deviations are then indicated by a shift of this threshold curve as shown by figure 3. Within a small range of temperature values the threshold hyperbolas (Fig. 2) can be approxi-

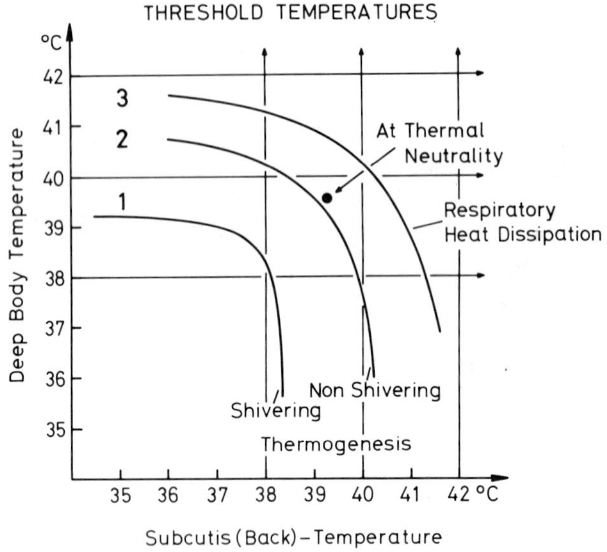

Fig. 2. *Threshold curves for shivering and nonshivering thermogenesis and for respiratory evaporative heat dissipation as a function of body surface and deep body temperature. Curve 1: non cold-adapted guinea pigs (11), curve 2: newborn guinea pigs (10). Curve 3 is based on data from (13, 23). The shape of curve 3 is inferred from studies in the rabbit by Stitt (44). At conditions of thermal neutrality, body temperature assumes a value represented by the large filled circle.*

mated by a straight line. Hence, the effector action may be related to a single temperature, a weighted mean body temperature, T_b, which is used in some of the following figures.

Physiological adaptation has been defined as the modification of organs and functional systems, including regulatory processes, taking place during prolonged or repeated action of an environmental stimulus or stressor (1). Based on the characterization of the control system given above, adaptive changes in the thermoregulatory system may consist of:

1) changes of the capacity for the magnitude of response of the effector systems
2) changes in threshold (Fig. 1) and gain for the effector actions

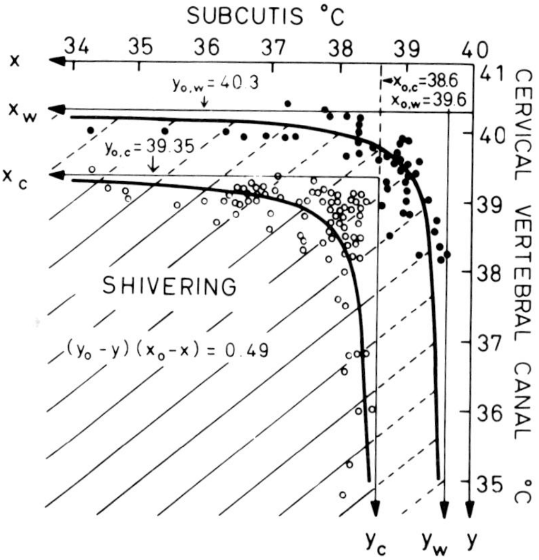

Fig. 3. *Shivering threshold curves for two groups of guinea pigs (aged 4 to 8 weeks) reared from birth at cold (5°C = o) or warm (28-30°C = •) environmental conditions. The values were obtained by independently changing temperature of the body surface and the cervical vertebral canal. The diagram shows, for instance, that at a body surface temperature of 37°C, shivering begins in warm-adapted animals if the hypothalamic temperature drops below 40°C. In the cold-adapted animals, however, shivering does not occur until hypothalamic temperature has reached a value slightly below 39°C. From (11).*

3) changes in the relative sensitivity of the static (proportional) and dynamic (time derivative) sensitivity of the control system

It is the adaptive changes in the capacity for response of the effector systems which has attracted the greatest interest during the past. Thus, on the cold side, the development of an additional mechanism of heat production, NST, has been described. An increase in the maximum sweat rate and in the maximum skin blood flow rate would be the corresponding capacity changes on the warm side. The range of environmental temperature at which an organism is able to maintain deep body temperature close to the value found under thermoneutral conditions ["tolerated ambient temperature range", (6)] may be considerably extended by capacity adaptation. A conceivable adaptive change in the dynamic and static sensitivity of the control system remains to be studied.

Adaptive changes in the threshold temperatures for the actuation of thermoregulatory responses will be the subject of the present paper. One way of viewing these adaptive changes would be as follows:

The threshold temperature curves (Fig. 2) may be deviated in such a way that the distance between the heat dissipation and heat production threshold curve is reduced - this would mean the precision of the regulation is increased. Under other conditions, however, the threshold shift may have the opposite direction and this will reduce precision of the regulation. The latter behavior would be the physiological basis for what is known as "tolerance adaptation", one example of which is hypothermic adaptation, as found in cold-adapted man. Another example is hibernation and torpidity. Hence the main phenomenon dealt with at this symposium could be reduced on a basic alteration in the thermoregulatory system: deviation of the threshold level for the actuation of cold defense reactions. The present paper aims to 1) give a survey on the experimental evidence for the occurrence of thermoadaptive threshold deviations in animals and man, and 2) show a possible neurophysiological correlate of such threshold deviation.

II. ADAPTIVE THRESHOLD DEVIATIONS IN THE GUINEA PIG

A. Heat Production

As shown in figure 3, in guinea pigs reared in the cold (+5°C) from birth, the shivering threshold hyperbola was about 1°C lower than in animals reared in neutral temperatures (28-30°C) (11). In order to interpret these results it must be known that shivering in the guinea pig is controlled by the combined action of body surface thermoreceptors and those located in the cervical part of the spinal cord. Thermosensors of the preoptic area influence NST, but with only a slight effect on shivering thermogenesis. Such a shift in shivering threshold is not confined to animals reared from birth in the cold. In another series of studies the same deviation was found in animals that had not been exposed to the cold (5°C) until their third month of life (55). In intermittently cold exposed animals (12 hr cold exposure per day at 5°C) shivering threshold was found to be in the range of permanently cold-adapted (CA) animals (13).

The threshold for NST is found in newborn as well as adult CA guinea pigs at a temperature level well above the shivering threshold, as indicated in figure 2. In the neonate this threshold is somewhat higher than in CA adult animals (50). In warm-adapted (WA) animals the maximum amount of NST is small in comparison with shivering; the threshold temperature for NST is practically identical with that for shivering (50).

B. Heat Dissipation

The guinea pig enhances heat dissipation by increasing respiratory frequency up to about 300 per min (heat polypnea). Respiratory frequency is determined by body surface temperature and by the temperature of the preoptic area and of the spinal cord. It is for this reason that respiratory rate in figure 4 is related to mean body temperature. If the respiratory rate of 200 per min is taken - operatively - as threshold frequency, the threshold is found at a level more than 1°C lower in CA than in WA animals. By contrast, animals exposed only intermittently to cold - 12 hr per day at +5°C for a period of 4 weeks - the threshold for heat polypnea is not shifted (Fig. 4).

C. Body Temperature

Deep body temperature (colonic) of CA animals has been found to be lower than that of WA animals by 0.5-1.0°C at environmental temperatures from 15-30°C (Fig. 5). Thus one may speak of a set-point deviation of the controlling system in these animals. By contrast, the intermittently cold-exposed animals do not seem to show such set-point deviation.

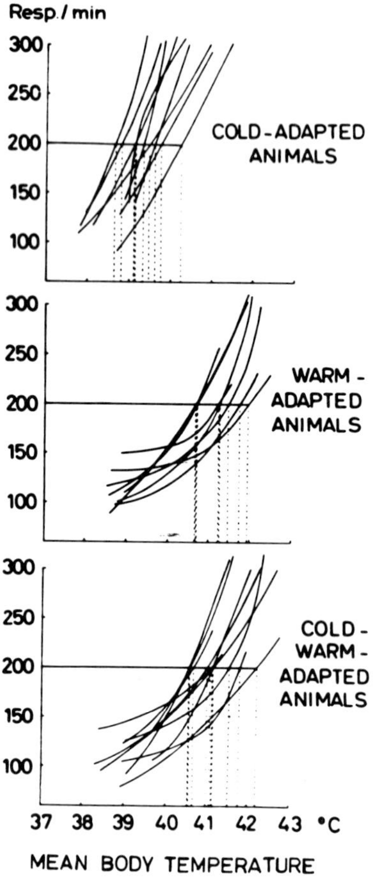

Fig. 4. *Respiratory frequency in relation to mean body temperature (arithmetic mean of body surface and colon temperature) in cold- and warm-adapted, and in intermittently cold-adapted (cold exposure 12 hours per day) animals. A respiratory frequency of 200 per minute is operatively defined as heat polypnea threshold. This threshold is shifted to lower values in the permanently cold-adapted animals. (From 13).*

D. Biological Significance of Threshold Deviations in the Guinea Pig

A mean body temperature reduction by 1-2°C would mean a measurable reduction in the metabolic cost for maintaining a constant body temperature, as may be seen from the following calculation: mean body temperature of a guinea pig may be maintained at 39°C while the ambient temperature is 10°C; this is brought about by a threefold increase in basal metabolic rate. After the cold-adaptive set-point deviation, mean body temperature may be maintained at 37°C under the same ambient conditions. The overall temperature difference between

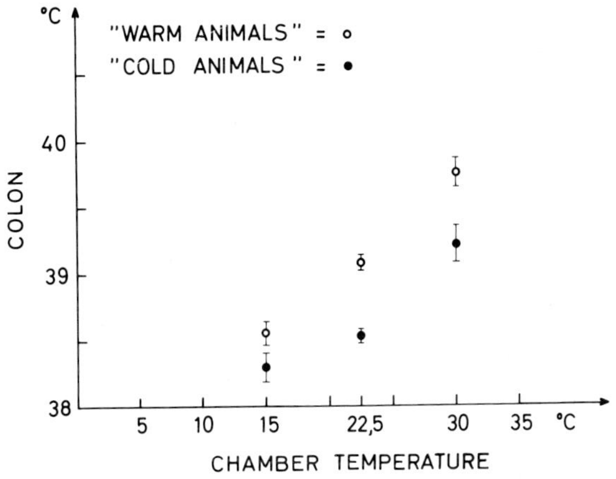

Fig. 5. *Comparison of colonic temperature in cold-adapted and warm-adapted guinea pigs exposed to three different ambient temperatures. For 22.5°C: 24 hr mean; for 15°C and 30°C: 2 hr mean. Bars indicate S.E. From (13).*

body core and environment is thereby reduced from 29 to 27°C, and this represents a reduction of the metabolic cost of about 7% or 20% of the basal metabolic rate. The lower threshold for heat polypnea in CA animals would be disadvantageous in a warm environment, since the animals would have to actuate and sustain respiratory heat dissipation at a lower body temperature than in WA animals. In the intermittently CA animals this disadvantage does not exist. Their mode of adaptation is characterized by a widening of the "interthreshold zone" (Fig. 1), within which neither thermoregulatory heat production nor respiratory heat dissipation is elicited. Hence, when intermittently CA animals are exposed to an environment with fluctuating temperatures (such as day-night alterations) they can make use of the heat storing capacity of their body mass to a much greater extent than permanently CA animals. Thus a widening of the interthreshold zone provides a basis for a more economical temperature regulation (7, 13) with less water and energy expenditure.

III. THRESHOLD DEVIATION IN MAN

A. Shivering Threshold Changes in Natural Adaptation

A number of field studies have indicated that the shift in shivering threshold - resulting in hypothermia - may be an essential modification in cold-adaptation. Thus LeBlanc (30) demonstrated a slight tendency to hypothermia in a group of soldiers living for several months in an arctic station. The most spectacular example of hypothermic adaptation was demonstrated by Scholander *et al.* (40) and Hammel et al. (21) in a group of Australian aborigines who slept without any substantial clothing in the open air at temperatures around +4°C. These people did not shiver when their mean body temperature had fallen to below 34°C in the morning, while body temperature in white controls fell less than in aborigines due to vigorous shivering. In the Amas, the Korean pearl divers who are used to diving several hr a day at water temperatures as low as 10°C, shivering was found to occur at much lower temperatures than in controls (25). The first example of an experimentally produced hypothermic adaptation was demonstrated by Davis (17). This study was originally designed to demonstrate the development of NST in man. For this goal, a group of students was observed over a period of 4 weeks, of which they spent 8 hr daily in a climatic chamber at about +8°C, wearing light clothing. NST did not develop but rectal temperature, heat production and electrical muscle activity were all reduced at the end of the study, i.e. the subjects had developed a hypothermic type of adaptation.

B. Experimental Adaptation

In several types of studies it was our intention to systemise the adaptive threshold deviations in man, whereby we tried to answer the following questions:

1) Is the shift in shivering threshold a cold-specific adaptate [for this term cf. (1)]; does cross-adaptation occur?

2) Are the shivering threshold deviations coupled with, or independent of, sweating threshold deviations?

1. Repeated Short-Term Exposures

In a series of studies, shivering threshold deviations were observed in the course of repeated short cold exposures (9). The subjects, wearing only shorts, were exposed to the following thermal conditions: 1) 30 min at 28°C, 2) followed by a linear decrease (0.5°C/min) of temperature down to between +5 and -5°C unless the subject wanted to end the cold-exposure earlier, and 3) quick rewarming. Such single cold-exposures were repeated 5 to 7 times over a 1 to 2 week period. As in the guinea pig, shivering is determined in man by deep body temperature and mean skin temperature ($\overline{T}_{sk}$), which is the average from skin temperatures of the chest, upper arm, thigh and lower leg. The weighted mean body temperature ($\overline{T}_b$), which is calculated from $\overline{T}_{sk}$ and deep body (esophageal) temperature according to the following equation, was used for threshold determination: $\overline{T}_b = 0.87\ T_{es} + 0.13\ \overline{T}_{sk}$. In two thirds of the subjects the shivering threshold, related either to ambient temperature or to $\overline{T}_b$, was shifted to lower values (Fig. 6). The oxygen uptake, the experience of thermal discomfort and the sensation of cold were also shifted to lower $\overline{T}_b$ (Fig. 6). Esophageal temperature was also significantly lower in the environmental temperature range of 0 to 28°C in the final exposure. The difference amounted to 0.2 to 0.3°C, on the average.

2. Endurance Training

In 7 long-distance (42 km or more) runners shivering threshold was found to be considerably decreased when compared to those young people who had undertaken no special training program (2). This difference is particularly pronounced when shivering threshold is related to a measure of body surface/mass ratio (Fig. 7). The $\overline{T}_b$ and esophageal temperature measured at neutral temperature were found to be lower in long-distance runners. The sweating threshold, as to be expected, was decreased in the long-distance runners. By contrast, in the cold-adaptation series we did not find any change in sweating threshold (9). With respect to the shivering threshold, cross-adaptation is evoked by the stressors cold and physical training. With respect to sweating threshold, however, no such cross-adaptation is produced by the two stressors.

3. Repeated Heat Exposure

Shivering as well as sweating thresholds were determined in a series of exposures to different temperatures in six resting subjects on six consecutive days (41). On day one and six the subjects were exposed to a standard cold test, on day two through five, to a standard heat test in which T_{es} reached levels of between 37.5 and 38.0 °C. Shivering threshold related to $\bar{T}_b$ was found to be unchanged in all six subjects on day six (Fig. 8). Sweating threshold, however, was shifted to $\bar{T}_b$ and T_{es} levels of about 0.3 °C lower on day five in four of the six subjects (Fig. 8). Moreover, in the course of the four heat exposures, the initial and final T_{es} decreased by 0.2 and 0.25 °C on the average. Due to this decreased T_{es} level and the unchanged shivering

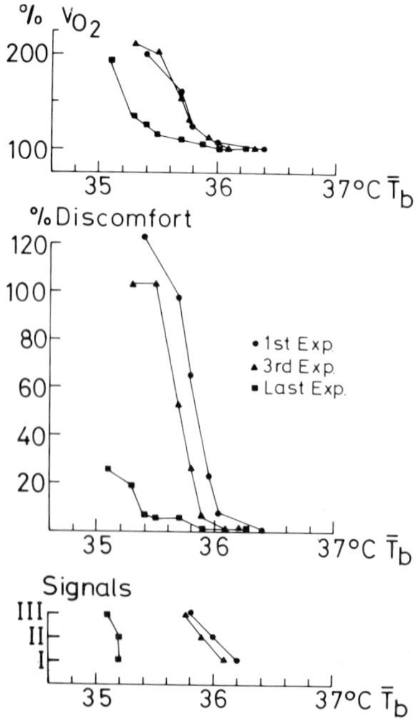

Fig. 6. *Study in a subject who was cold-exposed on six different days. Abbreviations: $\bar{T}_b$ = mean weighted body temperature; signals I = cool, II = cold, III = very cold. Discomfort means thermal discomfort as signalled by the subject using a slide-resistor. $\dot{V}O_2$ is oxygen uptake. Note: metabolic rate, cold sensation (signals) as well as the sensation of thermal discomfort were shifted to lower $\bar{T}_b$ in the course of repeated cold exposures. From (9).*

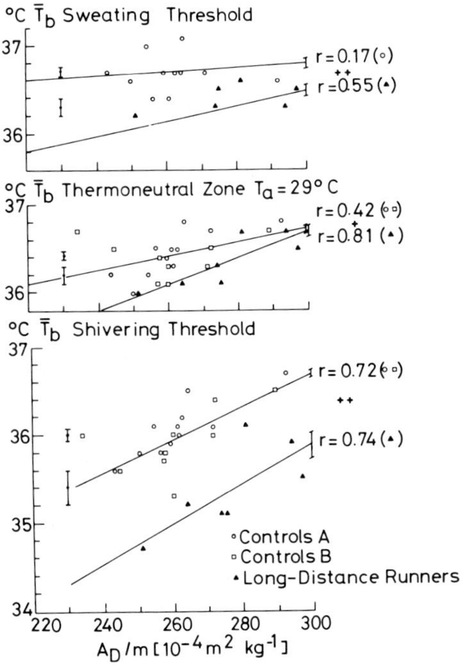

Fig. 7. *Comparison of long-distance runners with controls. Mean weighted body temperature ($\bar{T}_b$) as measured at shivering threshold, at neutral temperature and at the beginning of sweating, respectively. T_b is plotted against surface area/body mass quotient, A_D/m (the larger values correspond to thin men). Note shivering threshold as well as sweating threshold is shifted to lower T_b in long distance runners. Bars on right end of lines indicate S.E. of the regression lines, numbers are correlation coefficients. Probability of identity of the regression lines: + indicates $P<0.35$; ++: $P<0.05$. From (2).*

threshold, the subjects began to shiver sooner in the second cold test (41). The results from experimental manipulations in man indicate that shivering threshold deviations as found in the course of cold-adaptation and after endurance training are not to be considered a stressor-unspecific modification, and that the interthreshold zone, which was found to be extended after intermittent cold-exposure in the guinea pig and in man, can also be narrowed (after heat exposure).

C. Summary of Adaptive Threshold Deviations in Man

Figure 9 summarizes the adaptive threshold deviations as observed so far in man. Intermittent cold-adaptation is an example of a widening of the "interthreshold zone". Intermittent heat-adaptation is an example of a narrowing of the "interthreshold zone". In long-distance runners both threshold temperatures are

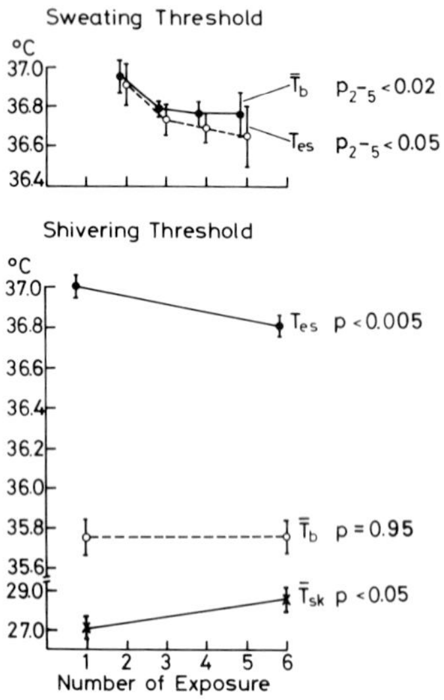

Fig. 8. *Summary of a study in 6 subjects who were repeatedly heat exposed (exposures 2 to 5) and cold exposed before (exposure 1) and after (exposure 6) the heat exposure series. Top diagram: average mean body temperature, $\bar{T}_b$, and esophageal temperature, T_{es}, at which sweating began. Bottom diagram: average T_{es}, $\bar{T}_b$ and mean skin temperature, $\bar{T}_{sk}$, at which shivering began. (P_{2-5} indicates null hypothesis - based on intrapair differences - between exposures 2 and 5).*

shifted to lower levels and the deep body temperature at neutral temperature is shifted to lower values. In this case one may speak of a set-point deviation of the control system. There is a major difference between intermittent heat- and cold-adaptation: In the latter, tolerance adaptation (hypothermic adaptation) is possible but the characteristic of intermittent heat-adaptation is that body temperature is more effectively defended against any tendency to become higher. A recent study (39) of sojourners in a hot climate showed tolerance adaptation against heat loads, i.e. deep body temperature was found to be more than 0.5°C higher, but no sweating was induced at this temperature. It appears from this that, depending on the exposure time or other exposure conditions, qualitatively different types of heat adaptation develop.

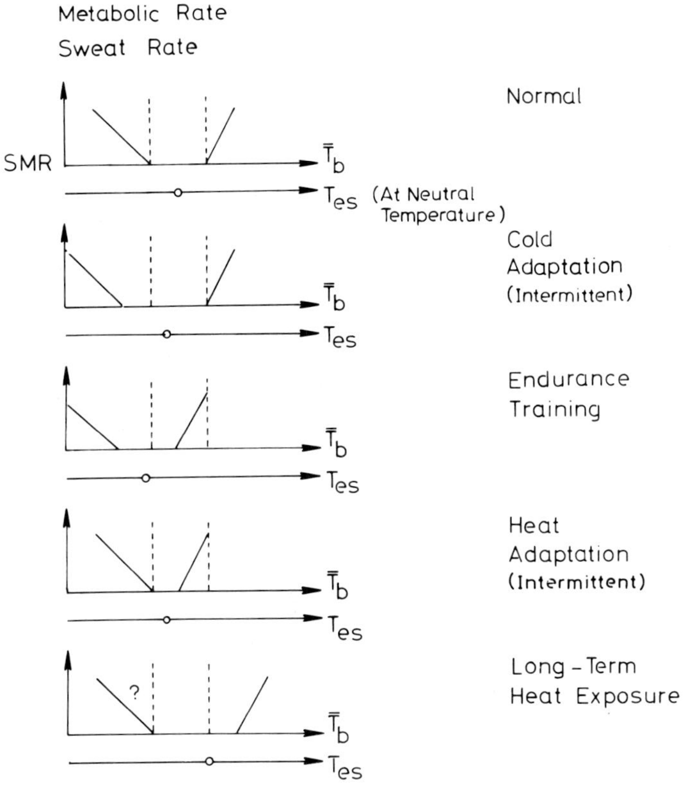

Fig. 9. *Schematic survey on the threshold deviations for the elicitation of shivering and sweating observed so far in man.*

D. Biological Significance of Threshold Deviations in Man

The decrease of sweating threshold temperature after intermittent heat exposures and endurance training may be seen as a means of preventing, or at least retarding, an increase in deep and mean body temperature to a critical level during exercise or heat exposure. This can only be a benefit, of course, if the water loss by sweating can be made up for by ingestion of sufficient water. As mentioned above, a downward shift of sweating threshold alone would narrow the interthreshold zone. Up to a point, this may render temperature regulation more precise, but beyond that point it could cause instability of the regulatory system. One may visualize a situation in which heat and cold loads quickly alternate, for instance in a vehicle moving from shadow into sunshine and vice versa. Here sweating and shivering could be actuated almost simultaneously. The concomitant decrease in shivering threshold in the runners may thus be seen as a means of preventing such instability. Besides the decreased sweating threshold, the lowered resting body temperature (set-point shift) in the long-distance runners (Fig. 9) is a second factor which prolongs the time period until a dangerous body temperature level is reached during heavy exercise (cf. 2). The hyperthermic type of heat adaptation (Fig. 9) would allow people to stay quite a while in the heat without losing a great amount of water by evaporation. This can only be advantageous, of course, in a thermally non-uniform environment with respect to time or space which provides the possibility of giving off the stored heat. The same would be true for the hypothermic type of cold-adaptation as demonstrated above.

IV. A POSSIBLE NEUROPHYSIOLOGICAL CORRELATE OF THRESHOLD DEVIATION

The following mechanisms may be discussed as possibly underlying the observed adaptive threshold deviations in man and animals:

1) Alteration in the function of cutaneous cold receptors.
2) Functional alteration of internal thermosensors.
3) A change in some non-thermosensitive central neurons which are determinants for the asymptotes of the hyperbolas which characterize the relation of the threshold curves to the body temperature (cf. Fig. 3).

In studies in cats adapted to +5°C, no indication of a functional change in facial cutaneous thermoreceptors was obtained (22). No results are yet available which demonstrate a conceivable adaptive change in internal thermosensitivity. A number of results are available, however, which speak in favour of the third possibility. They are all based on the observation that intrahypothalamic (i.h.) injection of NA may cause changes of the threshold temperatures for thermoregulatory metabolic reactions which resemble those found in the course of cold-adaptation (Sections II and III). The neurophysiological analysis of this NA effect will be described in the following sections.

A. Thermo- and NA-Sensitive Area of the Hypothalamus

By means of simultaneous implantation of 4 sets of probes into different frontal planes of the upper brainstem (52), it was possible to distinguish between a NA-sensitive and a thermo-sensitive area (53, 54). The sets of probes consisted of 2 electrodes for radio-frequency heating, a thermo-couple and a guide cannula for the microinjection of NA. As shown in figure 10, oxygen uptake, which was increased by external cooling (ambient temperature 21.5°C, which is a relatively low temperature for the restrained and outspread NB guinea pig) could be almost totally suppressed by local heating of plane III (preoptic area) but not of planes II and I. Heating of plane IV caused a concomitant increase in the temperature of plane III and this may have caused a small inhibitory effect produced by heating plane IV.

On the other hand, microinjection of NA (1 µg in 1 µl) was only effective when made in plane II, i.e. caudally to the thermosensitive preoptic area. When injected under thermoneutral conditions (30°C), the effect of NA is to increase oxygen uptake, either by evoking shivering or NST, or both.

B. Effect of i.h. Microinjection of NA on the Thresholds for Shivering and NST

For the determination of thresholds for shivering and NST, the temperatures are recorded from the sites of those receptive areas which control either shivering (body surface and spinal cord) or NST (body surface and preoptic area) (10, 11, 12). After i.h. microinjection of NA the threshold temperatures for shivering and NST increased (Fig. 11). As can be seen, the temperature pairs before injection are situated

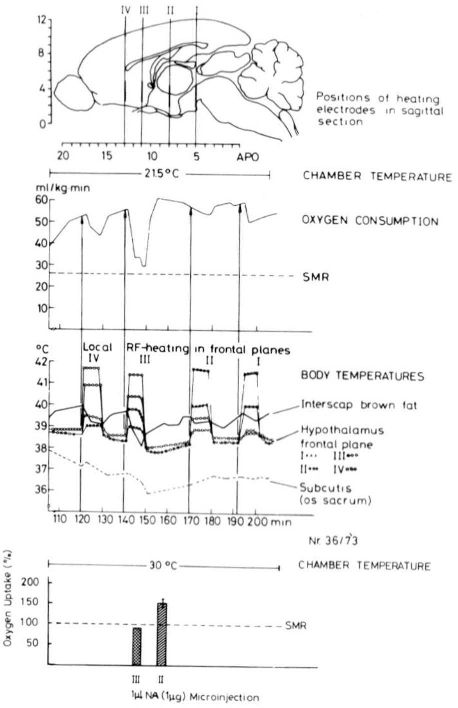

Fig. 10. *Effects of local radio frequency heating and of i.h. microinjection of noradrenaline in different regions of the brain in a newborn guinea pig exposed to a cool environment (21.5°C). The top diagram shows the projection of the four implanted electrodes on the saggital section of the brain. Only heating of plane III resulted in an almost complete suppression of the cold-induced NST (two middle diagrams). Noradrenaline administration was only effective when injected into plane II (noradrenaline-sensitive area, bottom diagram). Data from (52, 54).*

very close to the threshold curves determined in previous studies.

C. Effect of i.h. Injection of an Alpha-receptor Blocking Agent

The first indication that NA may naturally occur at the sites where its microinjection had caused effects on the thermoregulatory system was obtained from a series of studies (56) in which an alpha-receptor blocking agent, phentolamine, was injected into plane II (Fig. 10) of the hypothalamus. Depending on the dosage, shivering threshold was decreased,

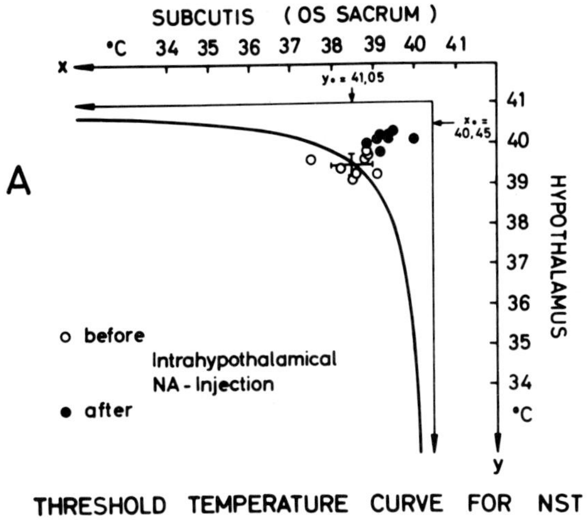

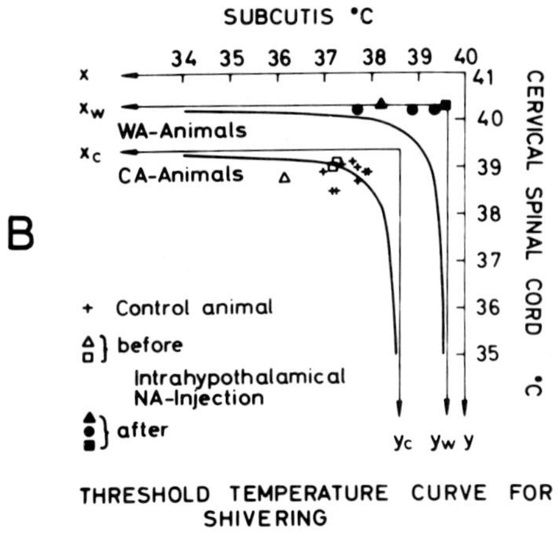

Fig. 11. *Threshold temperature for the onset of non-shivering thermogenesis (A) and shivering (B) before (open symbols) and after (closed symbols) intrahypothalamic noradrenaline injection in cold-adapted guinea pigs. The hyperbolas indicate average threshold temperatures as determined in previous studies (9, 10). From (53).*

i.e. an effect opposite to that produced by NA injection was obtained (Fig. 12). It was suggested by this result that special interneurons are located in plane II, the membranes of which contain adrenergic alpha-receptors.

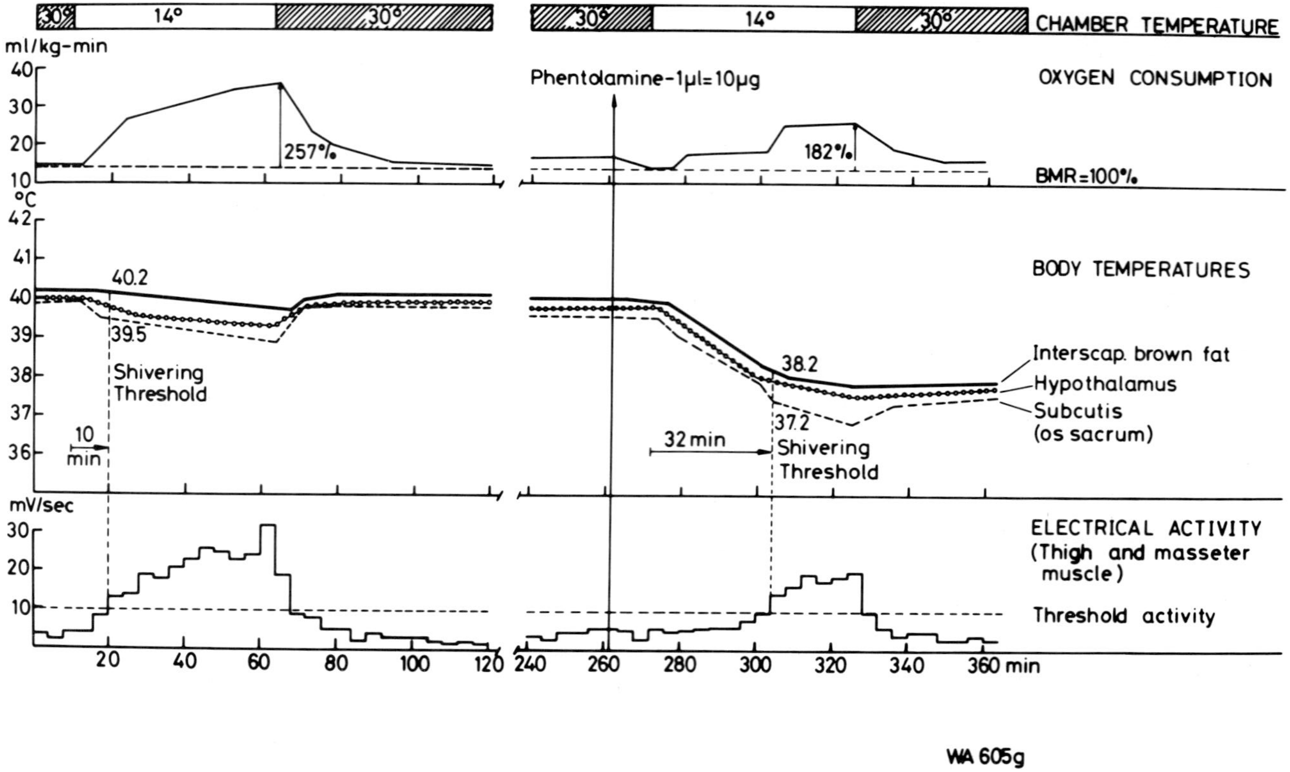

Fig. 12. *Thermogenic response to external cooling in a warm-adapted guinea pig after intrahypothalamic injection of phentolamine. From (55).*

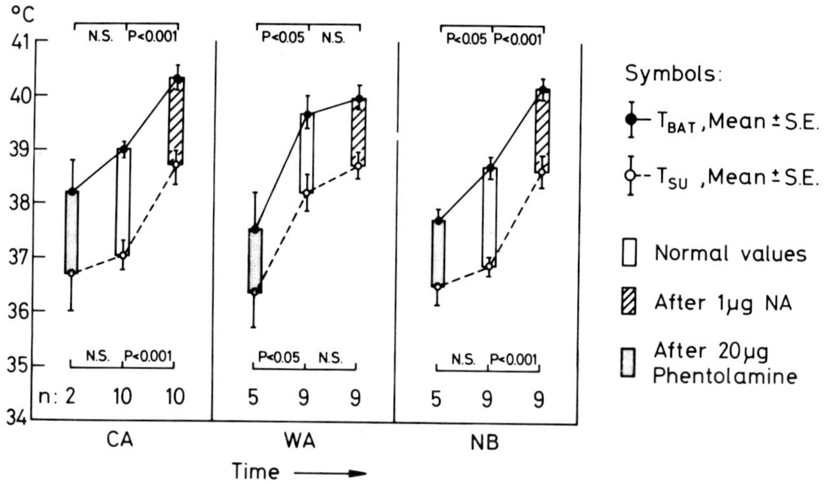

Fig. 13. *Changes of shivering threshold temperatures in cold-adapted (CA), warm-adapted (WA) and newborn (NB) guinea pigs after intrahypothalamic microinjection of noradrenaline (1 μg in 1 μl) or phentolamine (20 μg in 2 μl). Adapted from (55).*

D. Effects of NA and Phentolamine in Relation to State of Adaptation

Further evidence that hypothalamic catecholamines are determinants of thermoregulatory thresholds was obtained from the following experimental results (55). In WA animals in which the shivering threshold is found at higher temperatures than in CA animals, i.h. NA had no effect on shivering threshold; in CA animals, however, it shifted the shivering threshold upwards to the level normally found in WA animals (Fig. 13). Reversely, i.h. microinjection of phentolamine (20 μg) shifted the shivering threshold of WA animals downwards to a level found normally in CA animals; in CA animals this dosage had no effect. Only with a larger dosage could shivering threshold be shifted in both WA and CA animals to a level below that found in CA animals.

E. Catecholaminergic Pathway Projection to the Hypothalamic NA-sensitive Area

The above-described results suggest that there is a NA pathway projecting to the area II (Fig. 10); the terminals of this pathway would make synaptic contacts with the interneurons determining the threshold for shivering and NST. It has been assumed that such a pathway may be part of the system of ascending catecholaminergic pathways (ACP) described on the basis of fluoroscopic studies (16, 47). In order to substantiate this concept, a series of studies were conducted in guinea pigs using 2 sets of implanted probes, one in the hypothalamic area II (Fig. 10), the other in the midbrain; the latter consisted of a special multipolar electrode which allowed bipolar stimulation in four different depths of the brainstem (45). As shown in figure 14 electrical stimulation of a point in the midbrain increased oxygen uptake and the temperature of the brown adipose tissue (indicating NST), and evoked shivering. Thus such electrical stimulation produced the same effects seen after NA injection at neutral temperature. With repeated stimulation the effect vanished but could be restored by an i.h. injection of NA (Fig. 15, series A). These results are interpreted as due to a temporary depletion of NA stores in the terminals subsequent to the relatively long-lasting high frequency stimulation [for stimulating parameters see (45)]. The NA injection is thought to cause replenishment of the NA store in the terminals. The effect of electrical stimulation can be totally abolished by the adrenergic alpha-receptor blocking agent phentolamine (Fig. 15, series B).

In another series of studies, horseradish peroxidase (HRP) was injected through the guide cannula into area II of the hypothalamus to demonstrate, by means of a retrograde labelling, the presumed catecholaminergic pathway projecting to the hypothalamus (35). In these studies, relatively large karyocytes were labelled in an area of the midbrain reticular formation identical to that electrically stimulated in our studies described above. Furthermore, it was possible, using the fluoroscopic method, to demonstrate in this reticular formation NA-containing karyocytes that had the same shape and size as those labelled by HRP (34). Thus there is now physiological and morphological evidence for assuming the existence of a pathway projecting from the midbrain

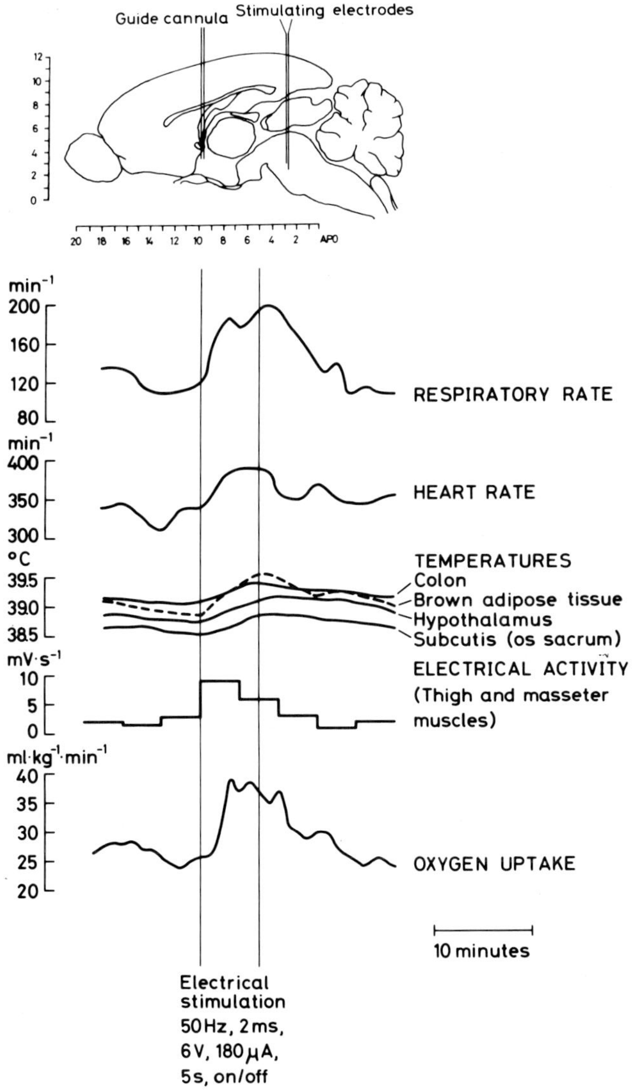

Fig. 14. *Response of young guinea pig to electrical stimulation of a site in the lower brain stem where the ascending noradrenergic pathways were presumed to be located. From (46).*

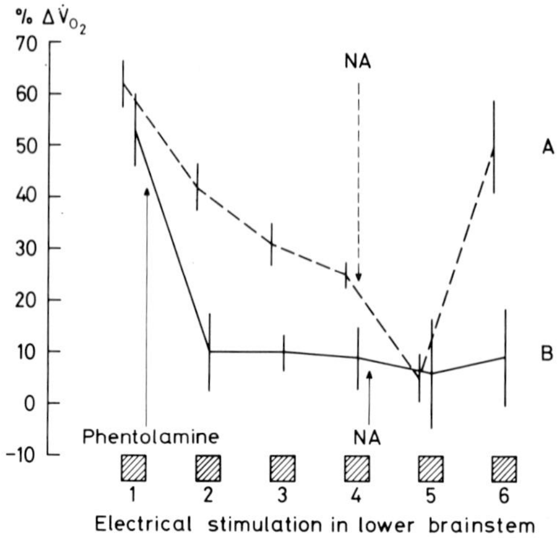

Fig. 15. *Comparison of the effects of brain stem stimulations repeated at intervals of ca. 30 minutes on the increase of oxygen uptake in two groups of guinea pigs. The broken line representing the first series (A) of 9 experiments in 9 animals shows the restoration of the fading effect of repeated electrical stimulation by intrahypothalamic noradrenaline injection. The solid line representing the second series (B) of 10 experiments in 10 animals shows the inhibitory effect of intrahypothalamic microinjection of phentolamine. The vertical bars indicate the S.E.M. From (46).*

reticular formation to hypothalamic neurons which play a role as interneurons in the thermoregulatory system.

F. A Neurophysiological Model of Temperature Regulation and Adaptation

The experimental results described in the foregoing paragraphs are summarized in a neuronal model depicted in figure 16. This model essentially consists of external and internal thermoreceptors, some effector neurons which govern the effector systems of the temperature regulatory system and some inhibitory interneurons which are inevitable in explaining the inhibitory effect on the metabolic effector systems of the warm receptors in the spinal cord and preoptic area. Thus this model

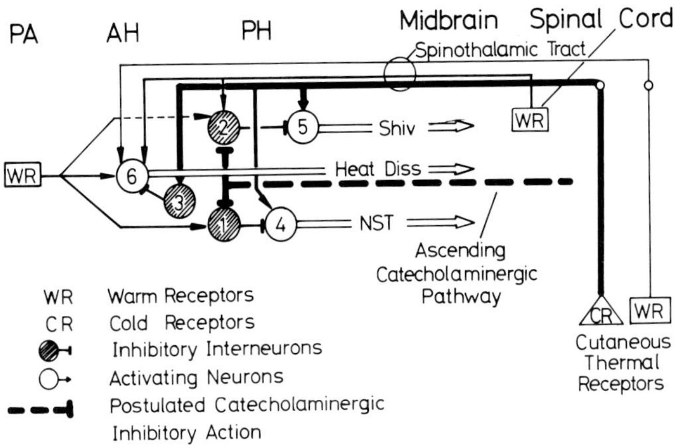

Fig. 16. *A neuronal model of the central integrative parts of the temperature control system. PA = preoptic area, AH = anterior hypothalamus, PH = posterior hypothalamus, WR = warm receptors, CR = cold receptors, NST = nonshivering thermogenesis. The dashed line pointing to neuron 2 indicates the minor influence of preoptic sensors on shivering in comparison with NST in the guinea pig. For further explanation see text.*

expresses the generally accepted view that cold defense reactions are actuated and sustained by the action of peripheral cold receptors and inhibited by deep body warm receptors; heat dissipation reactions, on the other hand, would be actuated and sustained mainly by internal warm receptors (and supported by cutaneous warm receptors) and inhibited by peripheral cold receptors (3, 12, 22, 38). The ACP described in section IV.E. would make synaptic contact with the interneurons nos. 1 and 2. On the basis of microiontophoretic studies, it must be assumed that NA has an inhibitory postsynaptic action (18, 26, 28). In particular, Jell (28) demonstrated so-called "warm-responsive NA-depressed neurons". The latter could be represented by interneurons nos. 1 and 2 in our model. The action of the ACP would then consist in inhibiting the interneurons nos. 1 and 2, thereby suppressing the inhibitory activity on the effector neurons. Shivering and NST could then be evoked with less drive from the cutaneous cold receptors, i.e.

at a higher skin temperature. In other words, the threshold for shivering and NST is shifted to a higher temperature.

It is to be noted that the ACP is by no means identical with the antero-lateral pathway (spinal thalamic tract) conveying the thermal information from cutaneous thermoreceptors to the diencephalon. This can clearly be derived from the following facts:

1) As mentioned above, NA acts as an inhibitory transmitter. If NA were the transmitter of the antero-lateral system, it should inhibit shivering and NST but the opposite was shown to be true.
2) There are some indications that 6-hydroxydopamine (6-OHDA) destruction of the noradrenergic terminals in the hypothalamus does not interrupt the cold defense reactions on external cold exposure (5). This is not what would be expected if the thermo-afferent pathways contained noradrenergic synapses.

We believe that the ACP depicted in figure 16 is to be seen as a system comparable to the ascending reticular activation system (ARAS), which is known from studies by Moruzzi and Magoun (36), to determine the state of vigilance. The ACP described here would adjust the autonomic control on the state of vigilance as determined by the ARAS, and on special environmental conditions. For the latter action one would have to assume further that the ACP is controlled by cortical elements (possibly of the limbic system) whose function is influenced by prolonged exposure to thermal stress. On the basis of studies reported in the following section, it also seems possible that long-term change in activity of the ACP is brought about by a change in the peripheral catecholamine level.

V. INTER-RELATIONSHIP BETWEEN PERIPHERAL AND CENTRAL NA

As shown in figure 13, the sensitivity of the hypothalamic plane II to NA injection is higher in CA and NB than in WA guinea pigs, while the sensitivity to phentolamine is the opposite. These results suggested that the activity of the ACP and, consequently, the concentration of the transmitter (NA) at postsynaptic sites in the hypothalamus is normally higher in WA than in CA or NB animals. In other words, the postsynaptic receptive sites in the hypothalamic area of the WA animals appear to be already saturated by endogenous NA and this may be the reason that the WA animals do not respond to the microinjection of exogenous NA. At first glance it is difficult to conceive that the hypothalamic release of catecholamines should be reduced in CA animals, since their peripheral release of catecholamines is known to be considerably increased (for lit. see 24).

To judge from the shivering threshold temperatures, the hypothalamic release of NA should also be reduced in NB animals (Fig. 13). Accordingly, the whole-brain NA concentrations are reported to be about 15-30% of the adult levels in the NB rat and 20-40% in the NB mouse. Adult brain NA concentrations are reached at the age of 4 weeks in the mouse and 6 weeks in the rat. In the guinea pig, which is born at an advanced stage of maturity, the brain NA concentration amounts to about 80% of the adult level, reaching adult concentrations at the age of 1-2 weeks (for lit. see 29). The peripheral release of NA seems to be increased in neonates, according to data from human neonates where the excretion of unmetabolized NA in urine amounts to about 80 ng/kg·h at an ambient temperature of 33°C (neutral temperature) and 180 ng/kg·h under cold exposure to 24°C (43). These values considerably exceed the adult NA excretion which does not amount to more than 30 ng/kg·h (19).

Thus the neonates not only have a similarly low shivering threshold temperature as CA animals, but they also have the same reciprocal relationship between central and peripheral NA levels. Here the question arises as to whether the high peripheral levels of NA reduce the activity in the ACP by some negative feedback action, resulting in a lowered hypothalamic NA level and a decrease in shivering threshold temperatures. In order to substantiate this hypothesis, a series of studies was performed in which the peripheral NA level was elevated by i.m. infusion and the effect of this manipulation on shivering threshold was studied (49). WA animals were used for these experiments as they presumably have a high hypothalamic NA level whereas

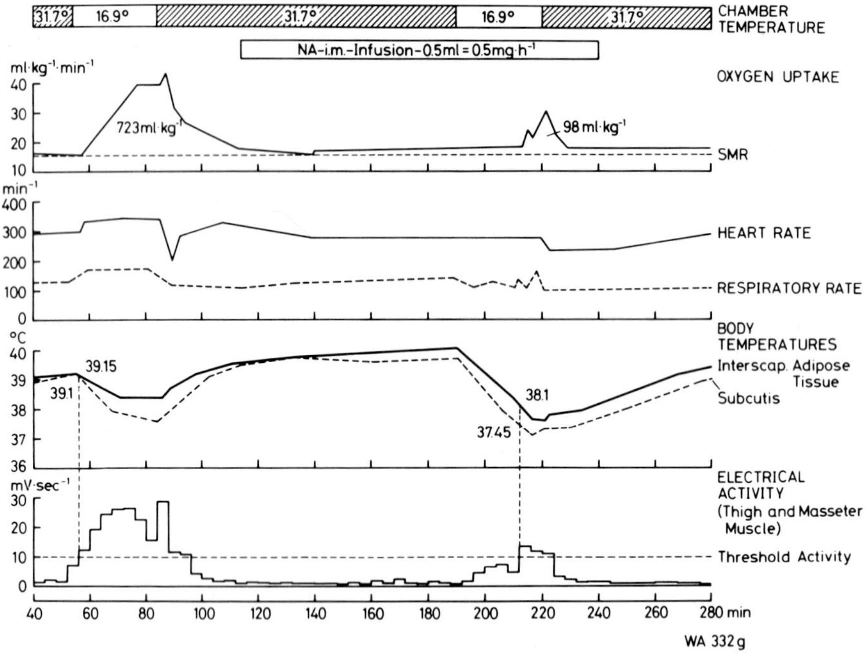

Fig. 17. *Metabolic rate, shivering and body temperature changes to cooling in a warm-adapted guinea pig showing the reduced metabolic response to cooling after an intramuscular noradrenaline infusion. From (57), by permission S. Karger AG, Basel.*

the release of catecholamines in the periphery is reduced. Figure 17 shows an example of such experiments. In the first part of the experiment the animal started to shiver immediately after the beginning of the control cooling at relatively high threshold temperatures. During the cooling period the animal used 723 ml more oxygen than at neutral ambient temperature. During the peripheral NA infusion, the cooling was repeated and the animal started to shiver at the end of the cooling period at lower body temperatures and used only 98 ml more oxygen than at neutral temperature.

The lowering of threshold temperatures for shivering during peripheral NA infusion in 10 WA animals is summarized in figure 18. Remarkably the shift in shivering threshold temperature which developed during the course of the infusion persisted for at least two hr after the end of the infusion (Fig. 20). With a dose of 1.3 mg/kg·h the shift of shivering threshold amounted to 1.2°C ($\bar{T}_b$) on average. In order to find out whether the shift was induced by a mechanism specifically mediated by NA, another series of studies was performed in which adrenaline (A) was infused into the muscles of a comparable group of WA guinea pigs. Figure 19 shows an example of such an experiment. In contrast to NA, comparable doses of A (1.4 mg/kg·h) did not decrease the threshold temperature for shivering. Figure 20 compares schematically the changes in shivering threshold temperature evoked in two groups of WA animals by i.m. infusion of NA or A. In contrast to NA, A infusion increased the response to cold exposure and the shivering threshold temperature. The interval between the onset of cooling and the beginning of shivering is shorter in the A-infused group than in the NA-infused group. Similarly the time period during which the animals continue to shiver after increasing the ambient temperature to 31°C is longer in the A-infused and shorter in the NA-infused group. These time changes are another indicator of the changes in the shivering threshold. With even higher doses of NA (6 mg/kg·h) it was possible to decrease the shivering threshold temperature by 2.5°C on average and this shivering threshold shift outlasted the end of the infusion at least 5 hr. With regard to the high doses of infused NA and A it must be taken into account that most of the intramuscularly infused NA is probably metabolized very rapidly; only a small fraction may reach the receptors. As judged from data on NA excretion reported by Leduc (32), the peripheral release of endogenous NA during severe cold exposure in the rat can amount to about 1/10 of the infusion dose used in our experiments (for calculations see 24). We assume that it is possible to cause the shivering threshold temperature to shift with doses of NA similar to the NA released endogenously under cold exposure, but it takes longer for the shift to develop. On the basis of these studies, we suggest that a high peripheral NA level reduces the NA release at the hypothalamic synapses of the ACP. One can only speculate about the nature of this negative feedback mechanism at present. The signals from peripheral baroreceptors

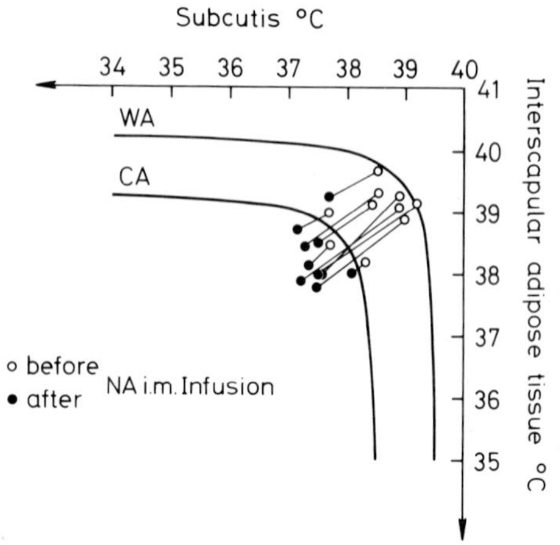

Fig. 18. *Shift of the shivering threshold to lower body temperatures evoked in 10 warm-adapted guinea pigs by continuous intramuscular noradrenaline infusion lasting 2 hours. From (57), by permission S. Karger AG, Basel.*

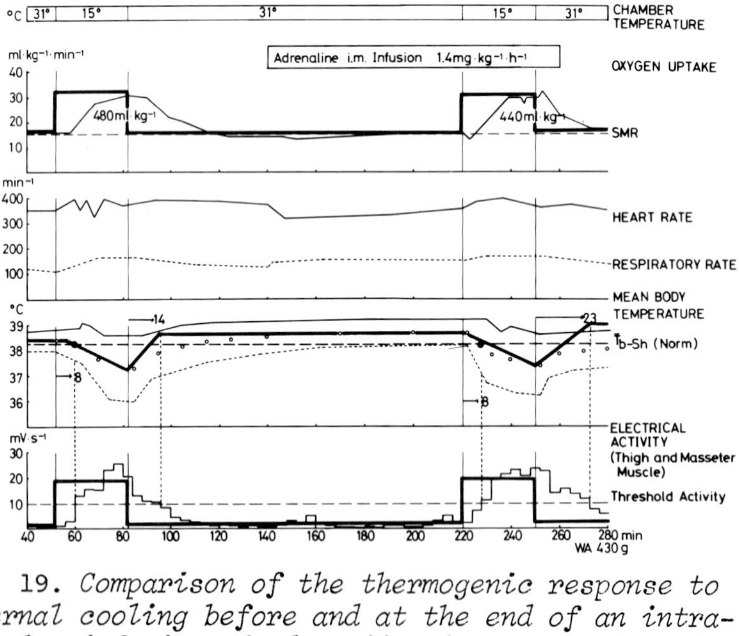

Fig. 19. *Comparison of the thermogenic response to external cooling before and at the end of an intramuscular infusion of adrenaline in a warm-adapted guinea pig. The thick lines schematically represent the cold-induced increase in oxygen uptake and electrical activity of the muscles as well as the changes in the mean body temperature, with respect to the beginning and end of shivering activity (numbers in the temperature record denote the time interval in minutes before the shivering begins or ends). From (51).*

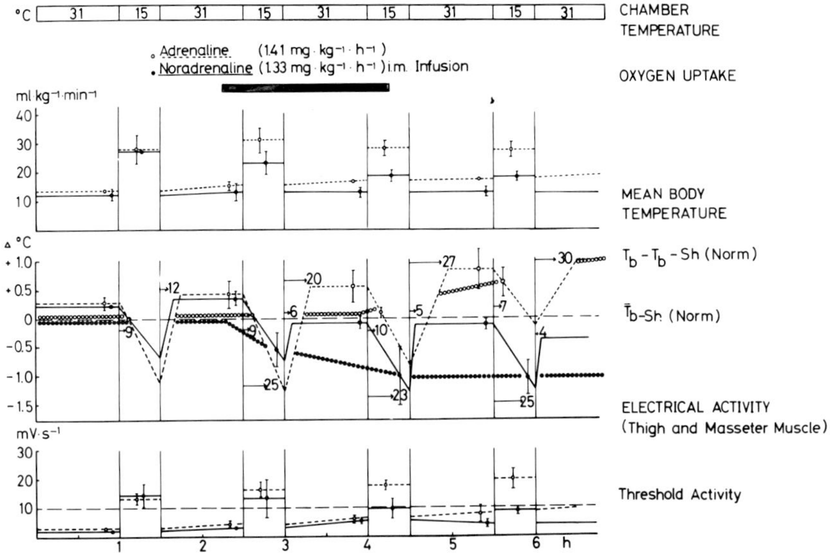

Fig. 20. *Thermogenic responses to external cooling before, during and after intramuscular infusion of noradrenaline (•) and adrenaline (○) in warm-adapted guinea pigs. Note that the shivering threshold temperature decreased in the noradrenaline-infused group, but increased in the adrenaline-infused group. Figures beside arrows indicate latency in occurrence of shivering after cooling or persistence of shivering after rewarming, respectively (in min). Horizontal black bar on top indicates time period of NA or A infusion. From (51).*

and chemoreceptors might be involved (37). On the other hand, it does not seem impossible that excessively high plasma NA concentrations may overcome the blood-brain barrier at special sites, e.g. in the area postrema, and act directly on the nerve cells in the brain stem. The same effect could be obtained if some metabolites of NA, even at lower concentrations, could pass the blood-brain barrier. The existence of a negative feedback mechanism between the peripheral and central NA levels is supported by findings of an increased NA turnover rate in the hypothalamus after adrenalectomy (20) and in the hypothalamus and lower brain stem after chemical sympathectomy with 6-OHDA (for review see 29). On the other hand, a decreased NA turnover rate in the brain has been found after acute elevation of the blood pressure in the rat (27), which is closely related to the increase in peripheral plasma NA concentration (14). The negative feedback hypothesis would also explain the original observations of Cannon *et al.* (15) on the responses of surgically sympathectomized cats to cold. These cats shivered violently, more than intact cats, when exposed to cold. It is possible to imagine that the decrease of peripheral NA level induced by sympathectomy removed the inhibition of shivering. An increase in shivering after chemical sympathectomy can also be inferred from recent experiments performed by Leduc (33). Both CA and non-acclimatised rats sympathectomized with 6-OHDA could produce as much heat as sham-operated rats, or even more when exposed to 6°C. Since the non-acclimated rats could not increase their oxygen uptake in any other way than by shivering, it is clear that the higher values of oxygen uptake in sympathectomized rats, compared with sham-operated animals, must be caused by the increase in shivering activity.

VI. A TENTATIVE BASIC CONCEPT OF COLD-ADAPTATION

Cold-adaptation is a complex process which appears to occur in several steps. Deviation of shivering threshold seems to belong to the early changes in this process. This shivering deviation quickly follows the acute response to cold exposure. In man (see III.B.1) a few consecutive cold exposures are sufficient to develop shivering threshold deviations. An essential feature of the acute cold exposure is the activation of the peripheral sympathetic system. The NA liberated from the sympathetic system reduces the heat loss due to peripheral vasoconstriction and also induces the NST in NB and CA animals. In WA animals (or in adult man) that do not possess the ability to produce heat without shivering, NA can promote the shivering activity by mobilisation of substrates for muscular contractions. During cold exposure, the amount of NA liberated peripherally probably exceeds the capacity for its metabolic destruction. This would explain the high excretion rates of unmetabolized NA during the early phase of cold-adaptation. Due to the accumulation of incompletely metabolized NA, suppression of the activity in the ACP of the lower brain stem can be induced. As a consequence, shivering threshold shifts to a lower mean body temperature. The animal is thus able to maintain the controlled variable (mean body temperature, cf. 11, 12) at a decreased level without shivering being activated: this is the hypothermic stage of adaptation. In the space of time following, brown adipose tissue and NST develop in small animals under the stimulation of the increased NA (and thyroxine) level (31). As the threshold temperatures for NST are well above those for shivering (Fig. 2), there is a tendency to re-establish normothermia. Whether or not complete normothermia is reached depends on some additional factors, in particular on the time course of cold exposure (see III.C.). In any case, shivering threshold remains decreased; this is to be expected as the peripheral catecholamine concentration is maintained at a high level under continuous cold exposure. In a further period lasting several months, body insulation is improved (e.g. by growth of fur). Now the peripheral release of catecholamines is reduced and shivering threshold returns to its normal level, thereby establishing a normothermic phase of adaptation.

The fact that the heat dissipation threshold can be shifted independently of the threshold for heat production in man (Fig. 9) as well as in guinea pigs suggests that there are two distinguishable neuronal networks in the hypothalamus which are concerned with the adjustment of the thresholds for heat production and heat dissipation, respectively. This concept is supported by recent stimulation and coagulation studies which led to the conclusion that the thermointegrative structures controlling heat polypnea are located in a more rostral part of the hypothalamus than the integrative structures for the control of shivering (48). The effect of NA or any other putative central transmitter on heat dissipation threshold in CA and WA guinea pigs remains to be studied.

ACKNOWLEDGMENTS

Part of the work underlying this article was supported by the Deutsche Forschungsgemeinschaft (SFB 122, Project B 1).

REFERENCES

1. Adolph, E. F. General and specific characteristics of physiological adaptations. Amer. J. Physiol. 184: 18-28, 1956.
2. Baum, E., K. Brück and H. P. Schwennicke. Adaptive modifications in the thermoregulatory system in long-distance runners. J. Appl. Physiol. 40: 404-410, 1976.
3. Benzinger, T. H. Heat regulation: Homeostasis of central temperature in man. Physiol. Rev. 49: 671-759, 1969.
4. Bligh, J. Thermoregulation: What is regulated and how? In: Proc. Lille Symposium on Thermoregulation. Guieu, J. D. and Y. Houdas (editors). XXVII Congress of the Int. Union of Physiol. Sci. (In press).
5. Bligh, J., A. J. Davis, D. F. Sharman and C. A. Smith. Unimpaired thermoregulation in the sheep after depletion of hypothalamic nor-adrenaline by 6-hydroxydopamine. J. Physiol. 265: 51-52, 1977.
6. Bligh, J. and K. G. Johnson. Glossary of terms for thermal physiology. J. Appl. Physiol. 35: 941-961, 1973.
7. Brück, K. Cold adaptation in man. In: Regulation of depressed metabolism and thermogenesis. Janský, L. and X. J. Musacchia (editors). Springfield, Ill.: Charles C. Thomas Publ. 1976, pp. 42-63.
8. Brück, K. Temperature regulation and catecholamines. Isr. J. Med. Sci. 12: 924-933, 1976.
9. Brück, K., E. Baum and H. P. Schwennicke. Cold-adaptive modifications in man induced by repeated short-term cold exposures and during a 10-day and night cold exposure. Pflügers Arch. 363: 125-133, 1976.
10. Brück, K. and H. P. Schwennicke. Interaction of superficial and hypothalamic thermosensitive structures in the control of non-shivering thermogenesis. Int. J. Biometeorol. 15: 156-161, 1971.

11. Brück, K. and W. Wünnenberg. Eine kälteadaptative Modifikation: Senkung der Schwellentemperaturen für Kältezittern. Pflügers Arch. 293: 226-235, 1967.
12. Brück and W. Wünnenberg. Meshed control of two effector systems: Non-shivering and shivering thermogenesis. In: Physiological and Behavioral Temperature Regulation. Hardy, J. D., A. P. Gagge and J. A. J. Stolwijk (editors). Springfield, Ill.: Charles C. Thomas Publ. 1970, pp. 562-580.
13. Brück, K. W. Wünnenberg, H. Gallmeier and B. Ziehm. Shift of threshold temperature for shivering and heat polypnea as a mode of thermal adaptation. Pflügers Arch. 321: 159-172, 1970.
14. Callingham, B. A. Catecholamines in blood. In: Handbook of Physiology, Section 7, Vol. VI. Greer, R. O., E. B. Astwood, H. Blaschko, G. Sayers, A. D. Smith and S. R. Geiger (editors). Washington, D.C.: American Physiological Society 1975, pp. 427-445.
15. Cannon, W. B., H. F. Newton, E. M. Bright, V. Menkin and R. M. Moore. Some aspects of the physiology of animals surviving complete exclusion of sympathetic nerve impulses. Amer. J. Physiol. 89: 84-107, 1929.
16. Dahlström, A. and K. Fuxe. Evidence for the existence of monoamine neurons in the central nervous system. Acta physiol. scand. 64: Suppl. 247, 1965.
17. Davis, T. R. A. Chamber cold acclimatization in man. J. Appl. Physiol. 16: 1011-1015, 1961.
18. Eisenman, J. S. Unit activity studies of thermoresponsive neurons. In: Essays on Temperature Regulation. Bligh, J. and R. E. Moore (editors). Amsterdam-London: North Holland Publ. Co. 1972, pp. 56-69.
19. von Euler, U.S. Noradrenaline. Chemistry, Physiology, Pharmacology and Clinical Aspects. Springfield, Ill.: Charles C. Thomas Publ. 1956.

20. Fuxe, K., T. Hökfelt, G. Jonsson and P. Lidbrink. Brain-endocrine interactions: Are some effects of ACTH and adreno-cortical hormones on neuroendocrine regulation and behavior mediated via central catecholamine neurons? In: Hormones and Brain Function. Lissak, K. (editor). New York: Plenum Press, 1973, pp. 409-425.
21. Hammel, H.T., R. W. Elsner, D. H. LeMessurier, H. T. Andersen and F. A. Milan. Thermal and metabolic responses of the Australian aborigine to moderate cold in summer. J. Appl. Physiol. 14: 605-615, 1959.
22. Hensel, H., K. Brück and P. Raths. Homeothermic organisms. In: Temperature and Life. Precht, H., J. Christophersen, H. Hensel and W. Larcher (editors). Berlin: Springer Verlag, 1973, pp. 503-732.
23. Herrmann, H. J. Die Wärmetachypnoe des Meerschweinchens in Abhängigkeit von der Temperatur des vorderen Hypothalamus, des Cervicalmarks, des Lumbalmarks und der Körperoberfläche. Inaugural Dissertation, Giessen, 1972.
24. Himms-Hagen, J. Role of the adrenal medulla in adaptation to cold. In: Handbook of Physiology, Section 7, Vol. VI. Greer, R. O., E. B. Astwood, H. Blaschko, G. Sayers, A. D. Smith and S. R. Geiger (editors). Washington, D.C.: American Physiological Society, 1975, pp. 637-665.
25. Hong, S. K. Comparison of diving and non-diving women of Korea. Amer. Soc. Exp. Biol. 22: 831-833, 1963.
26. Hori, T. and T. Nakayama. Effects of biogenic amines on central thermoresponsive neurones in the rabbit. J. Physiol. 232: 71-86, 1973.
27. Ito, A., K. Tanaka and T. Omae. Lowered turnover of norepinephrine in the brain associated with an acute elevation in blood pressure of rats. Jap. Heart Journal 16: 575-582, 1975.
28. Jell, R. M. Responses of rostral hypothalamic neurones to peripheral temperature and to amines. J. Physiol. 240: 295-307, 1974.

29. Lanier, L. P., A. J. Dunn and C. V. Hartesveldt. Development of neurotransmitters and their function in brain. In: Reviews of Neuroscience, Vol. 2. Ehrenpreis, S. and I. J. Kopin (editors). New York: Raven Press, 1976, pp. 195-256.
30. LeBlanc, J. Evidence and meaning of acclimatization to cold in man. J. Appl. Physiol. 9: 395-398: 1956.
31. LeBlanc, J. and A. Villemaire. Thyroxine and noradrenaline on noradrenaline sensitivity, cold resistance and brown fat. Amer. J. Physiol. 218: 1742-1745, 1970.
32. Leduc, J. Catecholamine production and release in exposure and acclimation to cold. Acta physiol. scand. 53: Suppl. 183, 1-101, 1961.
33. Leduc, J. Effect of sympathectomy on heat production in rats exposed to cold. Isr. J. Med. Sci. 12(9): 1099-1102, 1976.
34. Merker, G. and H. G. Hartwig. Thermosensitive und thermointegrative Areale des Hypothalamus: Struktur und Funktion. Anat. Anz. (In press).
35. Merker, G., E. Zeisberger, E. Bergheim-Hackmann and K. Brück. Horseradish peroxidase (HRP) tracing of ascending projections from lower brain stem to thermointegrative structures of the hypothalamus. Pflügers Arch. 368: R29, 1977.
36. Moruzzi, G. and H. W. Magoun. Brain stem reticular formation and activation of the EEG. Electroenceph. clin. Neurophysiol. 1: 455-473, 1949.
37. Mott, J. C. The effects of baroreceptor and chemoreceptor stimulation on shivering. J. Physiol. 166: 563-586, 1963.
38. Nadel, E. R., R. W. Bullard and J. A. J. Stolwijk. Importance of skin temperature in the regulation of sweating. J. Appl. Physiol. 31: 80-87, 1971.
39. Raynaud, J., J. P. Martineaud, O. P. Bhatnagar, H. Viellefond and J. Durand. Body temperatures during rest and exercise in residents and sojourners in a hot climate. Int. J. Biometeor. 20(4): 309-317, 1976.

40. Scholander, P. F., H. T. Hammel, J. S. Hart, D. H. LeMessurier and J. Steen. Cold adaptation in Australian aborigines. J. Appl. Physiol. 13(2): 211-218, 1958.
41. Schwennicke, H. P. and K. Brück. Thermoregulatory modifications in the course of repeated moderate heat-exposures in man. Pflügers Arch. 365: R27, 1976.
42. Simon, E. Temperature regulation: the spinal cord as a site of extrahypothalamic thermoregulatory functions. Rev. Physiol. Biochem. Pharmacol. 71: 1-76, 1974.
43. Stern, L., M. H. Lees and J. Leduc. Environmental temperature, oxygen consumption and catecholamine excretion in newborn infants. Pediatrics 36(3): 367-373, 1965.
44. Stitt, J. T. The regulation of respiratory evaporative heat loss in the rabbit. J. Physiol. 258: 157-171, 1976.
45. Szelényi, Z., E. Zeisberger and K. Brück. Effects of electrical stimulation in the lower brainstem on temperature regulation in the unanaesthetized guinea pig. Pflügers Arch. 364: 123-127, 1976.
46. Szelényi, Z., E. Zeisberger and K. Brück. A hypothalamic alpha-adrenergic mechanism mediating the thermogenic response to electrical stimulation of the lower brainstem in the guinea pig. Pflügers Arch. 370: 19-23, 1977.
47. Ungerstedt, U. Stereotaxic mapping of the monoamine pathways in the rat brain. Acta physiol. scand. Suppl. 367: 1-48, 1971.
48. Wünnenberg, W. Thermointegrative function of the hypothalamus. In: Regulation of depressed metabolism and thermogenesis. Janský, L. and X. J. Musacchia (editors). Springfield, Ill.: Charles C. Thomas Publ., 1976, pp. 26-41.
49. Zeisberger, E. Deviation of shivering threshold in warm-adapted guinea pigs induced by alteration of peripheral catecholamine level. Pflügers Arch. 362: R24, 1976.

50. Zeisberger, E. Ontogenic and thermoadaptive changes of thermogenic thresholds in the guinea pig. (In preparation).
51. Zeisberger, E. Comparison of the effects of peripheral noradrenaline and adrenaline infusion on shivering threshold. Pflügers Arch. (In press).
52. Zeisberger, E., E. Berger-Gaude and K. Brück. Drug effects on hypothalamic structures participating in the control of thermogenesis. In: Depressed Metabolism and Cold Thermogenesis Janský, L. (editor). Prague: Charles University, 1975, pp. 182-187.
53. Zeisberger, E. and K. Brück. Central effects of noradrenaline on the control of body temperature in the guinea pig. Pflügers Arch. 322: 152-166, 1971.
54. Zeisberger, E. and K. Brück. Effects of intrahypothalamically injected noradrenergic and cholinergic agents on thermoregulatory responses. In: The Pharmacology of Thermoregulation. Schönbaum, E. and P. Lomax (editors). Basel: Karger, 1973, pp. 232-243.
55. Zeisberger, E. and K. Brück. Alteration of shivering threshold in cold- and warm-adapted guinea pigs following intrahypothalamic injections of noradrenaline and of an adrenergic alpha-receptor blocking agent. Pflügers Arch. 362: 113-119, 1976.
56. Zeisberger, E. and K. Brück. The significance of central adrenergic alpha-receptive structures in the control of thermogenesis and in cold adaptation. Isr. J. Med. Sci. 12(9): 1103-1106, 1976.
57. Zeisberger, E., Z. Szelényi and K. Brück. Role of catecholaminergic brainstem pathways in thermoadaptive alteration of shivering threshold. In: Drugs, Biogenic Amines and Body Temperature. Cooper, K. E., P. Lomax and E. Schönbaum (editors). Basel: Karger, 1977, pp. 5-12.

STRATEGIES IN COLD

ADAPTATION OF MAN TO COLD

Jacques LeBlanc
Department of Physiology
School of Medicine
Laval University
Quebec (10), Quebec

ABSTRACT

In order to study the problem of adaptation to cold in men, it is necessary to try to relate human studies to the work which has been done on laboratory animals. Numerous investigations on the white rat suggest two types of adaptation to cold: one produced by *Continuous Exposure* to *Moderate Cold* (CM), the other by *Intermittant Exposure* to *Severe Cold* (IS). By the first method, the animals develop a non-shivering thermogenesis which is related to a hypertrophy of the brown adipose tissue and an enhanced response to noradrenaline. With the second method, the response to noradrenaline remains unmodified. In humans, adaptation to CM is probably non-existant. In the Australian aborigines, the Kalahari bushmen and the Ama, although shown to shiver less than the white subjects, there is no evidence that non-shivering thermogenesis exists. On the other hand IS are frequently experienced by humans. To quantify adaptation to IS, the responses of the autonomic nervous system to immersion of the hand in cold water and to blowing of the face by cold wind are measured. Adaptation to IS is characterized by a decreased sympathetic response, a diminution of pain sensation and an enhanced parasympathetic activity. It is interesting to note that similarities exist in hypothermic and autonomic responses between hibernators and non-hibernators adapted to cold: the night torpor of the aborigines and the decline in sympathetic activity and the increase of parasympathetic activity are examples. Exposure to cold produces shivering which is non-economical and mentally disturbing. In all animal species, adaptation to cold is manifested by a diminution of shivering and of cold pain sensation.

ISBN 0-12-734550-7

Various animal species have adapted to extremes of environmental conditions. Survival of Arctic mammals is made possible primarily through improved capacity for heat preservation provided by abundant fur and subcutaneous fat. This is particularly true for larger animals. Some animals cannot tolerate very cold temperatures because the larger surface area per unit of body mass in these smaller animals contributes to greater heat dissipation. Similarly, naked man, because of poor insulation, is very much at the mercy of cold temperature. The development of clothing and shelter proved sufficient for human beings to live comfortably even in colder regions of the earth. Independent of specific characteristics for tolerance to cold, all species seem to be able to improve their resistance through various mechanisms.

Figure 1 summarizes globally the perception and the reactions to stimulation by cold. The zone of induction deviates slightly from the comfort zone. This zone is well tolerated and represents an increased heat preservation through vasoconstriction. The reaction zone is where adaptation takes place. Depending on the

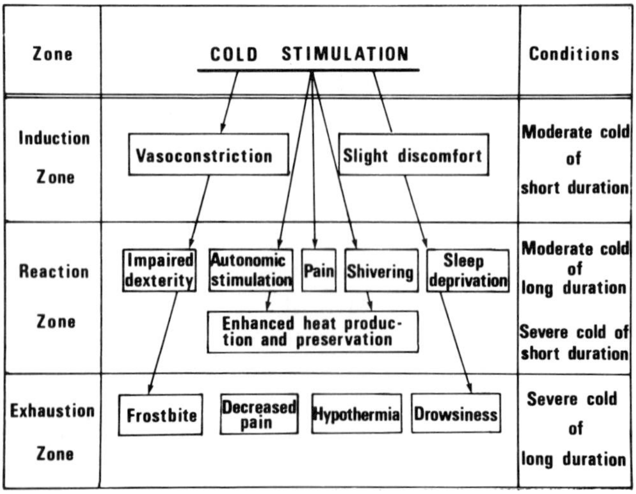

Fig. 1. *Responses of the organism to cold stimulation as affected by the duration and the severity of the exposure.*

type of exposure, various kinds of adaptation will develop. The cold environment is perceived differently on both duration of exposure and intensity of the stimulation. Various studies performed on laboratory animals suggest two types of adaptation (Fig. 2) (11, 12, 14, 15). One resulting from *Continuous Exposure to Moderate cold* (CM) and the other from *Intermittent Exposure to Severe cold* (IS). The prolonged exposure produces shivering initially and, as it will be discussed, adaptation that results from it is metabolic. The short exposure does not necessarily produce shivering but is painful especially on extremities and face and the adaptation that is observed is cardiovascular. In these two types of adaptation, the conscious disturbances which are experienced are 1) shivering which is disturbing and metabolically non-economical and 2) pain which may become quite stressful. Shivering and pain are the two perceptions which adaptation tends to alleviate. The different mechanisms by which these two types of adaptation (CM and IS) are produced will be discussed but it can be said immediately that the distinction between CM and IS adaptations has been proven to be real and useful at the same time.

A preliminary brief description of studies on laboratory animals will help clarify the problem of adaptation to cold in humans. Rats exposed for a few weeks at 6°C show a remarkable increase in heat production capacity which enables these adapted animals to withstand temperatures as low as -35°C compared to -18°C for non-adapted animals. A major break-through in the field of

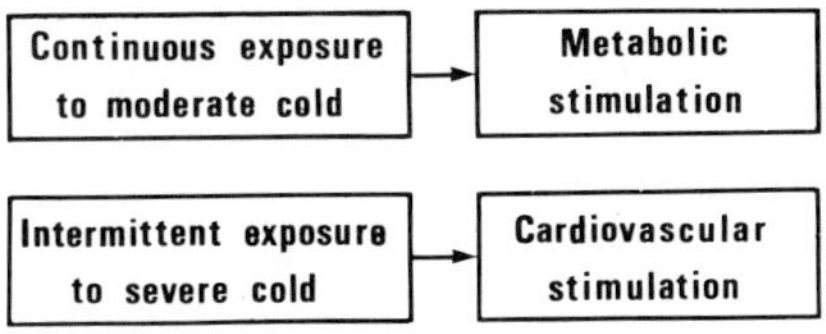

Fig. 2. *Prolonged exposure to moderate cold produces a metabolic type of adaptation and intermittent exposure to severe cold produces a cardiovascular adaptation.*

cold physiology was the discovery by Carlson and Hsieh (4) of a marked increase in sensitivity to noradrenaline in cold adapted animals (Fig. 3). This finding reported some 20 years ago has stimulated extensive research on the role of the autonomic nervous system in cold adaptation. Concomitant with this increased sensitivity to noradrenaline, it was found that the secretion of this hormone is significantly increased in cold exposed animals (7, 20) (Fig. 4). These results suggested that the increased secretion of noradrenaline, observed in cold adapted animals, produces an enhanced response to this amine and this might prove to be an important factor in cold adaptation. This hypothesis proved to be true since daily injections of noradrenaline to animals kept at room temperature (26°C) significantly increase the response to noradrenaline and also improved tolerance to cold (Fig. 5) (8, 13, 14, 15). Similar results were obtained with isoproterenol, a beta receptor agonist, indicating the importance of beta receptor in adaptation to cold (Fig. 6) (16). It is also important to note that, as these changes are taking place, shivering is gradually replaced by non shivering thermogenesis and this is the important characteristic of CM type of

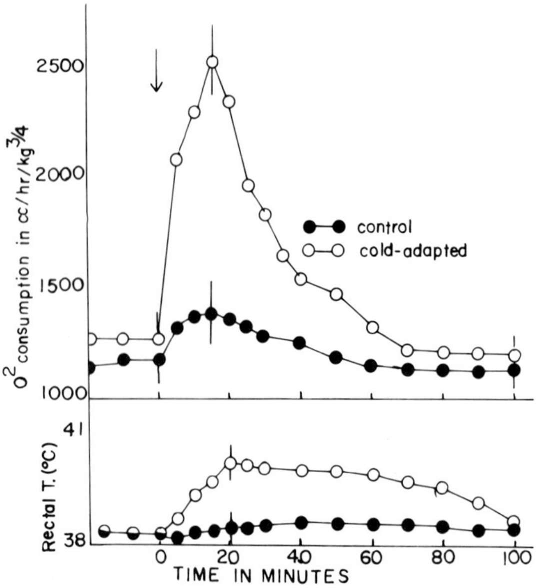

Fig. 3. *Effect of noradrenaline on oxygen consumption of normal and cold adapted animals showing the increased response caused by adaptation (4).*

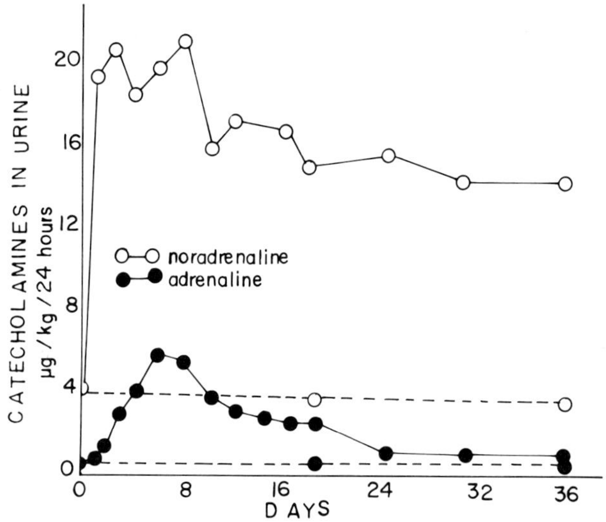

Fig. 4. *Urinary excretion of adrenaline and noradrenaline in rats continuously exposed to 5°C (20).*

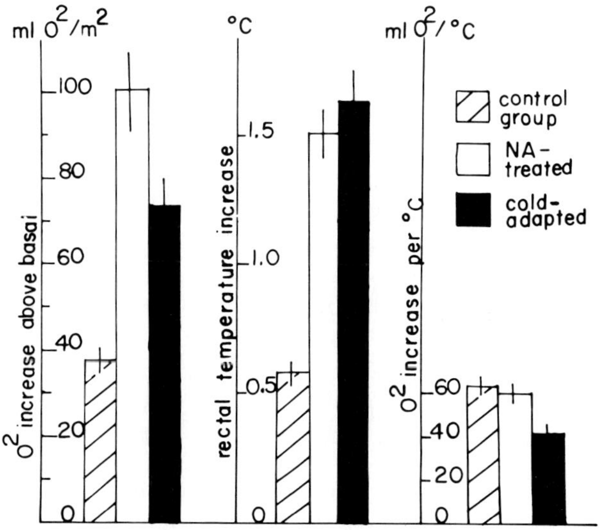

Fig. 5. *Repeated injections of noradrenaline produce an enhanced response to noradrenaline and a greater tolerance to cold (14, 15).*

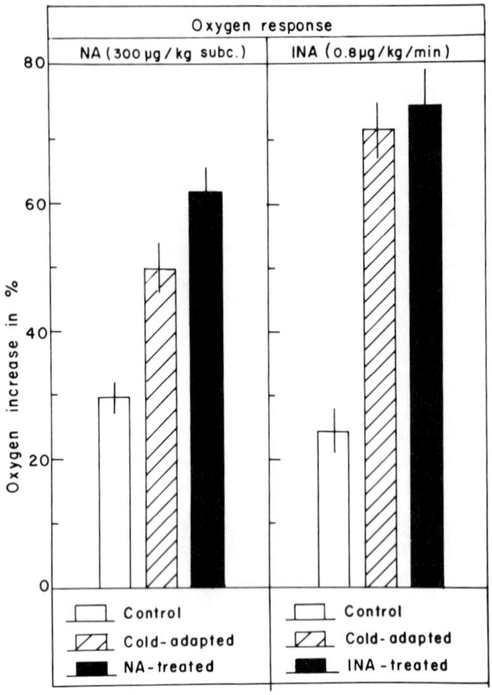

Fig. 6. *Repeated injections of isoproterenol resulted in greater tolerance to cold and enhanced response to catecholamines (16).*

adaptation. IS type of adaptation is different. The animal is placed in the cold for short periods of time 5 to 10 minutes but at colder temperatures (-5 to -20°C). The exposure is repeated at hourly intervals for a total 8 times in one day for 2 days. In all, the animals are placed about 3 hours in the cold and this represents very little shivering (11, 12). However, the degree of discomfort and pain is certainly high at these very low temperatures. This procedure also results in improved tolerance to cold as seen in figure 7. The mechanisms by which IS adaptation is acquired would seem to differ from those responsible for CM adaptation. Indeed the response to noradrenaline which was shown to be increased with CM adaptation, remains unchanged with IS adaptation (Fig. 8) (14).

What happens in humans? It has been said that man is a tropical animal. This can only mean that without clothing or shelter, we could only live comfortably in tropical zones; it could not mean that we cannot adapt

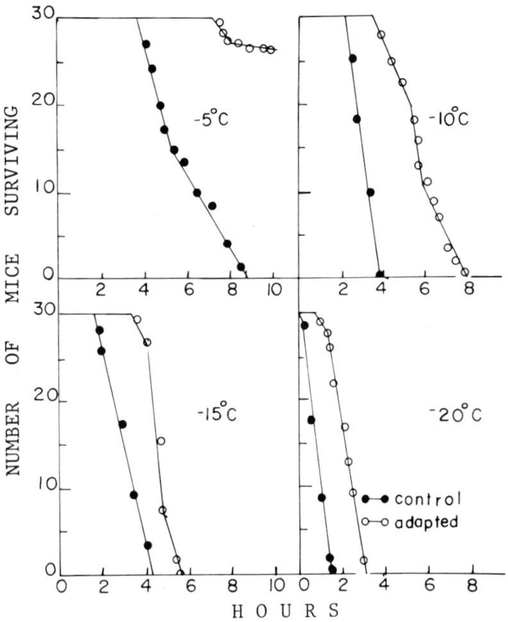

Fig. 7. *Improved tolerance to cold resulting from repeated short exposures to severe cold (11, 12).*

to cold temperatures. The question is to ask whether CM or IS types of adaptation are possible in man. Concerning CM adaptation, it would seem that very few populations really experience shivering for any length of time. The only three groups that can be cited are: The Australian aborigines (23), the Kalahari bushmen (3) and the Ama, the women divers of Korean Peninsula (22). Their responses to a standard cold test were compared to those of white subjects. Figure 9, 10 and 11 show a common behavior in all these groups. The response of the white subjects is characterized by shivering which contributes to maintain body temperature at a constant level throughout the test. In the adapted groups, an absence of shivering was noticed and this results in a slight fall of body temperature. In a sense, their reaction is comparable to a sort of overnight torpor from which recuperation is readily obtained when the atmosphere warms up at sunrise. As was observed, this suppression of shivering allows

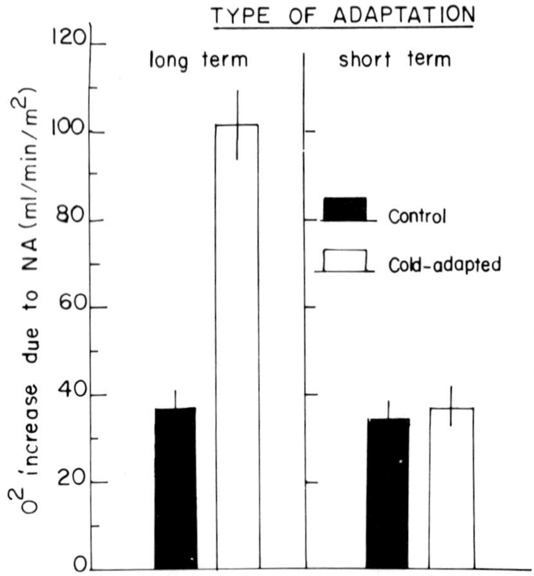

Fig. 8. *The response to noradrenaline which is increased by prolonged exposure to moderate cold, remains unchanged with intermittent exposure to severe cold (14).*

these populations to sleep which was impossible for the control subjects. These results are comparable to those observed in laboratory animals. Indeed adaptation to prolonged exposure to cold results in a decrease in shivering which is the cause for discomfort in this case. We have no evidence that shivering is replaced

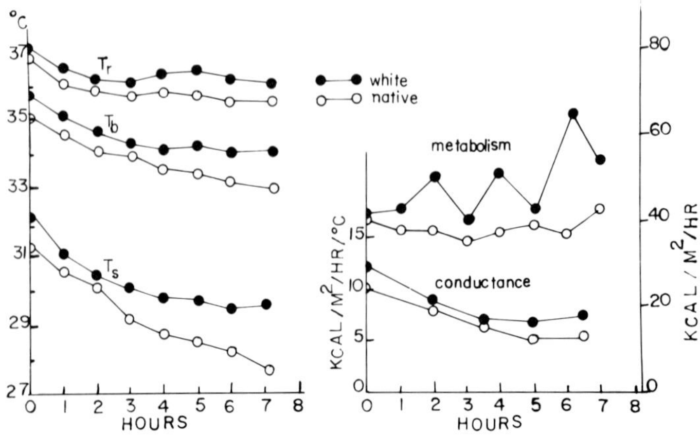

Fig. 9. *Responses to a standard cold exposure in Australian aborigines and control subjects (23).*

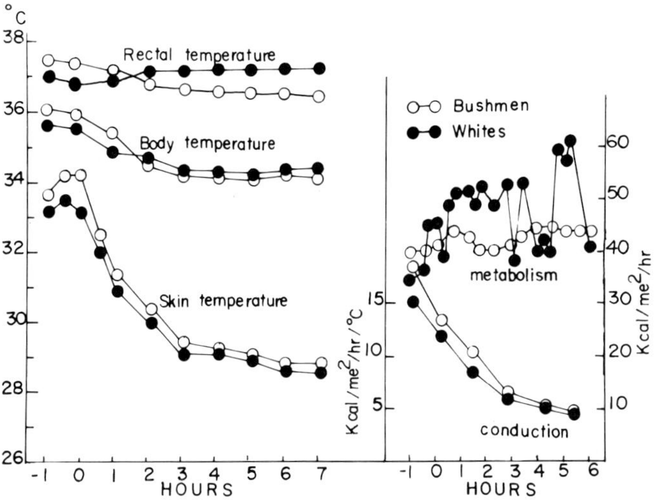

Fig. 10. *Responses to a standard cold exposure in Kalahari Bushmen and control subjects (3).*

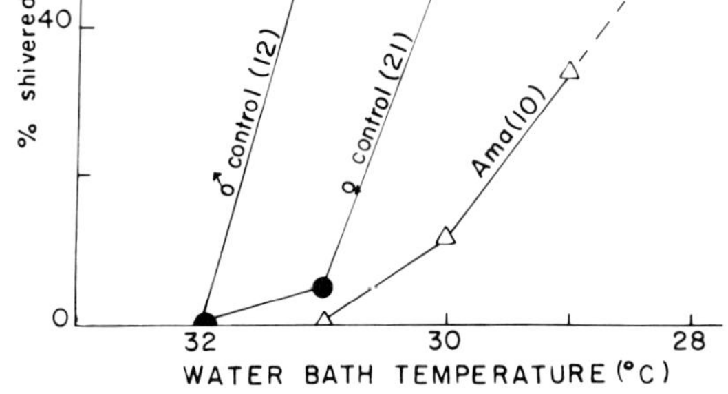

Fig. 11. *Responses to cold water baths in Korean diving women and control subjects (22).*

Reprinted from Federation Proceedings 22:831-834 1963.

by non-shivering thermogenesis in humans; in the present experiment, it is not; increased heat production does not take place and a slight fall in body temperature is observed. In laboratory animals, shivering is being replaced by non-shivering heat production. It must be emphasized here that the test on humans produces an increase of only 10 to 15% in oxygen consumption whereas the one for laboratory animals is of the order of 200%. As a consequence, when shivering disappears in the rat, it has to be replaced by another mechanism for heat production otherwise irreversible hypothermia would rapidly set in. Figure 12 illustrates these findings and permits an analogy between hibernators and non-hibernators with regard to their ways of resisting and adapting to a cold environment. In all cases, shivering disappears. In non-hibernators exposed to moderate cold, non-shivering thermogenesis gradually develops under the control of noradrenaline. Facing less severe cold expsoure, as in the three groups of humans just described, no extra heat is produced and slight hypothermia sets in. In hibernators, the responses are the same as in the humans with the difference that the degree of hypothermia is much more pronounced. These results do not mean that non-shivering thermogenesis could not develop in humans; on the contrary, indirect proof to that effect was obtained by some researchers (2, 5). Direct evidence will be difficult to obtain since there are no groups of humans that are exposed to temperatures as demanding as those used for laboratory

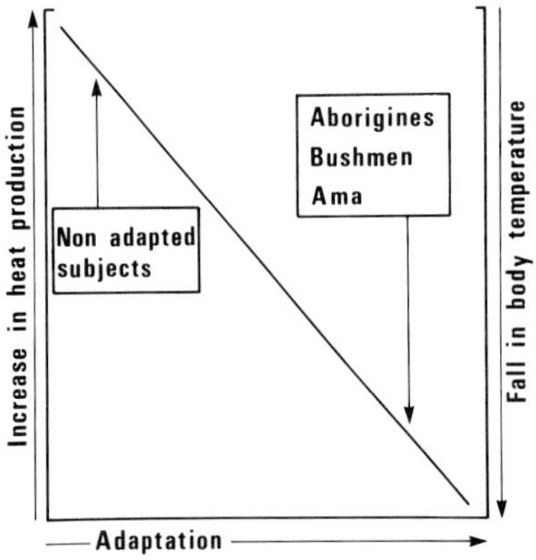

Fig. 12. *Effect of adaptation on body temperature and oxygen consumption upon exposure to cold.*

animals. In other words, for humans and possibly for many animal species living under natural conditions, the limits of their adaptative possibilities are not attained.

The best proof of this is found in the Eskimo. One might think that if adaptation to cold exists, it should certainly be found most developed among this population. This is far from reality. When tested in a cold room, they shiver as much as control subjects (1). The reason for this is that the Eskimos have developed proper clothing and shelter to protect themselves adequately. For my part, I have never seen an Eskimo shivering. On the other hand, living in the Arctic and even in moderate climate zone implies frequent exposures of the face and of the extremities to various combinations of temperature and wind. In other words, we are generally much more exposed to Intermittent Exposure of Severe cold than to Continuous exposure of

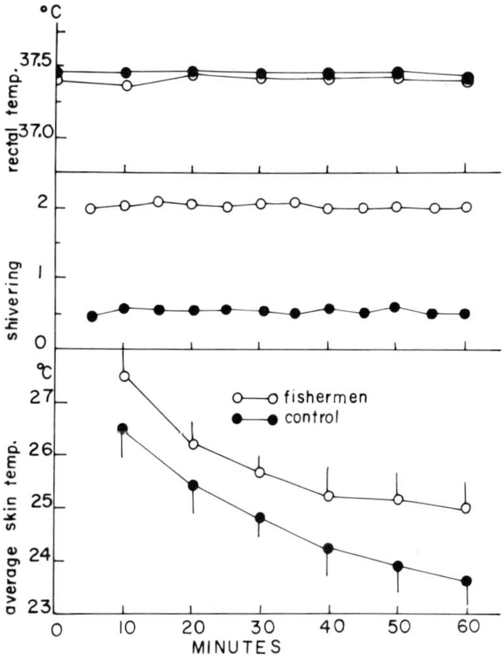

Fig. 13. *Responses to a standard cold exposure of Gaspé fishermen and control subjects (9).*

Moderate cold. Another example which illustrates this situation is that of the Gaspé fisherman. They immersed their hands for long periods in water varying between 0 and 10°C. When tested for metabolic adaptation, it was found that they shiver even more than the control subjects and their average skin temperature was higher (Fig. 13) (9).

We then proceeded to look for evidence of IS adaptation in humans by measuring the responses of both branches of the autonomic nervous system. Immersion of the hand into cold water causes an activation of the sympathetic as evidenced by an increase in systolic and diastolic blood pressure and by cardiac acceleration. Cold wind blown on the face produces the same effect on blood pressure but at the same time causes a reflex

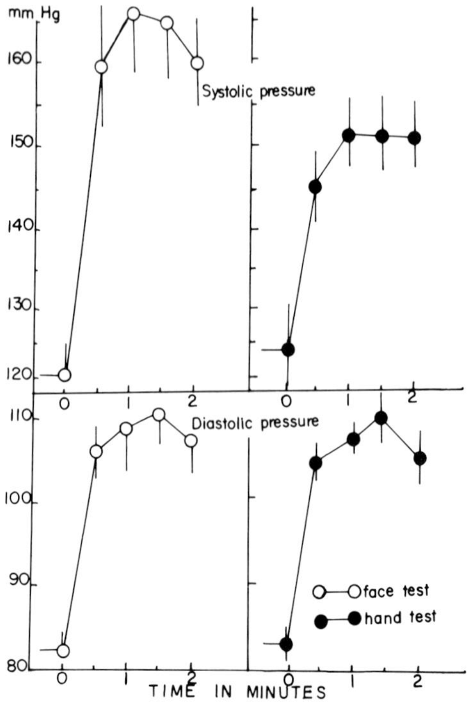

Fig. 14. *Blood pressure responses to immersion of one hand into cold water (5°C) and to a cold wind (0°C with 40 mph wind) blown on the face (17).*

bradycardia of vagal origin [Fig. 14, 15, 16 (17)]. Both these tests were used to test the sympathetic and parasympathetic activity of various populations. The sympathetic response was shown to be very significantly reduced in both Eskimos and Gaspé fishermen (Fig. 17, 18, 19) as well as in mailmen in the region of Quebec when tested at the end of the winter (Fig. 20) (18), or in the student population after a series of repeated exposures [Fig. 21 (10)]. The results on the bradycardia of cold face exposure show in general an activation of the response [Fig. 19 (19)]. Thus while the activity of the sympathetic nervous system is decreased with IS adaptation that of the parasympathetic seems to be increased in some cases. The decline in sympathetic response observed in adapted subjects has always been found to parallel a decrease in the subjective evaluation of the pain caused by cold stimulation (19).

The responses of adapted human subjects are in a way comparable to those of animals entering hibernation. In both cases, the activity of the sympathetic is

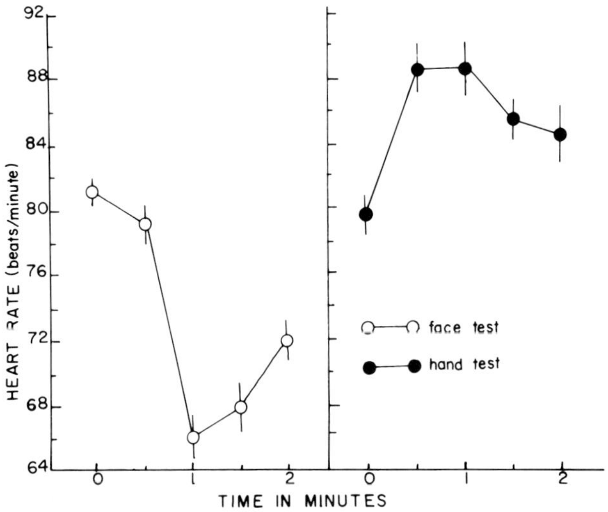

Fig. 15. *Heart rate responses to immersion of one hand into cold water (5°C) and to a cold wind (0°C with 40 mph wind) blown on the face (17).*

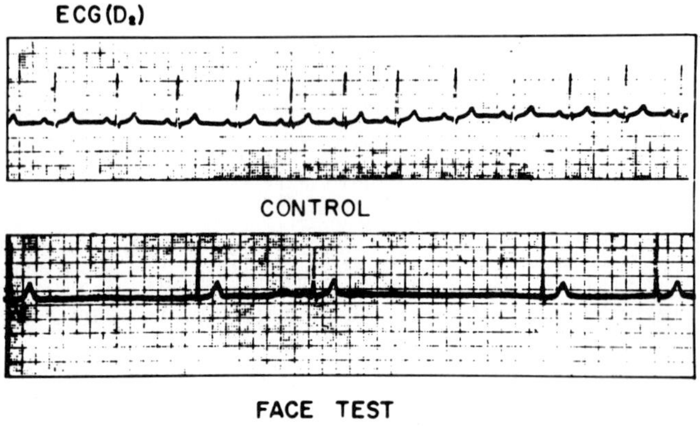

Fig. 16. *Illustration of the marked bradycardia observed in some subjects submitted to a cold face test (17).*

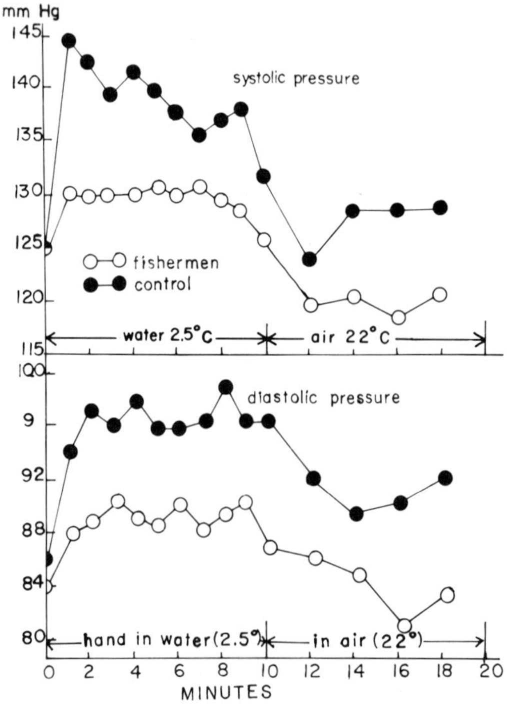

Fig. 17. *Blood pressure responses of Gaspé fishermen and of control subjects to a cold hand test (17).*

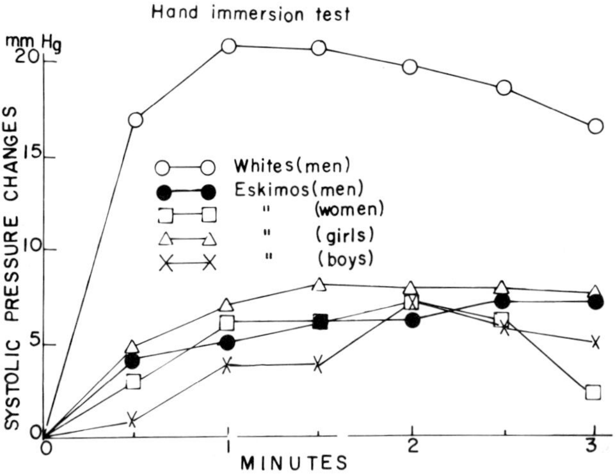

Fig. 18. *Comparison of blood pressure response of various groups of Eskimos and of white subjects to a cold hand test (18).*

attenuated while that of the parasympathetic is enhanced (18, 21). In adapted humans, the decreased response of the sympathetic seems to be related to habituation. Repeated exposures to a stress would produce through negative reflex conditioning a decline in autonomic stimulation accompanied by a marked decrease in subjective sensation. It is not known whether this interpretation could apply to hibernators. Obviously, something special happens in hibernators but it is interesting to note the similarities that exist between what happens in these animals and in other mammals. The night torpor of aborigines or the decline in activity of the sympathetic and the increase in that of the parasympathetic in adapted subjects may be given as examples. Similarly, a parallel may be drawn between the obese subject and the fat hibernator when placed in a cold environment. We have shown that subjects with a substantial subcutaneous fat layer will start shivering at colder room temperature and when the temperature of the skin has reached much lower levels than is observed with thin subjects (6). On the other hand, this analogy should probably not be carried too far because the

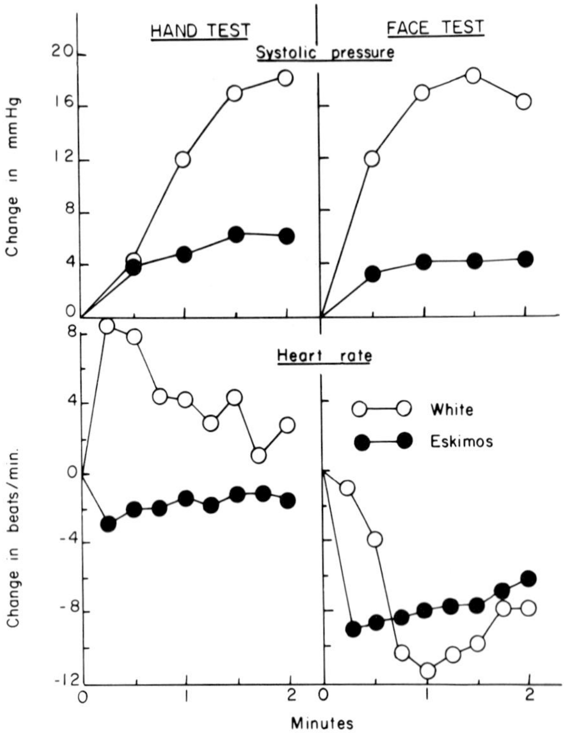

Fig. 19. *Blood pressure increases and heart rate modifications resulting from both cold hand and cold face tests in Eskimos and white subjects (17).*

situation is not completely comparable. For hibernators it is a question of economy of reserves as I understand it; in other mammals, it is a question of preserving dexterity and avoiding frost-bite, of avoiding shivering and at the same time hypothermia. The wisdom of the body is met in all animal species. The needs of the hibernator are different from those of the man, but within our flesh are found the means of resisting and adapting to the demand of the environment.

Whatever the specific requirements of various animal species as determined by the environmental conditions, there seems to be some common basic regulations. Figure 22 compares the responses of adapted and

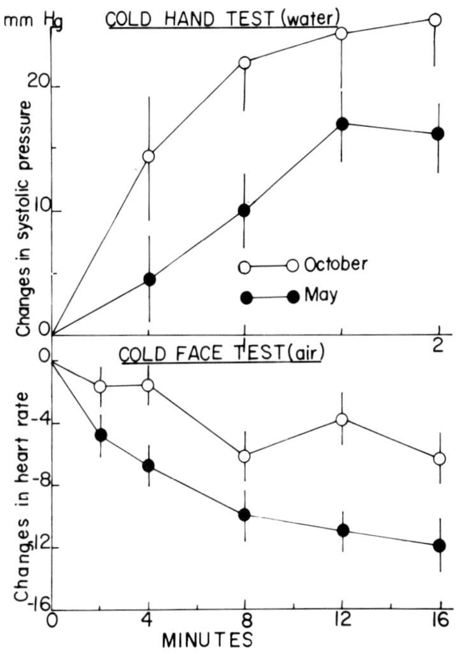

Fig. 20. *Blood pressure response to a cold hand test and heart rate changes due to a cold face test in mailmen of Quebec City. Results show significant differences between the responses in October and those in May (end of winter) (18).*

non-adapted animals. In non-adapted animals, exposure to cold produces shivering which by increasing heat production helps maintain body temperature near normal levels. In adapted animals (and hibernators may be included in this group), shivering disappears. In some cases, it is replaced by non-shivering thermogenesis. This happens when the demand of the environment is relatively great. For example, if a rat is placed at 5°C and shivering is prevented, irreversible hypothermia would soon develop. Non-shivering thermogenesis mediated by an enhanced sensitivity to noradrenaline is of course a major modification but it is essential if shivering is eliminated. If the demand of the environment is reduced, and this is what happens in the primitive groups that I have described, shivering does not take

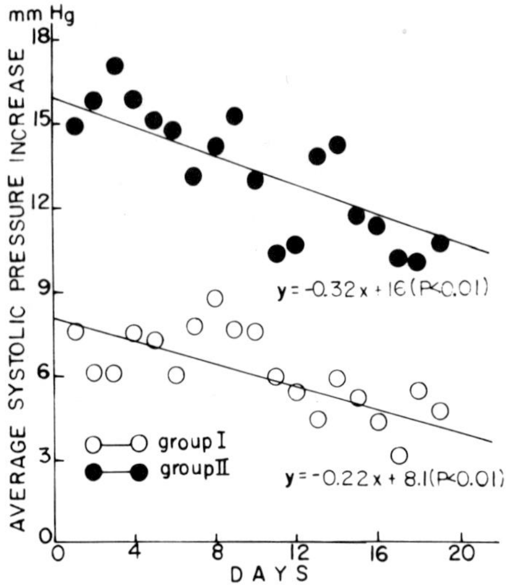

Fig. 21. *Evidence of adaptation to a cold hand test after repeated exposures in groups of medical students (10).*

Reproduced by permission of the National Research Council of Canada from the Canadian Journal of Physiology and Pharmacology, Vol. 44, pp. 287-293, 1966.

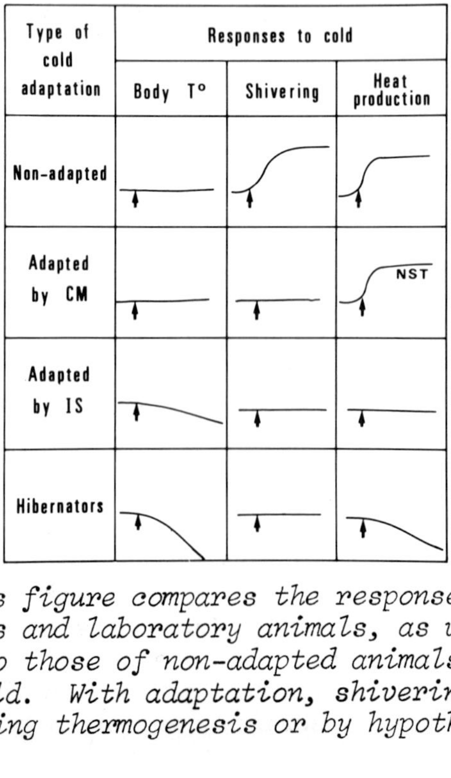

Fig. 22. *This figure compares the responses of cold adapted humans and laboratory animals, as well as of hibernators to those of non-adapted animals when exposed to cold. With adaptation, shivering is replaced by non-shivering thermogenesis or by hypothermia.*

place and it is not replaced by non-shivering thermogenesis. As a consequence, there is a small lowering of body temperature which is well tolerated. Finally, in the hibernator not only is shivering eliminated but at the same time basal heat production is lowered. The consequence is a more advanced degree of hypothermia.

These results are suggestive of a common reaction in organisms exposed to cold for a period of time. Indeed what seems important is to eliminate shivering. Shivering is non-economical and can be mentally very disturbing. The means by which shivering is depressed vary with the conditions of adaptation and also with some specific capacity of the animals such as being able to go into hibernation and come out of it without endangering survival. These conclusions do not seem to be very speculative and in any event they have the advantages of showing how the responses of human beings, living in contact with the environment, can fit into an overall scheme of nature.

REFERENCES

1. Adams, T. and B. Covino. Racial variations to a standardized cold stress. J. App. Physiol. 12: 9-12, 1958.
2. Davis, T. R. A. and S. R. Johnson. Seasonal acclimatization to cold in man. J. Appl. Physiol. 16: 231-235, 1961.
3. Hammel, H. T., J. A. Hildes, D. C. Jackson and H. T. Andersen. Thermal and metabolic responses of the Kalahari bushmen to moderate cold exposure at night. Tech. Rept. no. 62-44, Arctic Aeromedical Lab., Ladd AFB, 1962.
4. Hsieh, A. C. L. and L. D. Carlson. Role of adrenaline in chemical regulation of heat production. Am. J. Physiol. 190: 242-246, 1957.
5. Itoh, S. Physiology of cold adapted man. Sapporo, Hokkaido University School of Medicine, Japan, 1974.
6. LeBlanc, J. Subcutaneous fat and skin temperature. Can. J. Biochem. Physiol. 32: 354-359, 1954.
7. LeBlanc, J. and G. Nadeau. Urinary excretion of adrenaline and noradrenaline in normal and cold adapted animals. Can. J. Biochem. Physiol. 39: 215-217, 1961.
8. LeBlanc, J. and M. Pouliot. Importance of noradrenaline in cold adaptation. Am. J. Physiol. 207: 853-856, 1964.
9. LeBlanc, J., M. Pouliot and S. Rheaume. Thermal balance and biogenic amine excretion in Gaspé fishermen. J. App. Physiol. 19: 9-13, 1964.
10. LeBlanc, J. and P. Potvin. Studies on habituation to cold pain. Can. J. Physiol. Pharmacol. 44: 287-294, 1966.
11. LeBlanc, J. Adaptation to cold in three hours. Am. J. Physiol. 212: 530-533, 1967.
12. LeBlanc, J., D. Robinson, D.F. Sharman and P. Tousignant. Catecholamines and short term adaptation to cold in mice. Am. J. Physiol. 213: 1419-1423, 1967.
13. LeBlanc, J., A. Villemaire and J. Vallieres. Arch. Int. Physiol. Biochem. 77: 731-740, 1969.

14. LeBlanc, J. Hormonal control of thermogenesis. Proceedings Symposium on Nonshivering Thermogenesis. Prague, 1970.
15. LeBlanc, J., L. Lafrance, A. Villemaire, C. Roberge, J. Vallieres and S. Rosseau. Catecholamines and cold adaptation. Symposium on Environmental Physiology. Fed. Proc. Bioenergetics, 1971, pp. 71-79.
16. LeBlanc, J., J. Vallieres and C. Vachon. Beta receptor sensitization by repeated injections of isoproterenol and by cold adaptation. Am. J. Physiol. 222: 1043-1046, 1972.
17. LeBlanc, J. Man in the Cold. Springfield, Mass.: Charles C. Thomas, 1975.
18. LeBlanc, J., S. Dulac, J. Cote and B. Girard. Autonomic nervous system and adaptation to cold in man. J. App. Physiol. 39: 181-186, 1975.
19. LeBlanc, J., B. Blais, B. Barabe and J. Cote. Effects of temperature and wind on facial temperature heart rate and sensation. J. App. Physiol. 40: 127-131, 1976.
20. Leduc, J. Catecholamine production and release in exposure and adaptation to cold. Acta Physiol. Scand. 53 Suppl. 183: 1-101, 1961.
21. Lyman, C. P. and R. C. O'Brien. Autonomic control of circulation during the hibernating cycle in ground squirrels. J. Physiol. (London) 168: 477-499, 1963.
22. Rennie, D. W., B. G. Covino, B. J. Honell, S. H. Song and S. K. Hong. Physical insulation of Korean diving women. J. App. Physiol. 17: 961-966, 1962.
23. Scholander, P. F., H. T. Hammel, J. S. Hart, D. H. Lemesurier and J. Steen. Cold adaptation in Australian aborigines. J. App. Physiol. 13: 211-218, 1958.